TL 760 FIX

Fixed and Flapping Wing Aerodynamics for Micro Air Vehicle Applications

Fixed and Flapping Wing Aerodynamics for Micro Air Vehicle Applications

Edited by
Thomas J. Mueller
University of Notre Dame
Notre Dame, Indiana

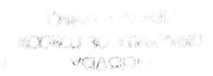

Volume 195
PROGRESS IN
ASTRONAUTICS AND AERONAUTICS

Paul Zarchan, Editor-in-Chief
MIT Lincoln Laboratory
Lexington, Massachusetts

Published by the
American Institute of Aeronautics and Astronautics, Inc.
1801 Alexander Bell Drive, Reston, Virginia 20191-4344

Progress in Astronautics and Aeronautics

Editor-in-Chief
Paul Zarchan
MIT Lincoln Laboratory

Editorial Board

Table of Contents

Thomas J. Mueller, *University of Notre Dame, Notre Dame, Indiana*; and
James D. DeLaurier, *University of Toronto, Downsview, Ontario, Canada*

Part I. Fixed Wing Aerodynamics

Mark Drela, *Massachusetts Institute of Technology, Cambridge,
Massachusetts*

Part II. Flapping and Rotary Wing Aerodynamics

Chapter 11   Thrust and Drag in Flying Birds: Applications to Birdlike Micro Air Vehicles 217

Jeremy M. V. Rayner, *University of Leeds, Leeds, United Kingdom*

Chapter 12   Lift and Drag Characteristics of Rotary and Flapping Wings 231

C. P. Ellington and J. R. Usherwood, *University of Cambridge, Cambridge, England*

Chapter 13   A Rational Engineering Analysis of the Efficiency of Flapping Flight 249

Kenneth C. Hall, *Duke University, Durham, North Carolina;* and Steven R. Hall, *Massachusetts Institute of Technology, Cambridge, Massachusetts*

Chapter 18   A Nonlinear Aeroelastic Model for the Study of Flapping Wing Flight **399**
Rambod F. Larijani and James D. DeLaurier, *University of Toronto, Downsview, Ontario, Canada*

Chapter 19   Euler Solutions for a Finite-Span Flapping Wing **429**
M. F. Neef and D. Hummel, *Technical University of Braunschweig, Braunschweig, Germany*

Part III. Micro Air Vehicle Applications

Chapter 23   Mesoscale Flight and Miniature Rotorcraft

Ilan Kroo and Peter Kunz, *Stanford University, Stanford, California*

Chapter 24   Development of the Black Widow Micro Air Vehicle . . .   519

Joel M. Grasmeyer and Matthew T. Keennon, *AeroVironment, Inc.,*
Simi Valley, California

Chapter 25   Computation of Aerodynamic Characteristics

Ravi Ramamurti and William Sandberg, *Naval Research Laboratory,*
Washington, DC

Preface

Recently, there has been a serious effort to design aircraft that are as small as possible for special, limited-duration missions. Vehicles of this type might carry visual, acoustic, chemical, or biological sensors. These aircraft, called micro air vehicles, are of interest because electronic detection and surveillance sensor equipment can now be miniaturized so that the entire payload weighs about 18 g. Although, the long-term goal of this project is to develop aircraft systems that weigh less than 30 g, have about an 8-cm wingspan, and can fly for 20 to 30 min at between 30 and 65 km/h, the current goal is to develop aircraft with a 15-cm wingspan that weigh less than 90 g. Because it is not possible to meet all of the design requirements of a micro air vehicle with current technology, research is proceeding on all of the system components at various government laboratories, companies, and universities. One of the areas of concern is the aerodynamic efficiency of various fixed wing, flapping wing, or rotary wing concepts as these vehicles are very small and must fly at very low speeds. The corresponding chord Reynolds number range for a 15-cm vehicle is from 50,000 to about 150,000. Very little, if any, information on the performance of various airfoil/wing shapes exists for this flight regime; however, there has been a long history of natural flight studies (e.g., insects and small birds) that may be helpful.

The design and evaluation techniques for fixed wings of relatively large aspect ratio and Reynolds numbers above 200,000 are reasonably well developed. Aerodynamic problems related to vortex lift on wings of aspect ratio of one and lower have made it difficult to extend existing techniques to lower Reynolds numbers. Further difficulties arise when one considers the possibility of designing flapping or rotary wing vehicles.

This volume is the collection of papers presented at the Conference on Fixed, Flapping and Rotary Wing Aerodynamics at Very Low Reynolds Numbers held 5–7 June 2000 at the University of Notre Dame. The conference was sponsored by the Department of Aerospace and Mechanical Engineering, College of Engineering, the Research Division of the Graduate School, and the Roth-Gibson Endowment of the University of Notre Dame, Notre Dame, Indiana. Over fifty active researchers in this field from Canada, Europe, Japan, and the United States were present. It is clear from the papers in this book that a great deal of progress has been made in the understanding the aerodynamics of fixed, flapping, and rotary wings. The ultimate goals of this understanding are the design and successful operation of a variety of micro air vehicles.

I am grateful to the authors and the reviewers of each paper for their efforts to ensure that this book is as good as it can be. I would also like to thank the staff of the AIAA for putting this book together.

Thomas J. Mueller
July 2001

Chapter 1

An Overview of Micro Air Vehicle Aerodynamics

Thomas J. Mueller*

University of Notre Dame, Notre Dame, Indiana

and

James D. DeLaurier†

University of Toronto, Downsview, Ontario, Canada

Nomenclature

AR	= full-span aspect ratio, b^2/S
b	= wingspan
C_D	= drag coefficient (three-dimensional)
C_d	= section drag coefficient
C_f	= skin friction coefficient
C_L	= lift coefficient (three-dimensional)
C_l	= section lift coefficient
$C_L^{3/2}/C_D$	= endurance parameter
C_l/C_d	= section lift-to-drag ratio
c	= root-chord length
Re_c	= root-chord Reynolds number
S	= wing area
t	= wing thickness
α	= angle of attack

Subscripts

max	= maximum
min	= minimum

*Roth-Gibson Professor, Department of Aerospace and Mechanical Engineering, Hessert Center for Aerospace Research. Associate Fellow AIAA.

†Professor, Institute for Aerospace Studies. Associate Fellow AIAA.

I. Introduction

T HERE is a serious effort to design aircraft that are as small as possible for special, limited-duration military and civil missions. These aircraft, called micro air vehicles (MAVs), are of interest because electronic surveillance and detection sensor equipment can now be miniaturized so that the entire payload mass is less than 18 g. The current goal is to develop aircraft with a 15-cm maximum dimension that have a mass of less than 90 g. In addition to being a compact system transportable by a single operator, MAVs have other advantages, including its rapid deployment, real-time data acquisition capability, low radar cross section, and low noise. The potential for low production cost is also an advantage. The primary missions of interest for fixed wing MAVs include surveillance, detection, communications, and the placement of unattended sensors. Surveillance missions include video (day and night), infrared images of battlefields (referred to as the "over the hill" problem) and urban areas (referred to as "around the corner"). These real-time images can give the number and location of opposing forces. This type of information can also be useful in hostage rescue and counter-drug operations. Because of the availability of very small sensors, detection missions include the sensing of biological agents, chemical compounds, and nuclear materials (i.e., radioactivity). MAVs may also be used to improve communications in urban or other environments where full-time line-of-sight operations are important. The placement of acoustic sensors on the outside of a building during a hostage rescue or counter-drug operation is another possible mission.

The requirements for fixed wing MAVs include a wide range of possible operational environments, such as urban, jungle, desert, maritime, mountain, and arctic. Furthermore, MAVs must be able to perform their missions in all weather conditions (i.e., precipitation, wind shear, and gusts). Because these vehicles fly at relatively low altitudes (i.e., less than 100 m) where buildings, trees, hills, etc. may be present, a collision avoidance system is also required.

Significant technical barriers must be overcome before MAV systems can be realized. These include issues in small-scale power generation and storage, navigation, and communications as well as propulsion, aerodynamics, and control. One of the most interesting and least understood aspects of small-scale flight is the aerodynamics. The combination of small length scale and low velocities results in a flight regime with very low Reynolds numbers. This places MAVs in a regime totally alien to conventional aircraft. The gross mass of MAVs and other flying objects vs Reynolds number is shown in Fig. 1. The payload mass vs wingspan for MAVs and unmanned air vehicles (UAVs) is presented in Fig. 2.

The long-term goal is to develop aircraft systems with a mass of less than 30 g, a wingspan of about 8 cm, and an endurance of 20 to 30 min at speeds between 30 and 65 km/h. It appears that flapping or rotary wing designs may have advantages over fixed wing MAVs at this small size. One of these advantages is in the generation of lift and thrust without excessive size or weight. The interest in achieving insectlike or birdlike flight performance has focused attention on their wing dynamics and unsteady aerodynamics. However, our understanding of the aerodynamics of bird and insect flight is limited. Biologists have studied bird and insect flight empirically for quite some time. One thing that is clear from these investigations is that all of these creatures use two specific mechanisms to overcome the small-scale aerodynamic limitations of their wings: flexibility and

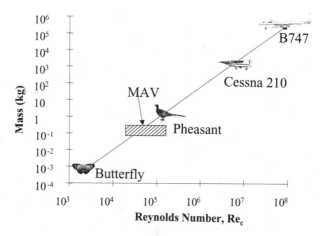

Fig. 1 Reynolds number range for flight vehicles.[1]

flapping. Birds and insects exploit the coupling between flexible wings and aerodynamic forces (i.e., aeroelasticity) such that the aeroelastic wing deformations improve aerodynamic performance. By flapping their wings, birds and insects effectively increase the Reynolds number seen by the wings without increasing their forward flight speed. Although some progress has been made in mimicking avian aeromechanics, more is necessary if an operational flapping wing MAV is to be successful.

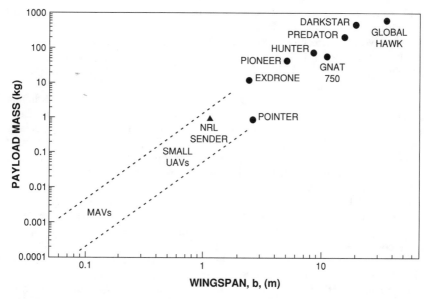

Fig. 2 UAV payload vs wingspan.[2] **(Reprinted with permission of MIT Lincoln Laboratory, Lexington, Massachusetts.)**

II. Fixed Wing Vehicles

The design requirements cover a wide range of parameters when one considers the diversity of possible applications for micro air vehicles. The MAV must be designed as a system consisting of airframe, propulsion, payload, and avionics. Although much smaller than operational fixed wing UAVs, electrically powered MAVs will have approximately the same weight fractions, that is, 21% for the airframe, 11% for the engine, 30% for the battery, 21% for the payload, and 17% for avionics and miscellaneous items (private communication with R. J. Foch, Naval Research Laboratory, Washington DC, 1996). Minimum wing area for ease of packaging and prelaunch handling is also important.

The airfoil section and wing planform of the lifting surface occupy a central position in all design procedures for flying vehicles. Also, as with all aircraft, MAVs share the ultimate goal of a stable and controllable vehicle with maximum aerodynamic efficiency. Aerodynamic efficiency is defined in terms of the lift-to-drag ratio. Airfoil section $C_{l_{max}}$, $C_{d_{min}}$, and $(C_l/C_d)_{max}$ as a function of Reynolds number are shown in Figs. 3a, 3b, and 3c for two-dimensional airfoils (i.e., infinite wings). It is clear from this figure (adapted from Ref. 3) that airfoil performance deteriorates rapidly as the chord Reynolds number decreases below 100,000. Performance of three-dimensional wings (i.e., finite wings), as measured by $(C_l/C_d)_{max}$, is less than that for airfoils except when the wing aspect ratio is reduced below a value of 2. While the maximum lift-to-drag ratio for most low-speed fixed wing aircraft including many UAVs ($U_\infty < 50$ m/s) is greater than 10, values for insects and small birds are usually less than 10. Furthermore, to achieve these values for MAVs at low Reynolds numbers, the wings must emulate bird and insect airfoils and be very thin (i.e., $t/c < 0.06$) with a modest amount of camber.

Requirements for a typical propeller-driven fixed wing MAV, for example, include long flight duration (i.e., high value of $C_L^{3/2}/C_D$ at speeds up to 65 km/hr at chord Reynolds numbers from about 45,000 to 180,000 and altitudes from 30 to 100 m). Because these vehicles are essentially small flying wings, there is a need to develop efficient low Reynolds number, low aspect ratio wings that are not overly sensitive to wind shear, gusts, and the roughness produced by precipitation. Furthermore, confidence that the operational vehicle will perform as designed is important in all applications.

Although design methods developed over the past 35 years produce efficient airfoils for chord Reynolds numbers greater than about 200,000, these methods are generally inadequate for chord Reynolds numbers below 200,000, especially for very thin airfoil sections. In relation to the airfoil boundary layer, important areas of concern are the separated regions that occur near the leading and/or trailing edges and transition from laminar to turbulent flow if it occurs. It is well known that separation and transition are highly sensitive to Reynolds number, pressure gradient, and the disturbance environment. Transition and separation play a critical role in determining the development of the boundary layer, which affects the overall performance of the airfoil/wing. The aerodynamic characteristics of the wing and other components, in turn, affect the static, dynamic, and aeroelastic stability of the entire vehicle. Therefore the successful management of the sensitive boundary layer for a particular low Reynolds number vehicle design is critical.

Laminar separation bubbles occur on the upper surface of most airfoils at chord Reynolds numbers above about 50,000. In principle, the laminar separation bubble and transition can be artificially controlled by adding the proper type of disturbance at the proper location on the airfoil. Wires, tape strips, grooves, steps, grit, or

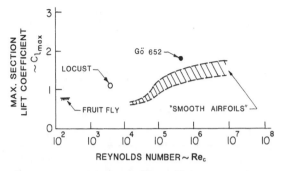

a) Maximum lift coefficient

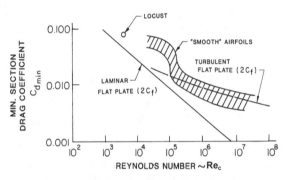

b) Minimum drag coefficient

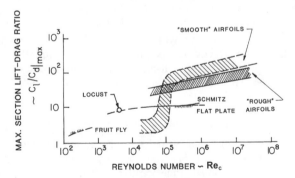

c) Maximum lift-to-drag ratio

Fig. 3   Airfoil performance.

bleed-through holes in the airfoil surface have all been used to have a positive influence on the boundary layer in this critical Reynolds number region. The type and location of these so-called turbulators and their actual effect on the airfoil boundary layer has not been well documented. Furthermore, the addition of a turbulator does not always improve airfoil performance. In fact, how the disturbances produced by a given type of turbulator influence transition is not completely understood.

The discussion of airfoils operating at low Reynolds numbers presented above is based on a large number of experimental and analytical studies performed over the past two decades. This understanding of airfoil performance has been very

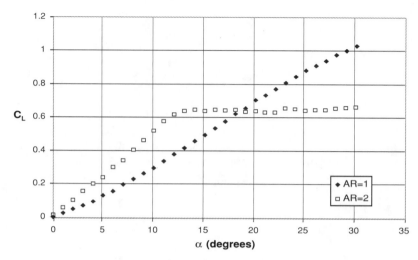

Fig. 4   Lift coefficient vs angle of attack for elliptical planform flat-plate wings, at $Re_c = 100,000.$

useful in the design of small UAVs with wingspans from 50 cm to 1 m (i.e., from about 200 g to 4 kg total mass) and chord Reynolds number well above 200,000. These UAVs have wings of large aspect ratio and thickness, which is characteristic of numerous successful vehicles. Recently, it has been found that for vehicles with aspect ratios around 1 and below,[4] thin wings cannot be designed using existing airfoil data. For these low aspect ratio wings the larger component of lift is produced by the tip vortices, which influence the majority of the wing surface. For example, at a chord Reynolds number of 100,000, the lift coefficient for an elliptical planform flat-plate wing with $AR = 2$ reaches its maximum value at an angle of attack of about 12 deg. As seen in Fig. 4, at the same Reynolds number for an $AR = 1$ elliptical planform flat-plate wing, the lift coefficient continues to increase past 30 deg. This behavior is similar to that of a thin delta wing.

III.   Flapping Wing Vehicles

The recent interest in micro air vehicles has generated renewed interest in mechanical flapping wing flight. This is motivated by the notion that flapping wings at small scales may offer some unique aerodynamic advantages over a fixed wing solution. In that respect, this runs counter to the historical solution for successful large-scale mechanical flight.

Humanity's earliest concepts for mechanical flight were inspired by birds and bats, leading to the assumption that flapping wings are required for propulsion and lift. Such a formidable challenge for replicating animal flight was bypassed by Sir George Cayley in 1799 when he introduced the concept of a fixed wing airplane with a dedicated propulsion system.[5] This innovation separated the functions of lift and propulsion into two distinct components, which proved to be the key for successful human-carrying flight. Soon thereafter mechanical flapping wing flight became marginalized within the aeronautical community, becoming the province of a few ornithopter inventors. Such efforts were generally hapless, with the notable exception of Alexander Lippisch's research in the late 1920s with a human-powered ornithopter.[6] What distinguished this work was that it was complemented with an analysis of flapping wing flight. Likewise, noted aerodynamicists such as von Karman and Burgers,[7] Garrick,[8] and Kuechmann and von Holst[9] performed detailed analytical studies of the physics of flapping wing propulsion. In more recent years, the topic of animal flying and swimming attracted the attention of Sir James Lighthill and his research group at Cambridge University.[10] There are other current examples, such as the bird-flight studies by Jeremy Rayner[11] and Geoff Spedding.[12] However, such interest has been more the exception than the rule among aerodynamicists, and one will find that the most consistent research in flapping wing flight has been among biologists and zoologists.

There is a fundamental difference between an ornithopter designer's interest in flapping wing flight and that of a biologist or zoologist. The primary motivation for studying animal flight is to explain the physics for a creature that is known to fly. That is, the fact that it achieves successful thrust and propulsion is given. Therefore, various analytical models are adjusted to match measured results, and conclusions are reached regarding energetics, migration capabilities, etc. An ornithopter designer, in contrast, is trying to develop a flying aircraft, and its ability to achieve this is no given fact. What such a person needs is a design-oriented analysis, which is not what is offered from animal-flight studies.

For a designer with an aerospace engineering background, publications from the animal-flight community can be fairly puzzling. That is, they may seem either too simplistic, involving inappropriate quasi-steady aerodynamic models, or too qualitative, presenting hypotheses regarding the supposed benefits of various free and bound vortices. What makes an engineer most comfortable is a comprehensive analysis verified with rigorous experiments over a wide range of parameters. Moreover, it is particularly appealing if other independent researchers successfully use the same analysis. However, it is also only fair to note that animal-flight researchers have not been provided with such an analysis for flapping wing flight from the aerospace community. What should be acknowledged is that there are a wealth of measurements, observations, and flow visualizations from animal-flight studies that are particularly appropriate to this renewed interest in mechanical flapping wing flight. For example, it is clear that flying creatures achieve successful flight in a variety of ways. Even among animals operating within the same Reynolds number regime, the solutions to flight widely differ.

At larger scales, such as that of large birds and bats, the aerodynamic modeling can draw upon familiar concepts from aerospace engineering. At sufficiently high aspect ratios, for instance, one can assume attached chordwise flow along the wing, with each section's aerodynamic characteristics being determined by the local angles of attack as influenced by wake effects. For nonflapping wings, the wake would be modeled by the familiar downwash concept. However, flapping wings require an unsteady shed wake, such as is familiar to researchers in advanced flight dynamics and aeroelasticity. The resulting sectional aerodynamic forces can then be resolved and integrated to give thrust and lift behavior. Such a model may ignore certain important nuances in animal flight, such as spanwise wing folding during the upstroke and tip-feather spread on raptors; however, this approach has been sufficiently accurate to be successfully applied to the design of large ornithopters.

It should be noted that because animals are constrained by their muscle actuation to achieve flight with flapping wings, they are making a virtue out of a necessity.

It was due to Cayley's brilliant observation that this need not be followed for mechanical flight that airplane inventors were freed from a slavish imitation of Nature. Moreover, there are observed wing motions on large birds, such as the previously mentioned wing folding, that are a consequence of muscle actuation cycling between tension and repose. This is not a constraint for an engine-powered flapping wing aircraft, where the design can take advantage of full power during both upstroke and downstroke. However, despite this advantage, and the fact that the propulsive efficiencies of large ornithopters can match the best that propellers can give, the additional complexity of such aircraft do not currently make them competitive with fixed wing designs for any function except mimicking animal flight.

Small-scale flapping wing flight is a different situation and is a rich topic for research. What motivates this is that unsteady aerodynamic features may provide advantages over comparable fixed wing designs. Separated flow in the form of vortices seems to be a given characteristic, and the primary interest is to understand and quantify how these work to enhance lift, thrust, and efficiency. The situation is further complicated by the fact that not all creatures of comparable size appear to use the same aerodynamic and configurational solutions to achieve flight. In some cases, a single pair of highly deformable wings generates strong leading-edge vortices for either translational flight (i.e., a Hawkmoth) or hovering (i.e., a hummingbird). In other cases, a pair of forward wings shed vortices that impinge upon a pair of rearward wings, presumably in an advantageous fashion. There are several other examples of differing configurations, but the main point is that quantifying these goes well beyond aerodynamic examples found in standard references. At present, solutions are being sought from sophisticated computational fluid dynamics methods, as well as inviscid panel methods with imposed vortex behavior.

Another activity that is serving well to elucidate the aerodynamic behavior of small-scale flapping wing creatures is to build bench-top mechanical analogs. Such an apparatus allows measurements and flow visualization to be performed in a controlled and repeatable way, providing further guidance for the development of analytical models. It is necessary, of course, to accurately replicate the significant characteristics of the wings being studied. However, an important (and perhaps inadvertent) result from this activity is that it can provide a bridge to the realization of a successful MAV owing to the necessity of materials selection and mechanical design. An example of this was inspired by the observations of zoologist Torkel Weis-Fogh,[13] who found extraordinarily high lift coefficients for tiny hovering insects whose wings make contact during the flapping cycle. This effect, which allows for bound circulation beyond that normally imposed by the Kutta condition, is referred to as the Clap-Fling hypothesis. A bench-top mechanical model successfully reproduced and confirmed this behavior and has now led to a 12-in. span flying MAV. This illustrates how a symbiosis between zoology and aerospace engineering can work to the benefit of both disciplines.

IV.   Concluding Remarks

The papers presented in this book focus on the current understanding of aerodynamic issues related to micro air vehicle design and performance. It is clear that the development of fixed wing micro air vehicles is significantly more advanced than that of flapping or rotary wing vehicles. However, the existence of mission capable MAVs does not negate the need for further research to improve their performance and efficiency. It would be helpful, for instance, to collect data on existing MAVs and compare these vehicles to insects and birds on the basis of maximum dimension and endurance vs weight. These comparisons would provide researchers with a better understanding of how close or far current technology is from Nature's most efficient fliers.

For fixed wing vehicles, many questions have surfaced from studies reported at the Conference on Fixed, Flapping, and Rotary Wing Aerodynamics at Very Low Reynolds Numbers held 5–7 June 2000. It has been asked, for example, whether camber is really beneficial at low Reynolds numbers and whether an optimum combination of planform shape and camber exists for maximized performance and stability. Other questions include the location and size of control surfaces, the influence of the propeller flowfield on the lift and drag of MAVs, and the feasibility of designing wing tips that can control the low-frequency rolling characteristic of low aspect ratio wings. Some of these questions could be answered by further wind tunnel research as well as flight testing.

For the successful development of flapping wing vehicles, it will be important to determine which movements of insects and birds are truly necessary for micro air vehicle flapping flight. For example, is the back and forth movement of bird wings really important? How critical is the control of the magnitude and time history of geometric wing section pitch? How sensitive is the optimum to small variations? How does this affect thrust and lift? If the pitch control matters, can it be achieved passively by elastic deformation of the wing over the speed range of interest? The study of sail and membrane wings might be helpful in answering these questions. Further wind tunnel studies of the unsteady wing movement of larger birds and bats could provide the necessary quantitative data on wing motion and energy consumption.

A thorough understanding of the kinematics of successful ornithopters may yield more useful results than the exact mimicking of bird or insect motions. In this regard, key research areas include the accurate modeling of unsteady membrane wings and an assessment of their low Reynolds number efficiency vs rigid-chord flapping wings. Also, no one has determined what the optimum planform should be (especially for membrane wings). Finally, birds and bats fold their wings on the upstroke. Does this have aerodynamic significance? Other areas of interest include the aerodynamics of an accelerating surface that is inclined to the streamwise flow and the use of three-dimensional dynamic stall in the design of a flapping wing MAV with attached flow. Because many of the current studies are two-dimensional, one important question is whether two-dimensional theories and/or experiments tell us much about what really happens on and around real three-dimensional flapping or plunging wings. As far as the local control of separation is concerned, are the effects of spanwise flows similar in magnitude to chordwise flows? Furthermore, what are the effects of trailing tip vortices that are dominant in some birds? The answers to these questions will likely require further wind tunnel experiments followed by theoretical or numerical simulations. Although early flapping wing MAVs have relied on a point design, greater success in the long term will be achieved using system design methods when sufficient data are available.

References

[1]Pelletier, A., and Mueller, T. J., "Aerodynamic Force/Moment Measurements at Very Low Reynolds Numbers," *Proceedings of the 46th Annual Conference of the Canadian Aeronautics and Space Institute*, Montreal, 3–5 May 1999, pp. 59–68.

[2]Davis, W. R., Jr., Kosicki, B. B., Boroson, D. M., and Kostishack, D. F., "Micro Air Vehicles for Optical Surveillance," *Lincoln Laboratory Journal*, Vol. 9, No. 2, 1996, pp. 197–213.

[3]McMasters, J. H., and Henderson, M. L., "Low Speed Single Element Airfoil Synthesis," *Technical Soaring*, Vol. 2, No. 2, 1980, pp. 1–21.

[4]Torres, G. E., and Mueller, T. J., "Aerodynamics Characteristics of Low Aspect Ratio Wings at Low Reynolds Numbers," *Proceedings of the Conference: Fixed, Flapping and Rotary Wing Vehicles at Very Low Reynolds Numbers*, Univ. of Notre Dame, Notre Dame, IN, 5–7 June 2000, pp. 228–305 and Chapter 7.

[5]Gibbs-Smith, C. H., "Chapter VI, The First Half of The Nineteenth Century," *A History of Flying*, B. T. Batsford, London, 1953, pp. 108–112.

[6]Lippisch, A. M., "Man Powered Flight in 1929," *Journal of the Royal Aeronautical Society*, Vol. 64, July 1960, pp. 395–398.

[7]Von Karman, T., and Burgers, J. M., "Problems of Non-Uniform and Curvilinear Motion," *Aerodynamic Theory*, W. F. Durand, Editor-in-Chief, Vol. II, Division E, Julius Springer, Berlin, 1934, pp. 304–310.

[8]Garrick, I. E., "Propulsion of a Flapping and Oscillating Airfoil," NACA Report No. 567, 1936.

[9]Kuechemann, D., and von Holst, E., "Aerodynamics of Animal Flight," *Luftwissen*, Vol. 8, No. 9, Sept. 1941, pp. 277–282; translated by L. J. Baker for the Ministry of Aircraft Production, R. T. P. Translation No. 1672.

[10]Lighthill, J., "Some Challenging New Applications for Basic Mathematical Methods in the Mechanics of Fluids That Were Originally Pursued with Aeronautical Aims," *Aeronautical Journal*, Vol. 94, No. 932, Feb. 1990, pp. 41–52.

[11]Rayner, J., "A Vortex Theory of Animal Flight. Part 2. The Forward Flight of Birds," *Journal of Fluid Mechanics*, Vol. 91, Part 4, 1979, pp. 731–763.

[12]Spedding, G. R., "The Aerodynamics of Flight," *Mechanics of Animal Locomotion*, edited by R. McNeill Alexander, *Advanced Comparative Environmental Physiology*, Vol. 11, Springer-Verlag, Berlin, 1992.

[13]Weis-Fogh, T., "Unusual Mechanisms for the Generation of Lift in Flying Animals," *Scientific American*, Vol. 233, No. 5, 1975, pp. 80–87.

Part I.   Fixed Wing Aerodynamics

Higher-Order Boundary Layer Formulation and Application to Low Reynolds Number Flows

Mark Drela*

Massachusetts Institute of Technology, Cambridge, Massachusetts

Nomenclature

A_{ij} = reduced wake vorticity-influence matrix for surface vorticity
$\vec{A}_{ij}$ = reduced wake vorticity-influence matrix for velocity
a_{ij} = vorticity-influence matrix for normal wash
$\vec{a}_{ij}$ = vorticity-influence matrix for velocity
B_{ij} = reduced source-influence matrix for surface vorticity
$\vec{B}_{ij}$ = reduced source-influence matrix for velocity
$\underline{b}_{ij}$ = source-influence matrix for normal wash
$\vec{b}_{ij}$ = source-influence matrix for velocity
C_D = dissipation coefficient
C_f = skin friction coefficient
D = dissipation integral
E = entrainment velocity
H = shape parameter
H^* = kinetic energy shape parameter
M = number of airfoil surface panel nodes
m = BL mass defect
N = number of all panel nodes
$\hat{n}$ = panel-normal unit vector
P = assumed pressure defect profile
p = pressure
$\vec{q}$ = potential-flow velocity vector
Re_c = chord Reynolds number
$\hat{s}$ = panel-tangential unit vector
U = assumed velocity profile

*Professor. Associate Fellow AIAA.

u, v = BL velocity components
x, y = BL curvilinear coordinates
Y = normalized coordinate

Greek Symbols
Δ^* = displacement moment integral
Θ = momentum moment integral
Θ^* = kinetic energy moment integral
γ = vortex panel strength
δ = reference BL thickness scale
δ_{ij} = Kronecker delta ($\delta_{ij} = 1$ if $i = j$, $\delta_{ij} = 0$ if $i \neq j$)
δ^* = displacement thickness
η = normalized normal coordinate
θ = momentum thickness
θ^* = kinetic energy thickness
κ = streamline curvature
ξ = coordinate metric
ρ = density
σ = source panel strength
τ = shear stress

Subscripts and Superscripts
$(\ )_\infty$ = freestream quantity
$(\ )_{INV}$ = strictly inviscid flow quantity (BL absent)
$(\ )_o$ = first-order quantity (higher-order BL terms absent)
$(\tilde{\ })$ = wake quantity
$(\ )_i$ = quantity at panel node i
$(\ )_i$ = equivalent inviscid flow (EIF) quantity
$(\ )^*$ = reference EIF quantity at $y = 0$
$(\ )_w$ = BL wall quantity
$(\ )_e$ = BL edge quantity
$(\ )'$ = turbulent fluctuation quantity

I. Introduction

THE standard two-dimensional boundary layer (BL) approximations[1] neglect streamwise diffusion and normal momentum transport, under the assumption that the BL thickness δ is small compared to its streamwise extent x and its longitudinal radius of curvature $1/\kappa$. Specifically, the streamwise Navier–Stokes equation terms can be ordered by relative magnitude as follows:

    transport, normal diffusion: 1 (standard),
    curvature-induced transport: $\kappa\delta$ (retain),
    curvature-induced normal diffusion: $\delta/x\ \kappa\delta$ (discard), and
    streamwise diffusion: δ^2/x^2 (discard).

Only the "standard" terms are retained in first-order BL theory. The present work will retain terms up to the next higher order $\mathcal{O}(\kappa\delta)$. The rationale is that in many flows this scale can be 0.1 or greater in limited but important regions such as the trailing edge and separation bubbles.

Much of the present derivation of the flow-curvature effects follows that of Le Balleur[2] and the further analysis of Allmaras.[3] Also closely related to the present derivation is the work of Lock and Williams,[4] who model the curvature-related terms and incorporate them into Green's lag-entrainment method[5] for viscous/inviscid calculations. Cousteix and Houdeville[6] perform a comparable treatment on their basic entrainment integral method. They place more emphasis on accounting for the very large effects that flow curvature (and additional strain rates in general) is known to have on Reynolds stresses, as discussed by Bradshaw.[7]

The present derivation uses the dissipation-closure approach, which employs the integral momentum and kinetic energy equations. The was the closure method used by Le Balleur.[2] The current derivation also details second-order corrections to the integral thickness expressions and various closure relations, which apparently were not considered in any previous work. Derivations will be performed for the turbulent Reynolds-averaged equations, although applications will be shown for laminar flows only. The modeling of curvature effects on turbulent flows is not yet complete.

A high-order panel method is developed to allow accurate calculation of the flow curvature and to form the coupling relations with consistency. This is essential for obtaining a well-posed and stable system of the coupled viscous and inviscid equations. The fully simultaneous solution technique is employed to solve the overall viscous/inviscid system of equations.

II.   Curvilinear Coordinates and Equations

Orthogonal curvilinear coordinates x, y, and velocities u, v are defined as shown in Fig. 1, with the x axis having a local curvature κ. In effect, x, y are local polar coordinates r, ϑ and

$$\mathrm{d}x = \mathrm{d}\vartheta/\kappa, \qquad \mathrm{d}y = -\mathrm{d}r$$

The x axis is assumed to be located on the displacement surface, so that $\kappa = \kappa^*$. This surface is also a streamline of the equivalent inviscid flow (EIF), so that $u^* = u_\mathrm{i}(0)$ and $v^* = 0$ by definition. Lock and Williams[4] make the alternative choice of placing the x axis at the wall so that κ is the wall curvature, although κ^* still enters into their formulation. For a wake, the "wall" at $y = y_w$ is actually the nominal wake centerline (however defined). For generality, the derivation will not assume the no-slip condition at $y = y_w$.

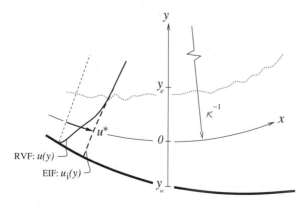

Fig. 1   Curvilinear thin shear layer coordinates.

The coordinate metric $\xi \equiv 1 - \kappa y$ gives an element of the physical length ℓ anywhere in the plane:

$$d\ell^2 = (\xi\,dx)^2 + dy^2$$

The Reynolds-averaged continuity, x-momentum, and y-momentum thin shear layer equations governing the real viscous flow (RVF) are then

$$\frac{\partial}{\partial x}(\rho u) + \frac{\partial}{\partial y}(\rho v\xi) = 0 \tag{1}$$

$$\frac{\partial}{\partial x}(\rho u^2) + \frac{\partial}{\partial y}(\rho uv\xi) + \frac{\partial}{\partial x}(p + \overline{\rho u'^2}) - \frac{\partial}{\partial y}(\tau\xi - \overline{\rho u'v'}\xi)$$
$$- \kappa\rho uv + \kappa(\tau - \overline{\rho u'v'}) = 0 \tag{2}$$

$$\frac{\partial}{\partial x}(\rho uv) + \frac{\partial}{\partial y}(\rho v^2\xi) + \frac{\partial}{\partial y}(p\xi + \overline{\rho v'^2}\xi) - \frac{\partial}{\partial x}(\tau - \overline{\rho u'v'}) + \kappa\rho u^2$$
$$+ \kappa(p + \overline{\rho u'^2}) = 0 \tag{3}$$

The laminar shear stress (and associated strain rate) is given by

$$\tau_{xy} = \mu\left(\frac{\partial u}{\partial y} + \frac{1}{\xi}\frac{\partial v}{\partial x} + \frac{\kappa u}{\xi}\right) \simeq \mu\left(\frac{\partial u}{\partial y} + \kappa u\right) \equiv \tau$$

with the normal stresses τ_{xx}, τ_{yy} being negligible in the second-order approximation.

Combining $2u$ [Eq. (2)] $+ 2v$ [Eq. (3)] produces the mean-flow kinetic energy equation, whose final form after discarding some of the $\mathcal{O}(\delta^2/x^2)$ terms is

$$\frac{\partial}{\partial x}(\rho u^3) + \frac{\partial}{\partial y}(\rho u^2 v\xi) + 2u\frac{\partial}{\partial x}(p + \overline{\rho u'^2}) + 2v\xi\frac{\partial}{\partial y}(p + \overline{\rho v'^2})$$
$$- \frac{\partial}{\partial y}[2u(\tau - \overline{\rho u'v'})\xi] + 2(\tau - \overline{\rho u'v'})\left(\frac{\partial u}{\partial y} + \kappa u\right)\xi = 0 \tag{4}$$

The $v\,\partial p/\partial y$ term is formally of higher order, but it is retained since its presence will actually simplify the integral form of this equation.

Assuming a variable density causes relatively few complications in the present development, and retaining it as a variable gives a physically clearer form to the equations. The incompressible assumption will be made only in the final integral closure development.

III.   Equivalent Inviscid Flow

The irrotational EIF streamwise velocity profile u_i (shown dashed in Fig. 1) is defined across the shear layer to leading order in κ as

$$u_i = u^*/(1 - \kappa y) \simeq u^*(1 + \kappa y)$$

$$\rho_i v_i\xi = -\frac{d(\rho^* u^*)}{dx}(1 + \kappa y)y$$

The $\rho_i v_i$ profile follows from the continuity equation (1).

IV. Entrainment Equation and Viscous/Inviscid Coupling

The continuity equation (1) put into defect form is simply

$$\frac{\partial}{\partial x}(\rho_i u_i - \rho u) + \frac{\partial}{\partial y}[(\rho_i v_i - \rho v)\xi] = 0$$

which is then integrated over the shear layer to give the following viscous/inviscid coupling condition:

$$\frac{d}{dx}(\rho^* u^* \delta^*) + \rho_{iw} E_{iw} \xi_w - \rho_w E_w \xi_w = 0 \tag{5}$$

The displacement thickness and entrainment velocity are defined as

$$\rho^* u^* \delta^* = \int_{y_w}^{y_e} (\rho_i u_i - \rho u)\, dy, \qquad E = \frac{u}{\xi}\frac{dy}{dx} - v$$

Physically, E is the volume flow per unit x length passing downward through the $y(x)$ boundary. The fictitious blowing mass flux $\rho V \cdot \hat{n}$ needed to model the influence of the BL on the potential flow is therefore given by

$$\rho V \cdot \hat{n}\, ds = -\rho_{iw} E_{iw} \xi_w\, dx = d(\rho^* u^* \delta^*) + \rho_w E_w \xi_w\, dx \tag{6}$$

which is the usual wall-transpiration coupling condition with a physical wall-blowing term added. For the case of zero physical wall blowing $E_w = 0$, and with ρ^*, u^*, and κ assumed to vary slowly in x, the coupling condition (5) integrates to

$$y_w = -\delta^* \left(1 + \tfrac{1}{2}\kappa \delta^*\right) \tag{7}$$

This is the usual δ^*–offset coupling condition, but with a second-order curvature correction term.

V. Integral Momentum and Kinetic Energy Equations

The defect forms of the momentum equation (2) and kinetic energy equation (4) are integrated between y_w and y_e and combined with the entrainment equation (5) to produce the standard integral BL equations with additional higher-order terms (h.o.t.):

$$\frac{d}{dx}(\rho^* u^{*2}\theta) + \rho^* u^* \delta^* \frac{du^*}{dx} + (\tau_i - \tau + \overline{\rho u'v'})_w\, \xi_w \qquad \text{[standard } \mathcal{O}(1)]$$

$$+ \rho_{iw} E_{iw} \xi_w (u_{iw} - u^*) \qquad\qquad\qquad \text{[h.o.t. } \mathcal{O}(\kappa \delta)]$$

$$+ \frac{d}{dx}\int_{y_w}^{y_e}(p_i - p - \overline{\rho u'^2})\, dy + (p_i - p - \overline{\rho u'^2})_w \frac{dy_w}{dx} \qquad \text{[h.o.t. } \mathcal{O}(\kappa \delta)]$$

$$- \kappa \int_{y_w}^{y_e}(\rho_i u_i v_i - \rho u v)\, dy + \kappa \int_{y_w}^{y_e}(\tau_i - \tau + \overline{\rho u'v'})\, dy = 0 \quad \text{[h.o.t. } \mathcal{O}(\kappa \delta)]$$

$$\tag{8}$$

$$\frac{d}{dx}(\rho^* u^{*3}\theta^*) + 2\rho^* u^{*2}(\delta^* - \delta_u)\frac{du^*}{dx} - 2(D_V + D_{12}) \qquad \text{[standard } \mathcal{O}(1)\text{]}$$

$$+ \rho_{iw} E_{iw}\xi_w(u_{iw}^2 - u^{*2}) - 2u_{iw}\tau_{iw}\xi_w \qquad \text{[h.o.t. } \mathcal{O}(\kappa\delta)\text{]}$$

$$+ 2\frac{d}{dx}\int_{y_w}^{y_e} u(p_i - p - \overline{\rho u'^2})\,dy + 2E_w\xi_w(p_i - p - \overline{\rho u'^2})_w \quad \text{[h.o.t. } \mathcal{O}(\kappa\delta)\text{]}$$

$$- 2(D_{11} + D_{22}) = 0 \qquad \text{[h.o.t. } \mathcal{O}(\kappa\delta)\text{]}$$

$$(9)$$

The various thicknesses and stress integrals in these equations are defined as follows:

$$\rho^* u^{*2}(\theta + \delta^*) = \int_{y_w}^{y_e}(\rho_i u_i^2 - \rho u^2)\,dy$$

$$\rho^* u^{*3}(\theta^* + \delta^*) = \int_{y_w}^{y_e}(\rho_i u_i^3 - \rho u^3)\,dy$$

$$\rho^* u^{*2}\frac{du^*}{dx}\delta_u = -\int_{y_w}^{y_e}(u_i - u)\frac{\partial p_i}{\partial x}\,dy = \int_{y_w}^{y_e}(u_i - u)\rho_i u_i\frac{\partial u_i}{\partial x}\,dy$$

$$D_V \equiv \int_{y_w}^{y_e}\left[\mu\left(\frac{\partial u}{\partial y} + \kappa u\right)^2 - \mu_i\left(\frac{\partial u_i}{\partial y} + \kappa u_i\right)^2\right]\xi\,dy$$

$$D_{12} \equiv \int_{y_w}^{y_e} -\overline{\rho u'v'}\left(\frac{\partial u}{\partial y} + \kappa u\right)\xi\,dy$$

$$D_{11} + D_{22} \equiv \int_{y_w}^{y_e} -\overline{\rho u'^2}\frac{\partial u}{\partial x}\,dy + \int_{y_w}^{y_e} -\overline{\rho v'^2}\frac{\partial v}{\partial y}\xi\,dy$$

$$\simeq \int_{y_w}^{y_e}(\overline{\rho u'^2} - \overline{\rho v'^2})\frac{\partial v}{\partial y}\xi\,dy$$

These definitions all reduce to the standard BL forms if the curvature κ is neglected. The usual total dissipation defect integral $D = D_V + D_{12}$ is split here into viscous-stress and Reynolds shear-stress components. The approximation for $D_{11} + D_{22}$ assumes a negligible velocity divergence.

It should be noted that the effects of the flow curvature enter virtually all the terms in Eqs. (8) and (9) via the local metric ξ, not just the h.o.t. terms.

VI. Turbulent Transport Equation

The integral form of the turbulent kinetic energy transport equation is used to model the evolution of the turbulence scale C_q, which is related to the Reynolds shear stress by the assumed constant a_{12}:

$$\overline{(u'^2 + v'^2 + w'^2)} = u^{*2}C_q, \qquad -\overline{u'v'} = a_{12}u^{*2}C_q$$

This is very similar to the "lag equation" formulation of Green.[5] This development will not be presented here, although the turbulence quantities will be shown as they appear.

VII.   Real Viscous Flow Profiles

The RVF profiles, defined via a suitable set of assumed profile families, are necessary for constructing the closure relations for Eqs. (8) and (9). Lock and Williams assume that the original first-order relations remain unchanged in the second-order treatment. Arguably, this assumption is inconsistent since it neglects possible contributions of $\mathcal{O}(\kappa\delta)$ to their shape parameter, skin friction, and entrainment closure relations. Here, all higher-order effects on the correlations will be retained.

A.   Assumed Profiles

The assumed profile of the RVF is the same normalized profile shape $u/u_e = U$ used in a typical first-order formulation, except that the constant edge velocity u_e is replaced by u_i, which is linear in y. The incompressible approximation is also invoked at this point:

$$u/u_i = U(y)$$

$$\rho/\rho_i = 1$$

Using the EIF profile expression (5), the complete RVF profile is given by

$$u = u^* U (1 + \kappa y) \tag{10}$$

and the total defect profiles follow similarly, with all terms $\mathcal{O}(\kappa^2\delta^2)$ and higher being discarded. The incompressible assumption is also invoked at this point:

$$\rho_i u_i - \rho u = \rho^* u^* [1 - U](1 + \kappa y)$$

$$\rho_i u_i^2 - \rho u^2 = \rho^* u^{*2}[1 - U^2](1 + 2\kappa y)$$

$$\rho_i u_i^3 - \rho u^3 = \rho^* u^{*3}[1 - U^3](1 + 3\kappa y)$$

$$(u_i - u)\rho_i u_i \frac{\partial u_i}{\partial x} = \rho^* u^{*2} \frac{du^*}{dx}[1 - U](1 + 3\kappa y)$$

B.   Pressure and Shear Stress Profiles

The laminar stress is given to second order as

$$\tau\,\xi = \mu\left(\frac{\partial u}{\partial y} + \kappa u\right)\xi \simeq \mu^* u^* \frac{\mu}{\mu^*}\left(\frac{dU}{dy} + 2\kappa U\right) \tag{11}$$

The pressure defect profile is estimated from the defect form of the y-momentum equation (3). Only the leading-order "centrifugal" term needs to be retained, and so

$$p_i - p = \rho^* u^{*2} \kappa\delta\,P + \overline{\rho v'^2} \tag{12}$$

The normalized pressure defect profile P can be approximated by

$$P(y) = \int_y^{y_e} \frac{\rho_i u_i^2}{\rho^* u^{*2}}(1 - U^2)\frac{dy}{\delta} \simeq \int_y^{y_e}(1 - U^2)\frac{dy}{\delta} \tag{13}$$

where δ is some convenient shear layer thickness scale, which is not necessarily the actual thickness of the shear layer. The simplification discards terms $\mathcal{O}(\kappa\delta)$. But because P itself always appears multiplied by $\kappa\delta$, this is in fact an $\mathcal{O}(\kappa^2\delta^2)$ approximation consistent with the present development.

C. Integral Thicknesses

It is convenient to introduce the normalized y distance function Y and the normalized integration coordinate η:

$$Y = \frac{y}{\delta}, \qquad \eta = \frac{y - y_w}{\delta}$$

where δ is the shear layer thickness scale introduced earlier. Again, this is not necessarily the actual thickness of the shear layer.

In the interest of forming more concise relations for the various integral thicknesses, it is convenient to define the following "basic" thicknesses devoid of curvature effects:

$$\bar{\delta}^* = \delta \int [1 - U]\, d\eta$$

$$\bar{\theta} = \delta \int [U - U^2]\, d\eta$$

$$\bar{\theta}^* = \delta \int [U - U^3]\, d\eta$$

It is also convenient to define the incompressible "thickness moment" integrals:

$$\Delta^* = \delta^2 \int [1 - U]Y\, d\eta$$

$$\Theta = \delta^2 \int [U - U^2]Y\, d\eta$$

$$\Theta^* = \delta^2 \int [U - U^3]Y\, d\eta$$

$$\Delta_P = \delta^2 \int P\, d\eta$$

$$\Delta_{PU} = \delta^2 \int PU\, d\eta$$

All the thickness ratios such as $\bar{\delta}^*/\delta$ and moment ratios such as Δ^*/δ^2, depend only on the fundamental parameter or parameters of the basic $U(\eta)$ profile family. Hence, these ratios can be precomputed and stored as functions in profile-parameter space. All the compressibility and higher-order curvature corrections are performed separately and do not affect these correlations.

Using all the definitions above, the full compressible thicknesses can now be expressed as follows:

$$\delta^* = \bar{\delta}^* + \kappa[\Delta^*]$$

$$\theta = \bar{\theta} + \kappa[2\Theta + \Delta^*]$$

$$\theta^* = \bar{\theta}^* + \kappa[3\Theta^* + 2\Delta^*]$$

$$\delta^* - \delta_u \equiv \delta^{**} = \kappa[-2\Delta^*]$$

The density flux thickness δ^{**} appearing in the kinetic energy equation (9) normally vanishes for incompressible flows, but not if curvature effects are significant. The κ terms in these relations effectively account for profile deformation due to flow curvature.

D. Laminar Skin Friction and Dissipation Integral

The RVF laminar skin friction term is obtained by simply evaluating the shear stress coefficient at the wall:

$$\frac{\rho^* u^* \delta}{\mu^*} \frac{\tau_w \, \xi_w}{\rho^* u^{*2}} \equiv Re_\delta \frac{C_f}{2} = \left(\frac{dU}{d\eta}\right)_w$$

Similarly, the laminar dissipation defect integral is expressed in terms of the assumed profiles as

$$\frac{\rho^* u^* \delta}{\mu^*} \frac{D_V}{\rho^* u^{*3}} \equiv Re_\delta \, C_{D_V} = \int \left(\frac{dU}{d\eta}\right)^2 d\eta \, + \, 2\kappa\delta$$

As with the various thicknesses, the integrals can be precomputed and stored as functions of the profile parameters. The turbulent dissipation integrals are not tied uniquely to the velocity profile and will be treated in a later section.

VIII.   Profile Families

For laminar flow, the particular choice for $U(\eta; \bar{H})$ used here is the Falkner–Skan profile family for incompressible flows, with $\bar{\delta}^*(x)$, $\bar{\theta}(x)$ being chosen as the fundamental variables, and $\bar{H} = \bar{\delta}^*/\bar{\theta}$ being used as the profile parameter. For $\bar{H} > 5.0$, nonsimilar profiles generated by finite-difference boundary layer calculations are used in lieu of the Falkner–Skan profiles. These have smaller velocities in the reversed-flow region, which is more representative of profiles found in separation bubbles.

A natural choice for the shear layer thickness scale is $\delta = \bar{\theta}$, so that $\eta = (y - y_w)/\bar{\theta}$. The auxilliary relations generated from this profile family then have the following functional dependencies:

$$\bar{\theta}^*/\bar{\theta} \equiv \bar{H}^* = f_1(\bar{H}), \qquad U'_w = f_2(\bar{H}), \qquad P_w = f_3(\bar{H}), \qquad C_{D_V} = f_4(\bar{H}) \quad \dots$$

The higher-order curvature terms also require the following moment integral correlations:

$$\Delta^*/\bar{\theta}^2 = g_1(\bar{H}), \qquad \Theta/\bar{\theta}^2 = g_2(\bar{H}), \qquad \Theta^*/\bar{\theta}^2 = g_3(\bar{H}) \quad \cdots$$

All the correlation functions are generated numerically from the assumed profile family and approximated with curve fits.

IX. Higher-Order Corrections

The profile correlations developed above already contain correction terms that are $\mathcal{O}(\kappa\delta)$. It is of course also necessary to determine the actual higher-order terms that appear in the integral Eqs. (8) and (9). The results here are consistent with those of Lock and Williams,[4] if allowance is made for the different coordinate systems used.

A. Entrainment Terms

The RVF wall entrainment terms containing E_w in Eqs. (8) and (9) will typically be prescribed and do not need to be modeled. The only exception occurs at a wake centerline, whose definition provides the necessary equation for E_w.[8]

The EIF wall entrainment terms are smaller than the standard terms by a factor of $\mathcal{O}(\kappa\delta)$ and are readily estimated from the known EIF profiles (5) and the approximate δ^*-y_w relation (7). For the momentum term we have

$$\rho_{iw} E_{iw} \xi_{iw}(u_{iw} - u^*) \simeq \frac{d(\rho^* u^* \delta^*)}{dx} u^* \kappa \delta^* \tag{14}$$

and likewise for the corresponding kinetic energy term we have

$$\rho_{iw} E_{iw} \xi_{iw}\left(u_{iw}^2 - u^{*2}\right) \simeq \frac{d(\rho^* u^* \delta^*)}{dx} 2u^{*2} \kappa \delta^* \tag{15}$$

Terms that are $\mathcal{O}(\kappa^3\delta^3)$ have been dropped in the approximations.

B. Normal Flux Integral

The momentum normal-flux integral term in Eq. (8) is approximated to second order as

$$\int_{y_w}^{y_e} (\rho_i u_i v_i - \rho u v)\, dy \simeq 0 \tag{16}$$

Aligning the coordinates with the displacement surface nearly eliminates the normal-flux integral (16). That we can neglect this integral is supported by finite-difference boundary layer calculations for the Simpson separating diffuser flow.[9] For a flat wall, where $\kappa = d^2\delta^*/dx^2$, these calculations indicate that the integral is typically $0.02\,\rho_e u_e^2 \kappa \delta^*$ or less, and never exceeds $0.10\,\rho_e u_e^2 \kappa \delta^*$ in any case. Hence, its relative contribution is usually less than 2% of the other higher-order terms, and assuming that it is zero is quite reasonable.

C. Pressure Defect Integrals

At the wall (or wake centerline), the pressure defect takes on the particular value

$$\Delta p_w \equiv p_{iw} - p_w \simeq \rho^* u^{*2} \kappa(\theta + \delta^*) + (\rho \overline{v'^2})_w \qquad (17)$$

so that the wall pressure term in Eq. (8) can be immediately estimated by

$$(p_i - p - \rho \overline{u'^2})_w \frac{dy_w}{dx} \simeq -\left[\rho^* u^{*2} \kappa(\theta + \delta^*) + (\rho \overline{v'^2})_w - (\rho \overline{u'^2})_w\right] \frac{d\delta^*}{dx} \qquad (18)$$

Likewise, the wall pressure work term in Eq. (9) can be estimated as

$$E_w(p_i - p - \rho \overline{u'^2})_w \simeq E_w\left[\rho^* u^{*2} \kappa(\theta + \delta^*) + (\rho \overline{v'^2})_w - (\rho \overline{u'^2})_w\right] \qquad (19)$$

which is normally zero at a solid wall. The Reynolds stress terms are of course also zero at a solid wall and usually can be neglected at wake centerlines.

The pressure-defect integrals in Eq. (8) can be obtained directly using the pressure-defect profile (13):

$$\int_{y_w}^{y_e} (p_i - p - \rho \overline{u'^2}) \, dy \simeq \rho^* u^{*2} \left(\kappa \Delta_P - C_q^2 \delta_\pi\right) \qquad (20)$$

The pressure-defect thickness moment Δ_P is approximated by

$$\kappa \Delta_P \simeq \tfrac{1}{2} \kappa(\delta^* + \theta)^2 \qquad (21)$$

as can be seen by examining the area between the p_i and p curves sketched in Fig. 2. This approximation was assumed by Lock and Williams.

The pressure-defect work integrals in Eq. (9) can likewise be obtained using the pressure-defect profile:

$$\int_{y_w}^{y_e} u(p_i - p - \rho \overline{u'^2}) \, dy \simeq \rho^* u^{*3} \left(\kappa \Delta_{PU} - C_q^2 \theta_\pi\right) \qquad (22)$$

The normal Reynolds stress flux thickness θ_π is treated in the next section.

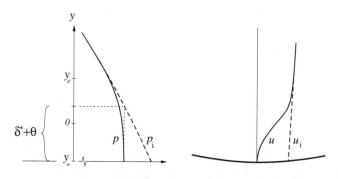

Fig. 2   RVF and EIF pressure and velocity profiles.

D. Shear Stress Defect Integral

The shear stress defect integral in Eq. (8) is estimated to first order as

$$\kappa \int (\tau_i - \tau + \rho \overline{u'v'})\, dy \simeq -\rho^* u^{*2} \kappa \delta \left\{ \frac{1}{Re_\delta} + C_q^2 \frac{\delta_\tau}{\delta} \right\} \qquad (23)$$

The EIF wall shear can be similarly expressed as

$$\tau_{iw} \xi_w = \mu_{iw}\, 2 \kappa u_{iw}\, \xi_w \simeq \rho^* u^{*2}\, 2 \kappa \delta \frac{1}{Re_\delta} \qquad (24)$$

The two higher-order shear stress terms above both appear in the momentum equation (8) and have the same form, and so it is convenient to lump them into one coefficient $C_{f\kappa}$, which is in effect a curvature-related correction term to the RVF skin friction coefficient C_f:

$$\frac{C_{f\kappa}}{2} \equiv -\frac{\tau_{iw}\xi_w}{\rho^* u^{*2}} - \frac{\kappa}{\rho^* u^{*2}} \int (\tau_i - \tau + \rho \overline{u'v'})\, dy = \kappa \delta \left\{ -\frac{1}{Re_\delta} + C_q^2 \frac{\delta_\tau}{\delta} \right\} \qquad (25)$$

The EIF shear-work term that appears in the kinetic energy equation (9) is readily estimated and expressed as a coefficient C_{D_κ}. This is in effect a curvature-related correction to the usual RVF viscous dissipation integral C_{D_V}:

$$C_{D_\kappa} \equiv \frac{u_{iw}\tau_{iw}\xi_w}{\rho^* u^{*3}} = \frac{\mu_{iw}\, 2\kappa u_{iw}^2\, \xi_w}{\rho^* u^{*3}} \simeq 2\kappa\delta \frac{1}{Re_\delta} \qquad (26)$$

X. High-Order Panel Method

The panel method employed in the present work is designed specifically to interface with the higher-order BL equations derived earlier. The major requirement is the unambiguous calculation of the velocity and flow curvature on the body and wake surface.

A. Discretization

All body surfaces and wakes are represented with panels having linear vortex sheet strength γ and linear source sheet strength σ, as shown in Fig. 3. The local strength at any surface point is then defined by the panel node sheet strengths γ_j, σ_j. The use of linear source strength is an improvement over the author's previous work,[10] which employed constant-source panels. Velocity $\vec{q}_i$ is evaluated at each field point $\vec{r}_i$ placed at the midpoint of panel $i \ldots i+1$:

$$\vec{q}_i = \vec{q}_\infty + \sum_{j=1}^{N} \vec{a}_{ij}\gamma_j + \sum_{j=1}^{N} \vec{b}_{ij}\sigma_j \qquad (27)$$

The summation is performed over all N panel nodes. The influence coefficients $\vec{a}_{ij}, \vec{b}_{ij}$ can be evaluated analytically for two-dimensional geometries. A numerical Romberg integration is necessary for axisymmetric and cascade geometries. The velocity $\vec{q}_i$ is evaluated right on the panel, with that panel's jump contribution

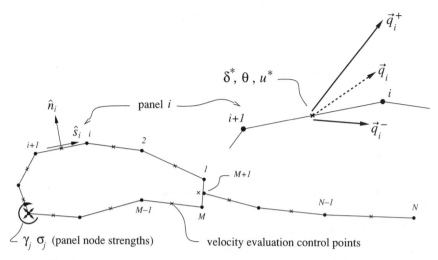

Fig. 3   Panel configuration, velocities, and viscous variable locations.

excluded. After the sheet jumps are added explicitly, the velocities $\vec{q}_i^{\pm}$ on the two sides of the panel are then given by

$$\vec{q}_i^{\pm} = \vec{q}_i \pm \tfrac{1}{4}(\gamma_i + \gamma_{i+1})\hat{s}_i \pm \tfrac{1}{4}(\sigma_i + \sigma_{i+1})\hat{n}_i \tag{28}$$

as shown in Fig. 3. The velocities defined by expressions (27) and (28) are used throughout the overall numerical formulation. The external velocity $\vec{q}_i^{+}$ will define the edge velocities u^* and streamline curvatures κ appearing in the BL equations. The interior velocity $\vec{q}_i^{-}$ is used to impose Neumann boundary conditions for the panel system. Both $\vec{q}_i^{+}$ and $\vec{q}_i^{-}$ are external velocities at a wake station; the average velocity $\vec{q}_i$ defines the trajectory of the wake. The panel midpoints are also the locations where the discrete BL variables are located. This differs from the previous work[10] which placed the BL variables at the panel nodes.

B. Corner Compatibility Relations

The linear panel vorticity and source distributions are continuous by construction. The velocity field representation is smooth, with the exception of very weak second-order logarithmic singularities at the panel nodes due to the slight corners at panel nodes. Finite panel corner angles, which typically occur at trailing edges, must have discontinuous panel strengths to preclude zeroth-order velocity singularities. This is accommodated by using two distinct panel nodes at the corner, as shown in Fig. 4. A smooth external velocity is obtained at the corner provided the following jump conditions on γ and σ between the two nodes are met:

$$\sigma_1\hat{n}_1 + \gamma_1\hat{s}_1 = \sigma_2\hat{n}_2 + \gamma_2\hat{s}_2 \tag{29}$$

This relation is used to set the source and vortex strengths of the panel spanning a blunt trailing edge and to construct the Kutta condition.

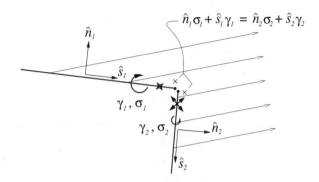

Fig. 4   Compatibility between panel strengths across a panel corner.

C.   Panel System Formulation

The $(M + 1)^2$ panel system is formed using simple Neumann boundary conditions applied at $M - 1$ panel midpoints $\vec{r}_i$ in the interior of a body. For a closed body a constant "leakage" normal velocity q_n must be introduced as the one additional $(M + 1)$th degree of freedom to keep the internal Neumann problem well-posed:

$$\vec{q}_i^- \cdot \hat{\boldsymbol{n}}_i = q_n$$

$$\sum_{j=1}^{M} a_{ij}\, \gamma_j = -\vec{q}_\infty \cdot \hat{\boldsymbol{n}}_i \; - \; \sum_{j=M+1}^{N} \tilde{a}_{ij}\, \tilde{\gamma}_j \; - \; \sum_{j=1}^{N} b_{ij}\, \sigma_j \qquad (30)$$

where

$$a_{ij} = \vec{a}_{ij} \cdot \hat{\boldsymbol{n}}_i$$

$$b_{ij} = \vec{b}_{ij} \cdot \hat{\boldsymbol{n}}_i - \tfrac{1}{4}(\delta_{ij} + \delta_{i+1\,j})$$

and $\tilde{a}_{ij}$, $\tilde{\gamma}_j$ correspond to wake nodes. The discrete Kutta condition is derived by applying the jump condition (29) at the panel corners at the trailing edge, and dotting it along the first wake panel direction $\hat{\boldsymbol{s}}_{M+1}$. This gives the necessary Mth equation of system (30):

$$(\hat{\boldsymbol{s}}_{M+1} \cdot \hat{\boldsymbol{s}}_M)\, \gamma_M \; - \; (\hat{\boldsymbol{s}}_{M+1} \cdot \hat{\boldsymbol{s}}_1)\, \gamma_1 \; + \; (\hat{\boldsymbol{s}}_{M+1} \cdot \hat{\boldsymbol{n}}_M)\, \sigma_M \; - \; (\hat{\boldsymbol{s}}_{M+1} \cdot \hat{\boldsymbol{n}}_1)\, \sigma_1 \; = \; \tilde{\gamma}_{M+1} \qquad (31)$$

The vortex strength $\tilde{\gamma}_{M+1}$ at the start of the wake biases the Kutta condition and tends to reduce the overall circulation. It is related to the wake curvature as derived below, and hence it is zero in first-order BL theory. The source strength terms in the Kutta condition (31), present only in viscous calculations, are somewhat unexpected. They can cause additional significant reduction in circulation, most notably on thick airfoils.

The last $(M + 1)$th equation for the normal velocity variable q_n is best constrained by imposing a zero velocity component along the trailing-edge bisector $\hat{\boldsymbol{t}}$, just inside the trailing-edge point:

$$(\vec{q} \cdot \hat{\boldsymbol{t}})_{TE} = 0$$

The resulting q_n determined from the overall linear system is typically very small, $10^{-4}q_\infty$ or less, and simply reflects imperfect flow tangency imposition caused by panel discretization errors.

The source strengths of all panels and the vortex strengths of the wake panels are provided externally to the panel formulation, and hence they are placed on the right-hand side of Eq. (30). This system is then reduced by LU factorization into explicit expressions for the surface node vortex strengths and velocities:

$$\gamma_i = \gamma_{\mathrm{INV}_i} + A_{ij}\,\tilde{\gamma}_j + B_{ij}\,\sigma_j \tag{32}$$

$$\vec{q}_i^{\pm} = \vec{q}_{\mathrm{INV}_i} + \vec{C}_{ij}\,\tilde{\gamma}_j + \vec{D}_{ij}\,\sigma_j \pm \tfrac{1}{4}(\gamma_i + \gamma_{i+1})\hat{s}_i \pm \tfrac{1}{4}(\sigma_i + \sigma_{i+1})\hat{n}_i \tag{33}$$

The inviscid-case quantities and viscous-case influence matrices are computed via LU decomposition as follows:

$$\gamma_{\mathrm{INV}_i} = a_{ij}^{-1}\{-\vec{q}_\infty \cdot \hat{n}_i\} \qquad \vec{q}_{\mathrm{INV}_i} = \vec{q}_\infty + \vec{a}_{ij}\gamma_{\mathrm{INV}_j}$$

$$A_{ij} = a_{ij}^{-1}\{-\tilde{a}_{ij}\} \qquad \vec{C}_{ij} = \vec{a}_{ij} + a_{ik}A_{kj}$$

$$B_{ij} = a_{ij}^{-1}\{-b_{ij}\} \qquad \vec{D}_{ij} = \vec{b}_{ij} + a_{ik}B_{kj}$$

D. Panel/Viscous Variable Relations

The panel source strengths in Eq. (33) are related to the viscous mass defect distribution via the coupling equation (6). Because the interior flow is zero, the source strength for incompressible flow is equal to the exterior normal EIF velocity:

$$\sigma = \vec{q}^+ \cdot \hat{n}$$

In terms of the BL station values shown in Fig. 5, the node source sheet strengths are then given in terms of the viscous mass defects $m = u^*\delta^*$ by the discrete version of the coupling equation (6),

$$\sigma_i = \frac{m_i - m_{i-1}}{\Delta s} \tag{34}$$

so that the total volume outflow resulting from this nodal σ_i value exactly matches the mass defect change, as indicated in Fig. 6.

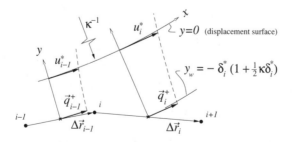

Fig. 5   Flow curvature and local x, y polar coordinates defined by velocities on two adjacent panels.

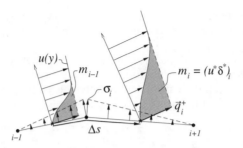

Fig. 6   Panel node source strength σ_i related to change in viscous mass defect.

The EIF velocity at the displacement surface, which is what is seen by the BL equations, then follows from the EIF profile (5), approximated to second order as

$$u_i^* = q_i^+(1 - \kappa y_w) = q_i^+\left[1 + \kappa\delta^*\left(1 + \tfrac{1}{2}\kappa\delta^*\right)\right] \simeq q_i^+ + \kappa m_i \qquad (35)$$

as indicated in Fig. 5. The geometry of the displacement surface does not need to be constructed.

Physically, the pressure profiles $p(y)$ of the upper and lower wake halves, defined by Eq. (12), must be equal at $y = y_w$. This gives the jump condition on the EIF pressures at the panel:

$$p_i^+ - p_i^- = \left[\rho^* u^{*2} \kappa \delta\, P(y_w)\right]^+ - \left[\rho^* u^{*2} \kappa \delta\, P(y_w)\right]^-$$

$$= \rho^* u^{*2}(\theta + \delta^*)\kappa$$

This EIF pressure jump implies a tangential EIF panel velocity jump, which in turn implies a nonzero wake vortex sheet strength $\tilde{\gamma}$. The relation is

$$\rho q \tilde{\gamma} = -\rho^* u^{*2}(\theta + \delta^*)\kappa \qquad (36)$$

as illustrated in Fig. 7. This nonzero wake vortex strength γ implies the presence of a load on the wake, which is manifested by the wake's total momentum defect $\rho^* u^{*2}(\theta + \delta^*)$ being deflected by the wake curvature.

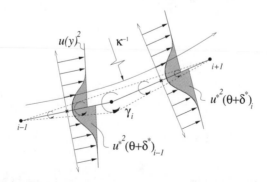

Fig. 7   Wake panel node vortex strength $\tilde{\gamma}_i$ related to curvature and momentum defect.

E.   Wake Trajectory

The wake trajectory follows the EIF average streamline, which is parallel to the control-point velocity $\vec{q}_i$ defined by Eq. (28), with the local wake panel strength terms excluded. The discrete defining relation is

$$\vec{q}_i \cdot \Delta \vec{r}_i = 0 \tag{37}$$

which implicitly determines the location of the wake nodes $\vec{r}_i$, $i = M+2, \ldots, N$. The first wake node $\vec{r}_{M+1}$ is fixed at the trailing-edge base midpoint. In practice it is rarely necessary to treat the wake implicitly during solution of the viscous problem. All higher-order quantities are computed with the velocity field directly, and the exact paneled wake shape has little influence on these velocities. The wake position is likely to be important only in cases such as multi-element flows where the wake is close to the surface of a trailing element.

XI.   Viscous/Inviscid System Formulation

A.   Newton System

The discrete integral momentum, kinetic energy, and turbulent kinetic energy BL equations become closed when expressions for the otherwise unknown velocities u_i^* and curvature κ are provided. The panel method provides these expressions in the form of Eqs. (35) and (38). Other supporting relations are also needed to provide auxiliary quantities.

The overall system is strongly nonlinear and is solved via a full Newton method, following the successful approach of the previous work.[10] The Newton variables are $\mathbf{U} = (m_i \ \theta_i \ C_{q_i})$. These then define the derived variables σ_i, $\vec{q}_i$, u_i^*, etc. The Newton residuals are the discrete boundary layer equations $\mathbf{R}(\mathbf{U}) = \mathbf{0}$. Newton iteration is then used to update $\mathbf{U}$:

$$\left[\frac{\partial \mathbf{R}}{\partial \mathbf{U}} \right] \delta \mathbf{U} = -\mathbf{R}$$

$$\mathbf{U}^{n+1} = \mathbf{U}^n + \delta \mathbf{U}$$

This higher-order system is only marginally more expensive to solve than the simpler first-order system described in Ref. 10. The additional work is due to the need to compute the wake vortex panel influence coefficients A_{ij} and C_{ij} and the need to represent the wake as two layers rather than one single layer. Computation of the Newton residuals and their Jacobians also now involves more terms, but these are relatively negligible.

B.   System Solution Procedure

The Newton system setup and solution proceeds incrementally, starting with the known, strictly inviscid panel solution γ_{INV}, $\vec{q}_{\text{INV}}$. These are also used to compute the initial wake trajectory and paneling. It is preferable to simply assume that this inviscid wake trajectory is fixed, since the BL source influence matrices can then also be precomputed and held fixed during the whole Newton iteration. Otherwise they must be recomputed every iteration with a several-fold increase in computational cost. In practice assuming a fixed inviscid wake trajectory produces only miniscule errors in the final viscous solution.

A Newton iteration cycle begins with the latest viscous variable iterates m_i, θ_i, and C_{q_i}. The corresponding panel node source strengths σ_i are then defined using relation (34).

Because the flow curvature κ is used only in second-order correction terms, it is sufficient (and in practice necessary) to compute it using only a first-order representation of the flowfield, denoted here by $(\ )_o$. This is provided by Eqs. (32) and (33), but with the higher-order wake vortex strengths $\tilde{\gamma}_j$ ignored:

$$\gamma_{o_i} = \gamma_{\text{INV}_i} + B_{ij}\,\sigma_j$$

$$\vec{q}_{o_i}^{\pm} = \vec{q}_{\text{INV}_i} + \vec{D}_{ij}\,\sigma_j \pm \tfrac{1}{4}\big(\gamma_{o_i} + \gamma_{o_{i+1}}\big)\hat{s}_i \pm \tfrac{1}{4}(\sigma_i + \sigma_{i+1})\hat{n}_i$$

$$u_{o_i}^{*} = q_{o_i}^{+} = |\vec{q}_{o_i}^{+}|$$

The flow curvature κ can be defined in terms of the velocity field $\vec{q}(\vec{r})$ via the differential relation

$$\kappa\,\vec{q} \cdot \mathrm{d}\vec{r} = \frac{1}{q}\,\hat{k} \cdot \vec{q} \times \mathrm{d}\vec{q} = \frac{1}{q}\,\hat{k} \cdot \vec{q} \times (\vec{q} + \mathrm{d}\vec{q})$$

where $\hat{k}$ is the unit normal to the plane. A convenient discrete version of this relation,

$$\kappa = \big(\vec{q}_{o_{i-1}}^{+} \cdot \Delta\vec{r}_{i-1} + \vec{q}_{o_i}^{+} \cdot \Delta\vec{r}_i\big)^{-1}\left(\frac{1}{q_{o_{i-1}}^{+}} + \frac{1}{q_{o_i}^{+}}\right)\hat{k} \cdot \vec{q}_{o_{i-1}}^{+} \times \vec{q}_{o_i}^{+} \qquad (38)$$

is used to define the average flow curvature over an interval between control points, as shown in Fig. 5. The curvature κ_i at the control point itself is obtained by using three-point central differences between the $i - 1$ and $i + 1$ points.

The nodal wake vortex strengths $\tilde{\gamma}_i$ are given to first order by the discrete form of relation (36), with the contributions of the upper and lower wake half quantities $(\ )^+$ and $(\ )^-$ added together:

$$\big(\vec{q}_{o_{i-1}} \cdot \Delta\vec{r}_{i-1} + \vec{q}_{o_i} \cdot \Delta\vec{r}_i\big)\,\tilde{\gamma}_i = -\hat{k} \cdot \vec{q}_{o_{i-1}}^{+} \times \vec{q}_{o_i}^{+}\big[(\theta + m/u_o^*)_{i-1}^{+} + (\theta + m/u_o^*)_i^{+}\big]$$

$$- \hat{k} \cdot \vec{q}_{o_{i-1}}^{-} \times \vec{q}_{o_i}^{-}\big[(\theta + m/u_o^*)_{i-1}^{-} + (\theta + m/u_o^*)_i^{-}\big] \qquad (39)$$

The value of $\tilde{\gamma}_{M+1}$ at the first point in the wake set by this equation directly influences the Kutta condition (31) and therefore has an effect on the airfoil lift.

With the flow curvatures κ and wake vortex strengths $\tilde{\gamma}$ defined in terms of the Newton variables, the Newton system for the full higher-order BL equations can be constructed. Its solution provides the Newton updates for m_i, θ_i, and C_{q_i} and thus completes one Newton iteration. The iteration is repeated with the updated viscous variables. Convergence is achieved when the Newton updates drop below some small tolerance.

XII. Results

A. Flat Plate Drag

The drag on a flat plate (NACA 0001) at zero angle of attack is computed down to a chord Reynolds number of 20. The theoretical drag based on analytical

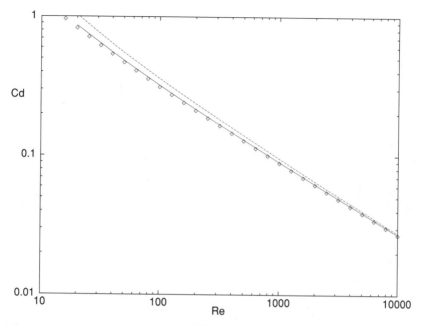

Fig. 8 Drag coefficient on a flat plate (NACA 0001) vs chord Reynolds number, for second-order method (solid line), first-order method (dashed line), and asymptotic analysis (symbols).

higher-order BL theory is given by Van Dyke[11] as

$$C_d = 2\left(\frac{1.328}{\sqrt{Re_c}} + \frac{2.326}{Re_c}\right)$$

Figure 8 compares this with computed values using the present second-order BL formulation and with the first-order BL formulation obtained by setting $\kappa = 0$ artificially. The latter result is essentially the same as that obtained with XFOIL code.[10] A marked improvement in accuracy is obtained at the lowest Reynolds numbers, where the shear layer thickness is comparable to the airfoil chord.

B. Very Low Reynolds Number Airfoil

The flow about a NACA 4404 airfoil is computed for a range of angles of attack at $Re = 1000$. The results are compared to full Navier–Stokes solutions computed with the INS2D code,[12] presented by Kroo and Prinz.[13] Figure 9 shows the extremely thick shear layers that form at these low Reynolds numbers. Figure 10 compares the drag polars and lift curves for the second-order, first-order, and Navier–Stokes calculations. Taking the Navier–Stokes result as effectively exact, we see that the second-order method shows a marked improvement over the first-order results. The computational effort required for both BL methods is still several orders of magnitude smaller than for the Navier–Stokes result.

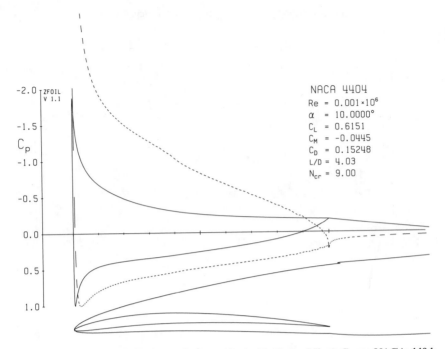

Fig. 9   Computed viscous (solid line) and inviscid (dashed line) C_p on NACA 4404 airfoil at $Re = 1000$.

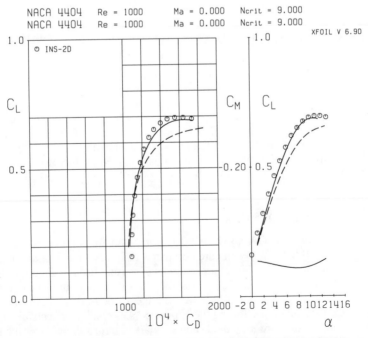

Fig. 10   Computed polars with present second-order method (solid line), first-order method (dashed line), and Navier–Stokes (symbols) on NACA 4404 airfoil at $Re = 1000$.

XIII. Conclusions

A higher-order boundary layer formulation has been strongly coupled to a high-order panel method, using the simultaneous solution technique via the full Newton method. Computed results indicate a marked improvement in the accuracy of the approach over the standard first-order BL formulation. The computational cost remains extremely modest relative to full Navier–Stokes solutions, making the method very attractive for design engineering work. Further development is needed to incorporate flow curvature effects into the turbulent transport equation to fully realize the advantages of the approach for high Reynolds number flows.

References

[1] Schlichting, H., *Boundary-Layer Theory*, McGraw-Hill, New York, 1979, Chap. VII.

[2] LeBalleur, J. C., "Strong Matching Method for Computing Transonic Viscous Flows Including Wakes and Separations," *La Recherche Aérospatiale*, Vol. 3, 1981, pp. 21–45; English ed.

[3] Allmaras, S. R., "Analysis of Le Balleur's Defect Formulation for Inviscid/Integral Boundary Layer Interaction," Boeing document, Sept. 1989 (unpublished).

[4] Lock, R. C., and Williams, B. R., "Viscous–Inviscid Interactions in External Aerodynamics," *Progress in Aerospace Sciences*, Vol. 24, 1987, pp. 51–171.

[5] Green, J. E., Weeks, D. J., and Brooman, J. W. F., "Prediction of Turbulent Boundary Layers and Wakes in Compressible Flow by a Lag-Entrainment Method, R & M Report 3791, Aeronautical Research Council, HMSO, London, 1977.

[6] Cousteix, J., and Houdeville, R., "Méthode intégrale de calcul d'une couche limite turbulente sur une paroi courbée longitudinalement," *La Recherche Aérospatiale*, Vol. 1, Jan.–Feb. 1977, pp. 1–13.

[7] Bradshaw, P., "Effects of Streamline Curvature on Turbulent Flow," AGARDograph 169, NATO, Neuilly Sur Seine, France, 1973.

[8] Drela, M., "Newton Solution of Coupled Viscous/Inviscid Multielement Airfoil Flows," AIAA Paper 90-1470, June 1990.

[9] Simpson, R. L., Chew, Y. T., and Shivaprasad, B. G., "The Structure of a Separating Turbulent Boundary Layer. Part 1. Mean Flow and Reynolds Stresses," *Journal of Fluid Mechanics*, Vol. 113, 1981, pp. 23–51.

[10] Drela, M., "Xfoil: An Analysis and Design System for Low Reynolds Number Airfoils," *Conference on Low Reynolds Number Airfoil Aerodynamics*, Univ. of Notre Dame, June 1989.

[11] Van Dyke, M., *Perturbation Methods in Fluid Dynamics*, Parabolic Press, Stanford, CA, 1975, pp. 137–139.

[12] Martinelli, L., and Jameson, A., "Validation of a Multigrid Method for the Reynolds Averaged Equations," AIAA Paper 88-0414, 1988.

[13] Kroo, I., and Prinz, F. B., "The Mesicopter: A Meso-Scale Flight Vehicle," NIAC Phase I Final Report, Stanford Univ., Stanford, CA, 1999.

Chapter 3

Analysis and Design of Airfoils for Use at Ultra-Low Reynolds Numbers

Peter J. Kunz* and Ilan Kroo[†]

Stanford University, Stanford, California

Nomenclature

a = geometric angle of attack in degrees
C_d = sectional drag coefficient
C_l = sectional lift coefficient
C_p = pressure coefficient
c = wing chord
D = drag force
L = lift force
L/D = lift-to-drag ratio
Re = Reynolds number based on chord length
V_∞ = freestream velocity

I. Introduction

INTEREST in very small aircraft operating in the $Re = 100,000$ to $150,000$ range has grown and many such vehicles are currently under development, but research and development at significantly smaller scales is in its infancy. Little experimental or computational work exists for aerodynamic surfaces operating at ultra-low Reynolds numbers below 10,000. The work presented here has been motivated by a microrotorcraft development program at Stanford University.[1] The goal of the Stanford project is to develop centimeter-scale powered air vehicles ranging in mass from 10 to 20 g. From a systems and manufacturing standpoint, technological advances in microfabrication techniques and in the miniaturization of electronics are beginning to make mechanical microflight vehicles feasible. The reduced scale and low flight speeds of these vehicles result in Reynolds numbers

*Doctoral Candidate, Department of Aeronautics and Astronautics.
[†]Professor, Department of Aeronautics and Astronautics. AIAA Fellow.

on the order of 1000. Although insects have been flying under these conditions for quite some time, this is a new flight environment for man-made aircraft.

The aerodynamics at these Reynolds numbers are considerably different from those of more conventional aircraft. The flow is laminar and dominated by viscous effects. Consequently, boundary layer thickness often reaches a significant fraction of the chord length and flow separation is an issue, even at low angles of attack. The study of flight under these conditions is only now becoming more than an academic problem. There is considerable literature on biological flight mechanisms, but very little detailed aerodynamic research is available. A lack of suitable manufacturing technologies, the absence of sophisticated computational analysis methods, and the difficulties associated with accurate experimental work at this scale have all restricted research. Advances in technology have reduced the significance of the first two issues, but the scarcity of experimental data remains to be addressed.

In the research described here, a series of airfoil sections are analyzed using a two-dimensional, incompressible, Navier–Stokes solver. Two-dimensional analysis allows a broad spectrum of parameters to be considered and provides a baseline for more detailed studies. The parameters investigated include thickness, camber, and the effects of leading and trailing-edge shapes. Depending on the particular application, three-dimensional flow effects may be significant, but the increased difficulty of the analysis would limit the scope of a parametric study.

Compared to performance at higher Reynolds numbers, airfoils exhibit an order of magnitude increase in the drag and a similar sized reduction in lift-to-drag ratios. Although the drag rapidly increases as the Reynolds number is reduced, significant lift coefficients are still attainable. Below approximately $Re = 10,000$, reducing the Reynolds number results in an *increase* in the maximum steady-state lift. As the Reynolds number is lowered, there is an alleviation of the leading edge suction peak, which results in less adverse gradients along the suction side of the airfoil. This delays separation and allows operation at higher angles of attack.

Examination of the geometry parameters reveals several trends. There is an expected drag penalty associated with increased thickness, but there is also a significant reduction in the lift curve slope. Given the benefits of reduced airfoil thickness, the remaining studies utilize 2% thick geometries. The performance of these thinner airfoils appears to be insensitive to the thickness distribution, but the magnitude and distribution of camber are still highly effective parameters. Building on the results of the geometry survey, several sections have been developed using an automated optimization method to maximize the two-dimensional lift-to-drag ratio at a given Reynolds number.

II. Computational Analysis Methods

The computational analyses make extensive use of the INS2d two-dimensional incompressible Navier–Stokes solver developed by Rogers.[2,3] This code utilizes the artificial compressibility method, first introduced by Chorin,[4] to deal with incompressible flows. Navier–Stokes solvers for the compressible flow equations generally require some form of preconditioning at very low Mach numbers. The artificial compressibility method offers a straightforward and efficient means of preconditioning to allow for the solution of an incompressible homogeneous flowfield.

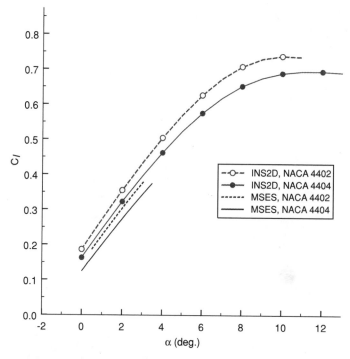

Fig. 1 MSES and INS2d predicted lift curves for the NACA 4402 and NACA 4404. Flow is fully laminar and $Re = 1000$.

Using integral boundary layer formulations in conjunction with inviscid flow-field solutions offers the potential for significant computational savings over viscous flow solvers. The MSES program developed by Drela[5] has been applied in this study with limited success. This is a two-dimensional Euler solver, coupled with an integral boundary layer formulation. It gives reasonable drag predictions over a narrow range of angles of attack, but the limitations of the boundary layer formulation cause the solution to diverge if significant regions of separated flow exist. This is a general limitation of coupled inviscid/boundary layer methods of this type.

A comparison of results from MSES and INS2d for the NACA 4402 and NACA 4404 airfoils at $Re = 1000$ are presented in Figs. 1 and 2. The upper end of each curve represents the maximum angle of attack for which a steady-state solution was attainable. Over the range in which MSES does converge to a solution, the trends in the results agree with INS2d, and in both figures the effects of increasing thickness are the same. Both analyses indicate similar reductions in the lift curve slope and increases in drag. The MSES solutions predict a lower lift curve slope and a slightly higher zero lift angle of attack, resulting in a deviation in predicted lift and approximately 5% lower drag than in the equivalent INS2d cases. That this method works at all is surprising, given the limitations of the boundary layer formulation, but under the restriction of low angles of attack, the much faster inviscid/integral boundary layer codes can provide a functional alternative to full viscous flowfield solutions.

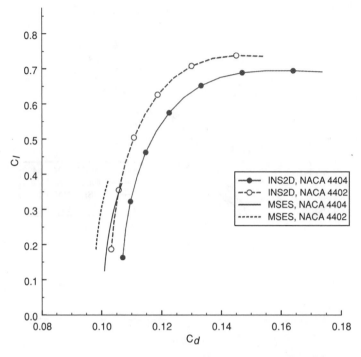

Fig. 2   MSES and INS2d predicted drag polars for the NACA 4402 and NACA 4404. Flow is fully laminar and $Re = 1000$.

III.   Flowfield Assumptions

The analyses make use of three assumptions about the flowfield: The flow is incompressible, fully laminar, and steady. Incompressibility is well justified for this application because the highest Mach number encountered in the associated vehicle development program was 0.3. For a broad range of applications, the Mach number will be considerably below this value and the flow is essentially incompressible.

The fully laminar flow assumption is more uncertain. In the absence of separation, the flow at these Reynolds numbers will be entirely laminar. Even slight to moderate separation will result in laminar reattachment for chord Reynolds numbers below 10,000 on a smooth airfoil. The degree of separation that might result in transition and the transition length are the unclear issues, but the alternatives are less satisfactory. The flowfield could be assumed fully turbulent, which is surely not the case, or transition could be artificially and rigidly imposed at a specified location. Of these three, the fully laminar assumption is the least restrictive and most physically accurate.

The steady-state assumption represents tremendous computational savings over time-accurate analysis and has been verified with time-accurate computations. Airfoil polars have been generated by increasing the angle of attack until the steady-state solution fails to converge. Analyses were completed at Reynolds numbers of 1000, 2000, 6000, and 12,000. For analyses at $Re = 6000$ and above, failure to converge is taken as an indication of unsteady phenomena in the flowfield. For the

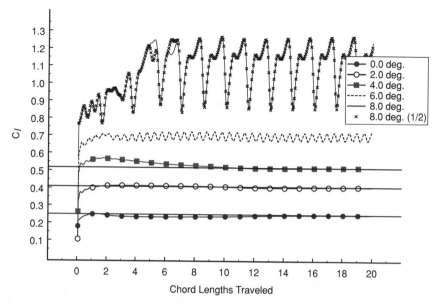

Fig. 3 INS2d time-accurate analysis of an impulsively started NACA 4402 at $Re = 6000$. Flow is fully laminar. The reference lines indicate steady-state results.

lower Reynolds number cases, lack of convergence can only be used as a definitive indicator of unsteadiness if the overall convergence rate is slowed considerably. For the $Re = 1000$ and $Re = 2000$ cases, the polars have been computed assuming a steady-state flowfield, but then a time-accurate analysis has been completed at or above the indicated maximum lift-to-drag ratio angle of attack. This ensures the absence of significant unsteady effects in the presented data but does not represent a rigidly defined upper limit. It is possible that higher angles of attack would still exhibit steady behavior.

The steady-state results for the NACA 4402 airfoil at $Re = 6000$ have been compared with data generated using the time-accurate mode of INS2d. The results of this study are illustrated in Fig. 3. The time-accurate computations consist of an impulsive start with 20 chord lengths of total travel. Each time step represents 0.02 chord lengths of travel. The steady-state solutions are indicated by the horizontal reference lines. The markers on the $\alpha = 8.0$ case represent a halving of the time step. These demonstrate that the 0.02 chord length time step is adequate for resolution of the temporal variations. Steady-state analysis failed to converge at $\alpha = 5.0$. Time-accurate solutions agree with the steady-state result past $\alpha = 4.0$. At $\alpha = 6.0$ small-amplitude periodic behavior is visible. The lift and drag of the section are still reasonably well defined. At $\alpha = 8.0$, the amplitude of the oscillations has increased considerably and the effects of multifrequency shedding are visible.

IV. Grid Topology

All calculations use a C-grid topology with either 256 by 64 cells or 512 by 128 cells in the case of the leading-edge studies and the optimized designs. The airfoil is paneled using 70% of the streamwise cells, with 10% clustered at the leading

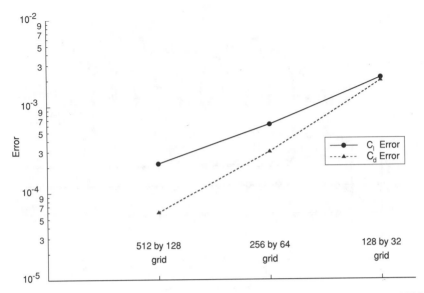

Fig. 4   **Grid-sizing study for the NACA 0002 at** Re **= 6000, with** α **= 3.0 and a 1024 by 256 reference grid.**

edge. Constant initial normal spacing provides for approximately 25 cells in the boundary layer at 10% chord for the 256 by 64 grids. The outer grid radius is placed at 15 chord lengths. The results of a grid-sizing study using the NACA 0002 are displayed in Fig. 4. Error values are based on analysis with a 1024 by 256 grid. The error convergence with grid size is close to quadratic and values for lift and drag are essentially grid independent with a 0.2% variation in C_l and a 0.7% variation in C_d over three levels of grid refinement.

V.   Comparison with Experiment

Although this chapter deals primarily with computational analyses at ultra-low Reynolds numbers for which experimental work is ongoing, there are a small number of relevant experiments in the literature that provide a reasonable basis for comparison. One interesting result from this computational study that is supported by experiment is an increase in attainable lift coefficient as the Reynolds number is reduced. Thom and Swart[6] tested a small R.A.F. 6a airfoil model in an oil channel and water channel at Reynolds numbers below 2000. They observed large increases in lift coefficient at fixed angles of attack as the Reynolds number was reduced from 2000 to almost 1.

Validation of the computational analyses is difficult owing to the almost complete absence of experimental data at relevant Reynolds numbers. The Thom and Swart experiment is based on a 1.24-cm chord airfoil with manufacturing deviations from the R.A.F. 6a. This small test piece was hand filed to shape causing the measured geometry to vary across the span. Although an exact validation is not possible because of the unknowns in the section geometry, comparison with computations for the R.A.F. 6 airfoil with a 256 by 64 grid shows reasonable

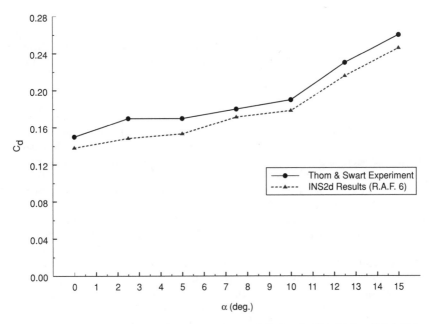

Fig. 5   Comparison of computed and experimental C_d for the R.A.F. 6 and R.A.F. 6a airfoils.

agreement with experiment. No coordinates for the R.A.F. 6a could be located, but the R.A.F. 6 appears to be nearly identical. The results are shown in Fig. 5. The Reynolds number varies from point to point and ranges from 650 to 810. The computed drag is on average 7.5% lower than that found by experiment, but the trends in C_d with angle of attack agree. Corresponding C_l data are only given for $\alpha = 10.0$. The computational result matches the experimental value of $C_l = 0.52$ within 3.0%.

VI.   Effects of Reynolds Number and Geometry Variations on Airfoil Performance

The effects of several airfoil geometry parameters at ultra-low Reynolds numbers have been investigated using the INS2d code and various members of the NACA four-digit airfoil family. The geometry definitions for these airfoils are parameterized for maximum thickness, maximum camber, and location of maximum camber. A wide range of variations is possible and section coordinates are easily generated.

A.   Reynolds Number

The most obvious effect of operation at ultra-low Reynolds numbers is a large increase in the section drag coefficients. Zero lift drag coefficients for airfoils range from 300 to 800 counts depending on the Reynolds number and geometry. The increase in drag is not reciprocated in lift. Lift coefficients remain of order 1, resulting in a large reduction in the L/D. Flight at these Reynolds numbers is much less efficient than at higher Reynolds numbers and available power is a limiting technological factor at small scales. It is important to operate the airfoil at its maximum L/D operating point, but this requires operating close to the maximum steady-state lift coefficient. Even small increases in the maximum lift coefficient are significant and generally translate to higher L/D.

Flow at ultra-low Reynolds numbers is viscously dominated, and as the Reynolds number is reduced, the effects of increasing boundary layer thickness become more pronounced. The definition of boundary layer at such low Reynolds numbers in a fully viscous flowfield is an inexact notion. Here, it is generalized as the low velocity flow adjacent to the body over which the pressure gradient perpendicular to the surface is close to zero. Regions of constant pressure extend a significant distance away from the surface of the airfoil and the effective geometry is significantly altered by the presence of the boundary layer.

In this operating regime the boundary layer has a dramatic effect on surface pressures, closer to that of a separation streamline. This is demonstrated by considering the inviscid and viscous pressure distributions on the NACA 0008 at zero angle of attack, shown in Fig. 6. Both the $Re = 6000$ and $Re = 2000$ cases are fully attached. As the Reynolds number is reduced, the value of the minimum pressure and the slope of the adverse gradient in the pressure recovery are reduced. This weakened pressure recovery does impact the pressure drag, but at a positive angle of attack, the largest effect is on lift. Figure 7 is a similar plot for the NACA 0002. Here again, all three cases are fully attached. For this 2% thick section, there is essentially no recovery of pressure in the $Re = 1000$ case. The flow behaves as

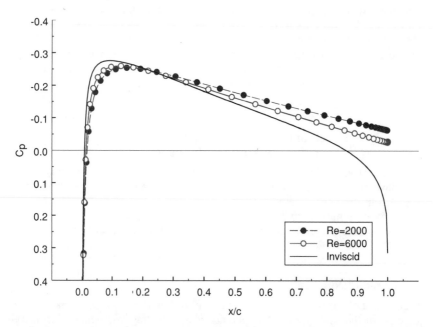

Fig. 6   Zero lift C_p distributions for the NACA 0008 at zero angle of attack. These viscous results are for fully laminar flow.

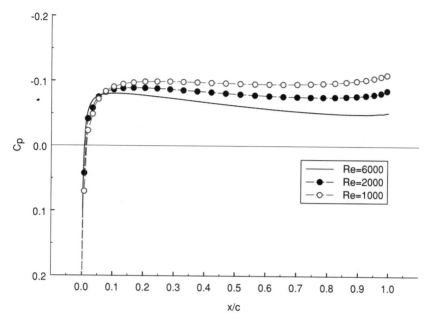

Fig. 7   Zero lift C_p distributions for the NACA 0002 at zero angle of attack for fully laminar flow.

if predominately separated. Also of note is the small increase in the magnitude of the minimum pressure, caused by the effective thickening of the section.

Viscous effects alleviate and smooth the high gradients present in the nose region of an airfoil. The reduction in the height of the leading-edge suction peak and the reduction in slope of the adverse pressure recovery gradient delay the onset of separation and stall. Leading-edge separation is delayed in thin sections, with trailing-edge separation delayed in thicker sections. The results are higher attainable angles of attack and higher maximum steady-state lift coefficients. Pressure distributions for the NACA 0008 airfoil at $\alpha = 2.0$ are presented in Fig. 8. The flow in the $Re = 6000$ case is on the verge of trailing-edge separation, but in the $Re = 2000$ case the flow does not separate until $\alpha = 3.5$. Lift coefficients for the two cases agree within 3.5%. The $Re = 2000$ case achieves the same amount of lift with a much weaker suction peak, a less adverse recovery gradient, and an additional margin of separation-free operation.

Reducing the Reynolds number affects the lift curve by reducing the slope in the linear range and extending the linear range to higher angles of attack. While operating within this range, the displacement effect of the boundary layer progressively reduces the effective camber of the section with increasing angle of attack. This change in the effective geometry increases as the Reynolds number is reduced. The delay in separation that accompanies reduced Reynolds numbers extends the linear range to higher angles of attack. Once the flow does separate, growth of the separated region is delayed by a reduction in the Reynolds number. The overall effect is a significant increase in both the maximum steady-state angle of attack and lift coefficient.

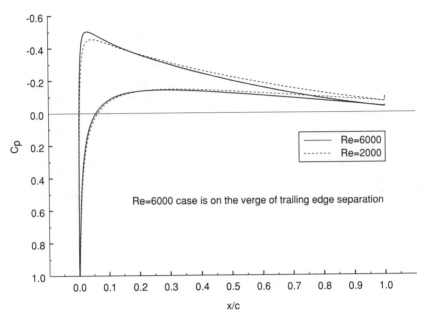

Fig. 8 C_p **distributions for the NACA 0008 at** $\alpha = 2.0$**. At** *Re* **= 6000, trailing-edge separation is imminent, but in the** *Re* **= 2000 case the flow remains fully attached.**

Lift curves for the NACA 0002 and NACA 0008 are presented in Fig. 9. The calculations are at $Re = 2000$ and 6000. The reduction of slope is most apparent for the NACA 0002 airfoil, but both sections exhibit the extension of the linear lift range. The NACA 0002 lift curves remain linear across its entire operating range, with the flowfield becoming unsteady owing to leading-edge separation. The $Re = 2000$ case reaches $\alpha = 5.0$ and attains a lift coefficient a full tenth greater than the $Re = 6000$ case. Similar gains occur for the NACA 0008 at $Re = 2000$.

Streamlines near the trailing edge of the NACA 0008 are displayed in Figs. 10 and 11 for $Re = 2000$ and 6000. The plots begin at the upper edge of the linear lift range for each Reynolds number. The streamlines originate from identical points in both figures. The onset of trailing-edge separation is pushed from $\alpha = 2.0$ at $Re = 6000$ to $\alpha = 3.0$ at $Re = 2000$. As the angle of attack is increased, the lower Reynolds number case achieves more than 2 deg higher angle of attack for similar amounts of trailing-edge separation.

B. Maximum Section Thickness

The effect of maximum thickness variations is investigated using uncambered NACA four-digit airfoils ranging from 2 to 8% thick in 2% increments. Performance estimates for each section were computed at $Re = 6000$ and at $Re = 2000$. Airfoil thickness variations appear to have two principal performance effects. A drag penalty, which is due to the pressure recovery attributable to increased thickness, is to be expected, but a strong reduction in the lift curve slope is also apparent.

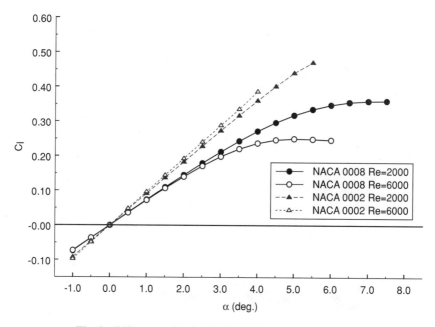

Fig. 9 Lift curves for the NACA 0002 and NACA 0008.

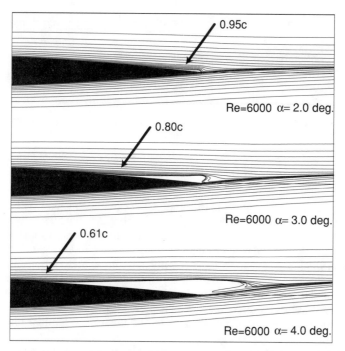

Fig. 10 Streamlines for the NACA 0008 at *Re* = 6000. The aft 45% of the airfoil is visible. The point of separation is indicated.

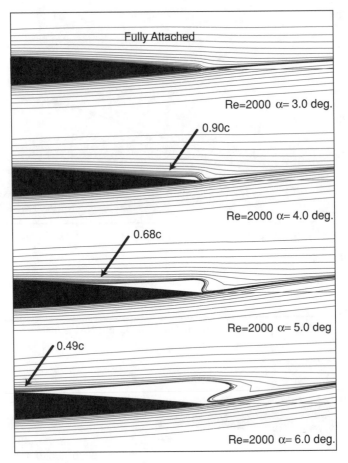

Fig. 11   Streamlines for the NACA 0008 at _Re_ = 2000. The aft 45% of the airfoil is visible. The point of separation is indicated.

1. Effect of Thickness on Drag

The variations in drag with section thickness are illustrated by the airfoil drag polars in Fig. 12. It is interesting to compare the calculated values with airfoils of the same family operating at higher Reynolds numbers and with the theoretical drag of a fully laminar flat plate. The experimental data for the NACA four-digit family is taken from Abbot and Von Doenhoff[7] at $Re = 6 \times 10^6$. Over a practical range of section thickness and for a fixed Reynolds number, the relationship between maximum thickness and the zero lift drag is well approximated by a linear function for this family of airfoils. The effect of thickness variations on the zero lift drag may be expressed as a reference zero thickness drag and a slope. The reference values and slopes are provided in Table 1. The theoretical laminar plate drag is also included for comparison.

The most apparent results are a general consequence of operation at ultra-low Reynolds numbers. Lift coefficients are similar to those seen at much higher

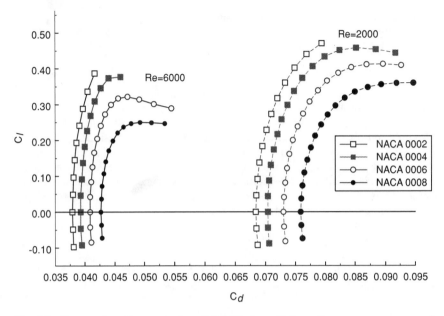

Fig. 12   Drag polars for uncambered NACA four-digit sections across a range of thickness.

Reynolds numbers, but the drag coefficients increase by an order of magnitude. This results in section lift-to-drag ratios of order 1 as opposed to order 100. Also notable is the large increase in the drag coefficient, nearly doubled, between the $Re = 6000$ and $Re = 2000$ results. In this regime, small changes in the Reynolds number result in large variations in the drag coefficient. The theoretical laminar flat-plate drag mimics these trends and provides a good estimate of the zero lift, zero thickness airfoil drag. The theoretical laminar flat-plate drag differs from the linear extrapolated zero thickness drag by only 5.3% at $Re = 6000$ and 9.5% at $Re = 2000$. This result is not surprising considering that the zero thickness reference *is* a flat plate. This supports the assumption of a linear relationship between thickness and zero lift drag for these airfoils.

The drag penalty associated with increasing thickness grows as the Reynolds number is decreased, but the rates of drag increase relative to the zero thickness drag are similar. Although the magnitudes of the drag coefficients increase dramatically,

Table 1   Effect of airfoil thickness ratio and Reynolds number on zero lift drag

Re	Zero thickness, zero lift C_d	C_d increase per % t/c	Laminar plate C_d
2000	0.0656	0.0013	0.0593
6000	0.0362	0.0008	0.0342
6.0×10^6	0.0040	0.0002	—

they are in line with simple laminar plate results. The variations with thickness also exhibit an order of magnitude increase, but the trends are consistent with those at higher Reynolds numbers.

2. Effect of Thickness on Lift

Within the linear range, the inviscid lift curve slope of an airfoil benefits from increased thickness, with thicker sections obtaining as much as a 10% increase in lift curve slope over the thin airfoil value of 2π per radian. At more conventional Reynolds numbers, viscous effects then degrade the lift curve slope, with the net result of lift curve slopes 5–10% below the inviscid thin airfoil value. The increased thickness of the upper surface boundary layer relative to the lower surface boundary layer at a positive angle of attack effectively reduces the camber of the airfoil. The inviscid gains and viscous losses typically cancel, resulting in lift curve slopes close to 2π per radian across a range of section thickness. This is not the case for Reynolds numbers below 10,000.

In this range of operation, the viscous boundary layer growth dominates and increasing thickness results in a significant decrease in lift curve slope in the linear range. The results in Fig. 13 show as much as a 35% reduction in lift curve slope for the 8% thick section. The 2% thick sections come closest to the inviscid thin airfoil value, showing a 15% reduction. The effect of reducing the Reynolds number from 6000 to 2000 is a further reduction in lift curve slope. In Fig. 14, the 2% section shows the greatest effect with a 5.2% reduction attributed to Reynolds number effects, with a decreasing effect with increasing thickness.

The decambering effect of the boundary layer is visualized by considering constant velocity contours in the flowfield. The area of reduced flow velocity is large and the validity of defining a reference edge value is debatable. The contours are

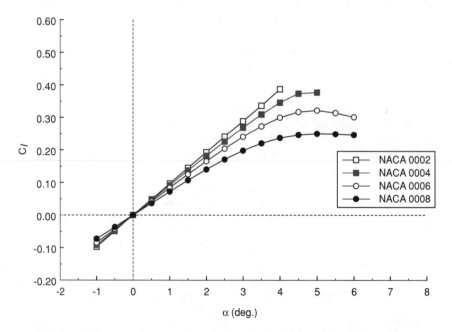

Fig. 13   Lift curves for uncambered NACA four-digit sections at $Re = 6000$.

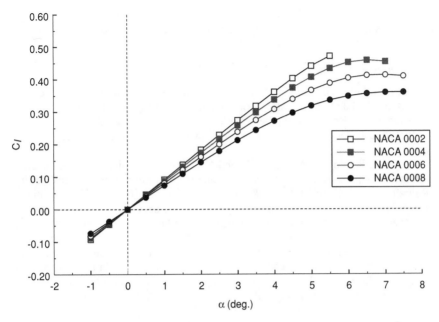

Fig. 14   Lift curves for uncambered NACA four-digit sections at $Re = 2000$.

chosen at a fixed fraction of the freestream velocity, low enough to be considered within the boundary layer, providing a qualitative notion of the boundary layer geometry. Several $0.2V_\infty$ contours are drawn for the NACA 0002 and NACA 0008 sections at $Re = 6000$ in Fig. 15. Three angles of attack, 0.0, 2.0, and 4.0 deg, represent the zero lift condition, the upper limit of the linear lift range for the NACA 0008, and a point within the nonlinear range, where trailing-edge separation comes into play. The boundary layer has little effect on the effective geometry of the NACA 0002, but the thicker upper surface boundary layer of the NACA 0008 significantly decreases the effective camber of the airfoil.

A second effect of increasing thickness is a more rapid reduction in the lift curve slope past the linear lift range, attributable to earlier and more severe trailing-edge separation. The flow over the NACA 0002 is fully attached up to stall for the same conditions that were presented in Fig. 10 for the NACA 0008. This plot begins at $\alpha = 2.0$, the edge of the linear range. The flow over the NACA 0008 is almost fully attached at $\alpha = 2.0$, with visible trailing-edge separation at 95% chord, but by $\alpha = 3.0$ there is significant separation at 75% chord. This moves to 60% at $\alpha = 4.0$. These separated regions result in a large displacement of the flow within

Fig. 15   NACA 0002 and NACA 0008 boundary layer development at $Re = 6000$. Several $0.2\,V_\infty$ velocity contours are drawn.

the aft boundary layer, increasing the decambering effect and resulting in larger reductions in the lift curve slope compared to the fully attached NACA 0002.

C.   Effect of Camber

The introduction of camber offers the potential for significant performance gains over a simple flat plate. The effects of camber do not differ significantly from those at much higher Reynolds numbers, but the fact that the detailed geometry is still an effective driver of performance at such low Reynolds numbers is itself a useful conclusion.

A comparison of the NACA 0002 and NACA 4402 airfoils indicates the gross effects of camber on performance. Lift curves and drag polars are provided for Reynolds numbers of 1000, 2000, and 6000 in Figs. 16 and 17. As at higher Reynolds numbers, the first-order effect on the lift curve is a translation toward lower zero lift angles of attack with increasing camber. The addition of 2% camber results in a 2.0–2.5 deg shift in the zero lift angle of attack. The maximum steady-state lift coefficients also increase. In the case of these two sections, there is a 30% increase in the maximum steady-state lift coefficient. Although the drag also increases, in this regime the ability to attain higher lift coefficients generally results in a net gain in lift-to-drag ratio. As the result of the introduction of camber, the maximum lift-to-drag ratio increases from 4.5 to 5.4 at $Re = 1000$ and from 9.3 to 11.0 at $Re = 6000$.

Within the linear range, the reduction in lift curve slope with a decrease in the Reynolds number is visible for the NACA 0002, but the effect of reducing the Reynolds number on the NACA 4402 is different. It appears as a drift

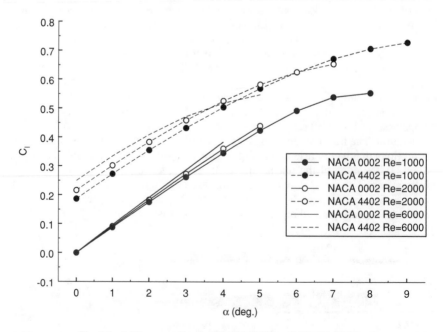

Fig. 16   Lift curves for the NACA 4402 and NACA 0002.

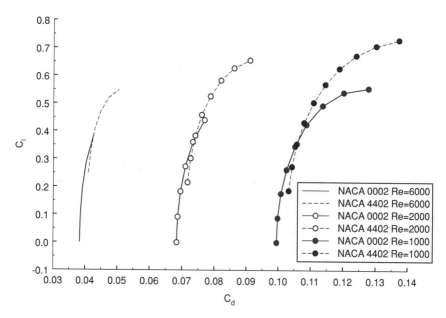

Fig. 17   Drag polars for the NACA 4402 and NACA 0002.

toward higher zero lift angles of attack as the reduction in Re uniformly reduces the effective camber of the section across the entire linear lift range. All but the $Re = 1000$ case eventually suffer from leading-edge separation, but the onset is delayed slightly. This is most likely attributed to the increase in the ideal angle of attack that comes with the introduction of camber.

The delay in trailing-edge separation attributed to reducing the Reynolds number is also visible in the NACA 4402 results. At $\alpha = 5.0$, the flow in the $Re = 6000$ case separates at 55% chord whereas in the $Re = 2000$ case it separates at 92% chord. Flow in the $Re = 1000$ case is still fully attached at this point, and at $\alpha = 7.5$ it is still attached up to 65% chord. Consideration of the drag polars in Fig. 17 reveals trends similar to those at higher Reynolds numbers. The addition of camber results in an increase in zero lift drag and an upward shift of the polars toward higher lift coefficients.

Further analyses investigate the possible benefits of varying the magnitude and distribution of camber. The design space is explored using nine airfoils spanning 2 to 6% maximum camber located at 30%, 50%, and 70% chord. All of the sections are 2% thick NACA four-digit profiles. All calculations were completed at $Re = 12,000$.

The lift curves provided in Fig. 18 are for 2% and 4% camber at all three of the chord locations. In both plots, the aft shift of maximum camber results in a less severe reduction of lift past the linear range, higher attainable lift coefficients, and higher lift-to-drag ratios. This correlates with reduced trailing-edge separation for a given angle of attack. The aft cambered sections exhibit separation at a lower angle of attack, but the growth of separation is retarded. As the angle of attack increases, the majority of the suction side experiences less adverse gradients than a similar section with forward camber. This contains the separation to aft of

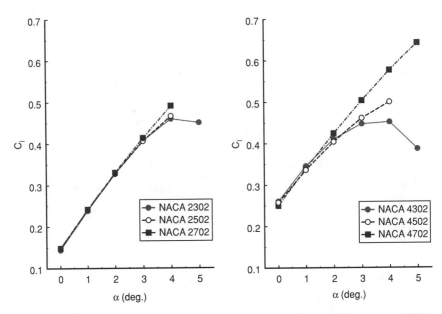

Fig. 18   Lift curves for 2% and 4% cambered NACA four-digit airfoils at $Re = 12,000$.

the maximum camber location by maintaining less adverse gradients ahead of it. The aft concentration of camber functions like a separation ramp in the pressure distribution.[8]

Comparison of the L/D of the nine sections reveals the complexity that is common in airfoil design. Only the 4% aft camber airfoil manages to outperform the 2% camber sections. The other two 4% camber cases are lower in L/D and roughly equivalent in lift coefficient. The maximum L/D values for all nine airfoils are provided in Table 2.

The effects of varying the amount of camber, while fixing the maximum camber location at 70% chord, are depicted in Fig. 19. The increase in camber causes a nonlinear penalty in drag for a given lift coefficient, but this is tempered by the ability to attain higher lift coefficients. For these three cases, the maximum L/D is attained by the 4% camber section. The effects of varying the location of maximum

Table 2   Effect of camber variations on L/D for a 2% thick airfoil

	L/D with max. camber at		
Max. camber (%)	0.3c	0.5c	0.7c
2	14.7	14.8	15.4
4	13.7	13.9	15.7
6	10.4	11.5	14.8

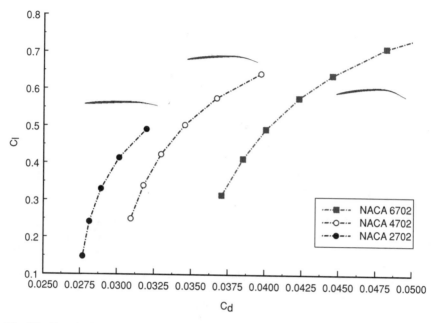

Fig. 19 Drag polars for NACA four-digit airfoils with varying amounts of camber. The maximum camber location is fixed at 70% chord and $Re = 12,000$.

camber for a fixed 4% camber is depicted in Fig. 20. The aft movement of camber results in significantly higher drag below $C_l = 0.45$, but this geometry is able to achieve higher lift coefficients within the steady-state operating limitation.

This simple nine-point test matrix indicates that aft camber is beneficial. Selection of the amount of camber is less clear, but for this particular camber definition, it should lie in the midrange of the values considered. This study is not meant to be a detailed indicator for design; the camber line is rather rigidly defined and the sampling is sparse. It is, however, indicative of the large variations in performance that exist within the design space and some of the physical trends responsible.

D. Effect of Leading-Edge Shape and Constant Thickness

Manufacturing limitations at small scales place minimum gauge constraints on the design of highly detailed cross sections. For very small chord lengths, the choice must be made between a traditional airfoil profile, at the expense of greater maximum thickness, or a constant thickness distribution at the minimum gauge. This tradeoff has been explored by comparing the performance of the NACA 0002 airfoil with 2% constant thickness plates. Two leading-edge shapes are considered for the plates: a leading edge radius and a nearly blunt leading edge. The trailing edge begins as a radius at 99% chord but is brought to a point to simplify grid generation. The three edge shapes are shown in Fig. 21.

The primary effect of substituting a constant thickness profile is a uniform increase in drag across the operating range of the section. The drag increase ranges

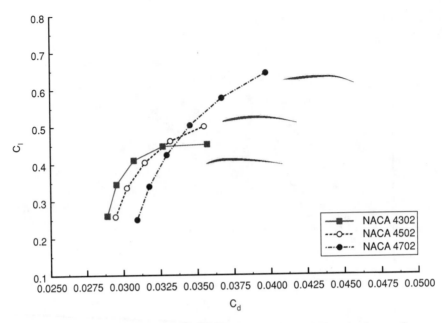

Fig. 20 Drag polars for NACA four-digit airfoils with variations in the maximum camber location and a fixed 4% camber at $Re = 12,000$.

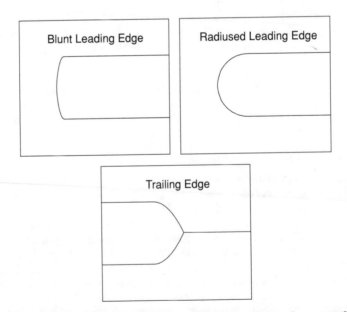

Fig. 21 Leading- and trailing-edge shapes used in the generation of constant thickness airfoils.

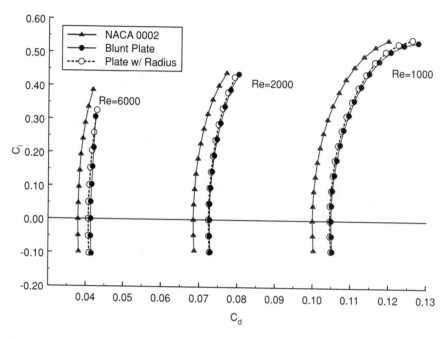

Fig. 22 Drag polars comparing the radiused and blunt constant thickness airfoils with the NACA 0002.

from 9% at $Re = 6000$ to 5% at $Re = 1000$. Drag polars for the NACA 0002 and both plate sections are shown in Fig. 22. This drag increase is comparable to increasing from a 2% thick airfoil to a 5% thick airfoil, but the plates suffer penalties in L/D that are minimal compared to an actual increase in thickness. An increase in section thickness results in a significant reduction in the lift curve slope and lower maximum steady-state lift coefficients, but the constant 2% thickness plates exhibit no reduction in lift curve slope relative to the 2% thick airfoil. An actual increase in thickness would result in a 20–25% reduction in maximum L/D. The penalty for the constant thickness plates is only 5% at $Re = 1000$ and $Re = 2000$, increasing to 18% at $Re = 6000$.

The leading-edge shape affects the formation of the leading-edge separation bubble. At $Re = 1000$, both plates exhibit fully attached flow past $\alpha = 4.0$. Once leading-edge separation occurs, the radiused plate gains less than 1/2 deg of angle of attack for equivalent bubble lengths. Leading-edge separation bubbles appear on the plates earlier than on the airfoil, but the leading-edge stall on the airfoil occurs very quickly. The net effect is a minimal penalty in lift for the plates. At $Re = 6000$, the leading-edge separation bubble forms almost immediately on the blunt section, but it does not form on the radiused plate until 1.5 deg later. Avoiding a blunt profile is advisable, but as the Reynolds number and maximum section thickness are reduced, the details of the thickness distribution become less relevant and the camber line becomes the dominant factor in performance.

VII. Airfoil Optimization

The optimization study makes use of the previous results to simplify the problem to its most essential elements. The maximum section thickness is fixed at 2% and the NACA four-digit thickness distribution is used. This should not affect the utility of the results and greatly facilitates automated grid generation. A specified thickness distribution reduces the number of variables considered but also simplifies the problem by removing minimum thickness constraints.

The free design element is the camber line. A wide range of variation is achieved by modeling the camber line with an Akima spline[9] anchored at the leading and trailing edges. Akima splines provide a curve that has the benefits of a tensioned spline, avoiding the undulations that can occur with simple cubic splines, but they are generally smoother across the control points and require no tuning to achieve a satisfactory interpolation. Four interior control points, or knots, are used to define the camber line. These are evenly distributed and their chordwise locations are fixed. The knots move perpendicular to the chord line constrained by upper and lower camber limits.

The optimization study utilizes a constrained simplex optimizer, a modified Nelder–Mead simplex, coupled with the INS2d code and a grid generator. With only four design variables, this simple optimization method is sufficient. In addition, each two-dimensional steady-state solution of the flow solver is relatively inexpensive. The simplex method is simple to implement and does not require (possibly noisy) gradient calculations. It is likely not the most efficient option, but the small problem size and inexpensive flow calculations make it a good solution.

Function evaluations are also designed to increase the robustness of the method. A coarse performance polar is generated for each function evaluation. Calculations begin at a low angle of attack where convergence is likely, regardless of geometry. The angle of attack is then increased until the L/D passes its maximum or the flow solver fails to converge. In either case, an objective function value is generated for that geometry and the process continues. This is computationally expensive and not particularly efficient, but it is robust and effective.

Two airfoils have been developed using this approach for $Re = 6000$ (R6) and $Re = 2000$ (R2). Both sections are shown in Fig. 23. The optimization runs were initialized with a flat-plate airfoil, but the converged solutions have been checked by restarting with a geometry near the upper camber limits. Both airfoils exhibit similar features with a prominent droop near the nose, well-defined aft camber, and distinct hump in the camber distribution that begins near 65% chord and reaches its maximum height at 80% chord. Since the spline knots are located between 20%

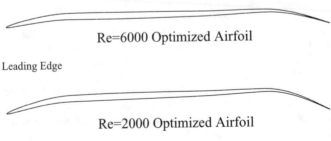

Re=6000 Optimized Airfoil

Leading Edge

Re=2000 Optimized Airfoil

Fig. 23   Optimized airfoils for $Re = 6000$ and $Re = 2000$.

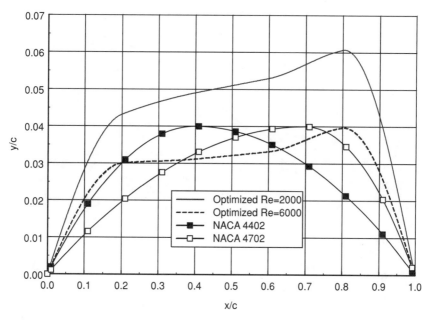

Fig. 24   Camber distributions for the 2% thick *Re* = 6000 and *Re* = 2000 optimized airfoils.

and 80% chord, the regions between the edges of the airfoil and the outermost control points are constrained to be nearly linear. The camber distributions are plotted in Fig. 24. The NACA 4402 and NACA 4702 are also plotted for reference.

Optimization at $Re = 6000$ resulted in a maximum camber close to 4%, but the R2 solution increases to 6% camber. This increase compensates for the larger reductions in effective camber at lower Reynolds numbers. The R2 airfoil achieves a maximum L/D of 8.2, 5% higher than the best four-digit section tested at this Reynolds number, the NACA 4702.

The 4% camber of the R6 airfoil is closer to the four-digit airfoils examined earlier and provides a better point for comparisons. This airfoil achieves an L/D of 12.9, 4% better than the NACA 4702 and a 16% improvement over the NACA 4402. Figure 25 shows the L/D vs geometric angle of attack for these three airfoils. The optimized section begins to show gains past $\alpha = 3.0$, increasing until the maximum L/D is reached at $\alpha = 5.0$.

Small improvements in lift and drag allow the R6 airfoil to outperform the NACA 4702. The majority of the gains in lift and drag are connected to 5% less trailing-edge separation on the R6 compared to the NACA 4702. The optimizer is attempting to exploit the benefit of limiting trailing-edge separation. The maximum camber is moved to the aft control point and this region is once again operating similarly to what has been described as a separation ramp.

At $\alpha = 4.0$, trailing-edge separation occurs at 88% chord, growing to 86% at $\alpha = 5.0$. Beyond $\alpha = 5.0$, the separation point almost immediately moves forward to 30% chord. Figure 26 shows the aft 60% pressure distributions at $\alpha = 4.0$ and $\alpha = 5.0$. At $\alpha = 4.0$, there is a distinct inflection in the pressure distribution at 82% chord. The reduced adverse gradients ahead of this location do not allow

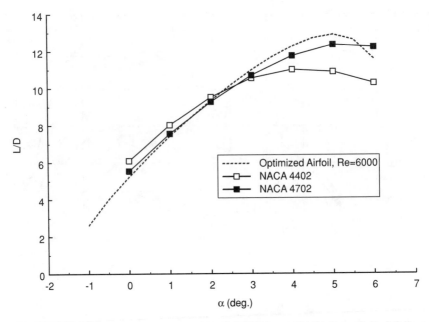

Fig. 25 L/D for the optimized Re = 6000 airfoil and two NACA four-digit airfoils.

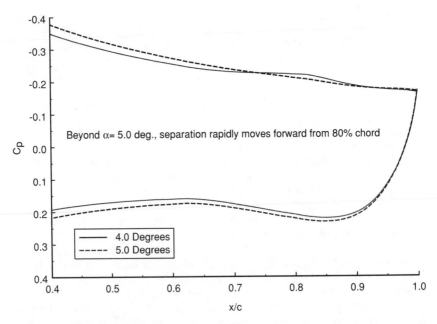

Fig. 26 C_p distributions for the Re = 6000 optimized airfoil.

separation to move forward, but the steeper gradients following the inflection point promote separation on the aft portion of the airfoil. This is the primary source of the drag penalty seen with aft camber at low angles of attack. At $\alpha = 5.0$, the inflection is no longer present. Beyond $\alpha = 5.0$, separation moves far forward and performance rapidly degrades.

The optimized design of these two airfoils highlights the ability of small modifications in geometry to be very effective in altering section performance. Additional degrees of freedom may be easily added to the problem by introducing more spline knots, but this simple four-variable problem succeeds in achieving significant performance gains over smooth formula-based camber lines. The lack of experience at ultra-low Reynolds numbers makes optimizers an effective and important tool, not only for design, but also for enhancing our understanding of this flight regime.

VIII. Conclusions

Achieving powered flight at ultra-low Reynolds numbers and true microaircraft scales requires a greater understanding of the relevant aerodynamics. Airfoils in this regime face unique operating conditions. The flow is dominated by viscous effects and the growth of thick boundary layers causes significant modifications to the effective geometry of the section. In addition to this effective loss of camber, flow separation occurs at low angles of attack, but the airfoil continues to operate in a steady-state manner while significant portions of the section experience separated flow, albeit with reduced performance. At more conventional Reynolds numbers, separation is usually a rapid precursor to stall, but under these conditions it is a common feature of the normal operating range of an airfoil.

Operating at ultra-low Reynolds numbers results in very high drag coefficients, but the increases agree with theoretical laminar plate results. Small changes in the Reynolds number cause large changes in drag, and as the Reynolds number is reduced, section L/D quickly falls to single digits. More interesting is the effect reducing the Reynolds number has on lift. Within the scope of this study, as the Reynolds number is reduced, the maximum steady-state lift coefficient generally increases. The study of NACA four-digit sections indicates that at ultra-low Reynolds numbers, geometry variations still have a strong effect on the aerodynamic performance of an airfoil. The impetus for design at these Reynolds numbers is further strengthened by the performance gains achieved with a simple optimization study.

References

[1]Kroo, I., and Kunz, P. J., "Development of the Mesicopter: A Miniature Autonomous Rotorcraft," *American Helicopter Society Vertical Lift Aircraft Design Conference*, AHS International, Alexandria, VA, Jan. 2000.

[2]Rogers, S. E., and Kwak, D., "An Upwind Differencing Scheme for the Steady-State Incompressible Navier–Stokes Equations," NASA TM 101051, Nov. 1988.

[3]Rogers, S. E., and Kwak, D., "An Upwind Differencing Scheme for the Time Accurate Incompressible Navier–Stokes Equations," *AIAA Journal*, Vol. 28, No. 2, 1990, pp. 253–262.

[4]Chorin, A. J., "A Numerical Method for Solving Incompressible Viscous Flow Problems," *Journal of Computational Physics*, Vol. 2, No. 1, 1967, pp. 12–26.

[5] Drela, M., and Giles, M. B., "ISES—A Two-Dimensional Viscous Aerodynamic Design and Analysis Code," AIAA Paper 87-0424, Jan. 1987.

[6] Thom, A., and Swart, P., "The Forces on an Aerofoil at Very Low Speeds," *Journal of the Royal Aeronautical Society*, Vol. 44, 1940, pp. 761–770.

[7] Abbot, I. H., and Von Doenhoff, A. E., *Theory of Wing Sections*, Dover, New York, 1959, pp. 148–157.

[8] Maughmer, M. D., and Somers, D. M., "Design and Experimental Results for a High-Altitude, Long Endurance Airfoil," *Journal of Aircraft*, Vol. 26, No. 2, 1989, pp. 148–153.

[9] Akima, H., "A Method of Univariate Interpolation That Has the Accuracy of a Third-Degree Polynomial," *ACM Transactions on Mathematical Software*, Association for Computing Machinery, Vol. 17, No. 3, 1991, pp. 341–366.

Chapter 4

Adaptive, Unstructured Meshes for Solving the Navier–Stokes Equations for Low-Chord-Reynolds-Number Flows

J. T. Mönttinen,[*] R. R. Shortridge,[†] B. S. Latek,[‡] H. L. Reed,[§] and W. S. Saric[§]

Arizona State University, Tempe, Arizona

Nomenclature

A	=	assembled fluid matrix; first node of the shortest segment in the front
B	=	assembled body-force vector; second node of the shortest segment in the front
C	=	third corner node of a newly generated element
C_D	=	drag coefficient, $F_D/(\frac{1}{2}\rho U^2 S)$
C_L	=	lift coefficient, $F_L/(\frac{1}{2}\rho U^2 S)$
C_p	=	pressure coefficient, $(p - p_\infty)/(\frac{1}{2}\rho U^2)$
D	=	domain where the governing equations are valid; diameter
F	=	general force vector
F_D	=	drag force
F_L	=	lift force
f	=	body force
K	=	general stiffness matrix
L	=	chord length
N	=	shape-function matrix
N^L	=	linear shape-function matrix
N^Q	=	quadratic shape-function matrix
n	=	number of nodes per general element; normal
n_i	=	shape function

[*]Graduate Research Associate, Mechanical and Aerospace Engineering. Student Member AIAA.
[†]Graduate Student, Mechanical and Aerospace Engineering.
[‡]Graduate Research Assistant, Mechanical and Aerospace Engineering. Student Member AIAA.
[§]Professor, Mechanical and Aerospace Engineering. Associate Fellow AIAA.

$nnodl$ = number of nodes in each linear element
$nnodp$ = number of nodes in each quadratic element
Re = Reynolds number based on the chord length, $\rho UL/\mu$
Re_D = Reynolds number based on the diameter, $\rho UD/\mu$
S = surface area
s = boundary variable
U = freestream velocity
$\bar{u}$ = vector of velocity variables, $[u\ v]^T$
$\tilde{u}, \tilde{v}, \tilde{p}$ = starting value at each iteration for u, v, and p, respectively
w = approximate solution to the general problem
z = exact solution to the general problem
α = angle of attack
Γ = element boundary
$\Gamma 1$ = boundary where essential boundary conditions are defined
$\Gamma 2$ = boundary where natural boundary conditions are defined
ε = error in the FEM approximation
λ = vector of unknown fluid variables, $[u\ p\ v]^T$
ρ = fluid density
σ = total stress tensor
Ω = element area

I. Introduction

IN recent years, both military and civilian organizations have expressed an increased interest in micro aerial vehicles (MAVs).[1] The goal of current MAV programs is to build a lightweight, quiet, and easy-to-use aerial vehicle with a maximum dimension of 6 to 10 in. for various surveillance missions. Ideally this vehicle would be invisible to radar and its noise would be impossible to distinguish from the background. Moreover, MAVs are required to operate under unsteady gust and lull conditions. Because of their size and unique flight conditions, these MAVs offer new challenges to aerodynamics, controls, and even basic design issues. To stabilize the vehicle, either winglets or fins are generally used in addition to active control mechanisms.[2] This leads to complicated geometries with various parameters to be optimized.

The development and evaluation of aerodynamic technologies for MAVs must include the effects of viscosity and three dimensionality that are inherent to their relatively low speeds and low chord Reynolds numbers (10,000 to 100,000). Because one expects local failure of classical inviscid aerodynamics in this flow regime it may be appropriately called *viscous aerodynamics*. Moreover, the flight environment is characterized as turbulence-free but with large unsteady gusts and lulls. Any design methodology and control schemes to be used on MAVs must be designed with these effects in mind.

For complex flows, advances in our prediction methods and in identification of basic mechanisms will come from those groups working hand-in-hand performing complementary computations and experiments on the same geometries. This kind of collaboration is critical for the following reasons:

1) Because of the heightened sensitivity of separation and transition to operating conditions and geometry in complex flows, computations provide validation of experiments and vice versa.

2) Detailed measurements are often difficult and costly. Here, computations can guide the experiments as to what effects are important and what needs to be measured.

3) When computations work with experiments, an explanation of the mechanisms at work is easier to determine and simpler models are thus developed. Each provides a different level of detail and a different perspective.

4) The process of reconciling computational and experimental results uncovers most of the hidden assumptions and flaws in each approach. It thus leads to results that are far more reliable. This is even more important in complex flowfields.

II. Approach

Our ongoing viscous aerodynamic studies include both experimental and numerical work, both conducted on the Arizona State University (ASU) main campus, and flight tests on a remotely piloted aircraft both on and off campus. Designs can be efficiently optimized through the computations. Even small changes to the model can lead to dramatic changes in the aerodynamic performance. The experimental work will be used to validate the computations and vice versa. In particular, initial flow visualization and force-balance measurements have been performed in the ASU Flow Visualization Tunnel (FVT). The baseline MAV studied is a flying-wing configuration using an Eppler 212 airfoil (Fig. 1). The ultimate goal of these force-balance measurements is to determine the effect of different wing-body-juncture and wing-tip designs. Preliminary flow-visualization studies have shown a significant sensitivity to the cant angle, span, and sweep angle of the winglet. The force-balance measurements will give quantitative results, which, guided by the computational results, will be used to find the ideal aerodynamic design for MAVs at low Reynolds numbers. By then testing the MAV under gusty conditions in the ASU Unsteady Wind Tunnel (UWT), the controls and trim may be configured before actual flights. Short flight tests can be done on the sports

Fig. 1   MAV configuration designed at ASU.

practice fields on our campus, and for longer missions either Williams Airport in Mesa, Arizona or the Electronic Proving Grounds (EPG) at Ft. Huachuca, Arizona is available.

Because the viscous effects are very significant, computationally we are solving the flowfield via direct numerical simulation (DNS). The chord Reynolds number Re is too low for one to use theoretical models commonly employed for solving problems at higher Re. DNS involves solving the full, incompressible Navier–Stokes equations without any assumptions or simplifications (e.g., turbulence modeling). At this Re range few well-documented data exist,[1] which adds extra challenges for the research. Computationally Ramamurti et al.[2] have developed a capability for designing MAVs both with and without the Baldwin–Lomax turbulence model. The goals of our current research are to determine basic mechanisms at work relative to the unsteady aerodynamics of these flows, suggest optimal configurations for MAVs, and produce a validated set of data from which simpler theoretical design methods can be developed.

In the early stages of the current work, the problem was investigated using structured, single-block grids that were sufficient for solving the two-dimensional problem accurately and predicting preliminary results for simple three-dimensional configurations such as a tapered wing. In Figs. 2 and 3 the comparison between experiment and computation is shown, illustrating the dominant tip vortex that affects on the order of 50% of the span of the MAV at $Re = 175,000$. Based on this result, the main emphasis of the current work is to minimize the effect of the tip vortex. Blending the wing and fuselage and adding winglets provides a reduction in the extent of these vortices as well as a refocusing of them away from the lifting surface. Adding a winglet to the model moved the vortex away from the airfoil in experiments, thus increasing the lift and also reducing the induced drag caused by this vortex. In Fig. 4 the velocity vectors at the trailing edge of a wing–winglet model are shown. It was observed that the vortex moved similarly as in experiments. However, the grid-generation scheme chosen was not able to handle these more complicated problems such as wing–winglet or wing–body

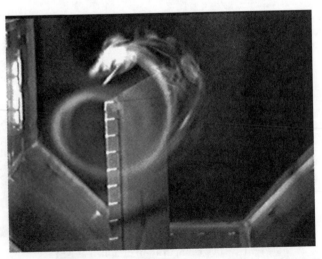

Fig. 2 Flow visualization of a dominant tip vortex on a tapered wing.

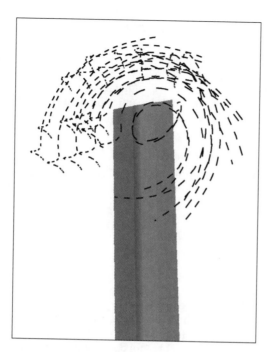

Fig. 3   Computational streaklines illustrating the tip vortex on a tapered wing.

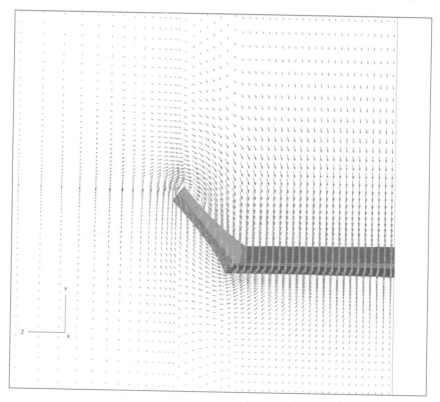

Fig. 4   Velocity vectors at the trailing edge of a wing–winglet model.

configurations to a satisfactory accuracy. To further investigate the problem, the possible choices for grid generation were to use either a structured multiblock grid or an unstructured grid. Adaptive refinement has been successfully used with unstructured grids for various fluid-dynamics problems[3-7] and was thus chosen for the current work along with an associated finite element method (FEM).

III.   The Finite Element Approximation

The FEM used is based on the method of weighted residuals (MWR), which involves the solution to a differential equation[8]

$$
\begin{aligned}
\nabla^2 z + f &= 0 & &\text{in } D \\
z &= g(s) & &\text{on } \Gamma 1 \\
\frac{\partial z}{\partial n} + \alpha(s)z &= h(s) & &\text{on } \Gamma 2
\end{aligned}
\tag{1}
$$

Here D is the domain of interest, $\Gamma 1$ is the part of the boundary where the unknown z has a specified value (i.e., essential boundary condition), and $\Gamma 2$ is the boundary where natural boundary conditions are defined. The domain is discretized into elements, each of which will have area Ω and boundary Γ. The solution of the above is approximated by $w(x, y)$ within each element, which leads to

$$
\frac{\partial^2 w}{\partial x^2} + \frac{\partial^2 w}{\partial y^2} + f = \varepsilon_n(x, y, w_1, w_2, \dots, w_n)
\tag{2}
$$

where n is the number of nodes in each element and $\varepsilon_n(x, y)$ is the error resulting from the fact that $w(x, y)$ is not an exact solution to the differential equation but an approximation of the form

$$
w(x, y) = \sum N_i(x, y)w_i \quad \text{for } i = 1, 2, \dots, n
\tag{3}
$$

The $N_i(x, y)$ are the nodal interpolation functions associated with the element being used and the w_i are the values at the nodes. The Galerkin method requires that the integral of the product of the error $\varepsilon_n(x, y)$ with each of the nodal interpolation functions $N_i(x, y)$ over the element Ω be equal to zero,

$$
\iint_\Omega \varepsilon(x, y)N_k(x, y)\, d\Omega = 0
\tag{4}
$$

for all $k = 1, 2, \dots n$. When this is applied to the governing equation

$$
\iint_\Omega \left[\frac{\partial^2 w}{\partial x^2} + \frac{\partial^2 w}{\partial y^2} + f \right] n_k(x, y)\, d\Omega = 0
\tag{5}
$$

results. Now, applying the definition of w and integration by parts, the final form can be written as

$$
\iint_\Omega \left(\frac{\partial n_k}{\partial x} \sum w_i \frac{\partial n_i}{\partial x} + \frac{\partial n_k}{\partial y} \sum w_i \frac{\partial n_i}{\partial y} \right) d\Omega + \int_{\Gamma 2} \alpha \sum w_i n_i n_k d\Gamma 2
$$

$$
= \iint_\Omega f n_k d\Omega + \int_{\Gamma 2} h(s) n_k d\Gamma 2 \quad \text{for } k = 1, 2, \dots n
\tag{6}
$$

which in matrix notation becomes

$$K_{nxn}w_n = F_n \tag{7}$$

All elemental contributions are finally assembled together to obtain the solution for the original problem.

IV. Fluid Solver

For viscous, incompressible flow the governing equations in primitive-variable form are

Conservation of Mass:

$$\frac{\partial u_i}{\partial x_i} = 0 \tag{8}$$

Conservation of Momentum, Navier–Stokes:

$$\rho \left(\frac{\partial u_i}{\partial t} + u_j \frac{\partial u_i}{\partial x_j} \right) - \frac{\partial \sigma_{ij}}{\partial x_j} + \rho f_i = 0 \tag{9}$$

for $i = 1, 2, 3$ and $j = 1, 2, 3$ and the sum over repeated indices is implied. To apply the Galerkin method in these equations, the unknowns are to be expressed in terms of the shape functions and the nodal values,[9]

$$\bar{u} = \sum_{i=1}^{nnodp} N_i^Q \bar{u}_i$$

$$p = \sum_{i=1}^{nnodl} N_i^L p_i \tag{10}$$

where superscripts Q and L refer to quadratic and linear interpolation, respectively. The $\bar{u}_i$ and p_i are the nodal values. The *nnodp* and *nnodl* are the number of nodes per element in quadratic and linear interpolation, respectively. A basic element in two dimensions is shown in Fig. 5 and in three dimensions in Fig. 6.

The FEM is to be used only in the spatial dimension and the Crank–Nicolson method is used in the temporal domain. Thus it is sufficient to apply the FEM formulation to the steady-state equations. The two-dimensional case leads to equations

$$\int_\Omega N_i^L \left(\frac{\partial N_j^Q}{\partial x} u_j + \frac{\partial N_j^Q}{\partial y} v_j \right) d\Omega = 0 \tag{11}$$

$$\int_\Omega \left[N_i^Q N_k^Q u_k \frac{\partial N_j^Q}{\partial x} u_j + N_i^Q N_k^Q v_k \frac{\partial N_j^Q}{\partial y} u_j + N_i^Q \frac{\partial N_l^L}{\partial x} p_l + \frac{1}{Re} \right.$$

$$\left. \times \left(\frac{\partial N_i^Q}{\partial x} \frac{\partial N_j^Q}{\partial x} u_j + \frac{\partial N_i^Q}{\partial y} \frac{\partial N_j^Q}{\partial y} u_j \right) \right] d\Omega - \int_{\Gamma1} \frac{1}{Re} N_i^Q \frac{\partial N_j^Q}{\partial n} u_j d\Gamma1 = 0 \tag{12}$$

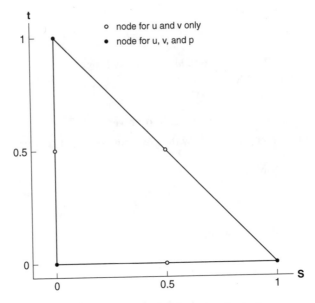

Fig. 5 Two-dimensional master element.

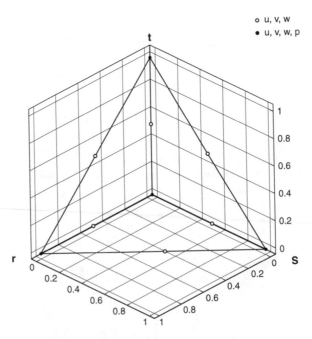

Fig. 6 Three-dimensional master element.

where both the mass and the momentum equations are shown. The $\Gamma 1$-integral is needed at the boundaries where the flow gradients are defined.

After summing the elemental contributions, the matrix equations can be written as

$$A\lambda = B \tag{13}$$

where λ is the vector of unknowns $[u \ p \ v]^T$ and B is the body-force vector. The ordering of unknowns is chosen in this way to make the extension to three dimensions as simple as possible, involving basically only the addition of the third velocity component into λ and using tetrahedral elements instead of triangular elements. The difficulty of this extension does not occur with the fluid solver but with the grid generation as will be discussed later.

The second-order-accurate fluid solver is based on the frontal-solution method developed by Irons in 1980 that has been proven to be very effective for solving positive-definite matrices and also works for unsymmetric problems that are often encountered in fluid mechanics.[9] The basic idea of this method is to work with only a small portion (front) of the domain at one time and solve for the unknowns in a node as soon as all the needed information is available. After this, that node is eliminated from the front and a new one is introduced. The procedure is illustrated in Fig. 7.

An iterative procedure[9,10] is used to solve for the unknowns. At first, uniform flow is assumed with the exception of boundary conditions and the unknowns are solved at each node. The starting values $\tilde{u}$ for the next iteration are obtained from the previous two iterations using a relaxation parameter ω

$$\tilde{u}_{n+1} = u_n + \omega(u_{n+1} - u_n) \tag{14}$$

and similarly for $\tilde{p}$ and $\tilde{v}$. Iterations are repeated until

$$\left| \frac{u_{n+1} - \tilde{u}_{n+1}}{u_{n+1}} \right| < 10^{-L}$$

where L is chosen appropriately for each problem. The computational results of this paper are obtained using $L = 6$.

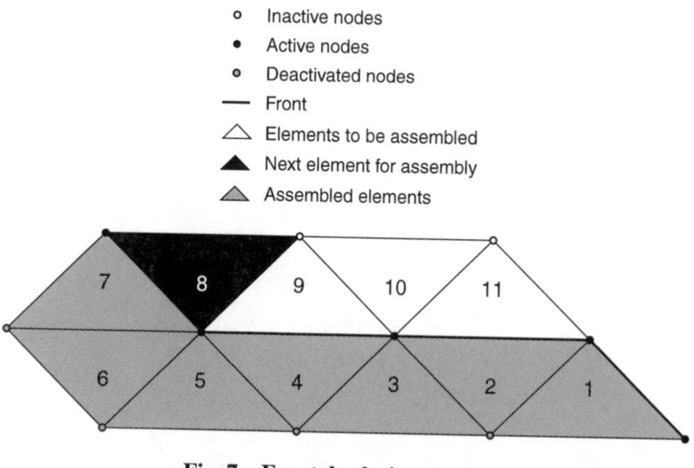

Fig. 7   Frontal solution method.

V.   Grid Generation and Adaptive Refinement

Unstructured, adaptive meshes have been successfully used for solving fluid-dynamics problems[3-7,11] mostly because of their ability to adjust according to variations in the flowfield. Triangular, unstructured meshes can also be easily fitted around complex geometries, which cannot be done effectively using structured, quadrilateral meshes. Parametric cubic splines[12] are used to define the geometries, as these can accurately model any geometry. Earlier a polynomial definition was used for an airfoil, but this led to stability problems at the leading edge where the two polynomials meet and caused the solution to lose accuracy.

The initial grid generation is based on a guessed mesh density, and the domain is filled using the Advancing Front method[13] starting from the shortest segment on the boundary[6,7] with endpoints A and B (Fig. 8). The ideal height of the triangle is found based on the mesh density. Then a preliminary, isosceles triangle is created using nodes A, B, and C, where C is called the ideal location. Next a search for existing nodes in the front within radius r from C is conducted and the points located within that radius are arranged according to the distance from C. The first node to satisfy the following two criteria is chosen to be the third node of the element:

1) The new triangle has positive area.
2) The new sides do not cross any existing segments.

If no valid location is found, the same checks are performed on C. If C is not a valid location, the search radius is extended to $5r$ and the above checks are repeated. If no valid location is found, the height of the triangle is scaled down by 5% at a time until an acceptable location is found. The front is updated after a new element is created.

Once the mesh is completed, the solution is calculated and the gradients of the solution are used to find the new mesh density in the whole domain and the new grid is generated following the above procedure. The initial mesh density is defined so that it would capture the details of the flow near the airfoil and thus give both a good approximate result and a good flowfield that can be used to refine the grid. The new mesh density is obtained from this solution by finding the maximum gradient of either velocity component in each node of the domain and allowing a maximum change of 5% of the freestream velocity within each element. A similar procedure can be used to refine the grid for unsteady problems as the previous solution can be interpolated to give values at the new grid-point locations. The criterion used to determine convergence and accuracy is the drag coefficient. If further refinement of the grid or enforcing a tighter convergence tolerance does not change the drag

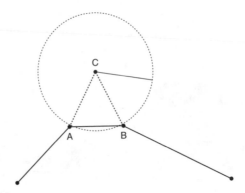

Fig. 8   Grid generation using the Advancing Front method.

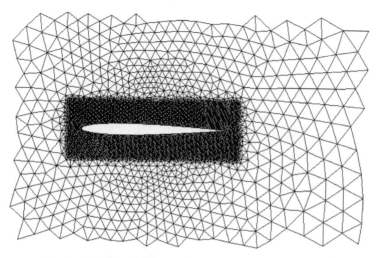

Fig. 9 Well-defined starting grid for a NACA 0008 airfoil.

coefficient, the solution is accepted. During these grid-convergence studies it was observed that two refinements are sufficient for a given Re value when starting from a well-defined grid. Consequently it was observed that these converged grids had at least 10 points in the boundary layer. A well-defined starting grid with 12,000 elements for the NACA 0008 airfoil is shown in Fig. 9 and another well-defined starting grid with 10,000 elements for the Eppler 212 is shown in Fig. 10. The choice of the starting grid depends on the Reynolds number for which the flowfield is to be solved. For higher Reynolds numbers smaller elements are required near

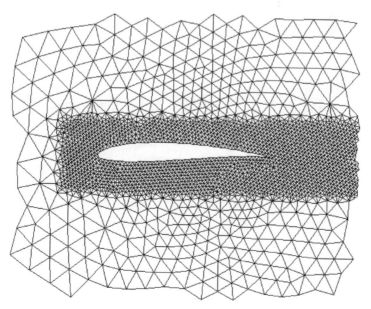

Fig. 10 Well-defined starting grid for an Eppler 212 airfoil.

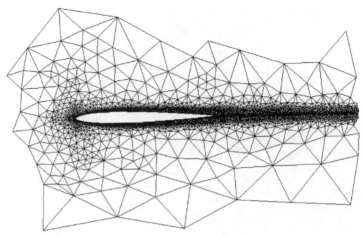

Fig. 11   Refined grid for NACA 0008 used to calculate the flowfield at $Re = 16{,}000$.

the airfoil to capture the details of the flow sufficiently well to provide information for grid refinement. A starting grid with small elements in the wake (Fig. 10) leads to convergent results in fewer iterations than one with larger elements (Fig. 9). This does not lead to savings in computational time though because the number of elements in the domain becomes much larger if the element size is kept small both near and behind the airfoil. Consequently more work is required to solve the system during each iteration. A converged grid for the NACA 0008 with 20,000 elements for $Re = 16{,}000$ is shown in Fig. 11 and one for the Eppler 212 with 19,000 elements for $Re = 4{,}000$ is shown in Fig. 12. It can be seen that the refined

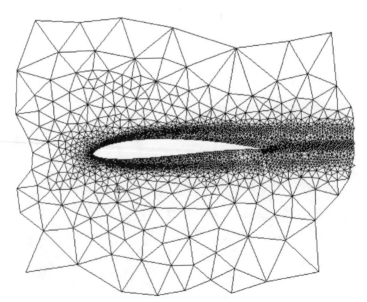

Fig. 12   Refined grid for Eppler 212 used to calculate the flowfield at $Re = 4000$.

grid has much smaller elements near the airfoil and in the wake of the airfoil but larger elements away from the airfoil.

VI. Results

At first, the solution method was tested for the well-documented[14–17] flow past a cylinder. A comparison of experimental data[16] with a simple curve-fit formula[17] for drag as a function of Reynolds number,

$$C_D \approx 1 + \frac{10}{Re_D^{2/3}} \qquad (15)$$

is shown in Fig. 13 and agreement can be seen for the investigated Re_D range. This fit (15) is in fair agreement with experimental results up to Reynolds number 250,000 but tends to give slightly higher values than obtained in experiments. In Fig. 14 the variation of the surface pressure on the cylinder at $Re_D = 40$ is shown together with two sets of earlier computational results[15] and these can be seen to agree.

We next provide a verification of our results with those of Kunz and Kroo[18] for the NACA 0008 airfoil at $Re = 2000$. In Fig. 15 a comparison is made for the pressure distribution for the zero-lift case and agreement can be seen. The lift coefficient as a function of the angle of attack is plotted in Fig. 16 and the comparison with Kunz and Kroo[18] shows excellent agreement. Finally, comparison of the two drag polars is shown in Fig. 17 and these agree as well for the investigated cases.

Further calculations for the NACA 0008 and other airfoils are currently underway. The airfoils being investigated are the Eppler 212,[19] which is the actual airfoil used for the current MAV design at ASU,[20] and the Eppler 61,[19] for which both

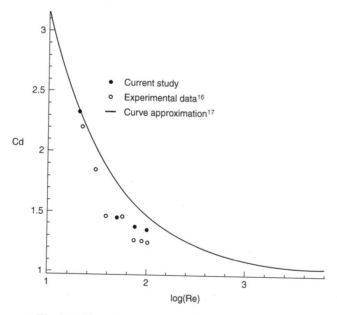

Fig. 13   Drag coefficient for the flow past a cylinder.

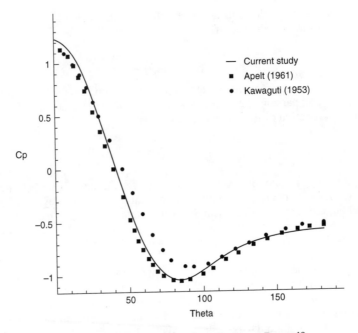

Fig. 14   Surface pressure of a cylinder at $Re_D = 40$.

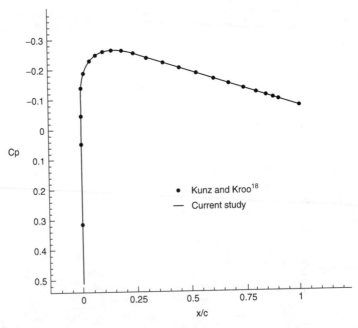

Fig. 15   Pressure distribution for the zero-lift case for NACA 0008 at $Re = 2000$.

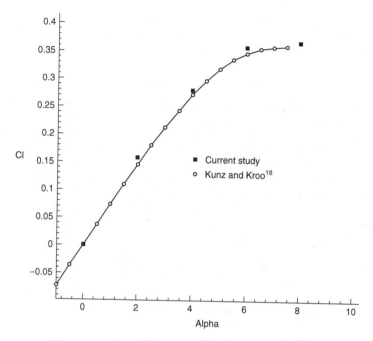

Fig. 16   Lift coefficient for NACA 0008 for *Re* = 2000.

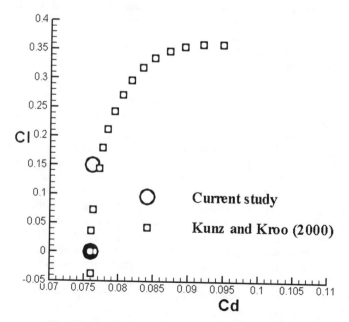

Fig. 17   Drag polar for NACA 0008 at *Re* = 2000.

experimental and computational data have been published earlier for Reynolds numbers of 40,000 and higher.[1] Further verification and validation can be done after completion of these computations.

VII. Database Validation

To provide useful data for future researchers, one has to be very critical of and complete in what one publishes. The computational database must be both verified with other computational results and validated with experimental data. The author must clearly explain the method used in such a way that the reader can repeat the same procedure in his or her research.

As mentioned before, the drag coefficient has been used in this study as one measure to determine the accuracy of the solution. This criterion chosen as the calculation of the drag coefficient involves both pressure and viscous terms and is thus sensitive to changes in either one. It was observed that matching the pressure distribution is much easier than matching the drag coefficient. One has to be careful when calculating the viscous drag as the gridlines are not perpendicular to the surface when using unstructured grids and additional postprocessing is necessary to determine the proper normal derivatives. A critical point is the definition of the geometry. Even small differences in geometry definition can lead to large differences in drag-coefficient calculations and thus give inaccurate results.

From the computational point of view, it is essential to document the boundary conditions used so that a future researcher can use the same conditions. In the current study, both the upstream and the freestream boundary are defined to be two chord lengths away from the airfoil and the downstream boundary is eight chord lengths away from the airfoil. These distances were noticed to be sufficiently large as further increasing the domain size did not change the flowfield. The freestream velocities are defined only at the upstream boundary and the no-slip condition is enforced on the airfoil. At all other boundaries, natural boundary conditions are automatically enforced by the FEM. This method was observed to be more reliable than defining the freestream conditions at the upper and lower boundaries of the domain.

VIII. Ongoing Work

As mentioned before, the extension of the fluid solver to three dimensions is straightforward as long as primitive variables are used. The information about the shape functions and Gauss points for numerical integration is well documented in the literature.[8,10] The fluid solver has already been extended to three dimensions and was first tested for flow in a channel. This example was used as it can be modeled using only a few elements and thus the resulting matrix system can be solved using direct solution methods or the frontal solver that still works for a small number of elements. Comparison of current computational results with the analytical results is shown in Fig. 18. As can be seen, the constant-velocity contours match exactly those obtained analytically.[17] However, practical three-dimensional problems include wing–winglet and wing–body configurations as well as solving the flowfield over the whole MAV. Figures 19 and 20 show such representative configurations of interest to the MAV development group at ASU.[20]

For these geometries the main problem is the grid generation. In two dimensions, triangular elements are used to fill the domain, starting by defining the node locations on the boundaries. In three dimensions, the starting point would be to create

Result of current work
—·—·— Analytical result

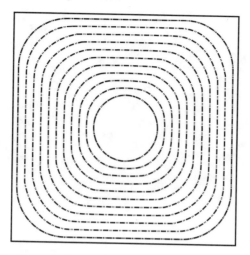

Fig. 18   Outflow velocity profile for three-dimensional channel flow.

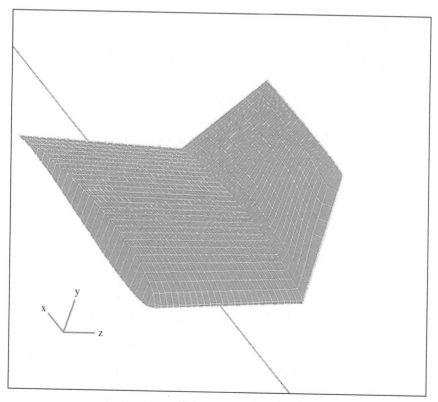

Fig. 19   Typical wing–winglet configuration.

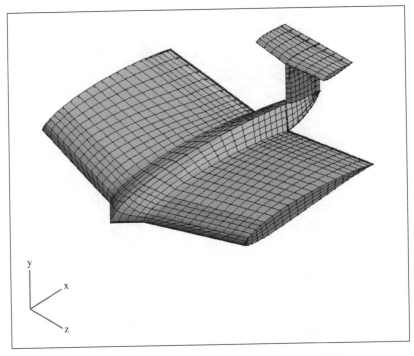

Fig. 20   Computational model of ASU's Mantis MAV.

triangular elements on each of the surfaces and then fill the domain with tetrahedral elements. To use adaptive refinement effectively, the surface grid must also be refined based on the flow near the surface. Also, directional refinement has to be applied to minimize the number of elements and still enable the accurate calculation of flowfield information. In directional refinement, both the size and the shape will be controlled based on the gradients in each direction, thus allowing long sides in directions where the gradients are small and small sides in the directions of strong gradients. A typical boundary-layer element after nondirectional and directional refinement is shown in Fig. 21. One can clearly see that directional refinement allows one to use a fewer number of elements for solving the problem and thus enables shorter computational times.[6] However, for transient problems, one has to compromise because directional refinement tends to be storage and CPU intensive.[4]

The number of elements required in two-dimensional problems is sufficiently low for one to use a single processor. However, with the number of elements increasing, both the size of the data files and computational cost have been seen to increase. This will become a greater problem in three-dimensional problems and thus parallel processing has to be implemented. The number of unknowns in three dimensions will be much higher as the number of variables increases from three to four and the number of grid points increases approximately by a power of 3/2. The frontal method has given satisfactory results thus far, but for larger problems a more effective solution method has to be applied. The frontal method also becomes inefficient in three dimensions as the front width increases as it has to accommodate variables of a whole plane rather than just those of a line as in

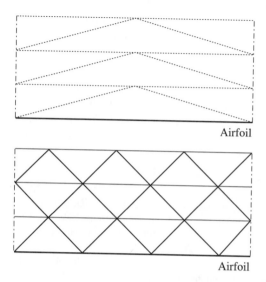

Airfoil

Airfoil

Fig. 21 Comparison of directional and non-directional refinement.

two dimensions. Methods such as $BiCGStab(l)$,[21] its modification $ASTAB(l)$,[22] and $PCGS$,[23] which use preconditioning, have been shown to be more effective than the frontal method. Most of the recent work using iterative methods for solving the Navier–Stokes equations has been done using $BiCGStab$ with preconditioning,[24,25] and so this can be applied in the current study as well. However, the choice of preconditioner is a difficult issue as only general guidelines can be found in the literature.[24–27] Also, most of the earlier work is only for Reynolds numbers less than 1000 and shows clearly that the effectiveness of a preconditioner depends on the Reynolds number. The methods used in this range are not necessarily effective for higher values of Re. One also needs to compromise, as some preconditioning methods such as incomplete LU-factorization can be as expensive as solving the original system.[26] A desirable alternative is element-by-element preconditioning[27] as it enables one to operate on small matrices rather than working with a large system matrix. Currently various preconditioning methods are being investigated, with the goal being to find one that would work for a wide Reynolds number range and be computationally efficient.

IX. Conclusions

Our research program in MAVs consists of complementary, experimental, computational, and flight-test elements. The goals of our research are to determine basic mechanisms at work relative to the three-dimensional unsteady viscous aerodynamics of these flows, suggest optimal configurations for MAVs, and produce a validated set of data from which simpler theoretical design methods can be developed.

Adaptive mesh refinement and the finite element method have been successfully applied for two-dimensional low-Reynolds-number flow calculations. Comparisons with existing computational data have shown good agreement. Further validations with available sets of experimental airfoil data for the desired Re range are underway. The extension of the fluid solver to three dimensions has proven to be straightforward. For these geometries the main challenges are the grid generation and the choice of preconditioning method.

Acknowledgment

This work is supported by the Air Force Office of Scientific Research under contract number F49620-99-1-0089.

References

[1]Mueller, T. J., "Aerodynamic Measurements at Low Reynolds Numbers for Fixed Wing Micro-Air Vehicles," presented at the RTO AVT/VKI Special Course on Development and Operation of UAV's for Military and Civil Applications, Belgium, 1999.

[2]Ramamurti, R., Sandberg, W., and Löhner, R., "Simulation of the Dynamics of Micro Air Vehicles," AIAA Paper 2000-0896, Jan. 2000.

[3]Löhner, R., and Morgan, K., "Improved Adaptive Refinement Strategies for Finite Element Aerodynamic Computations," AIAA Paper 86-0499, Jan. 1986.

[4]Löhner, R., "An Adaptive Finite Element Scheme for Transient Problems in CFD," *Computer Methods in Applied Mechanics and Engineering*, Vol. 61, No. 3, 1987, pp. 323–338.

[5]Löhner, R., and Martin, D., "An Implicit, Linelet-Based Solver for Incompressible Flows," *Advances in Finite Element Analysis in Fluid Dynamics, FED-Vol. 123*, American Society of Mechanical Engineers, Fairfield, NJ, 1991, pp. 9–20.

[6]Peraire, J., Morgan, K., Peiro, J., and Zienkiewicz, O. C., "An Adaptive Finite Element Method for High Speed Flows," AIAA Paper 87-0558, Jan. 1987.

[7]Peraire, J., Peiro, J., Formaggia, L., Morgan, K., and Zienkiewicz, O. C., "Finite Element Euler Computations in Three Dimensions," *International Journal for Numerical Methods in Engineering*, Vol. 26, 1988, pp. 2135–2159.

[8]Bickford, W., *A First Course in the Finite Element Method*, Richard D. Irwin, Burr Ridge, IL, 1994.

[9]Taylor, C., and Hughes, T. S., *Finite Element Programming of the Navier–Stokes Equations*, Pineridge, Swansea, Wales, U.K., 1981.

[10]Reddy, J. N., and Gartling, D. K., *The Finite Element Method in Heat Transfer and Fluid Dynamics*, CRC Press, Boca Raton, FL, 1994, pp. 163–178.

[11]Ramamurti, R., Sandberg, W., and Löhner, R., "Evaluation of a Three Dimensional Finite Element Incompressible Flow Solver," AIAA Paper 94-0756, Jan. 1994.

[12]De Boor, C., *A Practical Guide to Splines*, Springer-Verlag, Berlin, 1978, pp. 49–59.

[13]Lo, S. H., "A New Mesh Generation Scheme for Arbitrary Planar Domains," *International Journal for Numerical Methods in Engineering*, Vol. 21, 1985, pp. 1403–1426.

[14]Anderson, J. D., *Fundamentals of Aerodynamics*, McGraw-Hill, New York, 2001, pp. 256–264.

[15]Batchelor, G. K., *An Introduction to Fluid Dynamics*, Cambridge Univ. Press, New York 1994.

[16]Tritton, D. J., "Experiments on the Flow Past a Circular Cylinder at Low Reynolds Numbers," *Journal of Fluid Mechanics*, Vol. 6, 1959, pp. 547–567.

[17]White, F. M., *Viscous Fluid Flow*, McGraw-Hill, New York, 1991.

[18]Kunz, P., and Kroo, I., "Analysis, Design, and Testing of Airfoil for Use at Ultra-Low Reynolds Numbers," *Proceedings of the Conference: Fixed, Flapping and Rotary Wing*

Vehicles at Very Low Reynolds Number, edited by T. J. Mueller, Notre Dame, IN, 5–7 June 2000, pp. 349–372.

[19]Eppler, R., *Airfoil Design and Data*, Springer-Verlag, Berlin, 1990, pp. 222–223, 523.

[20]Latek, B., Reed, H., and Saric, W., "Implementation of Research Advances into an Operational MAV for the MAV Competition," *Proceedings of the Conference: Fixed, Flapping and Rotary Wing Vehicles at Very Low Reynolds Number*, edited by T. J. Mueller, Notre Dame, IN, 5–7 June 2000, pp. 478–488.

[21]Sleijpen, G. L. G., and Fokkema, D. R., "Bicgstab(l) for Linear Equations Involving Unsymmetric Matrices with Complex Spectrum," *Electronic Transactions on Numerical Analysis*, Vol. 1, 1993, pp. 11–32.

[22]Little, L. J., "A Finite Element Solver for the Navier–Stokes Equations Using a Preconditioned Adaptive BiCGStab(L) Method," Ph.D. Dissertation, Dept. of Mathematics, Arizona State Univ. Tempe, AZ, 1998.

[23]Saad, Y., *Iterative Methods for Sparse Linear Systems*, PWS Publishing, Boston, MA, 1995, pp. 214–217, 245–262.

[24]van der Vorst, H. A., and Sleijpen, G. L. G., "Hybrid Iterative Solvers and Their Application in CFD," *Solution Techniques for Large-Scale CFD Problems*, edited by W. A. Habashi, Wiley, New York, 1995, pp. 159–175.

[25]Dahl, O. and Wille, S. O., "An ILU Preconditioner with Coupled Node Fill-In for Iterative Solution of the Mixed Finite Element Formulation of the 2D and 3D Navier–Stokes Equations," *International Journal for Numerical Methods in Fluids*, Vol. 15, 1992, pp. 525–544.

[26]Saad, Y. "Preconditioned Krylov Subspace Methods for CFD Applications," Report UMSI-94-171, Minnesota Supercomputer Institute, Minneapolis, MN, Aug. 1994.

[27]Lee, H.-C., and Wathen, A. J., "On Element-by-Element Preconditioning for General Elliptic Problems," *Computer Methods in Applied Mechanics and Engineering*, Vol. 92, 1991, pp. 215–229.

Chapter 5

Wind Tunnel Tests of Wings and Rings at Low Reynolds Numbers

E. V. Laitone*

University of California, Berkeley, California

Nomenclature

A = aspect ratio = (span/chord) = $(b/\bar{c})$
C = length of circular cylinder of diameter D
C_D = drag force/(qS), $q = \rho V^2/2$
C_L = lift force/(qS)
$\bar{c}$ = average wing chord = S/b
Re = Reynolds number = $VC/\mathcal{V}$
S = wing planform area $(b\bar{c}) = \pi DC$ for ring
V = freestream velocity ahead of model
α = angle of attack of chord
$\Delta\alpha$ = angle of attack from zero lift
$\mathcal{V}$ = kinematic viscosity $\geq 1.464 \times 10^{-5}$ m^2/s = 1.576×10^{-4} ft^2/s

I. Introduction

T HE aerodynamic forces on wings and rings at low Reynolds numbers have become important because of the planned use of small remotely piloted aircraft, especially at high altitudes. For example, at 80,000 ft (24,400 miles) the kinematic viscosity has increased by a factor of 22 from its sea-level value; consequently, a sea-level Reynolds number (Re) of 880,000 has decreased to 40,000. Pilotless aircraft, and especially their aerodynamic control surfaces, are therefore left with inadequate wind tunnel data from the typical strain-gauge balance system. The balance described in Ref. 1 is a two-component beam balance with a sensitivity of ± 0.01 g ($\pm 10^{-4}$ N). With this, balance lift, and especially drag, can be accurately measured on small models at $Re < 10,000$.

The present tests are a continuation of those described by Laitone[1,2] and now include additional tests on the effect of aspect ratio and wing planform. Also, ring

*Professor Emeritus, Department of Mechanical Engineering. Fellow AIAA. Deceased.

and annular airfoils were tested at Re from 8000 to 50,000. These models were all cylindrical constant-diameter (D) tubes of thin (0.51 mm) metal cut to different lengths ($C < D$). The low turbulence wind tunnel had a square cross-section test area (0.66 m^2) with a tunnel empty turbulence level of 0.02%. The velocity range in these tests was from 10 to 20 m/s. The chords of the wing models ranged from 25 to 230 mm, while the ring models were approximately 10 to 50 mm in chord length.

II. Effect of Aspect Ratio and Planform on the Aerodynamic Lift and Drag

The wind tunnel tests ($Re < 50,000$) reported in Ref. 1 showed that a thin plate with 5% circular arc camber had the highest initial lift-curve slope (C_{L_α}) and the greatest lift-to-drag ratio (L/D) of all the same $A = 6$ aspect ratio wings with rectangular planforms. These included the NACA 0012 profile and other profiles with wedge shape or different camber. Also, it was the only wing that developed $C_L > 1$ after a very moderate stall at an angle of attack of $\Delta\alpha = 12.7$ deg, measured from $C_L = 0 = \Delta\alpha$. This surprising lift increase was ascribed to a two-dimensional momentum change produced by the curved lower surface of the wing profile acting as a turbine blade. To verify this conjecture other aspect ratios were tested at $Re = 20,700$, and Fig. 1 shows that the decrease in the span of the 5% circular camber profile for the $A = 6$ wing eliminated the two-dimensional momentum lift when $A = 2.18$ and 4. The momentum lift was restored by the wings with $A = 5$ and 8. The C_{L_α} measured for $A = 2.18$, 4, 6, and 8, were respectively, 0.065, 0.078, 0.098, and 0.109 ($\approx 2\pi/57.3$). These C_{L_α} values were all greater than the corresponding values given by Glauert's[3] calculations for rectangular planform wings with $A = $ span(b)/chord(c), as

$$C_{L_\alpha} = \frac{2\pi}{1 + (2/A)(1 + \tau)}, \qquad C_{D_i} = \frac{C_L^2}{\pi A}(1 + \delta) \qquad (1)$$

For example, when $A = 6$, Glauert[3] (p. 149) calculated $\tau = 0.163$ and $\delta = 0.046$, so that $C_{L_\alpha} = 4.528(0.07903)$ deg, which is less than the experimental value at $Re = 20,700$ of $C_{L_\alpha} = 0.098$, by a factor of 1.24! However, this remarkable increase in lift is counterbalanced by an even greater increase in skin friction and induced drag. As shown in Ref. 1 (Fig. 7 and text) the experimental data for an $A = 6$ wing with 5% camber at $Re = 20,700$ gives $C_{D_i} = 0.099\,C_L^2$ for $0.4 < C_L < 0.9$. This is 178% greater than C_{D_i} given by Eq. (1) [$= C_L^2(1.046/6\pi) = 0.0555\,C_L^2$]. Hence the L/D ratio is greatly reduced by low $Re = 20,700$. For example, $A = 2.18$, 4, 6, and 8 have corresponding L/D of 7.7, 11.4, 13.3, and 15.1. These are all much less than found at $Re > 10^6$.

The effects of planform and small aspect ratio ($A \le 4/\pi$) on the lift-to-drag ratio L/D are shown in Fig. 2. These models are all thin (0.62 mm) flat plates with wing chords from 25 to 150 mm with Re based on the chord from 20,000 to 50,000. The highest $L/D = 8$ for the $A = 6$ rectangular planform. Its $C_{L_\alpha} = 0.082$ is greater than the theoretical $C_{L_\alpha} = 0.0790$ from Eq. (1) with $\tau = 0.163$. Although the $A = 6$ elliptical planform had $C_{L_\alpha} = 0.087 > 0.0823$ [Eq. (1) with $\tau = 0$], its C_D was larger, and so its L/D was only 6.6. The $A \le 4/\pi$ all had lower $C_{L_\alpha} \le 0.048$, with a maximum for the square plate with $b = c$, $A = 1$, and $L/D = 5.4$. The $A \le 4/\pi$ could be useful for producing the largest forces when $\alpha > 30$ deg. The

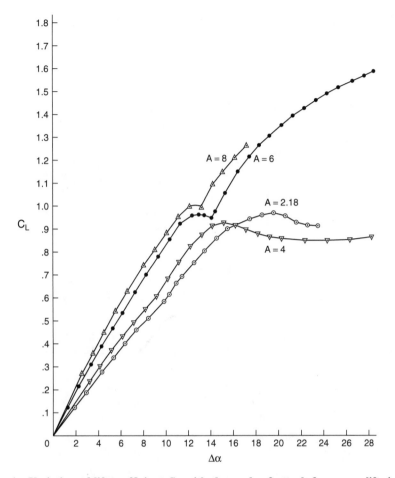

Fig. 1   Variation of lift coefficient C_L with the angle of attack from zero lift $\Delta\alpha$, for rectangular planform wings with different aspect ratios (A = span/chord), at Re = 20,700.

highest $C_L = 1.90$ was produced by the square plate at $\alpha = 38$ deg, but it also had the largest $C_D = 1.53$ and so $L/D = 1.24$. The circular disk had $C_{L_\alpha} = 0.044$ with $L/D = 5.7$ and $C_{Lmax} = 1.44$ at $\alpha = 35$ deg, but its L/D was only 1.39 because of its high drag $C_D = 1.04$. The "Delta wing," a triangle with $A = 1$, had $C_{L_\alpha} = 0.043$ with $L/D = 5.37$ and $C_{Lmax} = 1.72$ at $\alpha = 33$ deg and $L/D = 1.5$. However, the 5% camber, $A = 6$ rectangular planform wing (Fig. 1) had $C_{L\alpha} = 0.098$ with $L/D = (0.55/0.0415) = 13.25$ at $\alpha = 6.2$ deg and at $\alpha = 28$ deg, $L/D = (1.60/0.696) = 2.30$. This shows the advantage of 5% camber, but it also shows that more research is required to understand the increase in lift and the greater increase in drag for $Re \leq 20,000$ when $A = 6$. The data shown in Fig. 2 for $A \leq 4/\pi$ are for $Re = 50,000$ as they were less effected by the decrease to $Re = 20,000$.

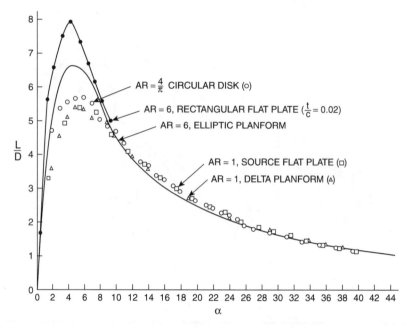

Fig. 2 Variation of L/D with the angle of attack from zero lift, $\Delta\alpha$, for various planforms with $AR = (\text{span}^2/\text{planform area})$. $Re = 20,000$ for $AR = 6$, and $Re = 50,000$ for $AR \leq 4/\pi$.

III. Effect of Low Reynolds Numbers on the Lift and Drag of Ring Airfoils

The second approximation to the lift-curve slope C_{L_α} of a thin-wall circular cylinder of diameter D and length or chord C is given by Weissinger[4] as

$$C_{L_\alpha} = \frac{\pi}{1 + (\pi C/2D) + (C/D)f(C/D)} = \frac{L}{q(\pi DC)\alpha}$$

$$f(C/D) = \arctan(1.2C/D) = (\pi/180)\tan^{-1}(1.2C/D) \tag{2}$$

The reference area πDC was used because it is one-half the wetted area and is equivalent to the planform area of a thin wing, so that at zero lift their C_D coefficients are comparable. Weissinger[4] showed that for $Re > 10^5$, Eq. (2) was satisfactory for all values of D/C, even the limiting case when $D \to 0$ because $f(C/D) \to \pi/2$ and Eq. (2) reduces to $C_{L_\alpha} = (D/C)/(1 + D/C\pi)$. This is equivalent to the momentum lift given by

$$L = \rho V(\pi D^2/4)(\alpha + \alpha_i)V = q(\pi DC)(D/C)(\alpha + \alpha_i)/2$$

$$\alpha_i \approx \alpha(1 - 0.24D/C), \quad C_{L_\alpha} \approx (D/C)(1 - 0.12D/C) \tag{3}$$

The α_i approximation was obtained from Belotserkovskii's[5] numerical calculations for $D/C \leq 3$, and so Eq. (3) is applicable for $D/C \leq 3$, as shown in Fig. 3.

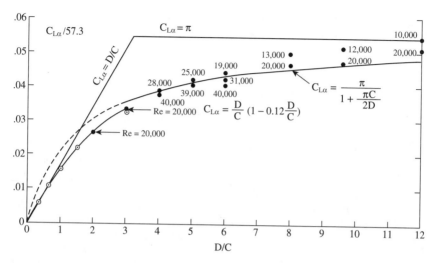

Fig. 3 Variation of the initial lift-curve slope C_{L_α} for thin-wall (0.51 mm) circular tubes of diameter (D) and length or chord (C). Text's experimental data ($\bullet$). Fletcher's[6] experimental data ($\odot$) at $Re \approx 10^6$.

Finally, for the other limit when $(D/C) \gg 1$,

$$f(C/D) = (1.2C/D) - (1.2C/D)^3/3 + O(C/D)^5$$

$$C_{L_\alpha} = \frac{\pi}{1 + (\pi C/2D)}[1 - 1.2(C/D)^2 + (1.2\pi/2)(C/D)^3 + O(C/D)^4] \quad (4)$$

If the higher-order terms are neglected this is identical to Ribner's[7] approximation after his reference area is changed to πDC. This, and Eq. (3) are shown in Fig. 3 where the wind tunnel data show that Eqs. (2) and (3) are valid for all Re when $D/C \le 3$, but for $(D/C) > 4$ Eqs. (2) or (4) can be less than the actual C_{L_α} for $Re \le 20,000$.

The theoretical induced drag was given for annular airfoils by Belotserkovskii[5] and Fletcher[6] as $C'_{D_i} = (C'_L)^2/2\pi D/C = D_i/qDC$, with the comment that this was one-half the induced drag of an elliptic wing. However, if the reference area is correctly increased to πDC, then one obtains the equation given by Weissinger,[4] since

$$C_{D_i} = \frac{C'_{D_i}}{\pi} = \frac{1}{\pi}\frac{(\pi C_L)^2}{2\pi D/C} = \frac{C_L^2}{2D/C} = \frac{D_i}{q(\pi DC)} \quad (5)$$

This shows that the induced drag force for an annular airfoil is greater than that of a wing that produces the same lift force. For example, if a thin flat plate of span πD and chord C is rolled into a circular cylinder with diameter D, then comparing Eq. (1) with Eqs. (2) and (5) gives

$$C_{D_i} = (1 + \delta)C_L^2/(\pi^2 D/C) < C_L^2/(2D/C)$$

$$C_{L_\alpha} = \frac{2\pi}{1 + (2C/\pi D)(1 + \tau)} > \frac{\pi}{1 + (\pi C/2D) + (C/D)f(C/D)}$$

Glauert[3] gives values of $\tau = 0.22$ and $\delta = 0.076$ for a rectangular planform wing with aspect ratio $\pi D/C = 3\pi$, and for $(D/C) = 3$, $(C/D)f = 0.1268$; consequently,

$$C_{D_i}(\text{wing}) = 0.0363 C_L^2 = C_{D_i}(\text{ring})/4.59$$
$$C_{L_\alpha}(\text{wing}) = 4.99(0.0871°) = 2.62 C_{L_\alpha}(\text{ring})$$

Weissinger[4] showed that Eqs. (2) and (5) closely approximated Fletcher's[6] data (C_{L_α} shown in Fig. 3) for annular airfoils having $D/C \leq 3$ at $Re \approx 10^6$. However, Fig. 3 shows a C_{L_α} greater than that of Eq. (2), or its first approximation, when $D/C > 4$ with $Re \leq 20{,}000$. Moreover, there is a rapid decrease in the L/D ratio as Re decreases. For example, Fig. 4, which gives the experimental data for $D/C = 8$, shows that decreasing the Reynolds number by a factor of 2 (from 16,400 to 8200) has decreased $L/D = C_L/C_D$ from 5.40 ($\alpha = 6.5$ deg) to 2.88 ($\alpha = 5.5$ deg), entirely as a result of the increase in the corresponding C_D from 0.0585 ($\alpha = 6.5$ deg) to 0.1320 ($\alpha = 5.5$ deg). This drag increase overcame the lift increase shown in Fig. 5 as a factor of 1.2 for the corresponding increase in C_L from 0.315 ($\alpha = 6.5$ deg) to 0.380 ($\alpha = 5.5$ deg). Hence L/D has decreased by a factor of $1.88 = (2.26/1.2) = (5.40/2.88)$. Figure 6 indicates that both the zero lift C_{D_o} and the C_{D_i} have increased more rapidly than the laminar flow skin friction factor given by $Re^{-1/2}$. Consequently, additional information would be required for the application of ring airfoils at $Re < 20{,}000$. The ring airfoil data for $D/C = 8$, shown in Figs. 4, 5, and 6, were obtained at $V \approx 40$ ft/s (12.2 m/s) and so the wind tunnel turbulence level was nearly constant. The circular cylinder for $Re = 16{,}400$ was a 6-in.-diam ($D = 8C$) metal tube with 0.02-in. (0.51-mm) thickness, and so its t/c was 0.026. For $Re = 8{,}200$ the 3-in.-diam ($D = 8C$) tube had a t/c of 0.052; this 200% increase in thickness ratio with a blunt leading could be responsible for the increase in C_{L_α} from 0.044 to 0.077 (Fig. 5) because of leading-edge suction

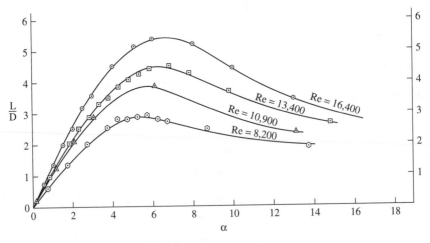

Fig. 4   Variation of L/D with angle of attack α for thin wall (0.51-mm) circular tubes with diameter (D) to chord (C) ratio, D/C, of 8 at a constant velocity of 12.2 m/s.

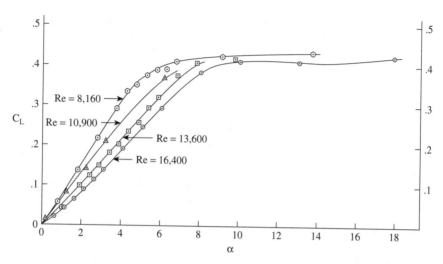

Fig. 5   Variation of lift coefficient C_L with angle of attack α for thin-wall (0.51-mm) circular tubes with diameter (D) to chord (C) ratio, D/C, of 8 at a constant velocity of 12.2 m/s.

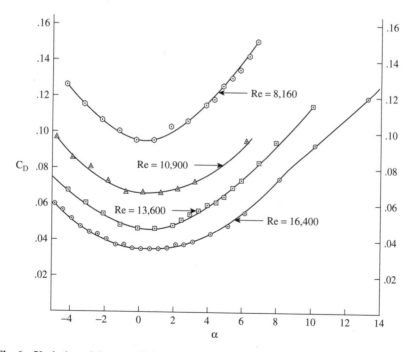

Fig. 6   Variation of drag coefficient C_D with angle of attack α for thin-wall (0.51-mm) circular tubes with diameter (D) to chord (C) ratio, D/C, of 8 at a constant velocity of 12.2 m/s.

(see Ref. 5, p. 165). This shows the need for further research at $Re < 20{,}000$, because the actual blunt leading edge is only 0.51 mm in thickness!

References

[1]Laitone, E. V., "Wind Tunnel Tests of Wings at Reynolds numbers below 70,000," *Experiments in Fluids*, Vol. 23, 1997, pp. 405–409.

[2]Laitone, E. V., "Aerodynamic Lift at Reynolds numbers below 7×10^4," *AIAA Journal*, Vol. 34, No. 9, 1996, pp. 1941–1942.

[3]Glauert, H., *Elements of Aerofoil and Airscrew Theory*, Cambridge Univ. Press, New York, 1947, pp. 145–150.

[4]Weissinger, J., "Elnige Ergebnisse aus der Theorie des Ringflügels in Inkompressibler Strömung," *Advances in Aeronautical Sciences*, Vol. 2, Pergamon, London, 1959, pp. 798–831.

[5]Belotserkovskii, S. M., "Theory of Thin Wings in Subsonic Flow," Plenum, New York, 1967, pp. 50 and 159.

[6]Fletcher, H. S. "Experimental Investigation of Lift, Drag and Pitching Moment of Five Annular Airfoils," NACA TN 4117, 1957.

[7]Ribner, H. S., "The Ring Airfoil in Nonaxial Flow," *Journal of Aeronautical Sciences*, Vol. 14, No. 9, 1947, pp. 529–530.

Chapter 6

Effects of Acoustic Disturbances on Low *Re* Aerofoil Flows

T. M. Grundy,[*] G. P. Keefe,[*] and M. V. Lowson[†]
University of Bristol, Bristol, England

Nomenclature

C_L = coefficient of lift
C_D = coefficient of drag
c = aerofoil chord
F = $2\pi f \nu U_o^{-2}$
f = frequency in Hz
M = Mach number
P = tunnel total pressure in atmospheres
Re = chord Reynolds number, cU_0/ν
SPL = sound pressure level dB re 0.00002 N/m²
TS = Tollmien–Schlichting
U_o = freestream velocity
ν = kinematic viscosity

I. Introduction

A COUSTIC disturbance is known to lead to aerofoil performance changes at low Reynolds numbers. This work was originally motivated by the substantial differences in results found by different authors in tests at low *Re*. A well-known example is the E61 aerofoil[1] shown in Fig. 1. Initial testing of the E61 aerofoil at low *Re* at Bristol demonstrated considerable variability in the results, particularly in the point at which the flow transitioned between states. This variability led to an exhaustive investigation of the possible errors in the experimental setup and procedure being used. No apparent reason was found for this variability. It was observed that the results fell into two groups, one generally repeatable set of results being obtained during the day and one in the evening.

---

[*]Student, Department of Aerospace Engineering.
[†]Professor, Department of Aerospace Engineering. Fellow AIAA.

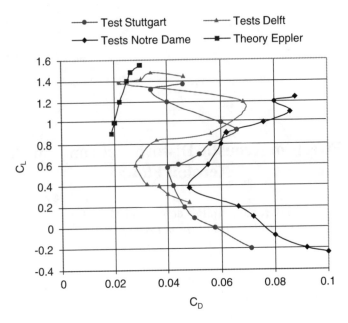

Fig. 1   **Eppler 61 lift polars at *Re* = 50,000.**

This remained a critical issue in the testing until it was observed that the external noise environment was having a major effect on the flow. The fast response balance used showed that any form of disturbance, such as remote machinery, a passing truck, conversation between the experimenters, minor contacts with the tunnel, or other people using neighboring areas, caused transient and, under certain conditions, permanent effects on the lift and drag observed. The work was therefore reoriented to provide more information on the effects of acoustic disturbance on the aerofoil performance.

Many authors have noted that acoustic energy can affect the measured lift and drag on an aerofoil at low Reynolds numbers. Schmitz[2] noted that blowing a whistle in the vicinity of the aerofoil causes a reduction in the critical Reynolds number (i.e., it promotes transition). Van Ingen and Boermans[3] present results obtained at Delft showing the beneficial effects of 104-dB 300-Hz sound beamed at the E61 aerofoil at *Re* = 80,000 and 145-Hz sound at *Re* = 50,000. In both cases the sound had the effect of making the drag polar straighter, removing some of the large variations in lift and drag with changing incidence. Mueller and Batill[4] found that a number of acoustic frequencies altered the aerofoil flowfield by changing the development of the boundary layer.

Ahuja et al.[5] demonstrated that sound at a preferential frequency can postpone the turbulent separation on an aerofoil in both pre- and poststall regimes. The optimum frequency was found to be $4U_0/c$. Goldstein[6] speculates that this effect may be due to enhanced entrainment promoted by instability waves that were triggered on the separated shear layers by the acoustic excitation.

Sumantran et al.[7] present a preliminary investigation into the effect of noise on a Wortmann FX-63-137 ESM aerofoil at a Reynolds number of 200,000 and showed that the stall hysteresis loops can be reduced in size by an acoustic disturbance introduced into the flow. The stalling angle and maximum lift coefficient were unaffected by the 5600-Hz tone, which had the effect of causing flows that were on the lower part of the hysteresis loop to "jump" back to their original values before stall had occurred. The ability of an acoustic tone to restore attached flow was found to be limited to a small range of angles of attack and the process was found to be more and more effective as Reynolds number decreased. In some cases, the flow remained in its restored state even after the tone was switched off. It was also found that even when a tone could not restore fully attached flow, it could alter the pressure distribution and temporarily increase the lift to some intermediate value.

Nash et al.[8] identified an aeroacoustic feedback process involving large-scale oscillations of the mean flow, extending back into the wake from the aerofoil trailing edge. The oscillations were involved in the selection of frequencies that comprise the tonal noise often radiated from aerofoils at modest Reynolds numbers. It is possible that the same amplification process may be utilized to affect the suction surface flow on an aerofoil, thereby modifying aerofoil performance. The hydrodynamic feedback process identified by Nash et al. interacts with the long separation bubble on the pressure surface of an aerofoil and the separating shear layer near the trailing edge to create an acoustic tone. It is reasonable to assume that an acoustic tone applied to the aerofoil will influence the feedback process and therefore alter the flow.

Although many authors have identified the effects of certain tones on certain aerofoils, there has not yet been a systematic study of the effects of various frequencies and sound pressure levels on an aerofoil, and there is no full explanation of the mechanisms by which the flow is affected.

The Eppler 61 (E61) aerofoil was designed to give high performance at low Reynolds numbers. Figure 1, based on a figure from Carmichael,[1] is a plot of C_L vs C_D (the lift-drag polar) for the aerofoil at a Reynolds number of 50,000, showing results obtained at Delft,[9] Stuttgart,[10] and Notre Dame.[11] The aerofoil shows sudden increases in lift and reductions in drag as a function of angle of attack, particularly at the lower Reynolds numbers. The results at various institutions show significant differences, with drag measured at Notre Dame up to 50% higher than at Delft and Stuttgart. Carmichael[1] presents a comparison of the differing turbulence intensities in the three test facilities, noting turbulence percentages of Delft: 0.025–0.085; Stuttgart: 0.08; Notre Dame: 0.1; and concludes that this does not explain the discrepancy.

Mueller reports results of experiments on thin flat and cambered plate aerofoils showing that lift and drag results are insensitive to changes in turbulence intensity below about 1%.[12] Although the E61 is a thicker aerofoil than those tested, it is relatively thin and sharp-edged and one might expect a similar result. Reporting the E61 results, Mueller[13] notes that the Delft and Stuttgart results were obtained using a wake rake to measure drag, whereas the Notre Dame results used direct force measurement through a strain-gauge balance. He suggests that the unsteady wake and the significant spanwise variations in the flow may cause the momentum method to be invalid at low Reynolds numbers. However, results reported by Mueller[12] using an improved balance arrangement have provided reasonably good agreement with Althaus's wake rake results at $Re = 40,000$. Mueller does not provide additional results for $Re = 50,000$ or $Re = 80,000$.

Donovan and Selig[14] investigated over 40 aerofoils in the low turbulence facility at Princeton and found good agreement with the Delft and Stuttgart results. They also used a wake rake for drag measurement but noted that there were significant spanwise variations in the wake at low Reynolds numbers. Up to 50% variation in the calculated drag was found depending on the location at which the wake was surveyed. This appears to support the view of Mueller and others that a wake rake may not adequately capture the total head in a three-dimensional, unsteady, separated wake.

Tests performed on an Eppler 387 aerofoil in extremely good low Reynolds number facilities at NASA Langley Research Center by McGhee and Walker[15,16] provide a direct comparison between wake rake measurements and strain-gauge balance results. Although agreement was good at Reynolds numbers greater than 100,000, below this value there were significant discrepancies. Balance data consistently measured higher drag than was suggested by the wake rake. Again, this was attributed to spanwise variations in the flow.

Recently, Lowson[17] suggested that the use of the wake rake may be invalid at low Reynolds numbers because of its upstream influence on the flow. Nash et al.[8] found that the introduction of a small hot-wire probe downstream of an aerofoil made a significant difference to measurements taken in a laminar separation bubble.

The present work seeks to examine the effect of various disturbances on the low Reynolds number flow about an E61 aerofoil. In particular it examines the effect of single-frequency acoustic disturbances and the acoustic background on the flow. It also examines the possibility that differing measurement techniques may influence the results.

The specific aims of the work were:

1) to replicate Mueller's experiments on the E61 aerofoil, measuring lift and drag using a strain-gauge balance,

2) to determine the nature of the flow over the E61 at low Reynolds numbers,

3) to investigate the wind tunnel environment, determining the influence of the acoustic background in the low Reynolds number range,

4) to perform a study into the effect of single-frequency acoustic disturbances on the flow and the resulting lift and drag behavior of low Reynolds number aerofoils, and

5) to examine the effect of a wake rake and other obstructions in the flow behind the aerofoil.

II. Experimental Arrangements

A. Tunnel

The tests were performed in a 2 × 2 ft horizontal, open circuit wind tunnel in the Department of Aerospace Engineering at the University of Bristol (Fig. 2). The tunnel inlet contains a honeycomb flow straightener and three antiturbulence screens with appropriate settling lengths between each. These are followed by a contraction with an area ratio of 9.7:1 leading to the test section. The octagonal test section is 1.07 m in length. A pitot-static tube at the entrance to the test section is connected to a Betz upright manometer, which allows determination of dynamic pressure heads to ±0.05 mm water.

The test section is followed by a 2.45-m-long diffuser with an area ratio of 1.24:1. The 10-bladed fan unit is isolated from the diffuser by a flexible collar to minimize transmission of vibrations to the test section. The unit is situated on rubber mounts, also to minimize vibration. After passing through the fan, the

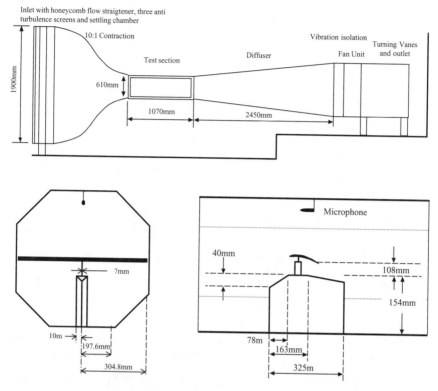

Fig. 2 Experimental setup.

flow is turned through 90 deg before being exhausted into the lab. The fan is driven by a variable-speed electric motor. Flow speeds of between 3.5 and 8.8 m/s were used in the work. Fixed tunnel speeds were selected to give the desired Re (e.g., for $Re = 50,000$ speed was set at 8.8 m/s). The temperature during all tests was close to laboratory ambient of 18°C. Error margins are believed to be better than 0.1 m/s.

B. Tunnel Environment Measurements

1. Turbulence Intensity

Turbulence intensity was measured using a DISA single component hot-wire anemometer. Figure 3 shows a plot of tunnel turbulence intensity against tunnel speed, expressed in terms of the equivalent chord Reynolds number corresponding to the model chord of 0.103 m used in this work. Note that although the turbulence intensity is high at low speeds, it soon drops to an almost constant 0.1%. Following Mueller,[13] the RMS output was passed through a 10-Hz high-pass filter to remove the low-frequency effects of large-scale recirculation in the lab (also see Section IV). The turbulence measurements were made without linearization.

"Turbulence" discussed in this chapter refers to the rotational hydrodynamic disturbance. Acoustic waves also include an irrotational velocity disturbance.

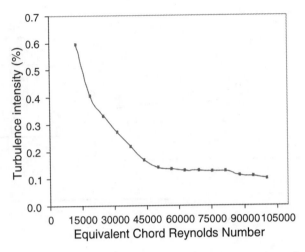

Fig. 3   Turbulence intensity in 2 × 2 ft open return tunnel.

2. *Acoustic Environment*

A Bruel and Kjaer (B&K) type 4134 0.5-in. condenser microphone was used for all acoustic measurements in this work. The standard diaphragm shielding was replaced by a highly polished nosecone to minimize aerodynamic noise generated by the flow.

Figure 4 shows the measured acoustic background in the wind tunnel with the tunnel not in operation during the day and evening. Note the broad range of acoustic frequencies present at levels above 50 dB. Other machinery in the lab and the movement and conversation of people all contribute to this background. During the evening when the lab is empty and other tunnels and machinery are silent, the background acoustic level is reduced. Noise levels in the tunnel are

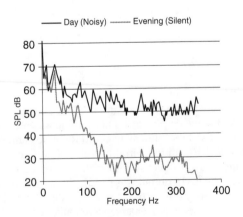

Fig. 4   Background noise levels.

not significantly greater than the tunnel-off values, particularly at low Reynolds numbers.

At the frequencies of interest (100–350 Hz) acoustic wavelengths are between 1 and 3 m. Under thése conditions sound within the present test section will be in phase over the whole aerofoil.

C.   New Strain-Gauge Balance

1.   Details

Lift and drag measurements were made using a new, two-component strain-gauge balance constructed specifically for use at low Reynolds numbers. The balance was constructed to a design used successfully at AeroVironment in the United States, following the kind provision of a full internal report on the balance by Grasmeyer. The balance uses the loadcells of two commercially available (ACCULAB VI-1200) strain-gauge balances and provides an overall capability of 600 g in lift and drag and a resolution of 0.1 g.

AeroVironment performed tests on loadcells of the same design as those used in this work and concluded that they are extremely insensitive to loading offsets (AeroVironment internal report). Weights of 200 g each were placed on the loadcell up to 12 in. away from the design loading point. Balance readings changed by a maximum of only 0.3 g in these tests.

2.   Model Mount

The balance has a single, central mounting point, which is actuated using a servomotor to give a range of approximately 15 deg in angle of incidence. The servomotor allowed the angle of incidence to be changed while tests were underway. Graduations on the servo controller dial allow the incidence to be changed in steps of 0.3 deg. The dial was calibrated using an inclinometer. The mount and servo arm are surrounded by a faired shroud, dimensions of which are provided in Fig. 2.

3.   Balance Calibration

The balance was calibrated using precision weights loaded into pans connected to the balance mounting point by wires running through low-friction pulleys. The wires were aligned to the horizontal and vertical using spirit levels and by sighting the wires against known vertical and horizontal features of the tunnel.

D.   Data Acquisition

Lift and drag measurements were available in real time on the commercial balance displays. However, to reduce errors, the strain-gauge voltages were amplified and sent to an AD2814 data acquisition card with 12-bit resolution in a 4DX266 PC. Two Pascal programs were written to sample the voltages. The first program was used to measure average voltages over an interval. The second provided a time history of the individual lift and drag voltages that were sampled over a period of time.

The following corrections have been applied to the data: buoyancy, solid and wake blockage, and streamline curvature. These followed the recommendations of Pope et al.[18]

E. Aerofoil Models

A model E61 aerofoil of chord 0.103 m and span 608 mm was used for these tests. The model was constructed to an accuracy of ± 0.2 mm.

The span of 608 mm was chosen to match the need to ensure two-dimensional flow behavior over the aerofoil with the need to prevent the model coming into contact with the tunnel walls. Because of the relatively flexible nature of the central model mounting point, it was necessary to leave a gap of approximately 1 mm between the model and the wall to prevent contact. Pope et al.[18] suggest that a gap of no more than 0.005 times the model span be left if two-dimensional flow is required. For the current model, this implies that a gap of less than 3 mm will ensure that the flow remains two-dimensional. Thus the 1-mm gaps left should provide satisfactory results.

F. Flow Visualization

1. Smoke Wire

The smoke-wire technique was used to gain a qualitative understanding of flow behavior. Unfortunately, owing to the problems of adequately lighting the smoke, no photographs of the flow visualization are available to date.

2. Laser Doppler Anemometer

The department's three-component laser Doppler anemometer (LDA) was used to obtain accurate, high-resolution, nonintrusive measurements of the flow over the E61 in the tunnel described above. This effectively allowed detailed flow visualization of the boundary layer across the aerofoil. The LDA results could also be used to gather information about the frequency content of any disturbances in the flow.

The LDA system consists of a 5-W argon-ion laser that emits a single beam to a transmitter box where it is split into three pairs of beams of distinctive wavelengths. These are transmitted by fiber optics to two optic heads. One optic head emits two pairs of orthogonal beams; the third pair is emitted by the second optic head. Both heads are mounted on a three-axis traverse mechanism. The pairs of beams are focused onto a small measurement volume of the flow. The measurements took advantage of the calibration technique developed by Swales et al.[19] which provides major improvements in resolution (0.05 mm) for the three velocity components measured.

III. Results

A. Basic Aerodynamic Tests on E61

Figure 5 presents the lift-drag polar obtained for the E61 at various institutions at a Reynolds number of 50,000 and includes the results found in this work. The trend of the Bristol results is seen to be the same as those obtained at Delft and Stuttgart. Particularly in the low-C_L range, the Bristol results closely follow the Stuttgart data. However, the maximum lift coefficient obtained at Bristol is almost exactly that obtained at Notre Dame. The correlation between the Bristol results and those obtained at Delft and Stuttgart is poor at high values of C_L.

The results appear to be adequately consistent with results obtained by other workers. However, the present results did show significant variability resulting

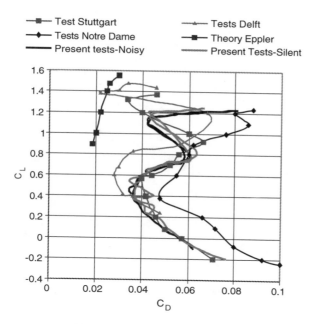

Fig. 5   Present lift-drag polars.

from acoustic inputs. As already noted in the Introduction this effect became the principal target for the present investigations.

B.   Aerodynamic Trends with Reynolds Number

Figure 6 shows results for lift vs incidence for the E61 aerofoil at a Reynolds number of 25,000. This figure provides a characteristic plot that may be used in conjunction with the discussion below. Figures 7a, 7b, and 7c show the full results for the E61 aerofoil at Reynolds numbers of 25,000; 35,000; 50,000; and 60,000; both for silent and noisy conditions. "Silent" conditions (S) refer to evening measurements when special efforts were made to maintain a minimum level of disturbance in all respects. "Noisy" conditions (N) refer to measurements made without special precautions, in which the ordinary background noise of a busy laboratory was present. It can be seen that all curves also have a generally similar shape, and the peak lift and minimum drag recorded for all cases are similar. At lower angles of attack, lift and drag generally follow similar trends for each case. As incidence increases above zero, lift increases slowly but drag shows a strong increase. This is the characteristic sign of a long laminar bubble.

At sufficiently high incidence the flow undergoes a sudden change, usually associated with the transition to a reattaching short bubble. Lift increases significantly and drag reduces. As incidence is further increased lift and drag increase slowly. If incidence is then reduced the high lift is retained but drag reduces considerably (counterclockwise hysteresis on the lift-drag polar). Generally, the curves are coincident at low and high values of C_L. However, the details of the process are extremely Reynolds number dependent. The point at which the long separation

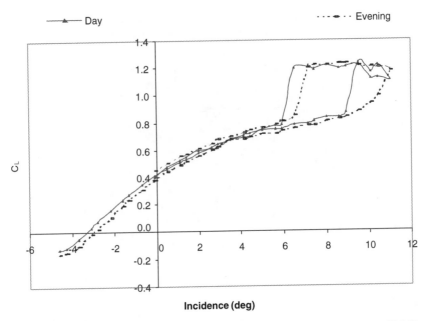

Fig. 6 Eppler 61 lift vs incidence effect of level of background noise at *Re* = 25,000.

bubble collapses when lift suddenly increases and the drag drops changes substantially with Reynolds number. As Reynolds number reduces, the laminar separation bubble persists longer. The point at which lift rises and drag drops to rejoin the tripped case is delayed and the maximum lift-to-drag ratio is reduced. The hysteresis loops become more evident at the lower Reynolds numbers.

Figure 6 shows that at a Reynolds number of 25,000 under noisy conditions, the lift remains low until an angle of incidence of about 9 deg is reached. At this angle, the lift suddenly increases. On reducing incidence, however, the lift does not immediately drop to its original value. This only occurs at the much lower incidence of 6 deg. Thus, the loop is followed in an anticlockwise sense. A hysteresis loop is seen for the drag on Fig. 7c at *Re* = 25,000 N. The increase in lift observed on Fig. 6 at 9 deg is matched by a drop in drag. The drag remains low until an incidence of 6 deg, where it increases to its original value. Thus, a clockwise movement on the drag curve matches the anticlockwise movement on the lift curve. This translates to an anticlockwise movement on the lift-drag polar of Fig. 7a. Similar hysteresis loops are observable at the other Reynolds numbers considered, but their extent reduces as Reynolds number increases.

The LDA was used to investigate the flow at 10.5 deg at a *Re* of 25,000. The mean velocity plots show a large amount of reverse flow, even quite close to the aerofoil's leading edge. RMS data show correspondingly high turbulence intensities, which peak 3 mm from the aerofoil surface. Use of a smoke wire showed that the separated flow does not reattach at all at this incidence. The aerofoil has stalled as a result of the point of reattachment moving past the trailing edge of the aerofoil. An extremely large, recirculating wake extends several chord lengths downstream of the aerofoil, the result of which is a very high drag.

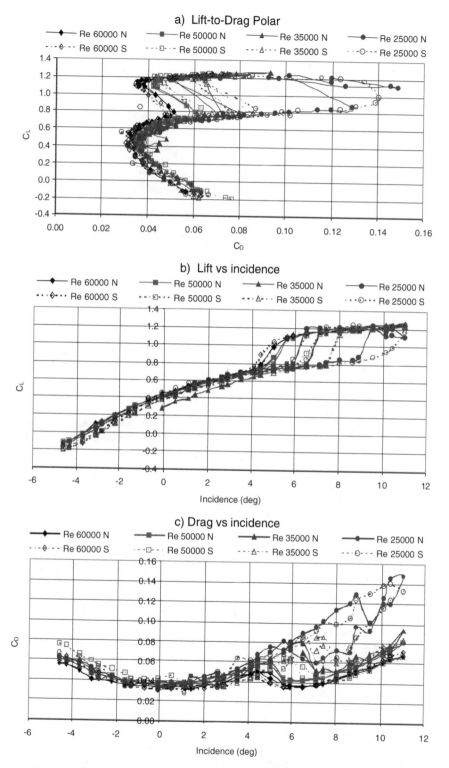

Fig. 7 Performance of Eppler 61 under silent (S) and noisy (N) conditions.

C.   The Effect of Background Acoustic Level

Clearly, the general trends observed in the noisy and silent conditions are similar. However, the two sets of data differ in detail—most notably in the angle of incidence at which the lift suddenly increases and the drag suddenly drops.

Figures 7b and 7c show that the acoustic background has little effect on the lift and drag behavior at high and low angles of incidence. It is in the region of the hysteresis loop that effects are seen. When the background acoustics are removed, the flow remains in the low-lift state until a much higher angle of incidence. Transition to the high-lift, low-drag state is much more gradual. Once this state is achieved, the flow returns to the low-lift state at a higher angle of incidence than is the case in the presence of higher background noise.

Essentially the same trends are observed at all Re, but the overall effect reduces as Re increases, until at $Re = 60,000$ the effects at the high angles of attack are not observed. Variations are within experimental error. But even at a Reynolds number of 60,000 the lift at negative angles of incidence is reduced in silent conditions. Generally, acoustic disturbances are having an effect similar to that of increasing Reynolds number in that hysteresis loops are reduced or eliminated.

D.   Key Comparison Issues

Analysis of the figures referred to above highlights the following main differences in the fundamental flow behavior over the E61 in both acoustic conditions described:

a) Maximum lift coefficients do not appear to be affected by changes in the acoustic background.

b) The transition process between the low- and high-lift states in silent conditions is found to be gradual. This same process was found to occur as a sudden step change in the presence of background acoustic disturbances. This is particularly noticeable at $Re = 25,000$ where seven individual sample data points can be observed during the transition between the low- and high-lift states during silent conditions, whereas only one is observed under daytime conditions.

c) Acoustic disturbances not only induce a sudden tripping process but also encourage it as the flow is found to change states at a lower incidence than would be expected in silent conditions. For example at $Re = 35,000$ transition occurs at approximately ~1 deg earlier under daytime testing conditions.

d) Tests taken under silent conditions present larger hysteresis loops in the lift-drag polar. This is particularly noticeable at a Reynolds number of 25,000 where even the shape of the hysteresis loop is found to be significantly different.

e) Tests taken in silent conditions show higher drag and lower lift throughout the range of negative incidences.

f) All the effects mentioned above are accentuated at the lower Reynolds numbers considered, with most obvious differences found at a Reynolds number of 25,000.

E.   The Effects of Single-Frequency Acoustic Disturbances on the Flow

The results given in Section III.B show that the flow, in particular the hysteresis loops, can be affected by acoustic disturbances. Acoustic disturbances cause the flow to transition more easily from the low-lift to the high-lift state. Single-frequency acoustic disturbances were used to investigate these effects further. This

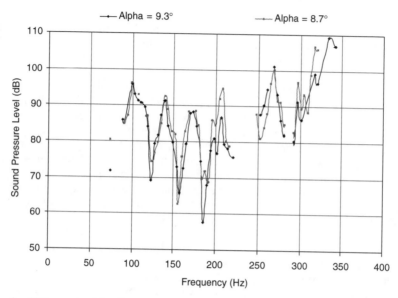

Fig. 8 SPL required for flow transition between states vs frequency for the E61 at *Re* = 25,000.

was done at a Reynolds number of 25,000, with the aerofoil incidence set in the region of the hysteresis loop. In this way, for each individual frequency, the gain required to induce the flow to change from the low-lift to the high-lift state could be recorded. The fact that transition had occurred could be inferred from the sudden, discontinuous increase in the lift readout on the balance display.

Figure 8 shows the sound pressure level at the position of the aerofoil inside the tunnel that was required to cause transition between states. The experiment was repeated with the aerofoil at two distinct angles of incidence; both curves are presented in Fig. 8. This shows that the SPL required to promote transition is strongly dependent on the frequency of the acoustic disturbance. The curves for the two angles of incidence follow exactly the same trend and similar SPLs are required to trip the flow in both cases. Over most portions of the curve, the SPL required at 8.7 deg is slightly higher than that at 9.3 deg.

Frequencies above 350 Hz were not found to affect the flow. Missing portions of the curve on Fig. 8 correspond to regions in which the maximum output of the loudspeaker was insufficient to cause the flow to transition.

The initial postulate under investigation was that the transition between states may be induced more easily at some frequencies than at others. The separated boundary layer in a separation bubble is known to act as a broadband amplifier of disturbances. This aids the process of collapsing the long bubble by promoting the transition from laminar to turbulent flow in the separated shear layer. Following results found by Nash et al.[8] it was suggested that the attached boundary layer upstream of the bubble may have frequencies at which a very small input disturbance would be more amplified strongly by the flow. When this amplified disturbance reached the separated shear layer it would be amplified even more strongly, causing a disproportionate effect.

As shown on Fig. 8, the sound pressure levels required to cause transition exhibit several maxima and minima. This would be expected if the boundary layer did indeed have resonant frequencies as described in the postulate above. The results obtained at two different incidences correspond closely and both show several sets of peaks and troughs. The troughs correspond to points where the flow trips easily. In other words, low sound pressure levels at these frequencies caused the flow to transition between states. At the peaks, however, the flow did not transition to the high-lift state until relatively high sound pressure levels were reached.

Although the curves for the two separate incidences are similar, there are important differences. Transition from the low-lift to the high-lift state under silent conditions occurs at an angle of incidence of about 10.5 deg. The aerofoil is further away from this natural transition when it is at 8.7 deg than it is at 9.3 deg. This is reflected in the higher sound pressure level that is in general required to trip the flow at the lower incidence. Despite this, the similarity of the curves suggests that the same mechanisms are at work in both cases.

It was also noted that even when acoustic disturbances had tripped the flow at 8.7 deg, removal of the disturbance caused the flow to revert to its original state (although there was often a delay of several seconds after removal of the disturbance before this occurred). At 9.3 deg, acoustic disturbances tripped the flow, which then remained tripped even when the disturbance was removed. This suggests that there is a region of the hysteresis loop where the high-lift state is unstable and will tend to revert to the low-lift state if insufficient disturbance is present. This is a similar effect to that discussed above, where it was shown that the high-lift state could be maintained at lower angles of incidence during the day when the background acoustic level was high.

There is an obvious possibility that the curves in Fig. 8 are related to acoustic resonances of the tunnel. This was measured by setting the external loudspeaker to a fixed level and measuring the acoustic response in the tunnel; the results are shown in Fig. 9. The curve from Fig. 8 corresponding to an angle of incidence

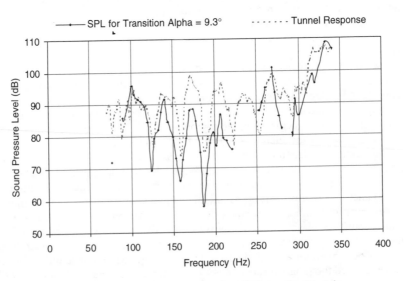

Fig. 9   Tunnel response compared to SPL required to trip.

of 9.3 deg is also presented on Fig. 9 for comparison. It can be observed that the shapes of the curves are remarkably similar.

The correspondence between the curves strongly suggests that there is a link between the tunnel response and the tripping process across the entire frequency range. It appears that the aerofoil response to acoustic disturbances is determined by the acoustic response of the wind tunnel.

Figure 9 shows that at frequencies where the tunnel is near a resonance (i.e., near a peak in the maximum gain plot), the flow requires a high gain to trip. Conversely, where the tunnel acts to reduce the sound pressure level at the aerofoil (at an antiresonance), only a small gain is required to trip the flow. This is again surprising since tunnel resonances serve to increase the sound pressure level at the aerofoil, creating large fluctuating acoustic pressures. These might be expected to affect the separated shear layer and promote transition to turbulence. Instead it appears that high sound pressure levels do not promote transition. The flow transitions more easily at the tunnel antiresonances.

In a resonant duct, peaks in sound pressure level correspond to minima in the induced fluctuating velocity level, and vice versa. Hence, Fig. 9 implies that the aerofoil responds most readily to frequencies that provide the maximum levels of fluctuating velocity. Thus it appears that the flow is affected through the velocity fluctuations that are imposed by the acoustic field. Figure 9 provides no obvious evidence that the boundary layer responds more readily to such fluctuations at specific frequencies.

The analysis above has two important implications:

1) The separated shear layer responds to the small velocity fluctuations generated by acoustic disturbances, not the pressure fluctuations.

2) Results of acoustic experiments would be expected to vary greatly among different wind tunnels, which will have different acoustic responses.

The importance of tunnel acoustic response on boundary layer response at low Reynolds numbers frequencies has been reported by Zaman et al.[20] although they did not relate this to the increased significance of external acoustic input as cause of premature transition.

F. Effect of Acoustics on Aerodynamic Forces

Further investigation of the effects of acoustics was performed by taking lift and drag measurements while generating single-frequency acoustic disturbances from the loudspeaker. The effects of these disturbances were most pronounced on the lift curves for the aerofoil, which will be presented here. All tests were performed at a Reynolds number of 25,000 with the laboratory silent other than the noise of the tunnel and the loudspeaker.

Figure 10 shows the effect of three such frequencies on the lift curve. Note that the low and negative incidence values remain almost the same or slightly lower as the values obtained with no sound. All of the frequencies generated a sound pressure level of 92 dB at the aerofoil. Note that all of the frequencies cause the flow to transition swiftly, eliminating the slow increase in C_L observed for the silent case. All inputs cause the hysteresis loop to be significantly reduced in extent, but different frequencies have different effects on the range of angles of incidence over which the hysteresis loop exists. However, the maximum lift obtained remains constant.

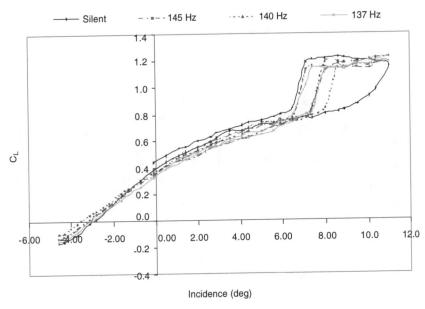

Fig. 10   Effect of various frequency inputs at 92 dB with _Re_ = 25,000.

The 140-Hz input causes very little reduction in the size of the hysteresis loop. However, although the 75-dB sound pressure level was the maximum obtainable for the 186- and 158-Hz inputs, a tunnel resonance allows the 140-Hz sound to be input at a higher level of 92 dB. The result of this input is shown on Fig. 11 along with the 140-Hz input at 75 dB for comparison. It can be seen that the increase to maximum gain causes the hysteresis loop to shrink in the same way as that seen on Fig. 10 for the effect of varying frequencies. However, once again, the range of angles of incidence over which the hysteresis loop exists is different.

Figure 9 suggests that preferential frequency ranges for acoustic tripping may exist between about 130 and 220 Hz and also around 300 Hz. The nature of the experimental procedure precludes making more specific analyses. Ahuja's suggestion of an ideal tripping frequency of $4U_0/c$ to suppress turbulent separation corresponds to 137 Hz ($U_0 = 3.5$ m/s; $c = 0.103$ m). This is at the lower end of the range found in the present experiments.

IV.   Discussion

The results of the present study demonstrate the sensitivity of the laminar separation bubble to disturbance generally. In broad terms, the result reported here at $Re = 25,000$ are very similar to those found by Sumantran et al.[7] at $Re = 200,000$. In both cases, it was found that appropriate sound inputs could cause either temporary or permanent changes in flow state, especially in regions where flow hysteresis was observed. This suggests that the general effects reported here may be found widely at low Re.

Transient effects of acoustic disturbances were also noted throughout the experiments. These have not been specifically recorded but are clearly relevant to

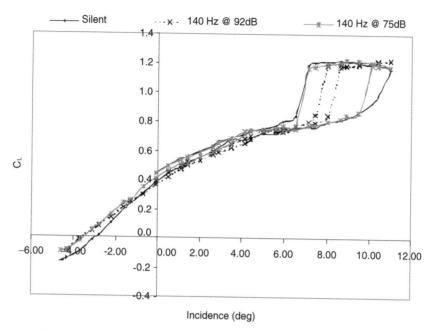

Fig. 11   Effect of gain with a single-frequency input with $Re = 25,000$.

any practical application. The most representative example of such transient effects occurred when the aerofoil was on the low part of the hysteresis loop. In this condition a slight increase in lift and reduction in drag was observed when any transient increase in the background noise level occurred or when there was other external disturbance. The effect was visible on the real-time balance displays while the disturbance took place.

Two different outcomes of the disturbance were observed:

1. The aerofoil response was truly transient and the flow returned to its original state once the disturbance was removed.

2. The disturbance would trip the flow over the aerofoil into its high lift "stable" state where it would remain even when the disturbance was removed. Under no circumstances did an acoustic disturbance cause the flow to change to a low-lift state from the high-lift condition.

Figure 12 shows results obtained at $Re = 60,000$ compared with results for a case in which the flow was tripped to turbulence using a trip wire. Comparisons of the three cases shows that generally lift is higher and drag is lower for the fully tripped case over a wide range of angles of attack. At the highest angles of attack beyond the transition to the short bubble, both lift and drag are found to be slightly higher in the tripped case.

Although no systematic experiments were undertaken, a number of tests were done that increased the turbulence level in the tunnel either locally or globally. It was found that this had little effect at the low Re of the present tests. Other tests on aerofoils at higher Re show considerable sensitivity to input turbulence. It appears that low Re flows are not particularly sensitive to turbulence input levels, consistent with Mueller's observation that turbulence intensities below 1% had

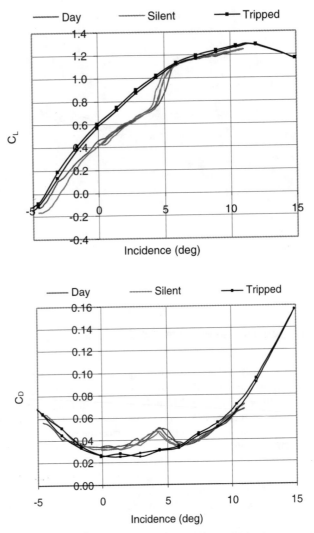

Fig. 12   Lift and drag vs incidence tripped and untripped at *Re* = 60,000.

little effect on the flow. Because the majority of the energy in the turbulence was below 10 Hz this provides partial post hoc justification for the use of the 10-Hz filter when measuring turbulence levels.

The sensitivity of the results to sound will have been enhanced by the two-dimensional nature of the present tests. Acoustic input at the frequencies examined here will lead to a disturbance that is in phase across the whole aerofoil and thus is particularly effective at generating two-dimensional instabilities. Indeed this technique has been explicitly used by Saric in his extensive series of studies on boundary layer receptivity (e.g, Saric et al.[21]).

In the majority of studies of low Re flow the frequencies will have sound wavelengths in relation to tunnel dimensions that to lead to two-dimensional disturbance. However, if the tunnel were of large scale then its response at the same frequency would not necessarily be two dimensional. This leads to a complication in extrapolating the results of the present tests to large-scale facilities.

Thus the sensitivity of the boundary layer to velocity disturbances caused by acoustic inputs is due to two separate effects. First, any velocity disturbance that is the result of an acoustic field will transmit in the tunnel without loss; indeed it may be amplified by acoustic resonances. Second, at typical test scales the acoustic disturbance can couple directly to two-dimensional disturbances on the aerofoil. In contrast, conventional hydrodynamic turbulence inputs will decay as they pass down the tunnel, especially at the relatively high frequencies of interest here, and will also be of three-dimensional rather than correlated two-dimensional form.

Saric and Reshotko[22] note that the key TS frequency bands are related to the dimensionless reduced frequency F defined by

$$F = 2\pi f v U_o^{-2}$$

where f is the physical frequency in Hz, v is the kinematic viscosity, and U is the freestream speed.

Saric and Reshotko give the expression

$$f = 0.116\,[F \times 10^{10}]PM^2$$

where P is the tunnel total pressure in atmospheres and M is the Mach number. Saric and Reshotko[22] give a value of $F = 30 \times 10^{-6}$ as a value in the middle of the TS band for lower Mach numbers.

Taking a total pressure of 1 atmosphere and a tunnel speed of 4.4 m/s corresponding to $Re = 25,000$ gives a value of f of 6 Hz. This figure is well below the frequencies found to be relevant in the present experiments. However, the transition process described here is not an attached TS transition. It is clearly due to transition in the separated shear layers, which have a far wider range of response frequencies. Thus it appears possible to conclude that attached TS transition processes have limited relevance to the effects found during the present work.

During one test an anomalous result was obtained. For this test there was no transition to the high-lift low-drag state at high angles of attack. Surprisingly, this was a test during which no specific levels of extra care were being taken to ensure disturbance-free conditions. This result was not duplicated in subsequent tests. Despite this, it is logical to assume that this represents a case in which the laminar bubble was not tripped. If so, then this result could be more representative of a completely undisturbed flow than those described in the rest of this chapter.

A number of other tests have also been undertaken during the present work. The suggestion that a wake rake could influence the flow under conditions of large-scale recirculation was examined by measuring the lift and drag with and without a dummy wake rake 3 chords downstream. The results were inconclusive. A modest difference was observed but it was not clear that this was outside the normal range of experimental error. It can be concluded that the wake rake has no large effects on the overall aerofoil performance. The possibility of a local upstream effect on the flow remains open.

Some of the tests involved the introduction of mounting devices causing blockages of up to 3% close to the tunnel wall. These had very strong effects on the results, and effects could still be observed even when the blockage was as far as 6 chords downstream. This was clearly due to change to local flow conditions at the aerofoil. This effect was considerably greater than expected and emphasizes the need for considerable caution in low Re tests.

As a result of this observation, additional tests were undertaken to measure the flow angles in the tunnel using the LDA. Blockage caused by the model mounting gave rise to an upwash of 1 deg locally at the center of the tunnel, although this reduced rapidly to zero away from the tunnel center. This will affect all of the results reported here. The effects are believed to be modest since all results are based on lift and drag measured on the whole wing.

It was also observed that, at negative angles of incidence at the lowest Re tested, lift and drag results showed large temporal variations. This was associated with resonances in the wing mounting structure and/or servo system. The effects were only noted at test conditions that were well away from the area of principal interest in the present work. Nevertheless, this raises the possibility of aerodynamic/dynamic coupling at negative angles of attack. There was some evidence that the response was coupled to a vortex shedding process, but time did not allow in-depth investigation of this effect.

V. Potential Use of Sound to Improve Performance

The results provided earlier in the chapter present possibilities for using sound to improve uninhabited air vehicle (UAV) or micro air vehicle (MAV) performance. Under normal circumstances, an MAV with an E61 wing flying at 5 deg incidence at a Reynolds number of 25,000 would be in the low-lift state. To achieve better performance it would need to increase incidence to above 7 deg to enter the high-lift state. Its incidence could then be reduced so that it could fly at the maximum lift to drag ratio. However, the MAV power requirement would have to be high to overcome the high drag encountered at 7 deg.

However, through careful choice of a particular tone to be generated by the MAV, the extent of the hysteresis loop could be reduced. The high-lift regime could then be entered at a much lower angle of incidence, thus reducing the maximum power required.

Any acoustic disturbance intended to improve performance should be chosen such that:

1) The energy in the disturbance is sufficient to close the hysteresis loop.

2) The frequency is chosen to provide a transition between states at the lowest possible incidence.

The individual frequency chosen could be generated by a small loudspeaker on the airframe. There may also be opportunities, for example, to tailor the propeller blade numbers and/or speed so that an acoustic disturbance of a beneficial frequency is generated. The practicability of this approach would also be strongly affected by aerofoil pitching moments, not recorded during the present experiments.

However, given the results obtained in this work, wind-tunnel tests to determine beneficial frequencies will have to be carefully interpreted to ensure that extraneous effects of wind tunnel response are eliminated. In effect the wind tunnel acoustic response process in the tests reported here has translated external disturbances into correlated two-dimensional disturbances on the aerofoil. Practical use of the effects found would require careful matching of disturbance to aerofoil geometry.

Figure 12 suggests that acoustic tripping will not be as effective in improving performance as conventional tripping using wires, etc.

VI. Conclusions

Tests have been performed on the E61 aerofoil in a 2×2 ft, open return, low turbulence intensity tunnel, using a fast response strain gauge balance to measure lift and drag. Aerofoil behavior at low Reynolds numbers was found to be strongly influenced by the background acoustic level and the presence of single-frequency tonal noise. Further, the response of the aerofoil to acoustic disturbances was dominated by the wind tunnel acoustic response. This finding implies that the consequences of acoustic disturbances at low Reynolds numbers will be sensitive to the facility in which the aerofoil is tested. This provides a new explanation for the variability in test results in different facilities reported by many investigators. It appears that the acoustic environment is more significant than the turbulence environment for low Re tests. These findings also offer new opportunities for performance modification of aerofoils by the use of selected noise inputs. The following specific conclusions can be reached from the results presented:

1) The flow about the E61 at this Reynolds number is dominated by the presence of a long separation bubble at positive angles of incidence. The bubble breaks down into a short bubble at high angles of incidence, causing a significant increase in lift and decrease in drag. The short bubble bursts into a long bubble at a lower incidence than that at which the reverse process occurs. Hence, two flow states exist over a range of angles of incidence.

2) High levels of background noise were found to promote the process of bubble breakdown and hence the transition between flow states. Under silent conditions, the bubble breakdown process was found to be gradual rather than sudden.

3) Background noise also reduced the size of the hysteresis loop.

4) Significant transient effects were observed in response to acoustic and other disturbances.

5) Aerodynamic tests at low Reynolds numbers should be undertaken in conditions that are as quiet as possible.

6) The effect of individual frequencies on the flow around the E61 is strongly dependent on the acoustic response of the wind tunnel. Hence, results quoted by different institutions are not likely to be comparable.

7) The mechanism by which acoustic disturbances affect the flow is through small velocity fluctuations rather than fluctuating pressure levels.

8) Individual frequencies over the range of 100–350 Hz promote breakdown of the laminar separation bubble at $Re = 25,000$ and hence cause transition between the two flow states. The sound pressure level required is related to the acoustic response of the wind tunnel.

9) Individual frequencies in this range promote transition at different angles of incidence. All of the frequencies tested appeared to reduce the size of the hysteresis loop.

10) The acoustic environment appears to be more significant than the turbulence environment for aerodynamic testing at low Re.

11) The possibility of using sound to improve UAV or MAV performance appears to justify study.

Acknowledgments

This work was carried out as part of an undergraduate project in the Department of Aerospace Engineering.[23] Additional support for the work was provided by a grant from DERA. The authors would like to thank Matt Crompton of the University of Bristol for his help in obtaining LDA data. Thanks are also due to Professor W. S. Saric for helpful comments on the interpretation of the results.

References

[1]Carmichael, B. H., "Update on High Performance Model Airfoils," *National Free Flight Society 12th Symposium*, 1979, pp. 34–41.

[2]Schmitz, F. W., "Aerodynamics of the Model Aeroplane," Redstone Arsenal Translation N70-39001, RSIC-721 Redstone Arsenal, Nov. 1967 (translation of 1942 original in German).

[3]van Ingen, J. L., and Boermans, L. M. M., "Research on Laminar Separation Bubbles at Delft University of Technology in Relation to Low Reynolds Number Airfoil Aerodynamics," *Proceedings of the Conference on Low Reynolds Number Airfoil Aerodynamics*, edited by T. J. Mueller, UNDAS-CP-77B123, Univ. of Notre Dame, Notre Dame, IN, June 1985, pp. 89–124.

[4]Mueller, T. J., and Batill, S. M., "Experimental Studies of Separation on a Two-Dimensional Airfoil at Low Reynolds Numbers," *AIAA Journal*, Vol. 20, No. 4, April 1982, pp. 457–463.

[5]Ahuja, K. K., Whipkey, R. R., and Jones, G. S., "Control of Turbulent Boundary Layer Flows by Sound," AIAA Paper 83-0726, 1983.

[6]Goldstein, M. E., "Generation of Instability Waves in Flows Separating from Smooth Surfaces," *Journal of Fluid Mechanics*, Vol. 145, 1984, pp. 71–94.

[7]Sumantran V., Sun, Z., Marchman, J. F., III, "Acoustic and Turbulence Influence on Low-Reynolds Number Wing Pressure Distributions," *Proceedings of the Conference on Low Reynolds Number Airfoil Aerodynamics*, edited by T. J. Mueller, UNDAS-CP-77B123, Univ. of Notre Dame, Notre Dame, IN, June 1985, pp. 323–334.

[8]Nash, E. C., Lowson, M. V., and McAlpine, A., "Boundary-Layer Instability Noise on Aerofoils," *Journal of Fluid Mechanics*, Vol. 382, 1999, pp. 27–61.

[9]deVries, J., Hegen, G. H., and Boermanns, L. M. M., "Preliminary Results of Wind Tunnel Measurements at Low Reynolds Numbers on Section Eppler 61," Delft Univ. Memo, 1980.

[10]Althaus, D., *Profilpolaren fur den Modelflug*, Neckar-Verlag Klosterring 1, 7730 Villingen-Schwenningen, 1980.

[11]Burns, T. F., "Experimental Study of Eppler 61 and Pfenninger 048 Airfoils at Low Reynolds Numbers," Graduate Thesis, Univ. of Notre Dame, Notre Dame, IN, 1981.

[12]Mueller, T. J., "Aerodynamic Measurements at Low Reynolds Numbers for Fixed Wing Micro Air Vehicles," presented at RTO AVT/VKI Special Course on Development and Operation of UAVs for Military and Civil Applications, 13–17 Sept. 1999.

[13]Mueller, T. J, "Low Reynolds Number Vehicles," AGARDograph No. 288, 1985.

[14]Donovan, J. F., and Selig, M. S., "Low Reynolds Number Airfoil Design and Wind Tunnel Testing at Princeton University," *Proceedings of the Conference on Low Reynolds Aerodynamics*, Univ. of Notre Dame, Notre Dame, IN, 1989.

[15]McGhee, R. J., and Walker, B. S., "Performance Measurements on an Airfoil at Low Reynolds Numbers," *Proceedings of the Conference on Low Reynolds Aerodynamics*, Univ. of Notre Dame, Notre Dame, IN, 1989, pp. 131–145.

[16]McGhee, R. J., and Walker, B. S., "Correlation of Theory to Wind-Tunnel Data at Reynolds Numbers below 500,000," *Proceedings of the Conference on Low Reynolds Aerodynamics,* Univ. of Notre Dame, Notre Dame, IN, 1989, pp. 146–160.

[17]Lowson, M. V., "Aerodynamics of Aerofoils at Low Reynolds Numbers," *Bristol UAV Conference 1999*, pp. 35.1–35.16.

[18]Pope, A., Rae, W., and Barlow, J., *Low Speed Wind Tunnel Testing,* 2nd ed. Wiley, New York, 1984.

[19]Swales, C., Rickards, J., Brake, C., and Barrett, R. V., "Development of a Pin-Hole Meter for Alignment of 3D Laser Doppler Anemometers," *Dantec Information,* Vol. 12, 1993, pp. 110–114.

[20]Zaman, K. B. M. Q., Bar-Sever, A., and Mangalam, S. M., "Effect of Acoustic Excitation on the Flow over a Low-Re Airfoil," *Journal of Fluid Mechanics,* Vol. 182, 1987, pp. 127–148.

[21]Saric, W. S., White, E. B., and Reed, H. L., "Boundary Layer Receptivity to Freestream Disturbances and Its Role in Transition," AIAA Paper 99-3788, July 1999.

[22]Saric, W. S., and Reshotko, E., "Review of Flow Quality Issues in Wind Tunnel Testing," AIAA Paper 98-2613, June 1998.

[23]Grundy, T., and Keefe, G., "The Effect of Disturbances on Stall at Low Reynolds Number," Dept. of Aerospace Engineering, Univ. of Bristol, Project No 929, Bristol, U.K., May 2000.

Chapter 7

Aerodynamic Characteristics of Low Aspect Ratio Wings at Low Reynolds Numbers

Gabriel E. Torres* and Thomas J. Mueller[†]
University of Notre Dame, Notre Dame, Indiana

Nomenclature

AR	= aspect ratio (b^2/S)
a_0	= two-dimensional lift-curve slope
b	= wingspan
C_D	= drag coefficient $[D/(\frac{1}{2}\rho U_\infty^2 S)]$
C_{D_0}	= drag coefficient at zero lift
C_L	= lift coefficient $[L/(\frac{1}{2}\rho U_\infty^2 S)]$
C_{L_α}	= lift-curve slope $(\deg^{-1})$
C_N	= normal force coefficient $(C_N = C_L \cos\alpha + C_D \sin\alpha)$
c	= wing chord
D	= drag force
K	= induced drag coefficient
L	= lift force
L/D	= lift-to-drag ratio
Re_c	= Reynolds number based on wing root chord $(\rho U_\infty c/\nu)$
S	= wing Area
U_∞	= freestream velocity
α	= angle of attack in degrees
$\alpha_{C_L=0}$	= zero-lift angle of attack
α_{stall}	= stall angle of attack
ν	= kinematic viscosity
ρ	= air density
τ	= Glauert parameter

*Graduate Research Assistant, Department of Aerospace and Mechanical Engineering.
†Roth-Gibson Professor, Department of Aerospace and Mechanical Engineering. Associate Fellow AIAA.

I. Introduction

INTEREST in the aerodynamics of flying vehicles at low Reynolds numbers has progressively increased over the past two decades. Most of the research during that time has been focused on applications such as remotely piloted vehicles (RPVs), man-powered airplanes, high-altitude aircraft, and radio-controlled model planes, to name a few. These vehicles, because of their wing size, cruise speed, or cruise altitude, normally operate at chord Reynolds numbers in the range of 10^4 to 10^6.[1]

During the past several years, the Reynolds number range of interest has shifted to significantly lower numbers as new applications have emerged. One of these applications is the design and development of what are called micro air vehicles (MAVs). MAVs are small, remotely piloted aircraft slated to be used for close-range reconnaissance or sensing missions. The military envisions flying vehicles that can be carried by soldiers in the field and can be deployed for the purpose of answering the question: "What's over the next hill?" For this purpose, MAVs must be small, easily controlled by soldiers, able to operate at short to medium ranges, and capable of carrying a variety of sensing and/or surveillance payloads. Some of the current goals for MAVs are a maximum dimension of approximately 6 in. (15 cm), autonomous or semi-autonomous control, operating range of 6.2 miles (10 km), average endurance of 30 min, and payload capability up to 0.63 oz (18 g).[2]

MAVs are expected to cruise at speeds ranging between 10 and 50 miles per hour (15 and 80 km per hour), thus yielding an operating Reynolds number range between roughly 50,000 and 250,000. Furthermore, the requirement of compactness dictates the use of very low aspect ratio (LAR) wings ($AR \approx 1$). Other applications of LAR wings at low Reynolds numbers are turbine blades at high altitudes and even an aircraft that would fly in the Martian atmosphere in future unmanned space flights.

The aerodynamics of low aspect ratio wings at low Reynolds numbers have essentially not been studied. LAR wings have been extensively researched at higher Reynolds numbers in the form of delta wings at subsonic, transonic, and supersonic speeds. Many of these studies are focused on the high-angle-of-attack aerodynamics of delta and other types of pointed LAR wings. For MAVs, delta wing planforms are not ideal because for a given maximum dimension, delta wings offer less lifting area (and therefore less payload capability) than a rectangular or circular wing of the same maximum dimension.

Some information is available, however, regarding nondelta LAR wings, with much of the research having been done between 1930 and 1950. Zimmerman,[3] Bartlett and Vidal,[4] and Wadlin et al.[5] performed experiments on LAR wings, although at significantly greater Reynolds numbers. Theoretical and analytical treatises of LAR wing aerodynamics have been performed by Bollay,[6] Weinig,[7] Bera and Suresh,[8] Polhamus,[9,10] and Rajan and Shashidhar,[11] to name a few.

Perhaps the most complete analysis and review of LAR wings was performed by Hoerner in his two-volume series on lift and drag.[12,13] Hoerner reviewed many of the theories developed for LAR wings of nondelta planforms. A variety of correlations as well as analytical methods were presented and compared with the available experimental data of the time. Although the information presented by Hoerner corresponds to higher Reynolds numbers than the ones considered in the present work, the aerodynamic theory still holds.

This theory correctly predicts that as a finite wing of a given aspect ratio generates lift, counter-rotating vortical structures form near the wingtips. These vortices strengthen as the angle of attack increases. For a low aspect ratio wing, the tip vortices may be present over most of the wing area and therefore exert great influence over its aerodynamic characteristics. Wings of AR less than or equal to 1 can be considered to have two sources of lift: linear and nonlinear.[13] The linear lift is created by circulation around the airfoil and is what is typically thought of as lift in higher AR wings. The nonlinear lift is created by the formation of low pressure cells on the wing's top surface by the tip vortices, such as that observed in delta wings at high angles of attack.[14] This nonlinear effect increases the lift-curve slope as α increases and it is considered to be responsible for the high value of α_{stall}.

II. Apparatus
A. Wing Models

A number of flat-plate models of varying planform shapes were tested. Figure 1 shows schematics of the planform shapes of these models. All models had a thickness-to-chord ratio of 1.96% and a 5 to 1 elliptical leading and trailing edge. Most models had a root chord of 8 in. (0.2032 m). The rectangular planforms had flat side edges whereas the nonrectangular ones had elliptical edges throughout. One particular shape used, called the "Zimmerman," is formed by joining two half-ellipses at either the quarter-chord location (for the Zimmerman) or the three-quarter-chord location (for the inverse Zimmerman).

B. Wind Tunnel

Wind tunnel experiments were performed using a low-turbulence, indraft, open-circuit wind tunnel located at the University of Notre Dame's Hessert Center for Aerospace Research. This wind tunnel has a contraction ratio of 20.6 to 1 and a

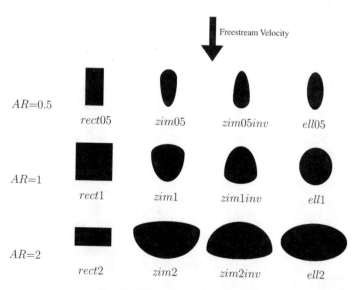

Fig. 1   Wing planform shapes.

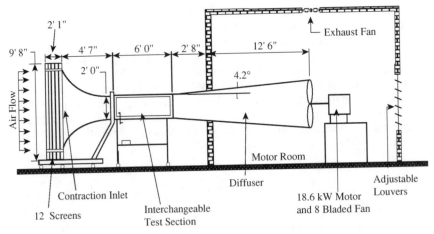

Fig. 2   Schematic of the low-speed wind tunnel.

rectangular inlet contraction cone designed for low-turbulence intensities. Over the speed range used for most experiments, the turbulence intensity in the test section has been determined to be approximately 0.05%.

A 6-ft-long (1.82-m) test section with a rectangular cross-sectional area of 2 × 2 ft (61 × 61 cm) is located directly downstream of the inlet. Downstream of the test section is a diffuser that slows the flow as it approaches the impeller, which is driven by a variable-speed electric motor. The tunnel exhausts into an enclosed area in a separate room, which can be ventilated by the use of louvers and exhaust fans when using flow visualization smoke. Figure 2 shows a schematic of the wind tunnel.

C. Force Balance

All lift and drag measurements were made using a specially designed three-component platform force balance. The balance is mounted externally on top of the wind tunnel test section. A sting connects the balance to the model. The forces are measured through two independent platforms (one for lift and one for drag). The platforms are attached to two flexures on which strain gauges are mounted in a full Wheatstone bridge configuration. The flexures are sized to allow a resolution of 1 g, which allows the balance to measure lift and drag forces of wings models down to a Reynolds number of approximately 20,000. The force balance has a limit of 26 oz (750 g) in the lift direction and 10 oz (300 g) in the drag direction. A computer-controlled stepper motor and associated gearing system is used for changing the angle of attack of the model.

The models were mounted vertically and attached to the force balance sting through a sting arm in the trailing edge of the model (see Fig. 3). A streamlined sting covering protected the sting from the freestream velocity, such that the only forces experienced by the force balance were the forces on the model. The forces acting on the horizontal sting arm were determined to be small enough to be neglected when calculating the forces on the wing model.

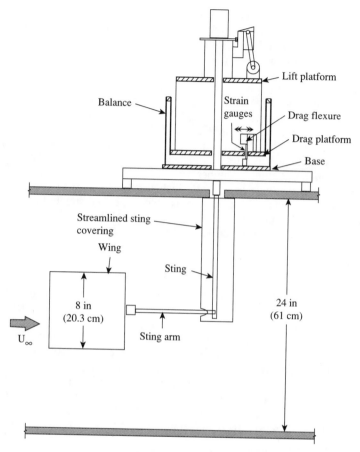

Fig. 3 Experimental setup: model mounting.

D. Data Acquisition

All experimental data were acquired using a PC-based data acquisition system running the LABVIEW® 5 graphical programming language. A 12-bit analog/ digital (A/D) card was used. Throughout the experiments, 4000 data samples were acquired quasi-simultaneously per channel at a frequency of 500 Hz. Both mean and standard deviation were calculated from these 4000 samples per channel. The standard deviation was used in the uncertainty analysis. For all the measurements an analog low-pass filter was necessary to reduce noise generated by the stepper motor used to change the angle of attack.

III. Procedures

The angle of attack of the model was varied from $\alpha = -15$ deg to a large positive angle between 20 and 40 deg, depending on the wing model being tested. For the larger wing models, the angle of attack had to be kept below a certain angle because the forces created by that model exceeded the force balance's load limits. The angle of attack was varied in increments of 1 deg using a computer-controlled stepper motor. The wing was then brought back to $\alpha = 0$ deg to determine whether hysteresis was present or not. No hysteresis was observed in any of the measurements.

At each angle of attack, the voltages of the lift and drag strain gauges were measured and recorded by the data acquisition system. The voltage of an electronic manometer connected to a pitot probe in the tunnel was also measured at each angle of attack to determine the dynamic pressure to be used for nondimensionalization of the forces at that angle of attack.

The chord Reynolds numbers presented are all *nominal* values (not corrected). For the experiments of lift and drag vs angle of attack, the freestream velocity U_∞ in the test sections was always adjusted with the model at $\alpha = 0$ deg to yield the desired nominal Reynolds number.

The lift and drag coefficients presented in this chapter have all been corrected for wind tunnel blockage (solid blockage, wake blockage, and streamline curvature) according to the techniques presented by Pankhurst and Holder.[15] Their techniques are equivalent to techniques described by Rae and Pope.[16]

IV. Uncertainty

The uncertainty in the angle of attack was determined to be of the order of $0.5 - 0.7$ deg. The two main sources of uncertainty for the lift and drag forces are the quantization error of the data acquisition card and the uncertainty arising from the standard deviation of a given mean strain-gauge output voltage. The Kline–McClintock[17] technique for error propagation was used to evaluate all uncertainties in the aerodynamic coefficients. This can yield very large percentage uncertainties for very small coefficients approaching zero. For larger coefficients (at $\alpha > 10$ deg), these percentage uncertainties are of the order of 6% for C_L and C_D. These percentages are for a Reynolds number of 70,000 and $AR = 1$. As Reynolds number increases, these uncertainties tend to decrease as a result of the larger forces involved.

V. Flow Visualization

Two forms of flow visualization techniques are included in this chapter: kerosene-smoke and surface flow visualization. For the kerosene-smoke technique, a constant stream of smoke was injected into the wind tunnel by dripping deodorized kerosene onto electrically heated filaments housed in four metal conduits. A blower cage fan forces the vaporized kerosene into a smoke rake, which has a series of heat-exchanger coils designed to bring the smoke to room temperature so as to minimize condensation. A filter bag inside the rake also filters out larger smoke particles. A detailed description of this system is given by Mueller.[18] The smoke generated using this device is very easily photographed either by direct illumination (from 300-W floodlights) or by use of a laser light sheet (generated by a 5-W argon-ion laser). All photographs were taken using a 35-mm Nikon® MF-16 camera with a 105-mm lens and either 3200-speed black and white film or 1600-speed color film.

Surface flow visualization was accomplished by use of a mixture of propylene glycol and water. A few drops of Kodak® Photoflow 200 were added as a wetting agent and a small amount of Fluorescein powder was dissolved in the mixture.

When illuminated by an ultraviolet light source, the mixture fluoresced. This fluid was liberally applied to the top of the model and the velocity in the tunnel was immediately brought to the desired speed. The patterns formed by the mixture were photographed from the top of the tunnel on color film.

VI.   Discussion of Results

A selection of the results obtained in wind tunnel experiments is presented in Figs. 4 through 21. The results correspond to experiments in the wind tunnel at Reynolds numbers of 70,000 and 100,000.

A striking feature of the C_L vs α curves for $AR = 0.5$ wings, especially at Re_c of 70,000 (Fig. 4), is the kink observable near $\alpha = 5$ deg. All four models of $AR = 0.5$ exhibit similar trends at the same low angle of attack. It was decided to study this matter in more detail using the rectangular $AR = 0.5$ wing at a Reynolds number of 70,000. A more refined step size in α was used for α between 0 and 16 deg. The angle of attack was changed in steps of 0.25 deg in the region of the kink and 0.5 deg elsewhere. The lift and drag coefficients were determined for both increasing and decreasing angles of attack to check for asymmetric behavior or hysteresis. Figures 22 and 23 show the results for the $rect05$ model. As can be seen from these figures, the variations of C_L with α for increasing and decreasing angles of attack are quite different. For increasing α, C_L and C_D follow the general shape of the kink but exhibit a large oscillation with respect to the average shape of this kink. For decreasing α, the C_L and C_D behave much more smoothly and do not exhibit any noticeable oscillation. These observations were found to be repeatable in several experiments conducted using the same conditions. It is noted that the kink is no longer apparent at the higher Reynolds numbers tested (at $Re_c = 140,000$, the kink is smoothed out and resembles the more typical nonlinear shape of the rest

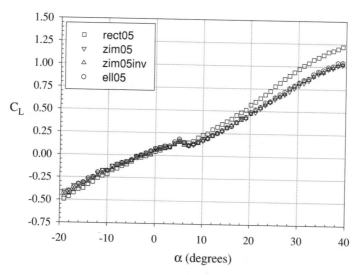

Fig. 4   Lift coefficient for $AR = 0.5$ models at $Re_c = 70,000$.

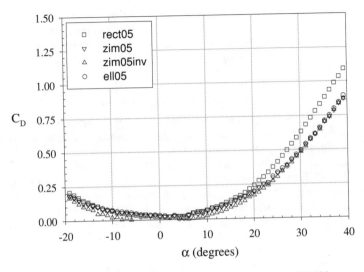

Fig. 5   Drag coefficient for $AR = 0.5$ models at $Re_c = 70{,}000$.

of the wing models). The cause for the occurrence of the kink is currently being studied.

Given the results presented in Figs. 4 through 21, some general conclusions can be drawn. In regards to lift, it can be stated that for $AR \leq 1.0$, the rectangular and inverse Zimmerman planforms generally exhibit advantages over the Zimmerman or elliptical ones. The difference in performance is particularly noticeable for the $AR = 1$ and $AR = 0.5$ wings. The drag of rectangular planforms,

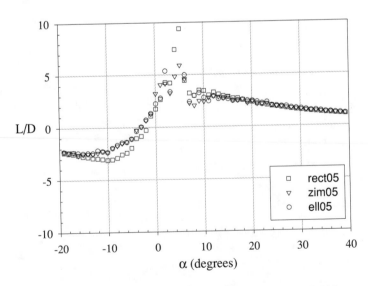

Fig. 6   Lift-to-drag ratio for $AR = 0.5$ models at $Re_c = 70{,}000$.

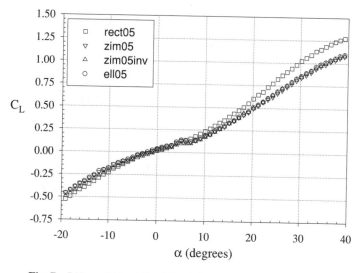

Fig. 7   Lift coefficient for $AR = 0.5$ models at $Re_c = 100,000$.

however, is higher than that of the inverse Zimmerman, Zimmerman, and elliptical shapes, specially at lower Reynolds numbers. The plots of lift-to-drag ratio show that for almost all aspect ratios below 2 and most Reynolds numbers, the inverse Zimmerman planform is most efficient at moderate angles of attack. Note that care must be taken when drawing conclusions from the L/D graphs at $Re = 70,000$ because the higher uncertainty in the aerodynamic forces at this Reynolds number can yield high uncertainty in the value of L/D. More valid conclusions can be determined from the $Re = 100,000$ graphs.

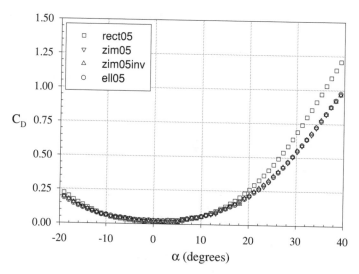

Fig. 8   Drag coefficient for $AR = 0.5$ models at $Re_c = 100,000$.

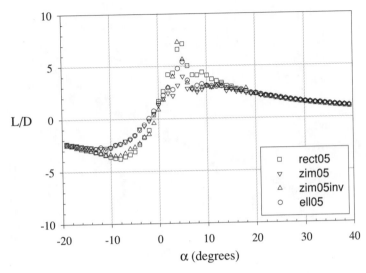

Fig. 9   Lift-to-drag ratio for $AR = 0.5$ models at $Re_c = 100,000$.

A.   Comparison of Lift

One way of comparing the different planform shapes is to compare their lift-curve slopes. The slope of the C_L vs α curve of each model at each Re_c was calculated by applying a least-squares linear regression to the curves of lift coefficient vs angle of attack for all models at all Reynolds numbers. Only the values of C_L corresponding to α between -10 and 10 deg were considered to assure that the lift curve behaved linearly. The assumption of linearity is not entirely accurate

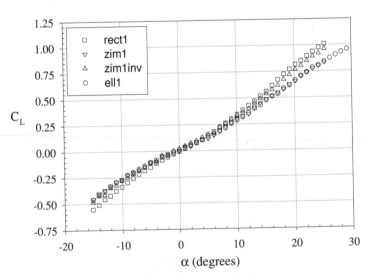

Fig. 10   Lift coefficient for $AR = 1.0$ models at $Re_c = 70,000$.

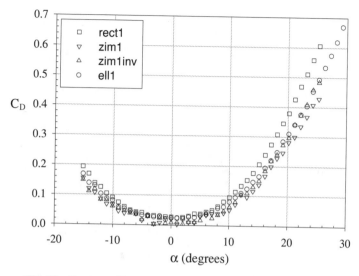

Fig. 11   Drag coefficient for $AR = 1.0$ models at $Re_c = 70,000$.

for the $AR = 0.5$ wings because of the inherent nonlinearity of the lift force. In addition, the kink in the C_L curve discussed earlier affects the lift-curve slope. Until a more thorough understanding of the drop in C_L for low angles of attack is obtained, C_{L_α} is determined in the same way as the higher AR wings without correcting for the kink in C_L.

A numerical technique was developed for obtaining the value of the lift-curve slope, C_{L_α}, and the uncertainty in this slope by taking into account the uncertainties

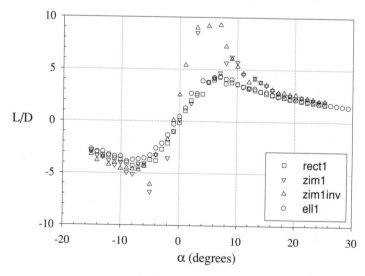

Fig. 12   Lift-to-drag ratio for $AR = 1.0$ models at $Re_c = 70,000$.

G. E. TORRES AND T. J. MUELLER

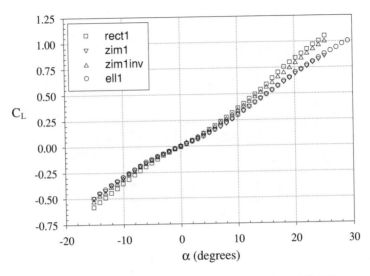

Fig. 13   Lift coefficient for *AR* = 1.0 models at *Re_c* = 100,000.

in the values of C_L and α. This task was accomplished by numerically generating values of C_L and α that differed from the mean values by a random fraction of the uncertainty at that data point. This procedure generated sets of data that simulated the occurrence of a new experiment by taking into account the known uncertainty at each point. The slopes of the least-squares linear regressions of 1000 sets of such generated angle of attack and lift data were averaged and their standard deviations determined the uncertainty in the lift-curve slope.

The values of C_{L_α} obtained as described above were compared to a number of theoretical predictions of the lift-curve slope. The first one is the classic

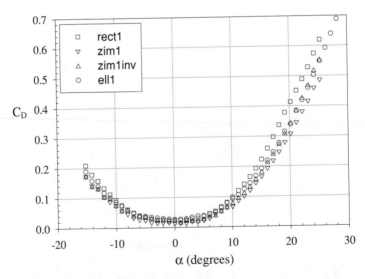

Fig. 14   Drag coefficient for *AR* = 1.0 models at *Re_c* = 100,000.

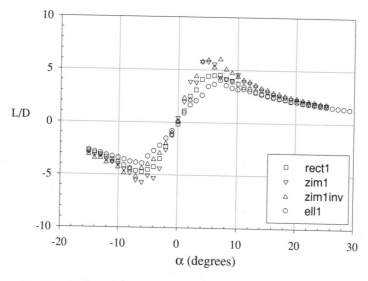

Fig. 15   Lift-to-drag ratio for $AR = 1.0$ models at $Re_c = 100,000$.

equation for C_{L_α},[19]

$$C_{L_\alpha} = \frac{a_0}{1 + \left(\frac{a_0*57.3}{\pi AR}\right)(1 + \tau)}. \tag{1}$$

The Glauert parameter τ is equivalent to an efficiency factor and varies typically between 0.05 and 0.25. The value of a_0 was taken to be $a_0 = 0.0938/\text{deg}$ (rather than the conventional value of $2\pi/\text{rad} = 0.10966/\text{deg}$) based on the average of experimentally determined two-dimensional slopes of flat-plate infinite wings

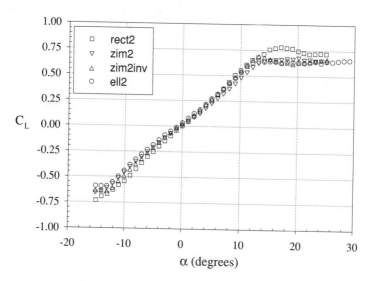

Fig. 16   Lift coefficient for $AR = 2.0$ models at $Re_c = 70,000$.

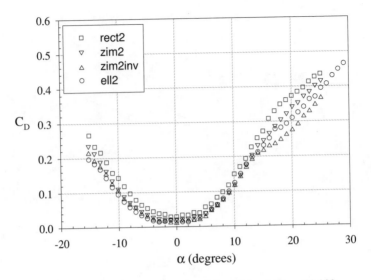

Fig. 17   Drag coefficient for $AR = 2.0$ models at $Re_c = 70,000$.

with the same thickness-to-chord ratio as the wings used in this work (Pelletier and Mueller[20]).

The second equation used is that proposed by Lowry and Polhamus,[21] which is supposed to be more accurate for small aspect ratios (less than 2):

$$C_{L_\alpha} = \left(\frac{1}{57.3}\right) \frac{2\pi AR}{2 + \sqrt{\frac{AR^2}{\eta^2}\left(1 + \tan^2 \Lambda_{c/2}\right) + 4}} \tag{2}$$

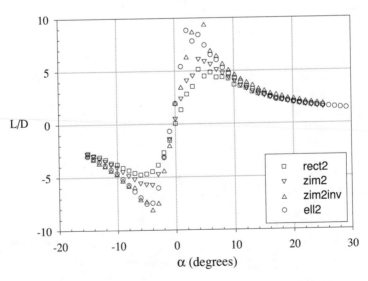

Fig. 18   Lift-to-drag ratio for $AR = 2.0$ models at $Re_c = 70,000$.

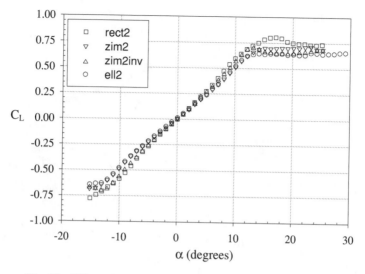

Fig. 19   Lift coefficient for $AR = 2.0$ models at $Re_c = 100,000$.

where

$$\eta = \frac{C_{L_\alpha,\text{airfoil}(\frac{1}{\text{rad}})}}{2\pi} \tag{3}$$

and $\Lambda_{c/2}$ is the sweep angle at midchord.

Finally, Hoerner and Borst[13] suggest that for thin rectangular plates of low AR (less than 2.5), the lift-curve slope is given by

$$C_{L_\alpha} = \left[\frac{36.5}{AR} + 2AR\right]^{-1} \tag{4}$$

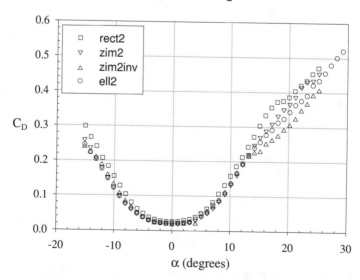

Fig. 20   Drag coefficient for $AR = 2.0$ models at $Re_c = 100,000$.

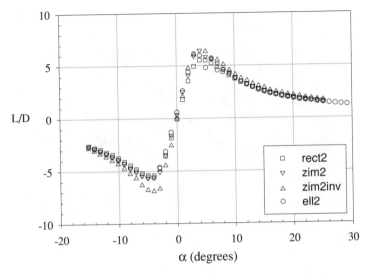

Fig. 21   Lift-to-drag ratio for $AR = 2.0$ models at $Re_c = 100,000$.

Figure 24 plots wind tunnel experimental C_{L_α} vs Re_c for all models. For all aspects ratios tested, the rectangular and inverse Zimmerman planforms have a higher C_{L_α} than the Zimmerman or elliptical planforms (note that the *rect* 2 and *zim* 2 models were not tested at $Re_c = 140,000$ but the trend from lower Reynolds numbers is expected to hold at higher ones). The value of C_{L_α} tends to increase with increasing Re_c.

Average values of C_{L_α} are plotted vs AR in Fig. 25. The values of C_{L_α} at $Re_c = 70,000$; $100,000$; and $140,000$ in the wind tunnel for a given model were averaged

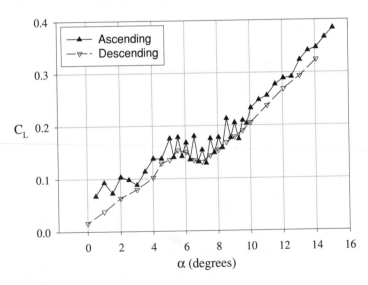

Fig. 22   Close-up of C_L vs α at $Re_c = 70,000$ for *rect* 05 model.

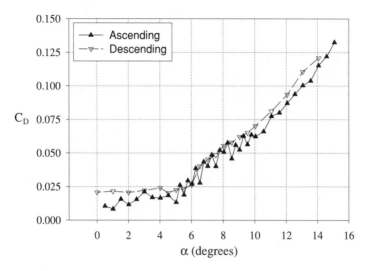

Fig. 23   Close-up of C_D vs α at Re_c = 70,000 for *rect*05 model.

and are assumed to represent a nominal lift-curve slope for that model for the Reynolds number range in question. Figure 25 also includes data from Pelletier and Mueller[20] of higher aspect ratio wing models. It should be noted that these models were rectangular flat-plate wings that had one end plate and whose semi-span-to-chord ratio determined the aspect ratio. The approximations of Eqs. (1) through (4) are also included in Fig. 25.

From Fig. 25, it can be concluded that Eq. (1) with $\tau = 0.05$ provides a good approximation of C_{L_α} for all wings of *AR* greater than 2 and for rectangular planforms of all aspect ratios. For round-tipped wings such as the Zimmerman and elliptical shapes, $\tau = 0.25$ gives a very good estimation of the experimental data.

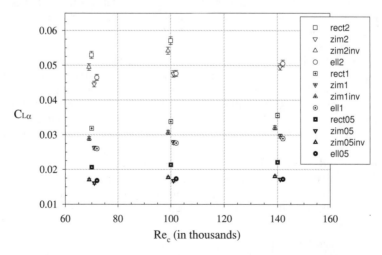

Fig. 24   Lift curve slope C_{L_α} vs Re_c for AR = 0.5, 1, and 2.

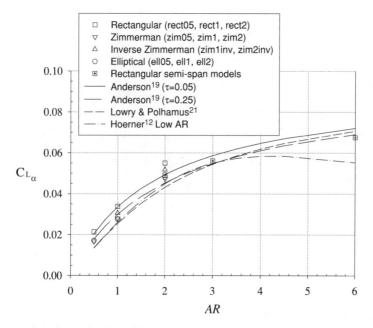

Fig. 25   Average lift-curve slope C_{L_α} vs AR.

The effect of wing planform shape on C_{L_α} for a given aspect ratio and Reynolds number can be further analyzed by plotting C_{L_α} vs the parameter $x_{\text{max span}}$. The variable $x_{\text{max span}}$ is the chordwise location (measured from the leading edge) of maximum wingspan, nondimensionalized by the root chord of the model. For the Zimmerman, elliptical, and inverse Zimmerman wings, $x_{\text{max span}}$ is 0.25, 0.50, and 0.75, respectively. For the rectangular wings, $x_{\text{max span}}$ is taken to be 1.0 rather than 0. In essence, $x_{\text{max span}}$ provides an indirect measure of the distance between the wingtip vortices as they develop over the wing and travel downstream. It was determined from flow visualization experiments that the distance between the wingtip vortices varied proportionally with the chordwise location of maximum span.

This can best be described by use of the sketch of Fig. 26. For wing shapes in which the maximum span is located upstream of the half-root-chord location ($x_{\text{max span}} \leq 0.5$), the tip vortices are seen to first develop at the location of maximum wingspan. The vortices then follow the outline of the wing up to a point and separate from the wing. In contrast, for wings with $x_{\text{max span}}$ greater than 0.5, the vortices are seen to separate from the wing at the location of maximum span. Thus the vortices of wings with $x_{\text{max span}} > 0.5$ are further apart than those of wings with $x_{\text{max span}} \leq 0.5$.

Hoerner and Borst[13] suggested that the lift performance of low aspect ratio wings improves as the distance between the wingtip vortices increases. Indeed, as can be seen from Fig. 27, C_{L_α} is seen to increase as $x_{\text{max span}}$ increases. Similar trends are evident at all Reynolds numbers tested.

B.   Flow Visualization

Surface flow visualization experiments were performed for three $AR = 1$ wings and one $AR = 0.5$ wing at several angles of attack. All experiments were conducted

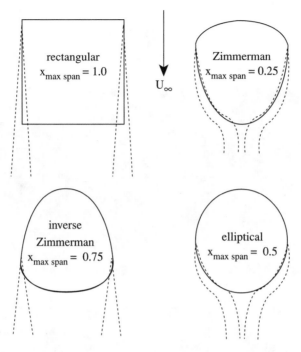

Fig. 26   Schematic of wingtip vortices for each wing planform shape.

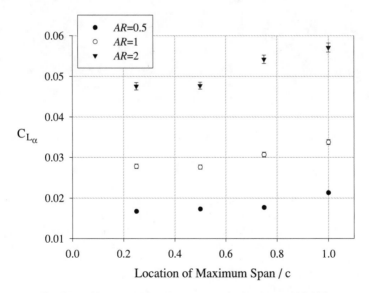

Fig. 27   Lift-curve slope C_{L_α} vs $x_{\text{max span}}$ for $Re_c = 100,000$.

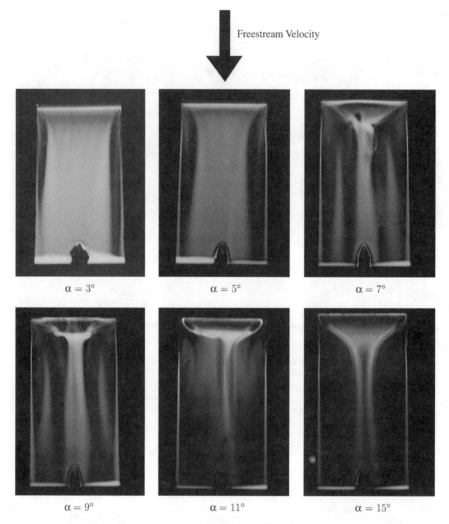

Fig. 28   *AR* = 0.5 **rectangular wing surface visualization at** Re_c = 70,000.

at a Reynolds number of 70,000. Figure 28 shows photographs of the *rect* 05 wing (*AR* = 0.5) at six angles of attack. These photographs illustrate a flow phenomenom that is typical of low Reynolds number flows: the separation bubble. A separation bubble occurs when laminar flow near the leading edge of a wing separates from the wing surface (see Fig. 29). A laminar free-shear layer forms and it is highly unstable if the Reynolds number exceeds a critical value. Small disturbances in the flow cause transition to turbulent flow in the free shear layer. The turbulent free-shear layer energizes the flow near the wall and in some cases reattaches in the form of a turbulent boundary layer. The width of the separation bubble can vary from a few percent of the chord to 20 or 30%.

In Fig. 28, a separation bubble exists near the leading edge for α between 7 and 11 deg. The bubble is apparent from the bright area, which is actually the area

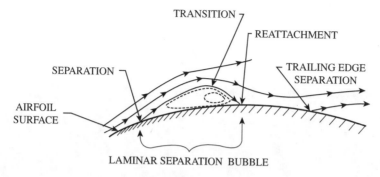

Fig. 29   Schematic of a separation bubble.

of recirculating flow defined in Fig. 29. The separation bubble is limited to the inboard section of the wing because, near the wingtips, the tip vortices energize the flow and eliminate the presence of the separation bubble.

Figure 30 shows a kerosene-smoke photograph of the separation bubble near the centerline of the *rect*05 model at $\alpha = 8$ deg and $Re_c = 70,000$. The locations of laminar separation and transition are quite clear from this photograph.

Figure 31 shows that a separation bubble is present in the *rect*1 model ($AR = 1$) for α between 5 and 7 deg, and to some extent it exists at $\alpha = 9$ deg. The proportion of the wing area affected by the tip vortices is apparent from the photographs at $\alpha = 11$ and 15 deg. The streaklines in the photograph at $\alpha = 9$ deg are caused by small sediments in the fluid used for flow visualization.

C.   Drag

It was desired to compare drag characteristics of different planform shapes of the same aspect ratio to determine which was most efficient. The approach taken

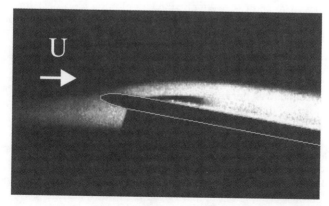

Fig. 30 $AR = 0.5$ rectangular wing smoke visualization at $\alpha = 8$ deg, and $Re_c = 70,000$.

Freestream Velocity

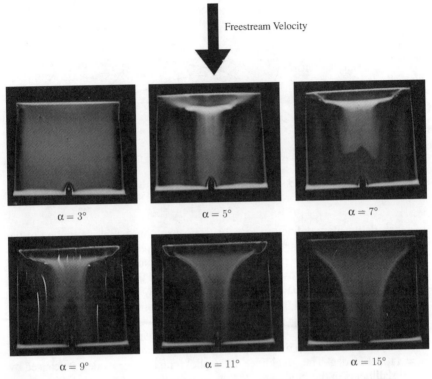

$\alpha = 3°$ $\alpha = 5°$ $\alpha = 7°$

$\alpha = 9°$ $\alpha = 11°$ $\alpha = 15°$

Fig. 31 *AR* = 1 rectangular wing surface visualization at *Re_c* = 70,000.

to attain this goal was to assume that the drag of a wing is given by

$$C_D = C_{D_0} + K C_L^2 \tag{5}$$

The values of K and C_{D_0} were obtained by plotting C_D vs C_L^2 for each model and Re_c and applying a least-squares linear regression to the data. Once again, only wind tunnel measurements corresponding to α between -10 and 10 deg were considered to ensure linear behavior. A computer program analogous to the one described in the lift section was used to calculate the mean value of K and its uncertainty for each model and Reynolds number. The relationship between C_D and C_L^2 was found to be linear except for the $AR = 0.5$ wings at Re_c of 70,000 and 100,000. Some of the nonlinearity of the lift of these wings at the low Reynolds number range is attributed to the kink in the C_L vs α curve discussed previously. The uncertainty in the drag coefficient, however, is relatively large for $AR = 0.5$ models (approximately 10%) and thus also contributes to the nonlinearity. For the purpose of comparison, a value of K is calculated for $AR = 0.5$ wings nevertheless. Note, however, that $AR = 0.5$ wings should not be compared solely on the basis of the value of K as the drag in these wing planforms seems not to be related to C_L^2 in a strictly linear manner.

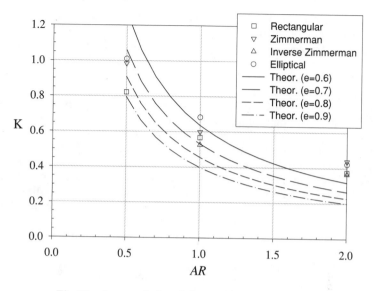

Fig. 32   Average induced drag coefficient K vs AR.

The classical theoretical approximation of K is (from Shevell[22])

$$K = \frac{1}{\pi AR\, e} \qquad (6)$$

where e depends on the shape, dihedral, and sweep angle of the wing, among other parameters. For high aspect ratio aircraft at high Reynolds numbers, e is usually given a value between 0.8 and 0.9. For small aspect ratios at low Re_c, however, e is much smaller. Hoerner[12] suggests $e = 0.5$ for flat plates. Figure 32 plots average values of K (average of K for a given model at three Reynolds numbers) vs AR as well as the predicted value of K from Eq. (6) for varying values of e.

It is apparent from this figure that the most appropriate value of e for small aspect ratios is approximately 0.6–0.7. The uncertainty in the values of K in Fig. 32 is on average ± 0.09, ± 0.05, and ± 0.03 for AR of 0.5, 1, and 2, respectively.

VII.   Vortex-Lattice Method

A modified vortex-lattice method (VLM) was implemented to predict the lift of low aspect ratio wings. To simplify the geometries used in the code, only rectangular wing planforms were considered. Zimmerman and elliptical planforms could be modeled as well but the gridding of the vortex elements for such complex geometries becomes cumbersome. The wing was modeled as an infinitely thin planar surface on which a grid of N horseshoe vortices was superimposed. The velocities induced by each vortex were computed by the Law of Biot–Savart. The velocity contributions of all vortices were added and formed a system of algebraic linear equations that must satisfy the condition of no flow through the surface of the wing (i.e., the component of velocity normal to the wing surface at the nth panel control point must be zero).

Following the classical VLM approach (summarized by Bertin and Smith[23]), the bound portion of the nth panel horseshoe vortex was located at the quarter-chord location of each panel. The control point of each panel was located at the three-quarter chord position, centered spanwise across the panel.

The orientation of the semi-infinite vortex elements with respect to the plane of the wing is extremely important for low aspect ratio wings. One approach commonly used in simplified VLM codes is to assume that the semi-infinite vortices (also called trailing vortices) convect downstream always in the plane of the wing. This approach simplifies the equations used to calculate induced velocities and yields an inherently linear prediction of lift. Another approach is to assume that the trailing vortices lie in a plane that is inclined at an angle Θ with respect to the wing plane. This inclination induces velocities in the chordwise and spanwise directions at each of the control points, which in turn lead to terms in the induced velocity equations that include $\sin^2$ and $\cos^2$. The predicted lift using this inclined assumption is nonlinear.

The latter approach was followed in this work, based mainly on the work of Bollay.[6] As a first approximation, the angle of inclination Θ for each trailing vortex pair corresponding to an nth horseshoe vortex is assumed to be a constant fraction of α for all N horseshoe vortices. This assumption represents the first step in application of a VLM method to a LAR wing. Future versions of the code will include routines that will time-step the inclination angle of each trailing vortex at each location in space. Bollay suggested that a good starting point for Θ is $\alpha/2$. This first approximation gives very good results as will be seen.

Once the required circulation strength of every horseshoe vortex is known, the lift is given by $L = \rho U_\infty \Gamma$, where Γ is the sum of the circulations of all panels. For low aspect ratio wings, the normal force coefficient C_N is commonly used

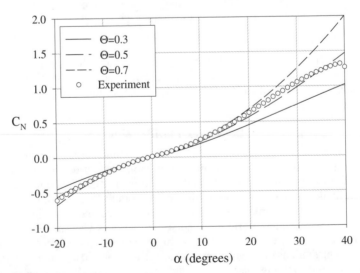

Fig. 33   Comparison of VLM and experimental results for $AR = 0.5$ and $Re_c = 140,000$.

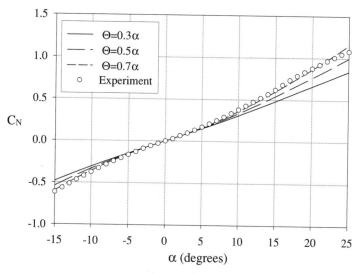

Fig. 34   Comparison of VLM and experimental results for $AR =1$ and $Re_c = 140{,}000$.

instead of the lift coefficient. Figure 33 shows the results for a rectangular wing of aspect ratio 0.5 with a grid of 6×6 panels ($N = 36$) and three values of Θ. The VLM results are compared with the experimental results of the *rect05* model at $Re_c = 140{,}000$. It is seen that if the assumption that Θ is a constant fraction of α is used, the best prediction is obtained with Θ having a value between 0.5α and 0.7α.

A similar comparison for an $AR = 1$ wing is shown in Fig. 34. The experimental results for the *rect1* model at $Re_c = 140{,}000$ are compared with VLM predictions. For this aspect ratio, $\Theta = 0.7\alpha$ gives a good approximation.

VIII.   Conclusions

The aerodynamic characteristics of uncambered wings with low aspect ratios and varying planform shapes were determined through wind tunnel experiments at chord Reynolds numbers between 70,000 and 140,000. Results showed that in this range of Reynolds numbers the lift curve becomes increasingly nonlinear as the aspect ratio decreases. The angle of attack for stall is also seen to increase as the aspect ratio becomes smaller. Furthermore, it was found that for aspect ratios less than or equal to 1.0, the rectangular and inverse Zimmerman planforms were generally most efficient, especially at $Re = 100{,}000$. A computer program that uses the vortex-lattice method to predict the normal force coefficient was also developed and applied to rectangular wings of $AR = 0.5$ and 1. The VLM results compare well with experimental data.

Acknowledgments

This research was sponsored by the U.S. Navy, Naval Research Laboratory, Washington, DC under Contract No. N00173-98-C-2025 and the Department of

Aerospace and Mechanical Engineering at the University of Notre Dame. The authors would like to thank Christopher Brown for his expert help in developing the flow visualization photographs, Alain Pelletier for his helpful comments during the preparation of this paper, and Joel Preston and Michael Swadener for their invaluable assistance in supporting the experimental facilities and electronic components used in this project. The authors also wish to acknowledge Rick Foch of the Naval Research Laboratory for his contributions to this paper.

References

[1]Mueller, T. J., "Low Reynolds Number Vehicles," AGARDograph No. 288, 1985.

[2]Dornheim, M. A., "Tiny Drones May Be Soldier's New Tool," *Aviation Week & Space Technology*, Vol. 148, No. 23, 8 June 1998, pp. 42–48.

[3]Zimmerman, C. H., "Characteristics of Clark Y Airfoils of Small Aspect Ratios," NACA TR 431, 1932.

[4]Bartlett, G. E., and Vidal, R. J., "Experimental Investigation of Influence of Edge Shape on the Aerodynamic Characteristics of Low Aspect Ratio Wings At Low Speeds," *Journal of the Aeronautical Sciences*, Vol. 22, No. 8, 1955, pp. 517–533.

[5]Wadlin, K. L., Ramsen, J. A., and Vaughan, V. L., Jr., "The Hydrodynamic Characteristics of Modified Rectangular Flat Plates Having Aspect Ratios of 1.00, 0.25, and 0.125 and Operating Near a Free Water Surface," NACA TR 1246, 1955.

[6]Bollay, W., "A Non-Linear Wing Theory and Its Application to Rectangular Wings of Small Aspect Ratio," *Zeitschrift fur Angewandte Mathematik und Mechanik*, Vol. 19, 1939, pp. 21–35.

[7]Weinig, F., "Lift and Drag of Wings with Small Span," NACA TM 1151, 1947.

[8]Bera, R. K., and Suresh, G., "Comments on the Lawrence Equation for Low-Aspect-Ratio Wings," *Journal of Aircraft*, Vol. 26, No. 9, 1989, pp. 883–885.

[9]Polhamus, E. C., "A Concept of the Vortex Lift of Sharp-Edge Delta Wings Based on a Leading-Edge-Suction Analogy," NASA TN D-3767, 1966.

[10]Polhamus, E. C., "Predictions of Vortex-Lift Characteristics by a Leading-Edge Suction Analogy," *Journal of Aircraft*, Vol. 8, No. 4, 1971, pp. 193–199.

[11]Rajan, S. C., and Shashidhar, S., "Exact Leading-Term Solution for Low Aspect Ratio Wings," *Journal of Aircraft*, Vol. 34, No. 4, 1997, pp. 571–573.

[12]Hoerner, S. F., *Fluid-Dynamic Drag*, Hoerner Fluid Dynamics, Brick Town, NJ, 1965, pp. 7-16–7-21.

[13]Hoerner, S. F., and Borst, H. V., *Fluid-Dynamic Lift*, Hoerner Fluid Dynamics, Brick Town, NJ, 1975, pp. 17-1–17-15.

[14]Hemsch, M. J., and Luckring, J. M., "Connection between Leading-Edge Sweep, Vortex Lift, and Vortex Strength for Delta Wings," *Journal of Aircraft*, Vol. 7, No. 5, 1990, pp. 473–475.

[15]Pankhurst, R. C., and Holder, D. W., *Wind-Tunnel Technique*, Pitman, London, 1952.

[16]Rae, W. H. J., and Pope, A., *Low-Speed Wind Tunnel Testing*, Wiley, New York, 1984.

[17]Kline, S. J., and McClintock, F. A., "Describing Uncertainties in Single-Sample Experiments," *Mechanical Engineering*, Vol. 75, No. 1, 1953, pp. 3–8.

[18]Mueller, T. J., *Fluid Mechanics Measurements*, edited by R. J. Goldstein, Taylor & Francis, Washington DC, 1996, pp. 367–450.

[19]Anderson, J. D. J., *Fundamentals of Aerodynamics*, McGraw-Hill, New York, 1991, p. 343.

[20]Pelletier, A., and Mueller, T. J., "Low Reynolds Number Aerodynamics of Low-Aspect-Ratio, Thin/Flat/Cambered-Plate Wings," *Journal of Aircraft*, Vol. 37, No. 5, Sept.–Oct. 2000, pp. 825–832.

[21]Lowry, J. G., and Polhamus, E. C., "A Method for Predicting Lift Increments due to Flap Deflection at Low Angles of Attack in Incompressible Flow," NACA TN 3911, 1957.

[22]Shevell, R. S., *Fundamentals of Flight*, Prentice-Hall, Englewood Cliffs, NJ, 1989.

[23]Bertin, J. J., and Smith, M. L., *Aerodynamics for Engineers*, Prentice-Hall, Englewood Cliffs, NJ, 1989.

Chapter 8

Systematic Airfoil Design Studies at Low Reynolds Numbers

Michael S. Selig,[*] Ashok Gopalarathnam,[†] Philippe Giguère,[‡]
and Christopher A. Lyon[§]
University of Illinois at Urbana-Champaign, Urbana, Illinois

Nomenclature

C_l = lift coefficient, chord = 1
H_{12} = boundary-layer shape factor δ_1/δ_2
α = airfoil angle of attack relative to chord line
α^* = segment design angle of attack relative to zero-lift line
α_z = airfoil angle of attack relative to zero-lift line
α_{zl} = airfoil zero-lift angle of attack
δ_1 = boundary-layer displacement thickness
δ_2 = boundary-layer momentum thickness
ϕ = segment arc limit used in conformal mapping

Subscripts

x_r = chordwise laminar bubble reattachment location
x_s = chordwise laminar-separation location
x_{tr} = chordwise transition location

I. Introduction

$\mathbf{F}$OR over 100 years, airfoil design has continued to capture the interest of practitioners of applied aerodynamics. The field is fueled by the ever-growing

---

*Associate Professor, Department of Aeronautical and Astronautical Engineering. Senior Member AIAA.

†Graduate Research Assistant, Department of Aeronautical and Astronautical Engineering; currently Assistant Professor, Department of Mechanical and Aerospace Engineering, North Carolina State University. Member AIAA.

‡Graduate Research Assistant, Department of Aeronautical and Astronautical Engineering; currently Senior Aerodynamicist, Enron Wind Corp. Member AIAA.

§Graduate Research Assistant, Department of Aeronautical and Astronautical Engineering; currently Software and Aerospace Engineer, Frasca International. Member AIAA.

combination of airfoil design requirements for unique applications, and this state of affairs is not likely to change. When one considers all possible permutations of the myriad airfoil design requirements, it quickly becomes apparent that the number of unique sets of requirements far exceeds the collection of existing airfoils. For this reason, the advancement and use of methods for airfoil design continues to be the economical solution. In contrast, the enrichment of airfoil "catalogs" for their own sake is felt to be of limited value.

The objective of this chapter derives from two topics. First, the alternative to our great legacy of airfoil design by geometric means guided by empirical study is to use an inverse method, and there are certain advantages to be had by adopting the latter while realizing that often geometric constraints must still be achieved. By adopting an inverse approach, the degree to which the aerodynamic performance can be controlled has reached a high level of sophistication, and this can be illustrated clearly by examples. Second, inverse design in the classic sense involves specifying a desired velocity distribution based on boundary-layer considerations. Taking this one step further by directly prescribing the desired boundary-layer characteristics is a step closer to controlling the desired outcome—the performance. Thus, employing an inverse boundary-layer-like approach can give the designer tremendous power in achieving the performance goals in the face of all the tradeoffs that one must consider in the process of airfoil design. These two aspects of a modern inverse airfoil design methodology form the subject of this chapter: design via boundary-layer considerations in an inverse sense.

To illustrate this approach, three series of low Reynolds number airfoils are presented. In each case, state-of-the-art tools for airfoil design[1,2] and analysis[3–5] were used. Although these airfoils were each designed for specific applications, the systematic and parametric studies show useful performance trends and tradeoffs in airfoil design at low Reynolds numbers. As will be shown, the overall design process has been validated through wind tunnel tests, and these results are presented together with the predictions.

II. Design Process

As an overview, the design process proceeded as follows. PROFOIL[1,2] was first used for rapid and interactive design. A new airfoil that appeared to meet the performance objectives was then screened through experience and analysis using the Eppler code[3,4] and/or XFOIL[5] to obtain the lift, drag, and pitching moment characteristics of the airfoil over a range of angles of attack. If at any state the candidate airfoil failed to meet the design goals, that additional experience was used to redesign the airfoil to more closely match the desired performance. This iterative process continued until a successful airfoil was designed, at which point the design was built and wind-tunnel tested to evaluate its performance. A more detailed description of each of these elements of the process follows.

A. PROFOIL

The PROFOIL code[1,2] embodies an inverse airfoil design method and an integral boundary-layer method for rapid analysis at the design points. The method draws on the pioneering work of Eppler[3,4,6,7] in inverse airfoil design and analysis through conformal mapping and integral boundary-layer techniques, respectively.

PROFOIL differs from the Eppler code in that laminar and turbulent boundary-layer developments can be directly prescribed through iteration on the velocity distribution. The method also allows for control over certain geometric constraints, such as the local geometry, maximum thickness, and thickness distribution. Additional differences are discussed in Refs. 1, 2, and 8. Both the boundary layer and thickness-constraint capabilities are used in the examples here. More details of the method will be given with the design examples to follow. A Web-based version of PROFOIL and further discussion is available online at http://www.uiuc.edu/ph/www/m-selig (cited September 2001).

B. Eppler Code

As mentioned, the Eppler code[3] was used for first-stage screening of candidate airfoils. Thus, only the analysis mode is discussed. In the analysis mode, the inviscid velocity distributions are determined by an accurate third-order panel method. Performance is then determined through the use of an integral boundary-layer method using the inviscid velocity distribution, which makes the analysis exceptionally fast (0.03-s elapse time on a 600-MHz PC per polar). In the version of the code used for this work, the drag caused by a laminar separation bubble is not calculated. Although the magnitude of the bubble drag is not determined, the method is invaluable when the user has had experience comparing its predictions with experiments. For example, the code has been used to design several successful low Reynolds number airfoils that can be found in the literature.[9–15]

C. XFOIL

XFOIL[5] is a design and analysis method for subcritical airfoils. In the data presented here, XFOIL has been used as a postdesign viscous/inviscid analysis tool. A linear-vorticity second-order-accurate panel method is used for inviscid analysis in XFOIL. This panel method is coupled with an integral boundary-layer method and an approximate e^n-type transition amplification formulation using a global Newton method to compute the inviscid/viscous coupling, requiring approximately 15 s of elapse time per polar on a 600-MHz PC. XFOIL has proven to be well suited for the analysis of subcritical airfoils even in the presence of significant laminar-separation bubbles.

The XFOIL analyses in this work were used primarily to study tradeoffs and effects of systematic variations in airfoils and for selecting the most suitable candidates before wind tunnel testing. For this purpose, the value $n_{crit} = 9$ was found to be quite suitable based on prior experience with comparisons between XFOIL results and the University of Illinois open-return subsonic wind tunnel data. This value $n_{crit} = 9$ was used for transition prediction in the data presented here.

D. Wind-Tunnel Tests

This section describes the wind-tunnel experiments. Because details of the method can be found in Refs. 16, 17, and 18, only a summary is given here. The experiments were performed in the University of Illinois open-return subsonic wind tunnel. The rectangular test-section dimensions are approximately 2.8×4 ft in cross section and 8-ft-long. To ensure good flow quality in the test section, the tunnel settling chamber contains a 4-in.-thick honeycomb and four antiturbulence screens, resulting in a turbulence level of less than 0.1% over the Reynolds number range tested.[16] The SA703x series models were made from hot-wire cut foam cores covered with fiberglass and resin under a mylar vacuum bag. The SG604x and S607x series models were from polyurethane RenShape®, milled using a numerically-controlled machine, then sanded and painted.

To isolate the ends of the airfoil model from the tunnel side-wall boundary layers and the outer support hardware, the airfoil models were mounted horizontally between two 3/8-in.-thick, 6-ft-long Plexiglas® splitter plates. Gaps between the model and splitter plates were nominally 0.05 in. All models had a 12 in. chord and 33 5/8-in. span. One side of the model was free to pivot. At this location, the angle of attack was measured using a potentiometer. The other side of the model was free to move vertically on a precision ground shaft, but it was not free to rotate. A loadcell restrained the motion of the model and measured the lift force. Linear and spherical ball bearings within the lift carriage helped to minimize any frictional effects.

The drag was obtained from the momentum method. To ensure that the wake had relaxed to tunnel static pressure, the wake measurements were performed 14.8 in. (approximately 1.25 chord lengths) downstream of the model trailing edge. Each vertical wake traverse consisted of between 20 and 80 total-head pressure measurements (depending on the wake thickness) with points nominally spaced 0.08 in. apart. Owing to spanwise wake nonuniformities,[19,20] wake profile measurements were taken at four spanwise locations spaced 4 in. apart over the center 12 in. of the model span. The resulting four drag coefficients were then averaged to obtain the drag at a given angle of attack.

The airfoil pitching moment was measured via a loadcell connected between two lever arms, one metric with the model and the other fixed to the lift carriage angle-of-attack adjustment plate.

The lift, drag, moment, and angle-of-attack measurements were corrected to account for the effects of solid blockage, wake blockage, and streamline curvature.[21] The velocity was not only corrected for solid and wake blockage but also for a "circulation effect" that is unique to setups that make use of splitter plates. For the current tests, the freestream velocity rather than being measured far upstream was measured between the splitter plates for higher accuracy. Because the pitot-static probe that was used to measure the freestream velocity was located fairly close to the model, the probe measurements were therefore corrected for airfoil circulation effects so as to obtain the true freestream test section speed. The details of this correction procedure can be found in Ref. 22.

Overall uncertainty in the lift coefficient is estimated to be 1.5%. The drag measurement error comes from three sources: accuracy of the data acquisition instruments, repeatability of the measurements, and the locations of the particular four wake profiles used to determine the average drag coefficient. Based partly on the error analysis method presented in Refs. 23 and 24, the uncertainty due to the instruments and measurement repeatability are less than 1% and 1.5%, respectively. Based on a statistical analysis (for a 95% confidence interval) of the spanwise drag results for the E374 airfoil[19] at $\alpha = 4$ deg, the uncertainties due to the spanwise variations were estimated to be approximately 1.5% at and above $Re = 200,000$. The current airfoils are expected to have approximately the same uncertainties. A more detailed discussion of this topic is presented in Ref. 20. For the angle-of-attack sensor, the uncertainty is estimated to be 0.08 deg.

To determine the accuracy of airfoil profiles, each model was digitized with a Brown and Sharpe coordinate measuring machine. Approximately 80 points were taken around each airfoil, and the spacing between points was approximately proportional to the local curvature. Thus, near the leading and trailing edges, the spacing was relatively small, whereas over the midchord it was no greater than 0.7 in. These measured coordinates were compared with the true coordinates using a two-dimensional least-squares approach (rotation and vertical translation), which yielded an average difference of approximately 0.010 in. or less for all airfoils discussed in this chapter.

Data taken on the E387 model for $Re = 200,000$ and $460,000$ are presented in Ref. 16 and compares well with data taken in the NASA Langley low-turbulence pressure tunnel (LTPT).[23] Moreover, surface oil-flow visualization taken to determine the laminar separation and oil-accumulation lines showed that the lines agreed with NASA Langley LTPT data to within 1–2% of chord.[25] This good agreement serves to validate the current experiments.

III.   Parametric Studies in Airfoil Design

In this section, three airfoil series are discussed with special emphasis given to the design process involving the aforementioned tools.

A.   SA 703x Series

This first series is for application to radio-controlled (R/C) model aircraft, specifically R/C soaring. The interest in designing such a series was motivated by the results of a survey taken at the 1996 AMA Nationals/Unlimited Thermal Soaring Competition in which flight duration and landing precision were emphasized. The survey showed that of the 101 responses (from nearly all of the participants) 40 of the pilots used the SD7037 airfoil[14] (shown in Fig. 1 together with its velocity distribution). Two other airfoils (S3021 and RG15) each numbered 8 in use, followed by the fourth (SD7080) with 6, and then 24 different airfoils were used on the remainder. Despite the overwhelming popularity of the SD7037, it was obvious that this airfoil could not be the optimum for a wide range of sailplanes with different sizes, wing loadings, and weather conditions. For example, in some situations pilots would benefit by having a faster version of the SD7037 (lower lift), and in others a slower version (higher lift) might be preferable. Thus, the SD7037 became the baseline for this first series.

For the series, the objective was to produce a range of similar performing airfoils that differed with respect to their lift range. One approach would be to make simple camber changes to the SD7037 to arrive at the new airfoils, but this process was not used because it is more attractive to control the performance by using the inverse capabilities in PROFOIL. In particular, PROFOIL was used to set the lower corner of the polar by specifying that the laminar boundary layer on the lower surface be close to separating at the lift coefficient at the lower corner of the polar. The shape of the polar above this lift coefficient was then obtained by tailoring the aerodynamics of the upper surface, which is discussed later.

Figure 2 shows the particular laminar boundary-layer development prescribed for the lower corner of the polar. This boundary-layer shape parameter H_{12} distribution, which is close to laminar separation, was specified to be achieved at the design angles of attack of 1.75, 2.15, 2.55, and 2.95 deg relative to the zero-lift

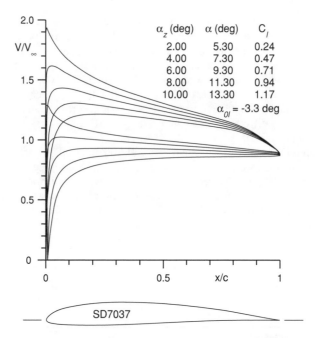

Fig. 1   SD7037 airfoil and inviscid velocity distributions.

line to produce each respective airfoil of the SA703x series shown in Fig. 3. (The SA7037 mimics the baseline SD7037.) In the figure, the velocity distributions are plotted at angle of attack increments equal to those that separate the aforementioned lower surface design angles of attack. The corresponding boundary-layer shape parameter developments as predicted by XFOIL for the lower surface design condition are shown in Fig. 4, and they are practically identical to the prescription shown in Fig. 2.

It is worth noting that the laminar boundary-layer H_{12} developments are plotted for a Reynolds number of 200,000; however, the laminar development to transition is independent of the Reynolds number, as discussed in Ref. 4. Also, although the

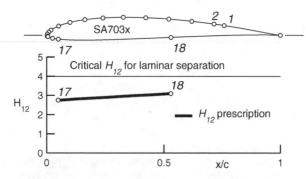

Fig. 2   Lower-surface laminar boundary-layer development prescribed for the SA703x airfoil series for the lower corner of the drag polar.

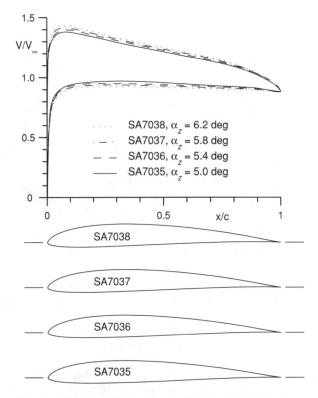

Fig. 3   SA703x airfoils and inviscid velocity distributions.

design specification was for there to be laminar flow on this segment at the design condition, in application the Reynolds number could quite well be off-design and cause transition before the end of this segment when operating at the design lift coefficient of the lower surface. Nevertheless, the prescription serves to define the lower-surface velocity distribution and corresponding geometry.

When each airfoil is operated below the respective design angle of attack for the lower surface, laminar separation and subsequent transition in the laminar-separation bubble quickly move forward and lead to higher drag at the end of the

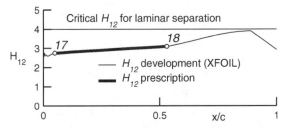

Fig. 4   Laminar boundary-layer development achieved for the SA703x airfoils and the agreement with the prescription.

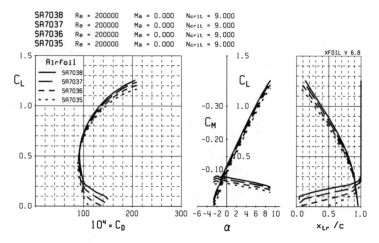

Fig. 5   XFOIL predictions for the SA703x airfoil series.

low-drag range. This result is found in the predictions of XFOIL shown in Fig. 5 and validated by the wind tunnel test results shown in Fig. 6. (In place of the SA7037, the baseline SD7037 was tested.) Additional data for these airfoils over the Reynolds number range from 100,000 to 300,000 can be found in Ref. 18, and tabulated data are available online at http://www.uiuc.edu/ph/www/m-selig (cited September 2001).

Of critical importance in the design of low Reynolds number airfoils is the upper-surface pressure distribution. The tendency of the flow to form a laminar-separation bubble can lead to a significant degradation in performance owing to

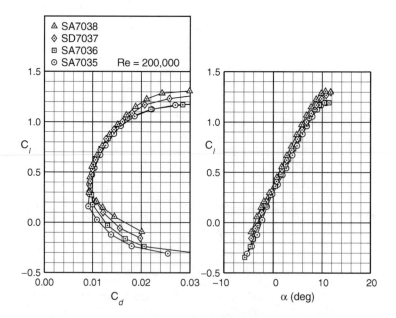

Fig. 6   Measured drag polars for the SA703x airfoil series.

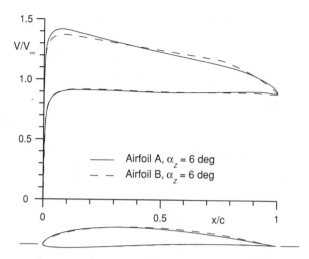

Fig. 7   Inviscid velocity distributions for airfoils A and B to study the different effects on drag.

the high bubble drag. To mitigate these adverse effects, a transition ramp in the pressure distribution is often employed to gradually bring the flow to transition in a thin bubble without a large pressure rise and high drag associated with an otherwise thick bubble. A general discussion of transition ramps can be found in Refs. 4 and 26, and additional details specific to low Reynolds number airfoils are discussed in Refs. 13, 14, 27, and 28.

The effect of the transition ramp is demonstrated using two example airfoils A and B. Figure 7 shows a comparison of the geometries and inviscid velocity distributions. These airfoils were designed using PROFOIL to each have a different transition ramp that is reflected in a different shape for the transition curve (C_l–x_{tr}/c curve) on the upper surface. The two airfoils were analyzed using XFOIL, and Fig. 8 shows the drag polars and upper-surface transition curves for a Reynolds number of 200,000. For the sake of discussion, the transition ramp is defined here as the region over which the bubble moves gradually as defined by the transition curve. (In this context, the transition ramp might be more aptly called a "bubble ramp."[14])

From the figure, it can be seen that airfoil A has lower drag than airfoil B at lift coefficients from around 0.3 to around 0.7, above which value airfoil B has lower drag. Also noticeable is the correlation between the drag polar and the shape of the upper-surface transition curve. For the C_l range from 0.3 to 0.7, where airfoil A has lower drag, the transition curve for airfoil A is shallower than that for airfoil B; that is, there is a larger change in the value of x_{tr}/c for airfoil A than for B. For values of C_l from 0.7 to 1.2 where airfoil B has lower drag, the transition curve for airfoil B is shallower than for A. This figure shows that the steepness of the transition curve is a direct indication of the bubble drag. By adjusting the shape of this curve, it is therefore possible to tailor the drag polar of an airfoil at low Reynolds numbers.

Figure 8 also includes an overlay of the variation of bubble size ($x_r - x_s$) with C_l. The size of the bubble for each C_l was obtained by determining the chordwise extent over which the skin friction C_f, as predicted by XFOIL, was less than

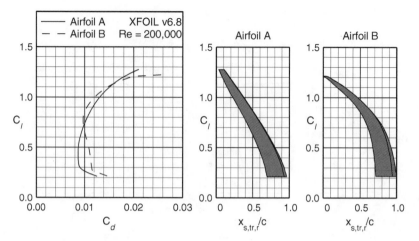

Fig. 8   XFOIL predictions for airfoils A and B to illustrate the effects of changes in the transition ramp on drag.

or equal to zero. Studying the bubble-size variation for the two airfoils further illustrates the connection between the shape of the transition curve and the bubble drag. The bubble is larger when the transition curve is steeper.

Figure 9 shows the inviscid velocity distributions for airfoil A at C_l values of 0.5 and 1.0 with the upper-surface bubble location marked in bold. A similar plot for airfoil B is shown in Fig. 10. Comparing the velocity drops across the bubble for the four cases, it can be seen that although airfoil A has a smaller velocity

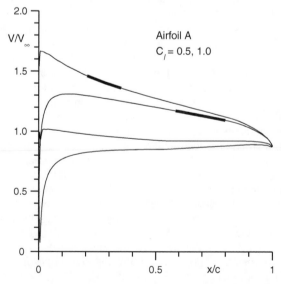

Fig. 9   Inviscid velocity distributions for airfoil A with the locations of the bubble marked.

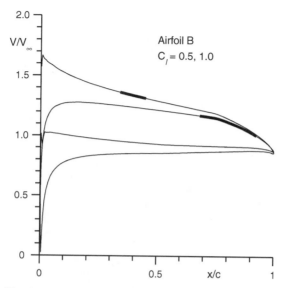

Fig. 10 Inviscid velocity distributions for airfoil B with the locations of the bubble marked.

drop than airfoil B at $C_l = 0.5$, the situation is reversed for $C_l = 1.0$. Because the pressure drag caused by the bubble increases with increasing velocity drop across the bubble, airfoil A has smaller bubble drag at the low C_l and larger bubble drag at the higher C_l. Thus, a steeper transition curve results in a larger bubble and also larger velocity drop across the bubble causing an increase in bubble drag.

Understanding the connection between the transition ramp in the velocity distribution, the C_l–x_{tr}/c transition curve, and the performance is the first step to obtaining optimum low Reynolds number airfoil design. The next step in design involves the implementation of these ideas. In studying the velocity distributions shown in Figs. 3 and 7 and the resulting behavior of the transition curves shown in Figs. 5 and 8, respectively, we see that the connection between the two is not straightforward owing to the subtle (yet important) differences. Moreover, it is worth adding that the differences in the airfoil shape are even less useful in guiding the design toward an optimum with respect to the transition ramp and its impact on drag.

In coping with this problem, a useful approach derives from an inherent feature of the Eppler theory for inverse airfoil design. Briefly, in the Eppler method, the designer can specify for a segment of the airfoil a design angle attack α^* (relative to the zero-lift line) over which the velocity is constant. For instance, the forward upper surface can be defined as one segment and given a design angle of attack of 10 deg ($C_l = 2\pi\alpha_z \sim 1$). When the resulting airfoil is then operated at 10 deg with respect to the zero-lift angle of attack α_{zl}, the velocity over that segment is then constant. For a higher angle of attack, the resulting pressure gradient is adverse and vice versa. Further discussion can be found in Refs. 4 and 29.

Because 1) the boundary layer responds to the pressure gradient, 2) the design angle of attack α^* for a segment has a direct effect on the pressure gradient, and 3) many such segments can be used to define an airfoil, these three aspects can be

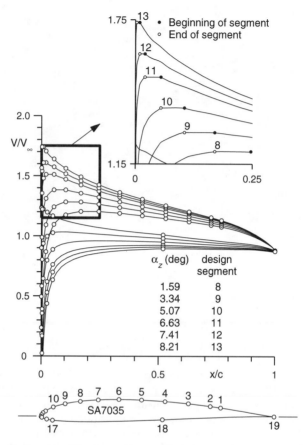

Fig. 11   SA7035 inviscid velocity distributions showing the zero pressure-gradient segments and corresponding design angles of attack.

connected to yield an elegant solution to having precise control over the C_l–x_{tr}/c transition curve. Figure 11 shows the velocity distributions for the SA7035 airfoil at several design angles of attack. As seen, for segments 8, 9, 10, 11, 12, and 13 the design angles of attack α^* are 1.59, 3.34, 5.07, 6.63, 7.41, and 8.21 deg, respectively. When the airfoil operates at these values of α_z, the velocity gradient over the respective segments is zero, as shown in the exploded portion of the figure. When these design angles of attack α^* are plotted vs the segment-endpoint arc limits ϕ used in the conformal mapping to generate the airfoil, the resulting curves shown in Fig. 12 mimic the corresponding transition curves (duplicated from Fig. 5) over the respective surface of the airfoil, thereby providing a means of controlling the transition ramp and resulting drag as was done also with the example of Figs. 7 and 8. For the SA703x series, the α^*–ϕ curve shown in Fig. 12 maps from leading edge back to segment-endpoint 1 or approximately 75% of the chord (see Fig. 11). It is over this region that the ramp is controlled by the α^*–ϕ curve, with the remainder being controlled by the final trailing-edge pressure recovery seen in Fig. 11. The essence of this technique has been employed in the design of most of the S*xxxx four-digit airfoils reported in the literature

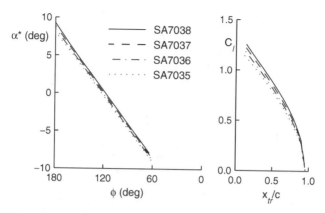

Fig. 12   Linkage between the $\alpha^*-\phi$ and C_l-x_{tr} curves for the SA703x airfoil series.

and archived online at http://www.uiuc.edu/ph/www/m-selig (cited September 2001) and in Refs. 14, 16, 17, 18, and 30.

B.   SG604x Series

The SG604x series of airfoils[31] as shown in Fig. 13 was designed for small variable-speed horizontal-axis wind turbines having a rated power of 1–5 kW. The operational Reynolds number for such machines is typically below 10^6. The focus here will be on the performance at a Reynolds number of 300,000, which represents rotors on the lower end of the range. In ideal conditions, variable-speed wind turbines operate at a constant tip-speed-ratio ($\Omega R/V_\infty$), which leads to the airfoil operating at a single angle of attack over a wide range of wind speeds. As a result, for optimum aerodynamic performance during variable-speed operation, the low-drag lift range (drag bucket) can be reduced in favor of having greater lift-to-drag ratios. However, to account for possible variations in the tip-speed-ratio caused by atmospheric turbulence and operational considerations, the best lift-to-drag ratio conditions should occur over a range of lift coefficients centered about the design lift coefficient. In rotor design, another factor deals with the tradeoff between the blade solidity and the design lift coefficient. With all else being equal, a high-solidity rotor requires an airfoil with lower lift than that required for a low-solidity rotor. Therefore, given the range of rotor designs, a family of airfoils covering a range of lift coefficients is desirable. These general considerations and others were taken into account in setting the design requirements, as detailed more thoroughly in Ref. 31. The current discussion is mainly concerned with the details of the design approach.

Again PROFOIL was used to prescribe the desired aerodynamic characteristics. In a manner similar to that used in the previously discussed SA703x series, the boundary-layer shape parameter was prescribed on the lower surface to control the lower corner of the low-drag range of the polar. In this respect, there were some minor differences in the prescriptions, but generally a similar behavior was obtained. The transition ramp on the upper surface differs from that of the SA703x series in much the same way that airfoil B differs from airfoil A in Fig. 7. Since the objective was to achieve a high lift-to-drag ratio with less emphasis being given to operation over a wide range, the ramp on the upper surface was made more shallow

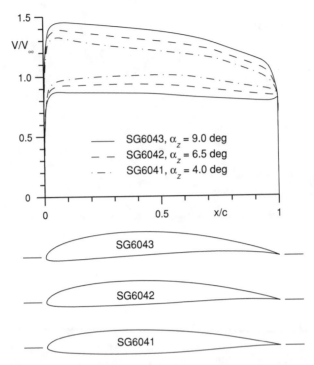

Fig. 13   SG604x airfoils and inviscid velocity distributions.

at the specific design lift coefficients of 0.6, 0.9, and 1.2 for the SG6041, SG6042, and SG6043, respectively. As before, the transition curve was controlled through the shape of the $\alpha^*-\phi$ curve shown in Fig. 14. Additional constraints included the pitching moment, which increased with the design lift coefficient, and also the airfoil thickness of 10%. It should be added that for ease of construction a finite trailing-edge angle was used, producing a zero trailing-edge velocity as seen in Fig. 13.

Performance predictions at a Reynolds number of 300,000 are shown in Fig. 15 and compare relatively well with the experimental results of Fig. 16. Most importantly, the trends in the predictions agree well with experiment, and also the behavior of the $\alpha^*-\phi$ curve is reflected in the transition curve.

Figure 17 shows the resulting experimentally determined lift-to-drag ratios at a Reynolds number of 300,000 compared with those for many previously existing airfoils. The SG6040 shown in the figure is the thicker companion root airfoil for the series of tip airfoils discussed here. Clearly, the objective of achieving high lift-to-drag ratios has been achieved. Given the high level of performance, these airfoils will likely find their way into applications beyond wind energy.

C. S607x Series

In contrast to the first two series, the S607x series is composed of three series of three airfoils each. The effort finally led to an airfoil for its intended application

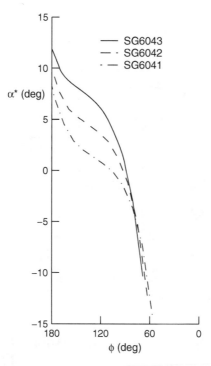

Fig. 14 SG604x airfoil series $\alpha^*-\phi$ curves for control over the C_l-x_{tr} curves.

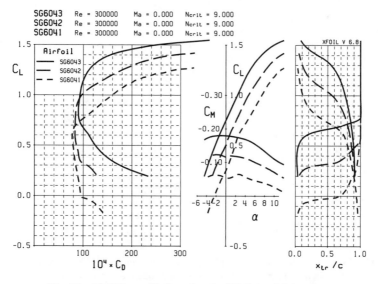

Fig. 15 XFOIL predictions for the SG604x airfoil series.

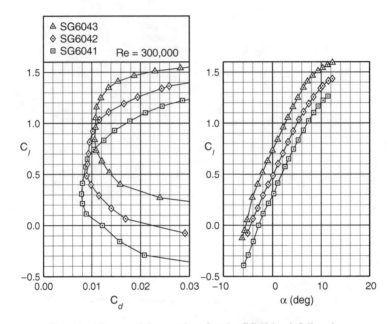

Fig. 16   Measured drag polars for the SG604x airfoil series.

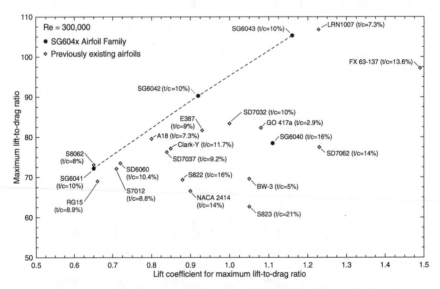

Fig. 17   Maximum lift-to-drag ratio vs the corresponding lift coefficient for the SG604x airfoil series compared with several previously existing airfoils.

that required a low pitching moment and optimum performance at a Reynolds number of 150,000 and lift coefficient near 1. Only data for a Reynolds number of 200,000 are presented here, however. The series evolved from a 9%-thick family to a 12%-thick one, and then permutations in the pitching moment were made while other improvements were incorporated. The series will be discussed in this order.

As in the prior examples, the lower surface of each airfoil was designed by prescribing the boundary-layer shape parameter development at its respective design angle of attack. For the first series S6071/2/3, the study centered on the upper-surface transition ramp. Figure 18 shows that small changes in the $\alpha^* - \phi$ curve affect the transition ramp, making it more shallow in going from airfoil S6071 to S6073. These differences result in small changes to the resulting velocity distributions shown in Fig. 19. As seen, the shallower $\alpha^* - \phi$ curve for airfoil S6073 results in a shallower pressure gradient over the forward upper surface. The differences in the airfoil shapes are so minute that a magnification of the y axis is needed to highlight the differences, as shown in Fig. 20. These small changes to the ramp, however, do have a significant effect on the bubble, as may be deduced by the drag predictions shown in Fig. 21. Wind tunnel tests depicted in Figs. 22 and 23 confirm the predicted trends.

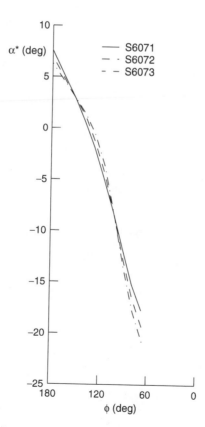

Fig. 18   S6071/2/3 airfoil series $\alpha^*-\phi$ curves for control over the C_l-x_{tr} curves.

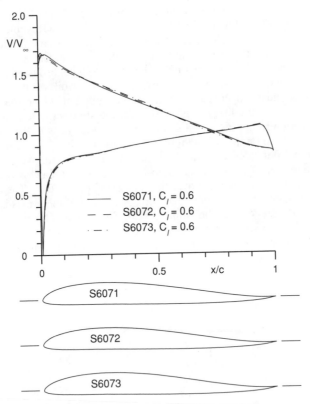

Fig. 19   S6071/2/3 airfoils and inviscid velocity distributions.

The next perturbation included largely a change in the airfoil thickness from 9% to 12%. Figure 24 shows the two bounding airfoils S6074/6 and their corresponding velocity distributions. Again the differences appear minor; however, as seen in Fig. 25 the trends are similar to the prior series, albeit with there being higher drag owing to the higher airfoil thickness. XFOIL predictions indicated similar increases in drag.

The wind tunnel tests of the S6074/5/6 airfoils revealed an undesirable feature of the series—stall hysteresis as shown Fig. 26, which was also found on airfoil S6071. Such hysteresis cannot be predicted by any method currently available. This type of hysteresis, however, has been found on other airfoils. As described in

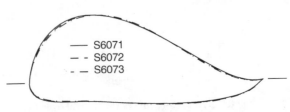

Fig. 20   S6071/2/3 airfoils with the thickness magnified to show the small differences.

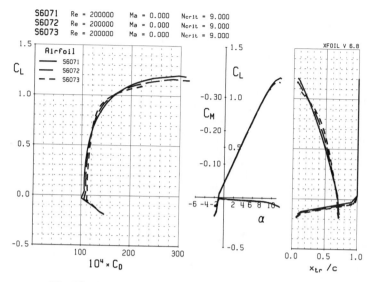

Fig. 21   XFOIL predictions for the S6071/2/3 airfoils.

Ref. 13, when the inviscid velocity distribution on the forward upper surface tends toward a concave shape, stall hysteresis of the type found here can be reduced. Unfortunately, no computational tool exists to quantify this effect, and the degree to which the pressure distribution (or velocity distribution) should tend in the concave direction has not been quantified. Nevertheless, the chief aim of the final series was to eliminate this hysteresis.

Figure 27 shows the two bounding airfoils S6077/9 designed to avoid stall hysteresis while satisfying the other constraints. A secondary purpose of this final

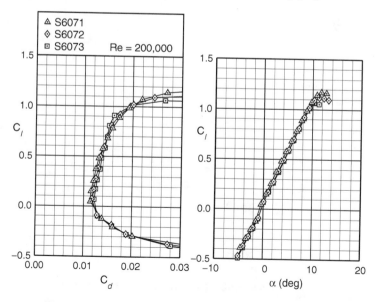

Fig. 22   Measured drag polars for the S6071/2/3 airfoils.

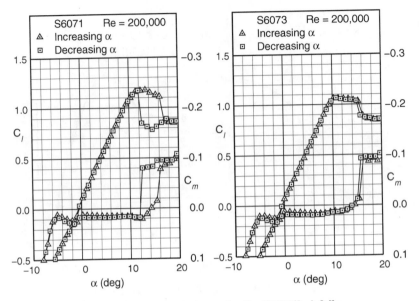

Fig. 23 Measured lift curves for the S6071/3 airfoils.

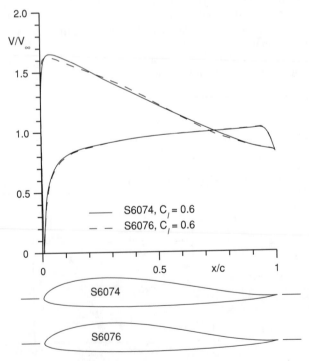

Fig. 24 S6074/6 airfoils and inviscid velocity distributions.

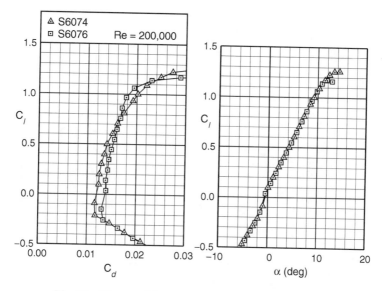

Fig. 25   Measured drag polars for the S6074/6 airfoils.

series was to examine the effects of a change in the pitching moment, which was specified using the inverse capabilities of PROFOIL. In comparing the velocity distributions shown in Fig. 24 with those in Fig. 27 it is seen that this current series has a more concave velocity distribution on the forward upper surface, and this difference is reflected in the shapes of the S6074/6 vs S6077/9 shown in Fig. 28. As seen in Fig. 29, this change in the velocity distribution is enough to eliminate the stall hysteresis. Finally, the performance is shown in Fig. 30. One feature of

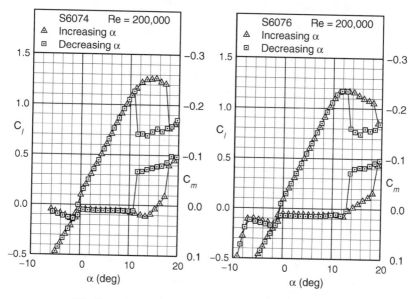

Fig. 26   Measured lift curves for the S6074/6 airfoils.

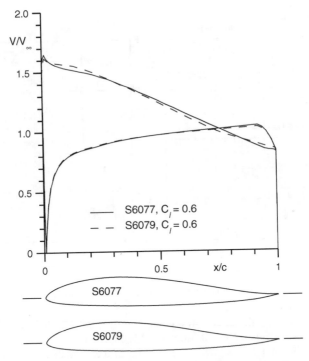

Fig. 27 S6077/9 airfoils and inviscid velocity distributions.

using a more concave distribution is that the $\alpha^* - \phi$ curve (not shown) becomes more shallow. As a result, the laminar-separation bubble drag is reduced at the upper corner of the low-drag range, yielding lower drag than the S6074/6 but also lower maximum lift as a tradeoff.

IV. Summary and Conclusions

In this chapter, three series of airfoils were designed to illustrate the power of modern computational tools for low Reynolds number airfoil design and analysis. Emphasis was placed on the design of the airfoils based on boundary-layer considerations. More specifically, the parameterization of the design problem centered around prescribing desirable boundary-layer features directly through an inverse

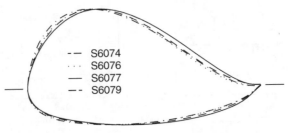

Fig. 28 S6074/6/7/9 airfoils with the thickness magnified to show the small differences.

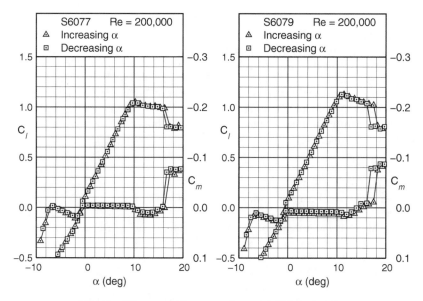

Fig. 29   Measured lift curves for the S6077/9 airfoils.

method. Formulating the design problem in this way offers the designer considerably more power than one would otherwise have using more traditional methods of inverse design (based on a single-point velocity distribution) and design by geometric perturbation. The design approach and philosophy can be used successfully to assess design tradeoffs with a high degree of control. Finally, wind-tunnel testing of low Reynolds number airfoils is, however, still needed to provide engineers with a necessary level of confidence required to make important engineering decisions.

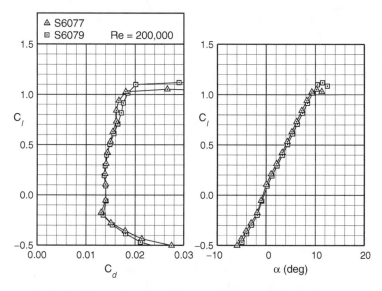

Fig. 30   Measured drag polars for the S6077/9 airfoils.

Acknowledgments

Funding for this work was provide through a variety of sources for which we are grateful, namely, private donations (see Refs. 16, 17, and 18 for individual listings), the National Renewable Energy Laboratory, and AeroVironment. Also, we thank all the model makers, Mark Allen, Allen Developments (SA7035, SG604x); Tim Foster and Frank Carson (SA7038); Jerry Robertson (SD7037); D'Anne Thompson (SA7036); and Yvan Tinel, Tinel Technologies (S607x series), for their meticulous and skillful efforts in model construction. The authors also wish to thank Mark Drela for providing the XFOIL code used in this work.

References

[1]Selig, M. S., and Maughmer, M. D., "A Multi-Point Inverse Airfoil Design Method Based on Conformal Mapping," *AIAA Journal*, Vol. 30, No. 5, May 1992, pp. 1162–1170.

[2]Selig, M. S., and Maughmer, M. D., "Generalized Multipoint Inverse Airfoil Design," *AIAA Journal*, Vol. 30, No. 11, Nov. 1992, pp. 2618–2625.

[3]Eppler, R. and Somers, D. M., "A Computer Program for the Design and Analysis of Low-Speed Airfoils," NASA TM 80210, Aug. 1980.

[4]Eppler, R., *Airfoil Design and Data*, Springer-Verlag, New York, 1990.

[5]Drela, M., "XFOIL: An Analysis and Design System for Low Reynolds Number Airfoils," *Low Reynolds Number Aerodynamics*, edited by T. J. Mueller, Vol. 54 of *Lecture Notes in Engineering*, Springer-Verlag, New York, June 1989, pp. 1–12.

[6]Eppler, R., "Direkte Berechnungvon Tragflugelprofilen aus der Druckverteilung," *Ingenieur-Archive*, Vol. 25, No. 1, 1957, pp. 32–57 (translated as "Direct Calculation of Airfoil from Pressure Distribution," NASA TT F-15, 417, 1974).

[7]Eppler, R., "Ergebnisse gemeinsamer Anwendung vo Grenzschicht- und Potentialtheorie," *Zeitschrift für Flugwissenschaften*, Vol. 8, No. 9, 1960, pp. 247–260 (translated as "Results of the Combined Application of Boundary Layer and Profile Theory," NASA TT F-15, 416, March 1974).

[8]Selig, M. S., "Multi-Point Inverse Design of Isolated Airfoils and Airfoils in Cascade in Incompressible Flow," Ph.D. Thesis, Pennsylvania State Univ., State College, PA, May 1992.

[9]Thies, W., *Eppler-Profile*, MTB 1, Verlag für Technik und Handwerk, Baden-Baden, Germany, 1981 (republished, Hepperle, M., MTB 1/2, 1986).

[10]Thies, W., *Eppler-Profile*, MTB 2, Verlag für Technik und Handwerk, Baden-Baden, Germany, 1982 (republished, Hepperle, M., MTB 1/2, 1986).

[11]Althaus, D., *Profilpolaren für den Modellflug—Windkanalmessung an Profilen im Kritischen Reynoldszahlbereich*, Neckar-Verlag, Villingen-Schwenningen, Germany, 1980.

[12]Althaus, D., *Profilpolaren für den Modellflug—Windkanalmessung an Profilen im Kritischen Reynoldszahlbereich*, Band 2, Neckar-Verlag, Villingen-Schwenningen, Germany, 1985.

[13]Selig, M. S., "The Design of Airfoils at Low Reynolds Numbers," AIAA Paper 85-0074, Jan. 1985.

[14]Selig, M. S., Donovan, J. F., and Fraser, D. B., *Airfoils at Low Speeds, Soartech 8*, SoarTech Publications, Virginia Beach, VA, 1989.

[15]Hepperle, M., *Neue Profile für Nurflügelmodelle*, FMT-Kolleg 8, Verlag für Technik und Handwerk GmbH, Baden-Baden, Germany, 1988 (republished, Hepperle, M., MTB 1/2, 1986).

[16]Selig, M. S., Guglielmo, J. J., Broeren, A. P., and Giguère, P., *Summary of Low-Speed Airfoil Data*, Vol. 1, SoarTech Publications, Virginia Beach, VA, 1995.

[17]Selig, M. S., Lyon, C. A., Giguère, P., Ninham, C. N., and Guglielmo, J. J., *Summary of Low-Speed Airfoil Data,* Vol. 2, SoarTech Publications, Virginia Beach, VA, 1996.

[18]Lyon, C. A., Broeren, A. P., Giguère, P., Gopalarathnam, A., and Selig, M. S., *Summary of Low-Speed Airfoil Data,* Vol. 3, SoarTech Publications, Virginia Beach, VA, 1998.

[19]Guglielmo, J. J., and Selig, M. S., "Spanwise Variations in Profile Drag for Airfoils at Low Reynolds Numbers," *Journal of Aircraft,* Vol. 33, No. 4, July–Aug. 1996, pp. 699–707.

[20]Guglielmo, J. J., "Spanwise Variations in Profile Drag for Airfoils at Low Reynolds Numbers," M.S. Thesis, Dept. of Aeronautical and Astronautical Engineering, Univ. of Illinois at Urbana–Champaign, Urbana, IL, 1995.

[21]Rae, W. H., Jr. and Pope, A., *Low-Speed Wind Tunnel Testing,* Wiley, New York, 1984.

[22]Giguère, P., and Selig, M. S., "Freestream Velocity Measurements for Two-Dimensional Testing with Splitter Plates," *AIAA Journal,* Vol. 35, No. 7, July 1997, pp. 1195–1200.

[23]McGhee, R. J., Walker, B. S., and Millard, B. F., "Experimental Results for the Eppler 387 Airfoil at Low Reynolds Numbers in the Langley Low-Turbulence Pressure Tunnel," NASA TM 4062, Oct. 1988.

[24]Coleman, H. W., and Steele, W. G., Jr., *Experimentation and Uncertainty Analysis for Engineers,* Wiley, New York, 1989.

[25]Lyon, C. A., Selig, M. S., and Broeren, A. P., "Boundary Layer Trips on Airfoils at Low Reynolds Numbers," AIAA Paper 97-0511, Jan. 1997.

[26]Wortmann, F. X., "Progress in the Design of Low Drag Airfoils," *Boundary Layer and Flow Control,* edited by G. V. Lachmann, Pergamon, London, 1961, pp. 748–770.

[27]Donovan, J. F., and Selig, M. S., "Low Reynolds Number Airfoil Design and Wind Tunnel Testing at Princeton University," *Low Reynolds Number Aerodynamics,* edited by T. J. Mueller, Vol. 54 of *Lecture Notes in Engineering,* Springer-Verlag, New York, June 1989, pp. 39–57.

[28]Drela, M., "Low Reynolds-Number Airfoil Design for the M.I.T. Daedalus Prototype: A Case Study," *Journal of Aircraft,* Vol. 25, No. 8, Aug. 1988, pp. 724–732.

[29]Gopalarathnam, A., and Selig, M. S., "Low-Speed Natural-Laminar-Flow Airfoils: Case Study in Inverse Airfoil Design," *Journal of Aircraft,* Vol. 38, No. 1, Jan.–Feb. 2001, pp. 57–63.

[30]Selig, M. S., *The Design of Airfoils at Low Reynolds Numbers, Soartech 3,* SoarTech Publications, Virginia Beach, VA, 1984.

[31]Giguère, P., and Selig, M. S., "New Airfoils for Small Horizontal Axis Wind Turbines," *ASME Journal of Solar Energy Engineering,* Vol. 120, May 1998, pp. 108–114.

Chapter 9

Numerical Optimization and Wind-Tunnel Testing of Low Reynolds Number Airfoils

Th. Lutz,* W. Würz,* and S. Wagner[†]
Institute for Aerodynamics and Gas Dynamics,
University of Stuttgart, Stuttgart, Germany

Nomenclature

A	=	amplitude of a Tollmien–Schlichting wave
c	=	chord length
c_d	=	drag coefficient
c_f	=	skin-friction coefficient
c_l	=	lift coefficient
c_p	=	pressure coefficient
FSD	=	full-scale deviation
H_{12}, H_{32}	=	shape factor
NC	=	numerically controlled
n	=	amplification factor
q	=	dynamic pressure
Re	=	Reynolds number
Re_{δ_2}	=	local Reynolds number based on δ_2
r_{le}	=	leading-edge radius
s	=	arc length
Tu	=	freestream turbulence level
t	=	airfoil thickness
U	=	local velocity
U_e	=	velocity at the boundary-layer edge
u_{pitot}	=	pitot reading
x	=	airfoil coordinate
α	=	angle of attack
α_i^*	=	design parameter
δ_1	=	displacement thickness
δ_2	=	momentum thickness

*Research Assistant.
[†]Professor, Head of Institute. Member AIAA.

δ_3 = energy thickness
γ = separation angle of the dividing streamline

Indices
crit   = critical value
max   = maximum value
R = reattachment point
S = separation point
T, tra   = transition point
US = upper surface
∞ = freestream condition

I. Introduction

FOR the Reynolds number regime of aircraft wing sections $(Re \gtrsim 1 \times 10^6)$ sophisticated direct and inverse methods for airfoil analysis and design are available.[1,2] In the hands of experienced users these methods allow a carefully directed airfoil design. Usually, specific airfoils are developed for a new aircraft to maximize the performance for the intended range of application.[2–6] Wind-tunnel tests mainly serve for verification of the predicted aerodynamic characteristics and are a necessary means to determine the stall behavior and to optimize turbulators or flap settings. Doubtless, the total number of design loops is significantly reduced as the result of the availability of reliable prediction methods.

This does not hold for airfoil design at low Reynolds numbers $(Re \lesssim 200,000)$. In this regime, transition usually takes place in a separated boundary layer if the clean airfoil is considered. Predicting the influence of laminar separation, of (unsteady) transitional separation bubbles, and of the viscous/inviscid interaction remains a challenge. This is especially true close to the critical Reynolds number where strong nonlinearities occur. In general, the uncertainty in the airfoil analysis increases with decreasing Re. For this reason, it is necessary to include the lessons learned from wind-tunnel tests in the design methodology of low Reynolds number airfoils. The most promising approach involves extensive experiments, physical interpretation of the results, and the transfer of the results into successively improved airfoil designs, as performed e.g. by Selig et al.[7,8]

Aware of these problems, an attempt was made to perform direct numerical shape optimizations of low Reynolds number airfoils $(200,000 \leq Re \leq 400,000)$. This approach is considered to be instructive for two reasons: First, the optimizer may find unconventional solutions that could inspire new ideas for an improved manual design, and second, weak points of the aerodynamic model can be identified if the optimized airfoils are examined in a wind-tunnel test. This in turn may lead to improvements of the prediction methods.

The intention of the present optimizations was to perform an unconstrained design of completely new airfoils rather than to modify existing ones. This necessitates considering a large number of design variables to enable a detailed representation of the airfoil shape and the pressure distribution. Furthermore, stochastic optimization algorithms should be applied to reduce the possibility of getting trapped in a local optimum. A high degree of freedom problem, in combination with a stochastic optimization strategy, however, requires a very large number of airfoils to be generated and analyzed until the optimization process converges. If the optimizations are to be performed with acceptable computation time, an efficient aerodynamic model is needed. For the present investigations, a potential-flow method coupled with an integral boundary-layer procedure, a simplified bubble model, and an e^n database method for transition prediction was applied. Viscous/inviscid interaction was neglected to minimize the computational effort, which actually is roughly two orders of magnitude below that needed for an XFOIL analysis. The complete airfoil optimization tool has been proven successfully for the design of high Reynolds number ($Re \geq 1 \times 10^6$) NLF airfoils[9,10] and is routinely applied for various design tasks.

In the following sections the ingredients of the method will briefly be described. Thereafter, optimization examples of low Reynolds number airfoils including wind-tunnel verification will be discussed.

In addition to this theoretical approach to airfoil design, fundamental wind-tunnel tests for very low Reynolds number airfoils ($Re \lesssim 50,000$) will be presented. The first detailed experimental investigations of airfoil characteristics for this Reynolds number regime were published by Schmitz.[11] He proved that thin airfoils featuring a sharp leading edge have a lower critical Reynolds number than "conventional" airfoil designs. In the experiments of Schmitz, the cambered plate (Gö 417a) shows the best performance up to $Re \approx 100,000$.

The problem of how to provoke boundary-layer transition by shaping to avoid extensive laminar separation was examined by Hamma[12] for $Re < 50,000$. Among other designs, a 9.7%-thick symmetrical airfoil with a sharp leading edge was tested in a water tunnel and showed promising results. However, very complex flow phenomena were observed. For example, laminar separation with laminar reattachment, followed by a transitional separation bubble occurred for a specific α and Re regime. Furthermore, the separation angle of the shear layer increased abruptly for a distinct angle of attack. Such complex flow phenomena cannot be predicted reliably with available theoretical models. The present authors, therefore, prefer an experiment-based approach to establish guidelines for the design of very low Reynolds number airfoils. In a recent test campaign, the influence of relevant geometric parameters, such as the leading-edge radius, on the aerodynamic performance is being examined. Initial results have been obtained and will be discussed.

II. Aerodynamic Model

For the present investigations, the determination of the aerodynamic characteristics during the optimization process is based on the airfoil design and analysis method of Eppler.[2] The method features an inverse conformal mapping procedure and a higher-order panel method in combination with an integral boundary-layer method. Major extensions with respect to boundary-layer computation and transition prediction were added to this tool.

Reliable transition prediction is of essential importance for a successful design and optimization of natural laminar flow airfoils. For the present investigations, the semi-empirical e^n method according to van Ingen[13] and Smith and Gamberoni[14] was implemented. The basic idea of the e^n method is that transition may be assumed when the most amplified frequency reaches a certain critical amplification factor $n_{crit} = \ln(A_{crit}/A_{initial})$. The value of n_{crit} depends on freestream conditions, the receptivity mechanism, and the definition of the transition "point." Therefore, n_{crit}

has to be correlated individually for each wind tunnel (or for free-flight conditions) taking the numerical model used for the airfoil analysis into account.

With the present e^n implementation, spatial disturbance growth is considered. Because stability analysis based on a direct solution of the Orr–Sommerfeld equation requires too much computational effort for the purpose of numerical airfoil optimization, a database method was implemented. This database contains the amplification rates for 36 shape factors (including separated flows) at 40 different Reynolds numbers and 40 different frequencies.

To account for the influence of transitional separation bubbles that are "short" according to the definition of Tani,[15] a new simplified bubble model was developed.[9,10,16] The objective was not to enable a detailed calculation of the flow properties inside the bubble but to efficiently determine realistic initial values for the turbulent boundary-layer calculation downstream of the bubble. Hence, a minimum number of superimposed empirical relations was aspired. Following the classical approach proposed by Horton,[17] a constant outer-flow velocity U_e is assumed in the laminar part of the bubble. Based on a Falkner–Skan separation profile successively shifted in wall normal direction, a new family of boundary-layer profiles has been introduced for this region. The wall distance of the dividing streamline and, accordingly, the magnitude of δ_1 and H_{12} are assumed to increase linearly with the arc length s, depending on the separation angle γ, which is determined by an empirical correlation.[9] Downstream of transition, a strong abstraction of the real physics is introduced. The reattachment point is assumed to coincide with the transition location, which results in an abrupt drop of the outer-flow velocity to the inviscid value. Furthermore, if $\delta_1 = constant.$ and $c_f = 0$ is assumed, the integral momentum equation can be solved analytically in the reattachment region, which yields the δ_2 increase over the whole bubble. The main advantage of this approach is that the use of uncertain dissipation laws in the unsteady recirculation region is avoided, making a direct, noniterative calculation possible.

For the calculation of turbulent boundary layers, the lag-dissipation method proposed by Drela[1] was implemented with a new shape-factor relation H_{12} (H_{32}, Re_{δ_2}).[10,16] The drag coefficient of the airfoil is finally determined from the boundary-layer properties at the trailing edge using the Squire–Young formula. If turbulent separation is predicted, the inviscid lift coefficient is corrected in a semi-empirical way as proposed by Eppler.[2] A more detailed description of the aerodynamic model along with validation examples can be found in other publications of the authors.[9,10]

III.   Experimental Setup

A.   Wind Tunnel

The experiments were carried out in the model wind tunnel (MWT)[18,19] of the Institute for Aerodynamics and Gas Dynamics (IAG). The MWT (Fig. 1) is an openreturn tunnel with a closed test section. It is placed inside a large test hall, which ensures stable operation conditions. The test section measures 0.375×0.6 m and is 0.9-m long. The two-dimensional airfoil models span the short distance of the test section. A thyristor-regulated 4-kW engine and an adjustable belt drive give control over the rpm of the nine-bladed fan. Because of the high contraction ratio of $19.6 : 1$ the turbulence level Tu, for the frequency range of 10–5,000 Hz, is between 2×10^{-4} and 8×10^{-4} depending on the tunnel speed (4 m/s < U_∞ < 30 m/s).

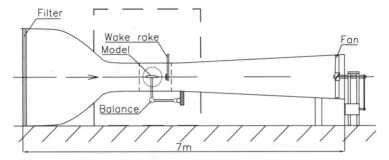

Fig. 1   The model wind tunnel of the IAG.

Typical chord lengths of the models are in the range of 0.12 to 0.2 m, which results in possible Reynolds numbers from 30,000 to 400,000.

The airfoil models are built of reinforced fiberglass in NC-milled molds to ensure the necessary surface accuracy. A special milling technique is used to reduce the remaining surface roughness of the mold to 0.02 mm. Therefore, only a small amount of work is necessary to smooth and polish the surface of the models. The airfoil models are mounted horizontally between two circular end plates, which are flush with the tunnel side walls. A special construction of the gap's geometry of the end plates reduces leakage effects caused by different static pressure inside and outside the test section. The end plates are connected to a balance with an HBM-Z6C2 high-precision loadcell with a linearity of $\leq \pm 0.1\%$. The output signal is amplified by an HBM-KWS3073 carrier frequency amplifier. The system is calibrated before every run using reference weights. The lift is determined directly by this force measurement, whereas the drag is measured by an integrating wake rake.[3] The wake rake is positioned 30% of the chord length downstream of the trailing edge and is traversed in spanwise direction to account for drag variations caused by longitudinal inherent structures in the boundary layer. Four Furness Controls Ltd. micromanometers (FCO01 and FCO14, $FSD \leq \pm 0.02\%$) are used to measure simultaneously the dynamic pressure, the mean pressure loss in the wake, the static pressure difference, and the maximum pressure loss in the center of the wake. The angle of attack is evaluated by a Megatron AL1720 high-precision potentiometer (linearity $\leq \pm 0.25\%$). The output signal is amplified by a differential amplifier to fit it to the input range of the analog/digital (A/D) converter. A typical overall accuracy of $\leq \pm 0.07$ deg in angle of attack is achieved. The complete measurement system is highly automated and controlled by a PC equipped with a 12-bit A/D converter. Standard wind-tunnel corrections[20] are applied and the results are monitored online.

B.   Correlation of the Critical Amplification Factor

Applying the e^n method requires the determination of a limiting n factor for the test facility (or free-flight conditions) to enable a realistic recalculation of the experiments. Different procedures to determine n_{crit} are possible. One direct approach is to use the measured turbulence level of the wind tunnel considered together with available correlations for n_{crit}.[21,22] The main disadvantage of this approach is that the integral value of the turbulence level is used and the spectral distribution of the velocity fluctuations is neglected. Typically, the unstable Tollmien–Schlichting (TS) waves are connected with a small frequency range in comparison to the whole range used for the evaluation of the turbulence level. The direct measurement of the velocity fluctuations in this critical frequency range is almost impossible, because of its small amplitude (for low-turbulence facilities), which is of the order of the electronic noise of the hot-wire equipment.

To avoid this problem, an indirect "calibration" is commonly preferred. The predicted transition positions for several test cases are compared to the experimentally determined location, and the n factor is adjusted until a good match is achieved. This procedure also takes small differences in the numerical evaluation of the n factor into account because the numerical model is directly involved.

To establish a well-defined test case, a special symmetrical airfoil that provides transition without a separation bubble at zero angle of attack was designed (see Fig. 2). To meet this objective, a long instability region with an almost constant shape factor of $H_{12} = 3.3$ was introduced between $x/c = 0.15$ and $x/c = 0.95$. Based on the inviscid velocity distribution, which represents a good fit to the measured pressure distribution, a boundary-layer calculation was performed and used as input for the spatial stability analysis. The resulting amplitude development is depicted in Fig. 3.

The experimental determination of the transition "point" is a serious problem because transition is a more or less continuous process. Therefore, the evaluated position depends strongly on the method used for transition detection. This again has a nonnegligible influence on the correlated value of n_{crit}. For the present investigations, a classical procedure[23] based on the change of the wall shear stress was applied. A miniature pitot tube with 0.25-mm opening is traversed on the surface of the model, and the pressure difference from a reference pressure is recorded (see Fig. 4). For $Re = 250,000$, the wall shear stress increases at $s/s_{max} = 0.65$

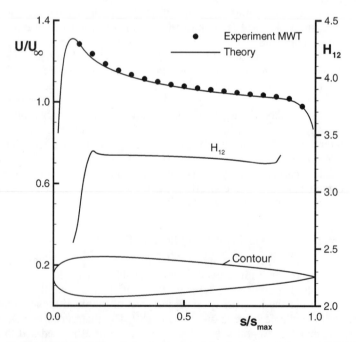

Fig. 2   Velocity distribution and development of the shape factor for airfoil GW98-109 at $\alpha = 0$ deg and $Re = 250,000$.

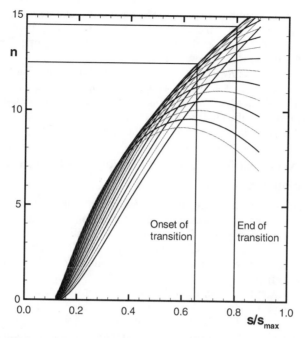

Fig. 3   Amplification curves for different TS frequencies for airfoil GW98-109 at $\alpha = 0$ deg and $Re = 250{,}000$.

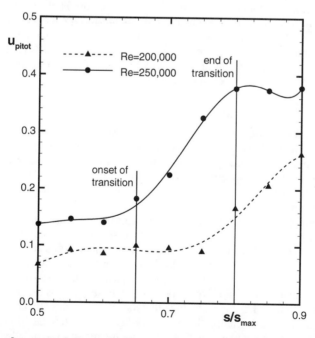

Fig. 4   Development of the wall shear stress in the transition region for airfoil GW98-109 at $\alpha = 0$ deg (arbitrary scale).

and reaches its maximum at $s/s_{max} = 0.8$. The corresponding n factors resulting from Fig. 3 are $n = 12.5$ and $n = 14.5$. The upstream position is similar to the station where the first sound can be heard with a stethoscope. The transition "position" was uniform in spanwise direction.

The comparisons for $Re = 200,000$ yield similar n factors. For the numerical model, a rounded value of $n_{crit} = 12$ is used to predict the location for the switch from laminar to turbulent closure relations. According to the above considerations, this position indicates the onset of transition and, therefore, the first station where the mean velocity profile is influenced by the transition process.

IV. Numerical Optimization of Low Reynolds Number Airfoils

A. Optimization Algorithm

With direct numerical optimization, an automated search for an optimal solution with respect to a user-specified objective function (e.g., minimization of the drag coefficient) is performed. This is done by an iterative variation of the design variables chosen. For airfoil optimizations, it is usually necessary to introduce geometric or aerodynamic constraints such as a limitation of the thickness or of the moment coefficient.

Regarding the optimization algorithm we can determine deterministic methods, like gradient-based approaches, and stochastic methods, like evolution strategy or genetic algorithm. In general, gradient methods converge fast for a simple topology of the objective function but may get trapped in a local optimum. As a result, the "optimized" airfoil can look very similar to the initial shape. Actually, this is true for most of the numerical airfoil optimizations published so far. Stochastic algorithms offer a greater chance to find a better optimum and, furthermore, can cope with complex topologies of the objective function. However, usually many more iterations and, therefore, airfoil analyses are required. Since the objective function considered in the present investigations shows multimodal behavior and strong nonlinearities[10] a hybrid optimizer was applied. The POINTER tool[24] chosen enables constrained optimization and consists of a combination of genetic algorithm, downhill simplex, and gradient method.

B. Airfoil Parameterization

To perform numerical optimizations, proper design variables have to be chosen (i.e., the airfoil has to be parameterized in an adequate way). Typically, geometric shape functions such as Legendre polynomials, Wagner functions, or bezier splines are applied with the respective coefficients used as design variables. The shape functions may directly define the airfoil contour or represent perturbations of a given basic shape.

Another approach was preferred for the present investigations. Instead of optimizing the contour in a direct way, the inverse conformal mapping procedure according to Eppler has been applied to generate the airfoil shape. The input parameters α_i^* of this approach were used as design variables for the optimization process. The α_i^* values represent the angle of attack (relative to the inviscid zero-lift line) for which the outer-flow velocity in a certain section i is intended to be constant. This approach has the advantage that the α_i^* values directly control the local pressure gradient and, finally, the boundary-layer characteristics. Furthermore, a spline representation of the sensitive leading-edge region is avoided.

C. Optimization Examples

The objective of the present investigations was to optimize airfoils with minimized average drag coefficient for a set of given angles of attack relative to the inviscid zero-lift line:

$$\sum c_d(\alpha_{\text{design}}, Re_{\text{design}}) \overset{!}{=} \min \tag{1}$$

$$\alpha_{\text{design}} = 0, 1, 2, 3, 4, 5, 6 \deg \text{ (relative to zero-lift line)} \tag{2}$$

From the specified α regime a design lift range of $0 \lesssim c_{l_{\text{design}}} \lesssim 0.66$ can be estimated. Prescribing fixed values for α was preferred to a direct specification of $c_{l_{\text{design}}}$ to avoid an α iteration or an interpolation of the predicted drag polar.

Three different Reynolds numbers were considered: 2×10^5, 3×10^5, and 4×10^5. Since an experimental verification in the MWT of the institute was planned, a value of $n_{\text{crit}} = 12$, representative for this facility, was specified for transition prediction. The NACA 66_2–415 section was used as initial airfoil for all optimization runs. The required α_i^* distribution of this airfoil was determined from the known contour using an automated redesign method.[25] The optimization process was driven solely by the objective to minimize the aerodynamic drag (Eq. 1) because no geometric or aerodynamic constraints were introduced at all. To enable a detailed representation of the airfoil shape, a large number of 34 α_i^* values were chosen as design variables. Almost 300,000 different airfoil shapes were generated and analyzed during one optimization run, which required almost 100 h of CPU time on a desktop workstation. To minimize the computational effort and to reduce the roughness of the objective function, an iterative boundary-layer coupling was neglected during the analysis. The resulting airfoil contours are depicted in Fig. 5. It is obvious that the shapes completely differ from the 15%-thick NACA airfoil. As expected, the airfoil thickness reduces with decreasing design Reynolds number. The absolute value of t/c is furthermore determined by the design lift range, which is identical for all present optimization examples.

The TL 133-O airfoil being optimized for $Re_{\text{design}_{\text{II}}} = 3 \times 10^5$ will now be discussed in more detail. In Fig. 6 the inviscid velocity distributions for different angles of attack (dashed lines) are given. On the lower side, the airfoil shows no significant pressure recovery, and for $Re_{\text{design}_{\text{II}}}$, the boundary layer is predicted to be laminar up to the trailing edge. In contrast, a steep almost linear flow deceleration was introduced on the suction side. Upstream of the pressure recovery, a remarkable valley can be observed in the inviscid velocity distribution. The upstream flank of

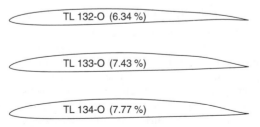

Fig. 5   Shapes of the optimized airfoils for $Re_{\text{design}} = 2 \times 10^5, 3 \times 10^5,$ and 4×10^5 (from top to bottom).

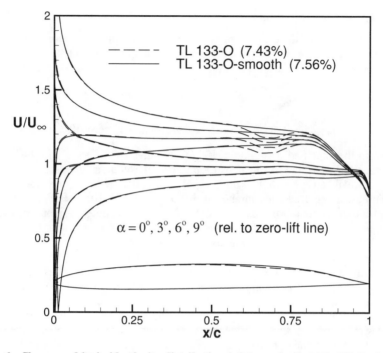

Fig. 6 Shapes and inviscid velocity distributions of the optimized TL 133-O airfoil and the modified section TL 133-O-smooth.

the valley causes laminar separation, and reattachment is predicted to occur at the end of the valley, just upstream of the main pressure recovery. By these means transition is fixed to the ideal position. Then, the turbulent boundary layer is able to overcome the pressure rise without separation up to the trailing edge.

For the specified design conditions, the optimizer obviously prefers transition initiated by a provoked separation bubble rather than a continuous destabilization of the attached boundary layer. However, this result may be attributed to the limited design range or to the simplified bubble model applied, which determines the δ_2 increase over the bubble only from *local* flow properties at the separation and the reattachment point, respectively. Therefore, with the present aerodynamic model the velocity distribution inside of the bubble does not directly affect the predicted drag.

D. Experimental Verification and Discussion

For the first wind-tunnel test it was decided to locally smooth the unusual valley-shaped velocity distribution (solid lines in Fig. 6) and apply an artificial turbulator upstream of the main pressure recovery. The measured drag polars are given in Fig. 7 along with the theoretical result. The drag polar shows a distinct laminar bucket, which coincides with the design lift region. A substantial drag increase results for the clean airfoil owing to a separation bubble on the suction side.

In a second test, the original airfoil with the dent was examined. Without a tripping device, the bubble drag is significantly reduced compared to the smoothed airfoil (see Fig. 8). However, reattachment still occurs in the main pressure recovery

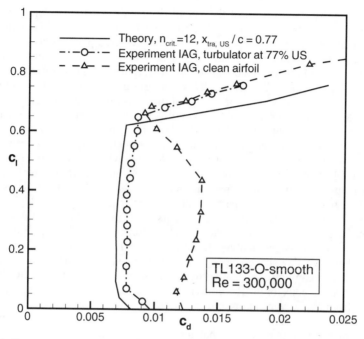

Fig. 7 Drag polars for the modified airfoil TL 133-O-smooth in the MWT at $Re = 3 \times 10^5$.

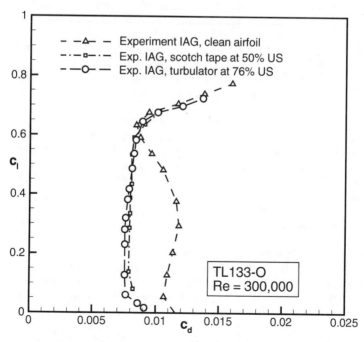

Fig. 8 Drag polars for the optimized airfoil TL 133-O in the MWT at $Re = 3 \times 10^5$.

region, slightly downstream of the intended location. Obviously, the numerical model overpredicts the destabilization effect of the surface dent. To simulate experimentally a "natural" transition close to the predicted position, the initial amplitudes of the relevant TS waves have to be increased. For a given freestream disturbance level this can be obtained by applying thin roughness elements (e.g., adhesive tape) to the airfoil surface. This increases the receptivity of the boundary layer and forces an upstream movement of the transition point but does *not* cause bypass transition as a usual turbulator. With an adequate application of Scotch® tape, the separation bubble can be adjusted to its optimum length for a considered lift region. Using a single height of the roughness element, however, it was not possible to obtain optimum performance within the whole laminar bucket.

To finally compare this airfoil to the smoothed section a bump tape, located slightly upstream of the main pressure recovery, was applied. Even though this turbulator was placed in the separated boundary layer, the bumps were high enough to trigger transition, and reattachment occurs at the intended location. Within the whole design lift region, the measured drag coefficient for this configuration is lower than for the smoothed airfoil with optimized turbulator (see Fig. 8). This shows that the valley-shaped velocity distribution resulting from the optimization is physically meaningful and that a provoked shallow transitional separation bubble can be advantageous for the Reynolds number regime considered. This conclusion differs from the design philosophy proposed by Pfenninger and Vemuru,[6] who preferred a distinct concave main pressure recovery in combination with transition control by means of a suitable turbulator.

To further examine the quality of the optimized airfoil, comparative wind-tunnel tests were performed for a manually designed airfoil, namely the RG-15 section (Fig. 9). This airfoil has proven to be very successful in model glider applications, but it was *not* especially developed for the design conditions considered in the present optimization. Therefore, it should be expected that this airfoil is inferior at least in the present design lift region according to Eq. (2). However, the comparison of the measured drag polars as depicted in Fig. 10 shows that the average drag coefficient in this region is identical for both airfoils ($\overline{c_d} = 0.008$). Despite its greater thickness, the RG-15 offers a lower minimum drag coefficient, whereas the optimized airfoil has some gain near the upper edge of the laminar bucket.

A further examination of the RG-15 shows that long but shallow separation bubbles occur on both airfoil sides in the minimum drag region. Whereas on the suction side transition is observed at $x_{tra}/c \approx 0.85$, on the pressure side the boundary layer remains laminar up to the trailing edge. To introduce such long and shallow bubbles extending almost into the wake seems to be advantageous compared to a design with distinct transition and a subsequent steep pressure recovery. However, because of the limited ability to model this type of bubbles with the present noninteractive aerodynamic method, the optimizer could not take advantage of this positive effect. To investigate this problem on a theoretical basis, it is necessary to account for viscous/inviscid interaction effects. Numerical optimizations considering these problems (e.g., by using the XFOIL code), perhaps for a reduced number of design variables, would therefore be valuable to obtain more insight into the problem of optimum shape design for low Reynolds number airfoils.

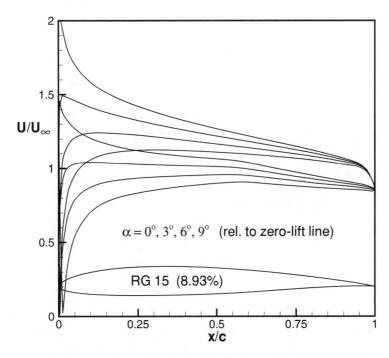

Fig. 9 Shape and inviscid velocity distributions of the RG-15 airfoil.

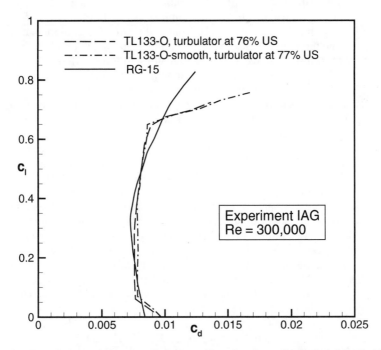

Fig. 10 Comparison of the measured drag polars for the optimized airfoil TL 133-O and the RG 15 section in the MWT at $Re = 3 \times 10^5$.

V. Experimental Investigations on Very Low Reynolds Number Airfoils

At Reynolds numbers above $Re = 1 \times 10^6$, the design of the airfoil leading edge is an important task because of the fundamental influence on the width of the low drag bucket and the stall behavior.[26] If the leading edge is too sharp, then a pronounced suction peak occurs directly at the leading edge for angles of attack slightly above the laminar bucket. This suction peak will cause an abrupt jump of the transition position toward the leading edge, which in turn produces a fast forward movement of the turbulent boundary-layer separation. The stall characteristics of such an airfoil design are poor (i.e., an abrupt breakdown in lift has to be expected). However, if the nose is too blunt, over-velocities result downstream of the leading edge, and the transition position moves upstream for lower angles of attack than with an ideal nose geometry. This reduces the width of the low drag bucket. However, the stall behavior will be more gentle compared to an airfoil with small nose radius. In any case, at higher Reynolds numbers a laminar-separation bubble near the leading edge should be prevented because of possible bubble "bursting,"[27] which will cause a sudden stall.

At low Reynolds numbers the situation is somewhat different. First, the influence of laminar-separation bubbles is much more pronounced. Second, the amplification rates of the TS waves are low; therefore, it is much easier to achieve a long laminar run. For Reynolds numbers below $Re \approx 100,000$ it becomes more and more difficult to obtain a turbulent boundary layer without extended laminar separation, which may cause significant additional drag. The common way to solve this problem is the application of an artificial turbulator on the suction side. In general, the lower surface is designed to be fully laminar. A fixed turbulator on the upper surface has one main disadvantage: If it has to be effective at higher lift coefficients, then the turbulator has to be placed far upstream. So, if a multipoint design is considered, at lower values of c_l, the turbulator will force earlier transition than achievable with a clean airfoil. A possible approach to avoid this disadvantage may be the controlled use of a transitional separation bubble near the leading edge. The bubble can be forced to occur at higher angles of attack by a steep pressure rise resulting from a "sharp" leading edge. For lower angles of attack, the suction peak is not present, and a longer laminar run results. A leading-edge bubble must not increase the drag significantly if no premature turbulent separation is caused. This can be seen in Fig. 11 where the relative δ_2 increase over a transitional separation bubble is depicted as it results from the present bubble model. The growth of δ_2, which finally determines the drag increase, is proportional to the momentum thickness δ_{2_s} at the laminar separation point S, which is small near the leading edge.

To investigate the effect of such a forced leading-edge bubble, three different airfoil shapes were examined in the MWT. Because of the intended low Reynolds number of 50,000, it was preferred to investigate the influence of systematic geometric modifications of a known airfoil rather than to design completely new shapes using theoretical methods. To reduce the number of parameters to be varied, a symmetrical airfoil (NACA 0009) was chosen as initial shape. In the present study, the influence of the leading-edge radius r_{le} was investigated. Two mutations of the NACA 0009 ($r_{le}/c = 0.0089$) were generated by specifying a reduced value for r_{le} but retaining the original contour for $x/c \geqq 0.1654$. The coordinates in between were obtained from a polynomial approximation spline, which ensures a smooth curvature distribution near the junction. The resulting airfoils show nose radii of $r_{le}/c = 0.00074$ (designated "sharp") and $r_{le}/c = 0.0036$ (designated "medium"),

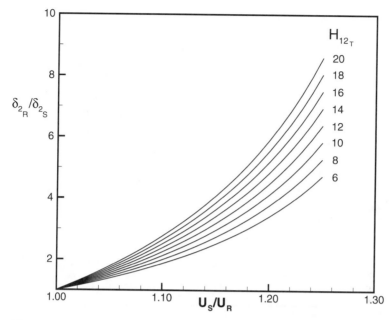

Fig. 11   The dependence of the relative δ_2 increase over a transitional separation bubble on the maximum shape factor H_{12_T} and the outer-flow velocity ratio U_S/U_R according to the present bubble model.

respectively. The shapes and the inviscid pressure distributions for the airfoils examined are presented in Fig. 12. The modified sections exhibit a small over-velocity near the junction and also the desired suction peak near the leading edge.

The wind-tunnel models with a chord length of $c = 120$ mm were built with 30 pressure orifices on one side of the airfoil in order to measure the pressure distribution. A special NC-milled template was used to drill the holes with a diameter of 0.3 mm exactly perpendicular to the surface. The pressure orifices were connected to a scanivalve system and a Furness Controls FCO01 micromanometer to obtain the pressure coefficient according to $c_p = 1 - q/q_\infty$. An integration time of 3 s was sufficient to achieve a reliable mean value. No wind-tunnel corrections were applied to the data. Figures 13, 14, and 15 show the measured distributions for different angles of attack.

The distributions show the same characteristics as the inviscid calculations for $\alpha \leq 2$ deg, despite the fact that there is a laminar flow separation without reattachment in the rear part. With increasing angle of attack, the laminar separation moves upstream and forms a closed separation bubble. This forward movement starts *earlier* for the original NACA 0009 and the medium contour. For $\alpha > 5$ deg the separation bubble begins just downstream of the leading edge. As expected, the total length of the bubble is the shortest for the sharp leading edge. However, because of the high over-velocity at the separation point, a higher pressure recovery is observed in the reattachment region. This increases the momentum thickness. It can also be seen that the gradient of the pressure recovery has become steeper for the modified leading edges. The measured distributions for $\alpha \approx 4$–5 deg (depending on the airfoil section) exhibit a very unusual feature (see $\alpha = 5$ deg, Fig. 13).

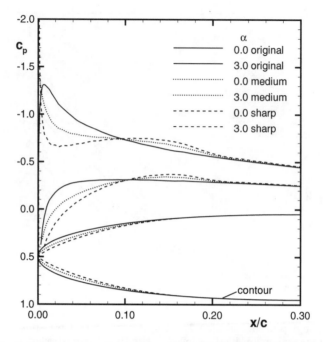

Fig. 12   Inviscid pressure distributions of the NACA 0009 and modifications.

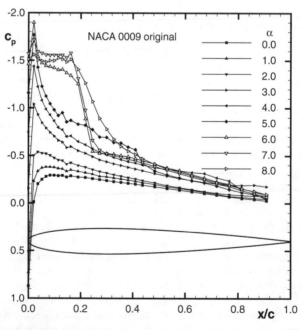

Fig. 13   Measured pressure distributions of the NACA 0009 on the suction side in the MWT at *Re* = 50,000.

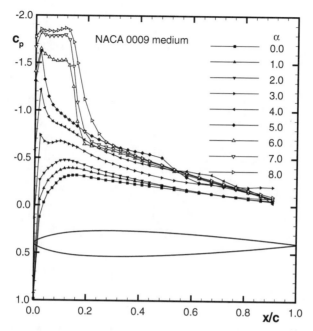

Fig. 14   Measured pressure distributions of the NACA 0009 medium on the suction side in the MWT at $Re = 50,000$.

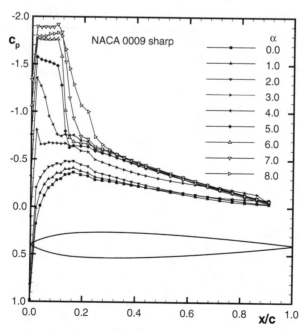

Fig. 15   Measured pressure distributions of the NACA 0009 sharp on the suction side in the MWT at $Re = 50,000$.

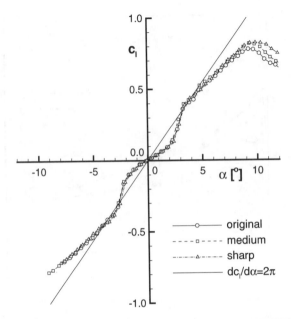

Fig. 16   Measured lift coefficients vs angle of attack for the NACA 0009 with and without modifications in the MWT at $Re = 50,000$.

From the general behavior, it looks as if an unsteady separation bubble exists between $x/c = 0.2$ and $x/c = 0.4$. This was checked for the original airfoil with a stethoscope. If the stethoscope was placed at $x/c = 0.3$ and at the boundary-layer edge, a "flapping," which indicates a rapid upstream/downstream movement of the separation bubble, could be heard. Because of the small diameter of the pressure orifices and the necessary tube length, unsteady effects cannot be resolved and the c_p distribution shows mean values.

Figure 16 shows the lift coefficient plotted vs angle of attack. As a result of the laminar separation without reattachment (subcritical state) at angles of attack below 2.5 deg, the gradient of the lift curve is much smaller than the theoretical inviscid lift slope ($dc_l/d\alpha = 2\pi$ for a flat plate). Above $\alpha = 3$ deg, the turbulent reattachment process gets started, (i.e., a separation bubble that moves upstream with increasing angle of attack is formed). A steep c_l increase results for this α regime. For $\alpha > 5$ deg, the lift slope again is clearly below 2π, which is caused by the turbulent boundary-layer displacement effect along with local turbulent separation close to the trailing edge. The $c_{l_{\max}}$ is the highest for the airfoils with the modified leading edge. The stall behavior is moderate for all configurations.

Figure 17 shows the measured drag characteristics. Despite the strong modifications in the leading-edge region, the drag polars are quite similar. The subcritical range of $-0.2 < c_l < 0.2$ shows no drag increase caused by separation effects. A small advantage of the medium airfoil can be observed for $c_l \approx 0.5$.

The investigations were supplemented by measurements for $Re = 30,000$; 80,000; and 120,000. It turned out that for the sharp and the medium leading edge the value of $c_{l_{\max}}$ is almost constant in this Re regime, whereas it decreases with Re for the blunt nose (see Fig. 18). The NACA 0009 medium shows the widest laminar

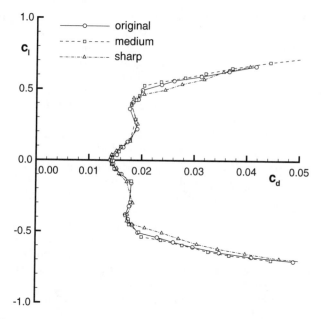

Fig. 17   Measured drag polars for the NACA 0009 with and without modifications in the MWT at $Re = 50,000$.

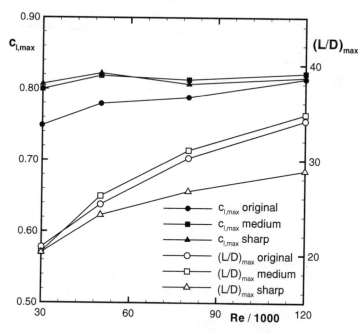

Fig. 18   Measured values of $c_{l_{max}}$ and $(L/D)_{max}$ vs Reynolds number for the NACA 0009 with and without modifications in the MWT.

bucket for $Re \gtrsim 50{,}000$ and the maximum L/D increases in strength with Re than for the sharp nose. An improvement of the performance within the laminar bucket could not be achieved by a variation of only one parameter, namely the leading-edge radius. Significant modifications of the inviscid pressure distribution can be nullified by the dominating viscous effects at such low Re. More combined theoretical and experimental investigations are necessary in the future to establish a straightforward design methodology for very low Reynolds number airfoils.

VI. Conclusion and Outlook

A numerical optimization tool was applied to the design of low Reynolds number airfoils ($200{,}000 \leq Re_{\text{design}} \leq 400{,}000$) with minimized drag for a specified lift region. The inviscid velocity distributions of the optimized airfoils exhibit a characteristic valley on the suction side to provoke a transitional separation bubble. Wind-tunnel tests showed that for the Reynolds number regime considered, this design approach is slightly superior to a smooth velocity distribution in combination with a turbulator upstream of the main pressure recovery. The predicted drag characteristics are in good agreement with the measured polar, and the resulting laminar bucket coincides with the design lift region.

To deduce guidelines for the design of very low Reynolds number airfoils, wind-tunnel tests were performed for $30{,}000 \leq Re \leq 120{,}000$, enabling the influence of relevant geometric parameters to be investigated. In a first campaign, the NACA 0009 plus two modifications with reduced leading-edge radii were examined. The airfoil with the sharp leading edge shows higher $c_{l_{\max}}$ values for $Re \leq 120{,}000$ than the original airfoil but also a smaller width of the laminar bucket, which results in lower values of $(L/D)_{\max}$ for $Re \geq 50{,}000$. A certain improvement could be achieved with the medium leading edge, which combines the high L/D of the original section with the high $c_{l_{\max}}$ of the sharp version. Measurements of the pressure distribution show that strong viscous/inviscid interaction effects occur along with unsteady flow phenomena resulting from large laminar-separation bubbles. To get more insight into the very complex flow physics, the leading-edge separation phenomena will be further studied by applying different flow visualization techniques. A careful interpretation and a comparison of measurement and theoretical prediction will be necessary to finally establish proper airfoil design criteria for this special Reynolds number regime.

References

[1]Drela, M., "Xfoil: An Analysis and Design System for Low Reynolds Number Airfoils," *Proceedings of the Conference on Low Reynolds Number Aerodynamics*, edited by T. J. Mueller, Univ. of Notre Dame, Notre Dame, IN, 1989, pp. 1–12.

[2]Eppler, R., *Airfoil Design and Data*, Springer Verlag, Berlin/Heidelberg/New York, 1990.

[3]Althaus, D., *Niedriggeschwindigkeitsprofile*, Vieweg, Braunschweig/Wiesbaden, 1996.

[4]Boermans, L. M. M., and van Garrel, A., "Design and Windtunnel Test Results of a Flapped Laminar Flow Airfoil for High-Performance Sailplane Applications," *Technical Soaring*, Vol. 21, No. 1, 1997, pp. 11–17.

[5]Drela, M., "Low-Reynolds-Number Airfoil Design for the M.I.T. Daedalus Prototype: A Case Study," *Journal of Aircraft*, Vol. 25, No. 8, 1988, pp. 724–732.

[6]Pfenninger, W., and Vemuru, C. S., "Design of Low Reynolds Number Airfoils—I," *Journal of Aircraft*, Vol. 27, No. 3, March 1990, pp. 204–209.

[7]Selig, M. S., Donovan, J. F., and Fraser, D. B., "Airfoils at Low Speeds," Soartech No. 8, Virginia Beach, VA, 1989.

[8]Selig, M. S., Guglielmo, J. J., Broeren, A. P., Giguère, P., Lyon, C. A., Ninham, C. P., and Gopalarathnam, A., *Summary of Low-Speed Airfoil Data*, Vols. 1–3, Dept. of Aeronautical and Astronautical Engineering, Univ. of Illinois at Urbana–Champaign, Urbana, IL, 1995, 1996, 1998.

[9]Lutz, Th., "Berechnung und Optimierung subsonisch umströmter Profile und Rotationskörper," Dissertation, Institut für Aerodynamik und Gasdynamik, Universität Stuttgart, Germany, 2000, pp. 113–140.

[10]Lutz, Th., and Wagner, S., "Numerical Shape Optimization of Subsonic Airfoil Sections," *Proceedings ECCOMAS 2000: European Congress on Computational Methods in Applied Sciences and Engineering, Barcelona, Spain, September 11–14, 2000*, The International Center of Numerical Methods in Engineering, Barcelona, 2000.

[11]Schmitz, F. W., *Aerodynamik des Flugmodells*, Luftfahrt und Schule/Reihe IV/Band 1, C. J. E. Volckmann Nachf. E. Wette, Berlin-Charlottenburg, 1942.

[12]Hamma, W., "Neu entwickelte Tragflügelprofile für $Re < 10^5$ mit den zugehörigen Polaren und Ergebnissen aus Grenzschichtbeobachtungen an diesen Profilen," Diploma Thesis, Institut für Aerodynamik und Gasdynamik, Technische Hochschule Stuttgart, Stuttgart, Germany, 1963.

[13]van Ingen, J. L., "A Suggested Semi-Empirical Method for the Calculation of the Boundary Layer Transition Region," Report V.T.H. 71, V.T.H. 74, Dept. of Aeronautical Engineering, Delft Univ. of Technology, The Netherlands, 1956.

[14]Smith, A. M. O., and Gamberoni, N., "Transition, Pressure Gradient and Stability Theory," Report ES 26388, Douglas Aircraft Company, 1956.

[15]Tani, I., "Low-Speed Flows Involving Bubble Separations," *Progress in Aeronautical Sciences*, Vol. 5, Pergamon Press, New York, 1964, pp. 70–103.

[16]Weigold, W., "Untersuchungen zur halbempirischen Ermittlung des Umschlags in laminaren Ablöseblasen und Programmierung eines vereinfachten Blasenmodells," Diploma Thesis, Institut für Aerodynamik und Gasdynamik, Universität Stuttgart, Stuttgart, Germany, 1998.

[17]Horton, H. P., "A Semi-Empirical Theory for the Growth and Bursting of Laminar Separation Bubbles," CP No. 1073, Ministry of Technology, Aeronautical Research Council, Queen Mary College, Univ. of London, London, 1969.

[18]Althaus, D., *Profilpolaren für den Modellflug I*, Neckar-Verlag, Villingen–Schwenningen, Germany, 1980.

[19]Althaus, D., *Profilpolaren für den Modellflug II*, Neckar-Verlag, Villingen–Schwenningen, Germany, 1985.

[20]Abbott, I. H., von Doenhoff, A. E., and Stivers, L. S., "Summary of Airfoil Data," NACA Rept. 824, 1945.

[21]Mack, L. M., "A Numerical Method for the Prediction of High-Speed Boundary-Layer Transition Using Linear Theory," NASA SP 347, 1975.

[22]van Ingen, J. L., and Boermans, L. M. M., "Research on Laminar Separation Bubbles at Delft University of Technology in Relation to Low Reynolds Number Airfoil Aerodynamics," *Proceedings of the Conference on Low Reynolds Number Airfoil Aerodynamics, UNDAS-CP-77B123*, edited by T. J. Mueller, Univ. of Notre Dame, Notre Dame, IN, 1985, pp. 89–124.

[23] Schubauer, G. B., and Skramstad, H. K., "Laminar Boundary Layer Oscillation and Transition on a Flat Plate," NACA Rept. 909, 1948.

[24] Synaps Inc., *Pointer Optimization Software Release 1.1 User's Guide,* Synaps Inc., Atlanta, 1997.

[25] Lutz, Th., "Automatisierter Nachentwurf von Tragflügelprofilen und Anwendung auf praktische Entwurfsprobleme," Diploma Thesis, Institut für Flugzeugbau, Unversität Stuttgart, Stuttgart, Germany, 1991.

[26] Althaus, D., "Effects on the Polar due to Changes or Disturbances to the Contour of the Wing Profile," *Technical Soaring,* Vol. 10, No. 1, 1986, pp. 2–9.

[27] Gaster, M., "The Structure and Behaviour of Laminar Separation Bubbles," *AGARD CP-4 Flow Separation Part II,* AGARD, 1966, pp. 813–854.

Chapter 10

Unsteady Stalling Characteristics of Thin Airfoils at Low Reynolds Number

Andy P. Broeren[*] and Michael B. Bragg[†]

University of Illinois at Urbana-Champaign, Urbana, Illinois

Nomenclature

C_l = mean lift coefficient
$C'_{l,\mathrm{rms}}$ = root-mean-square of the fluctuating lift coefficient
$C_{l,\mathrm{max}}$ = maximum mean lift coefficient (at stall)
c = airfoil chord
f = frequency
RMS = root mean square
Re = Reynolds number based on chord
St = Strouhal number, $fc \sin \alpha / U_\infty$
U_∞ = freestream velocity
x = chordwise location on airfoil
α = airfoil angle of attack
α_{stall} = stalling angle of attack

I. Introduction

THE very low Reynolds number operating regime of airfoils designed for micro air vehicles generally requires airfoils with small thicknesses and significant camber.[1,2] These airfoils generally provide optimum lift-to-drag ratio or other performance-based parameters. However, previous research has shown that thin airfoils often have undesirable stalling characteristics. This is manifest in the potential for large-scale unsteady flow during the onset of static stall. Indeed, as early as the 1930s, Jones[3] observed "violent fluctuations" of lift and drag on airfoil models operating near stall. This unsteady behavior was later investigated by Farren[4] using a "fast-response balance" in a related study. The pioneering

---

*Postdoctoral Research Scientist, Department of Aeronautical and Astronautical Engineering. Member AIAA.

†Professor and Head, Department of Aeronautical and Astronautical Engineering. Associate Fellow AIAA.

work of these researchers suggested that airfoil stall could be classified by three fundamental types. The authors further noted that the large-scale unsteady behavior only occurred for airfoils exhibiting two of the three fundamental stall types.

Stall type research was resumed again in earnest at the National Advisory Committee for Aeronautics in the 1950s. McCullough and Gault[5] conducted detailed testing and established the presently accepted definitions of stall type. *Trailing-edge stall* is preceded by movement of the turbulent boundary-layer separation point forward from the trailing edge with increasing angle of attack. *Leading-edge stall* has abrupt flow separation near the leading edge generally without subsequent reattachment. The abrupt separation usually results from a small laminar separation bubble that "bursts" at stall and usually causes a sharp decrease in lift. *Thin-airfoil stall* is preceded by flow separation at the leading edge with reattachment (a laminar-separation bubble) at a point that moves progressively downstream with increasing angle of attack. The authors point out that airfoil stall type is a function of several variables such as Reynolds number, surface roughness, or free stream turbulence. Therefore, any particular airfoil may exhibit a combination of stall types, or its stall type may change when flow conditions are changed.

More observations of low-frequency unsteady stalling characteristics for thin airfoils were reported throughout the course of this stall type research. For example, Gault[6] observed "a cyclic change between stalled and unstalled conditions" of the flow past a NACA 63-009 (9% thickness) at the maximum lift coefficient for a Reynolds number of 5.8×10^6. McCullough and Gault[7] reported "large and relatively low-frequency fluctuations of velocity associated with the separated boundary layer" for a NACA 64A006 (6% thickness) at angles of attack near stall. McCullough and Gault[5] also reported similar flow behavior for a 4.23%-thick double wedge airfoil. This airfoil had a sharp leading edge such that a separation bubble formed at the leading edge for any incidence other than 0 deg. These observations indicate that large-scale and low-frequency unsteady flow is by no means an anomaly for stalled airfoils. However, few detailed studies exist that quantify the magnitude of the force fluctuations and characterize the frequency of the unsteadiness as these may be related to stall type. Also, these Reynolds numbers are very high relative to the flight regime for micro air vehicles.

These aspects of unsteady stall phenomena were revisited again some 30 years later by researchers at NASA. Zaman et al.[8] described a very low frequency flow oscillation associated with the stall of an LRN-1007 airfoil operating at a Reynolds number of 75,000. This airfoil was thin (7.3%) and highly cambered (5.9%) and was originally designed for optimum low Reynolds number performance.[9,10] The authors performed a detailed investigation into this naturally occurring, quasi-periodic phenomenon. The flow oscillation frequencies measured in the airfoil wake converted to Strouhal numbers on the order of 0.02, approximately 10 times lower than those associated with bluff-body shedding or a von Kármán vortex street. Here, the Strouhal number was defined as $f c \sin \alpha / U_\infty$, where the cross-stream length scale, $c \sin \alpha$, is the projected airfoil height. The "low-frequency oscillation" occurred in the range of static stall, or maximum lift, from $\alpha = 14.5$ to 16.5 deg and involved a quasi-periodic switching of the flow between stalled and unstalled conditions. This resulted in large-amplitude force fluctuations, up to 50% of the mean lift coefficient 15 deg angle of attack. Curiously, the low-frequency oscillation completely diminished as the angle of attack was increased, with bluff-body shedding frequencies (i.e., $St = 0.2$) being measured at $\alpha = 18$ deg.

Zaman et al.[8] classified the LRN-1007 airfoil as having a combination thin-airfoil and trailing-edge stall. This observation was generally consistent with the results of the early researchers. Jones[3] described "violent fluctuations" in lift and drag that occurred for airfoils that would later be classified as having thin-airfoil and trailing-edge stall. The Reynolds number at which Jones made these observations was 110,000. Farren[4] performed more detailed measurements following Jones's observations and noted substantial normal-force fluctuations near stall on an RAF 28 airfoil. Estimates from the balance data indicated that the force fluctuations amounted to approximately 25% of the mean and occurred at a Strouhal number of approximately 0.019. Although not explicitly stated by either Jones or Farren, this airfoil could be considered "thin," having an approximate thickness of about 10%.

Research into this low-frequency oscillation on the LRN-1007 airfoil was subsequently performed by others[11−15] and the details of the unsteady features are well known. Bragg et al.[14] extended the Reynolds number range up to 1.25×10^6 and measured the oscillation frequency at twelve angles of attack from 14.4 to 16.6 deg. The Strouhal number varied from 0.017 to 0.030 and had very little dependence on Reynolds number but had a very strong dependence on angle of attack. The authors found no fundamental change in the flow oscillation over this large Reynolds number range. In the same paper, Bragg et al.[14] provided flow visualization results obtained for the LRN-1007 airfoil prior to the onset of the unsteady flow. There was both a leading-edge separation bubble and significant turbulent boundary-layer (or trailing-edge) separation features present on the upper surface. This confirmed the stall type classification made by Zaman et al.[8] In a later study, Broeren and Bragg[15] performed two-component laser-Doppler velocimeter measurements during the low-frequency oscillation on the LRN-1007 airfoil upper surface at $Re = 300,000$. These results showed an interaction between the leading-edge separation bubble reattachment and the tailing-edge separation as the airfoil stalls and unstalls. Since these flowfield features were observed on the airfoil prior to the unsteadiness, there may be a relationship between the airfoil stall type and the low-frequency oscillation.

This study considers the stalling of several airfoils at low Reynolds number that have different stall types as defined by McCullough and Gault.[5] Thin airfoils are of particular interest in this paper because they are more likely suited to very low Reynolds number applications. The objectives were threefold. The first was to quantify the magnitude of lift fluctuations near stall for airfoils of different stall types. The second was to determine if these fluctuations contained distinct low-frequency components and if there was a relationship between this unsteadiness and the stall type. The third objective was to identify any similarities in the unsteady flowfields of different airfoils. To accomplish these objectives, time-dependent lift measurements, wake-velocity measurements, and flow visualization were carried out for 12 different airfoils, encompassing five different mean stalling characteristics.

II.   Experimental Methods

All experiments in this study were performed at the University of Illinois Subsonic Aerodynamics Laboratory, using the low-speed, low-turbulence wind tunnel. The wind tunnel is a conventional indraft open-return type. The inlet settling chamber contains a 4-in.-thick honeycomb flow straightener and four

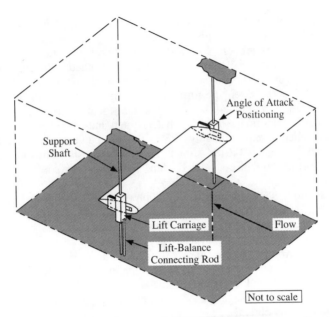

Fig. 1   General experimental apparatus (splitter plates not shown for clarity).

turbulence-reduction screens that reduce the empty test-section turbulence levels to less than 0.1% for all operating speeds (10 to 5000 Hz bandwidth).[16] The general experimental apparatus utilized for the time-dependent lift data acquisition is shown in Fig. 1. The airfoil model (12-in. chord by 33.63-in. span) was supported on each end and was isolated from the support structure by 3/8-in.-thick polycarbonate splitter plates. The gap between each end of the model and the splitter plates was nominally 0.05 in. One end of the model was actuated to adjust and measure the angle of attack. As shown in Fig. 2, the lift force was measured directly via a connecting rod from the lift carriage, which translated freely

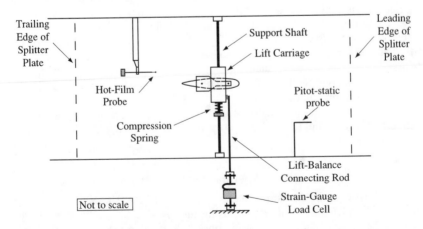

Fig. 2   Detailed schematic drawing of the lift-balance apparatus.

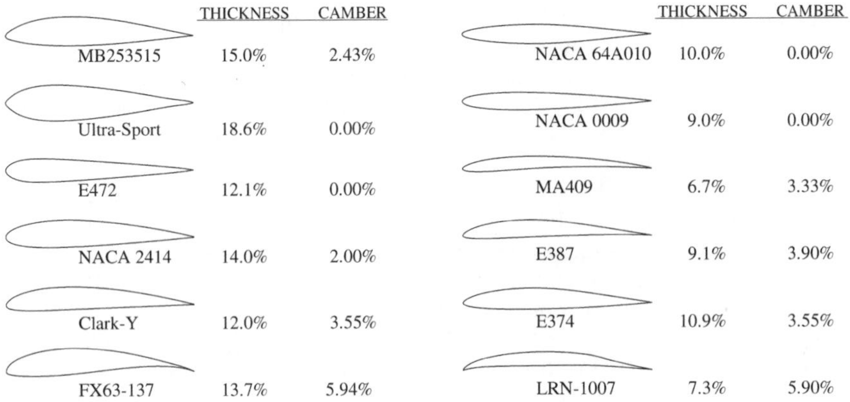

	THICKNESS	CAMBER		THICKNESS	CAMBER
MB253515	15.0%	2.43%	NACA 64A010	10.0%	0.00%
Ultra-Sport	18.6%	0.00%	NACA 0009	9.0%	0.00%
E472	12.1%	0.00%	MA409	6.7%	3.33%
NACA 2414	14.0%	2.00%	E387	9.1%	3.90%
Clark-Y	12.0%	3.55%	E374	10.9%	3.55%
FX63-137	13.7%	5.94%	LRN-1007	7.3%	5.90%

Fig. 3   The airfoils tested in this study.

on the support shaft using linear ball bearings. A compression spring was used to support the weight of the lift carriage and model. The lift force was measured with a strain-gauge load cell. For the present experiments, the load cell was routinely calibrated in its measurement position to determine the effects of mechanical hysteresis and/or other nonlinearities. The freestream dynamic pressure was measured upstream of the model between the splitter plates with a single pitot-static probe. Figure 2 also shows the hot-film probe, which was used to record the wake-velocity voltage.

A total of 12 different airfoils were tested in this study and are shown in Fig. 3 along with their geometric parameters. The airfoils were selected based upon their expected stall types and encompassed a large range of thickness and camber variations. References 17–19 provide the coordinates, more performance data (including drag and pitching moment), and contour accuracy for these airfoil models.

All of the ambient conditions and wind-tunnel data were acquired using a desktop computer equipped with an analog-to-digital conversion board. All data were acquired at a chord Reynolds number of 300,000 and over a 0 to 25 deg angle of attack range. The lift and wake-velocity signals were low-pass filtered using a 500-Hz cutoff. Detailed sets of time-dependent data were acquired for airfoils exhibiting unsteady stalling characteristics. Power spectra of both the lift and wake-velocity signals were performed where appropriate using a digital signal analyzer. The power spectra were acquired over a 0 to 100 Hz bandwidth with a resolution of 0.25 Hz.

Surface-oil flow visualization and smoke flow visualization were performed to reveal important features of the airfoil flowfields. The surface-oil flow visualization was performed at a Reynolds number of 300,000 and involved spraying a light coat of oil on the surface of the airfoil models. The oil was allowed to flow for 20 to 30 min with the tunnel on. The resulting patterns gave information regarding time-averaged boundary-layer separation, reattachment, and transition. These features were recorded for each airfoil as the angle of attack was increased into stall. The boundary-layer features were generally determined to within ± 0.02 x/c ($\pm 2\%$ chord). The smoke flow visualization was performed at lower Reynolds number (75,000 to 100,000) to facilitate video imaging. Using this method, a qualitative picture of the time-varying flowfield about the airfoil upper surface was obtained. Although these Reynolds numbers were substantially less than 300,000, the flow-fields were thought to be fundamentally similar based on wake-velocity measurements. Furthermore, small-scale changes in the unsteady flow over this Reynolds number range would have been undetectable with the smoke flow visualization method.

The time-dependent lift data were used to calculate the mean lift coefficient and the root mean square (RMS) of the fluctuating lift coefficient. The lift data were corrected for wind-tunnel interference effects using standard methods similar to those given by Selig et al.[17] and Giguere and Selig.[20] The magnitude of the corrections was significant. For example, a typical airfoil having an uncorrected C_l of 1.14 had a corresponding corrected C_l of 1.00. The time-dependent lift data were digitally low-pass filtered with a 20-Hz cutoff to remove the unwanted contribution to the RMS of the model and lift-balance natural frequencies. The digital filtering used a zero-phase delay, forward-reversing algorithm to ensure that no phase lag was imposed on the time-dependent data. The lift-balance frequency response had very little magnitude attenuation or phase delay over this bandwidth. Filtering the data had the effect of lowering $C'_{l,\mathrm{rms}}$; however, this effect was small. Furthermore, the filtering process ensured that the RMS data from each airfoil were comparable because the structural contamination was removed. Broeren[16] and Broeren and Bragg[21] provide more detailed information on the lift-balance frequency response and data reduction.

Strouhal numbers were calculated from the power spectra of both the lift and wake-velocity signals. The dimensional frequency was determined as the midpoint of the -3 dB bandwidth of a spectral peak. This frequency and the tunnel conditions were then used to calculate the Strouhal number. The uncertainty in the peak frequency was estimated to be ± 1.5 dB. Therefore, the uncertainty in the Strouhal number was higher for broader low-frequency peaks. The typical relative uncertainty in the Strouhal number was $\pm 3\%$.

The experimental uncertainty was computed for the reduced quantities following the method presented in Coleman and Steele.[22] The calculated uncertainties only included bias errors based upon 20:1 odds. The uncertainty in the angle of attack was ± 0.08 deg and the relative uncertainty in the freestream velocity was $\pm 0.80\%$ at 50 ft/s. The uncertainty in the mean lift coefficient is shown in Fig. 4 along with data from other facilities. The error bars are ± 2 to 3% of the mean C_l. Agreement among the data is very good within the linear range. The divergence after stall may be due, in part, to the large-scale unsteadiness that occurs for the E387 airfoil. This is especially true for the data from Ref. 23, since the lift values were obtained from the integration of surface pressures. Broeren[16] provides more detailed information on the experimental methods and data uncertainty analysis.

III. Results and Discussion

A. The Influence of Stall Type on Lift Fluctuations

A summary of the time-dependent lift data is shown in Table 1, which lists the airfoils tested and classified by stall type. The stall type was determined from interpretation of the lift curves and flow visualization data. Here the stalling angle of attack is equal to the angle of attack at maximum mean lift, and so these terms are used interchangeably. The data show that the airfoils having thin-airfoil stall and a combination of thin-airfoil and trailing-edge stall had the highest

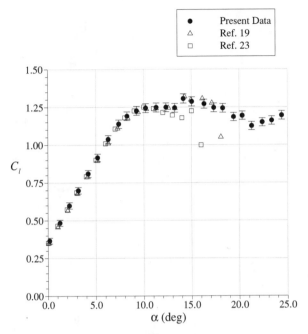

Fig. 4   Comparison of present data with data from other facilities for the E387 airfoil at $Re = 300,000$.

$C'_{l,\text{rms}}$ levels at maximum lift, indicative of large-scale unsteady flow. Of these, the $C'_{l,\text{rms}}$ values for the combination stall type were nearly twice that for the pure thin-airfoil stall cases. It should be noted that the maximum value of the RMS lift coefficient did not necessarily coincide with the maximum mean lift coefficient, but the values are representative of the general trends. The trailing-edge stall, leading-edge stall, and their combination stall type exhibited much steadier characteristics.

Table 1   Summary of time-dependent lift data

Stall type	Airfoil	α_{stall}(deg)	$C_{l,\text{max}}$	$C'_{l,\text{rms}}$
Trailing-edge	MB253515	8.1	1.06	0.002
	Ultra-Sport	10.1	0.90	0.004
Leading-edge	E472	13.3	1.26	0.003
	NACA 2414	15.2	1.25	0.005
Trailing-edge/leading-edge	Clark-Y	14.0	1.37	0.006
	FX63-137	17.1	1.79	0.010
Thin-airfoil	NACA 64A010	10.1	0.89	0.060
	NACA 0009	10.8	0.92	0.060
	MA409	11.9	1.14	0.070
Thin-airfoil/trailing-edge	E387	13.9	1.31	0.120
	E374	12.9	1.15	0.160
	LRN-1007	16.1	1.50	0.180

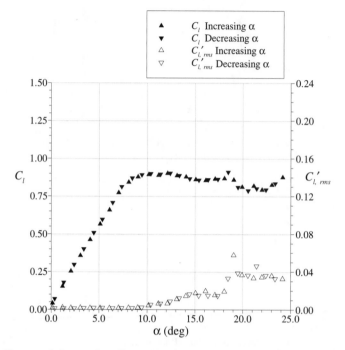

Fig. 5   Mean and fluctuating lift coefficient variation with angle of attack for the (trailing-edge stall) Ultra-Sport airfoil at _Re_ = 300,000.

Of particular interest here are the thin-airfoil and combination thin-airfoil and trailing-edge stall types, as these sections are more representative of very low Reynolds number airfoils. For example, the MA409 airfoil was designed for use in free-flight model competition with chord Reynolds numbers as low as 30,000.[24] However, it is useful to consider the trailing-edge stall and leading-edge stall characteristics to provide a baseline for the other cases.

Classic trailing-edge stall behavior was exhibited by the Ultra-Sport airfoil and is summarized in Fig. 5. The lift-curve slope remained constant until a few degrees prior to the stall. Past the first lift-curve peak, the lift remained approximately constant, or decreased only slightly. There was a steady rise in the RMS lift coefficient beyond stall, which was representative of the trailing-edge stall airfoils. The $C'_{l,\text{rms}}$ values at $\alpha \approx 20$ deg plateau at about 0.040, which was a trend observed for all of the airfoils tested. This high angle of attack range (i.e., $\alpha > 20$ deg) was always characterized by bluff-body shedding, as determined from the power spectra of both the lift and wake-velocity signals. Surface-oil flow visualization performed for this airfoil confirmed the trailing-edge stall classification as the boundary-layer separation point moved forward from the trailing edge as the angle of attack was increased into stall.

The behavior of the trailing-edge stall airfoils is contrasted with the leading-edge stall E472 airfoil. The lift data, Fig. 6, illustrate the sharp decrease in lift at stall that is a trademark of this stall type. The RMS lift also changed abruptly with the loss of lift, increasing by an order of magnitude from 0.003 to 0.030. It is also

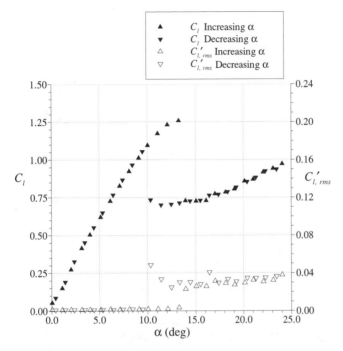

Fig. 6   Mean and fluctuating lift coefficient variation with angle of attack for the (leading-edge stall) E472 airfoil at *Re* = 300,000.

interesting to note that the $C'_{l,\text{rms}}$ remained at this high level until the mean lift was recovered as the angle of attack was decreased below stall. This hysteresis in the lift is fairly well known (e.g., see Ref. 25) and the present data suggest that the fluctuating lift exhibits similar behavior.

The mean and fluctuating lift characteristics of thin-airfoil stall airfoils are markedly different from the previous two cases. The NACA 64A010 airfoil exhibited the classic characteristics of thin-airfoil stall, as shown in Fig. 7. There was a distinctive decrease in the lift-curve slope at $\alpha = 4$ deg associated with the significant (streamwise) growth of a separation bubble and a gentle stall with maximum lift occurring at $\alpha = 10.1$ deg, followed by a gradual lift recovery with increasing incidence. The variation in $C'_{l,\text{rms}}$ was very different from the previous stall types as there was a significant increase as maximum lift was attained. Also, these values of $C'_{l,\text{rms}}$, approximately equal to 0.050 to 0.070, were an order of magnitude higher than those of the previous stall types. As shown in Fig. 7, the peak values of the RMS lift coefficient were actually in the stall region, that is, just beyond the first mean-lift-curve peak. The RMS lift coefficient decreased as the angle of attack was increased further. Here again, the similar value of 0.040 for the $C'_{l,\text{rms}}$ was seen for angles of attack in this range (i.e., $\alpha > 15$ deg). The $C'_{l,\text{rms}}$ spectra near stall contained low-frequency components as described for the LRN-1007 airfoil in Section I.

The MA409 airfoil had a more representative geometry for typical low Reynolds number airfoils than the NACA 64A010. Despite these differences in geometry, the mean and fluctuating lift were very similar, as illustrated in Fig. 8. The mean

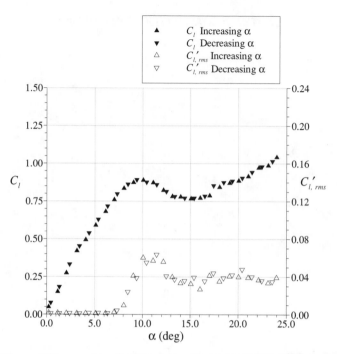

Fig. 7   Mean and fluctuating lift coefficient variation with angle of attack for the (thin-airfoil stall) NACA 64A010 airfoil at *Re* **= 300,000.**

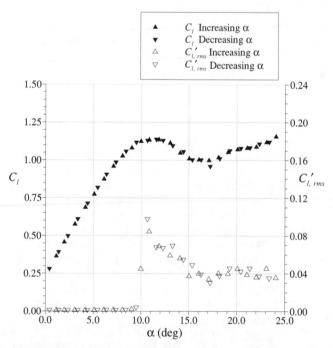

Fig. 8   Mean and fluctuating lift coefficient variation with angle of attack for the (thin-airfoil stall) MA409 airfoil at *Re* **= 300,000.**

lift variation was analogous to the NACA 64A010, except that the maximum lift coefficient (1.14) was higher, owing to the increased camber. The RMS lift variation was slightly higher ($\approx$0.080), having reached its peak slightly before the maximum mean lift coefficient. The fluctuating lift spectra in this range also exhibited distinct low-frequency components.

The highest levels of the fluctuating lift were observed for the combination thin-airfoil/trailing-edge stall airfoils. Two of these were the Eppler sections, E387 and E374, which are well known in the low Reynolds number airfoil community. The third is the somewhat unconventional LRN-1007 airfoil, which had a unique profile (see Fig. 3). Despite these differences in airfoil contour, the stall types were very similar and so were the unsteady characteristics of the stall. The E374 airfoil is considered first. The mean lift curve (Fig. 9) shows characteristics of thin-airfoil stall, such as the distinctive reduction in the lift-curve slope (at $\alpha \approx 6$ deg), caused by the significant (streamwise) growth of a leading-edge separation bubble. This conclusion was determined from the flow visualization results, which also showed substantial boundary-layer separation. The latter is a common feature of the trailing-edge stall type; thus the stall of the E374 was classified as a combination of thin-airfoil and trailing-edge stall. The RMS lift data had a narrow band, coincident with the maximum mean lift, where the levels were extremely intense, peaking at $C'_{l,\mathrm{rms}} = 0.160$. The RMS lift coefficient dropped off rapidly as the angle of attack was increased beyond maximum lift and the usual value of approximately 0.040 was attained. The highest levels of the fluctuating lift were characterized by quasi-periodic oscillations in the lift signal that were very similar in character to previous data [8,11–14] acquired for the LRN-1007 airfoil.

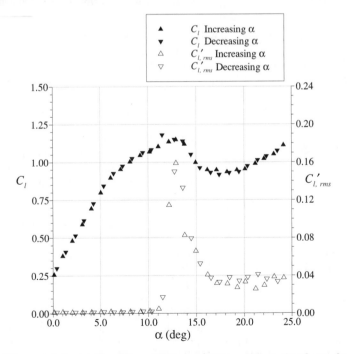

Fig. 9   Mean and fluctuating lift coefficient variation with angle of attack for the (combination thin-airfoil/trailing-edge stall) E374 airfoil at $Re = 300{,}000$.

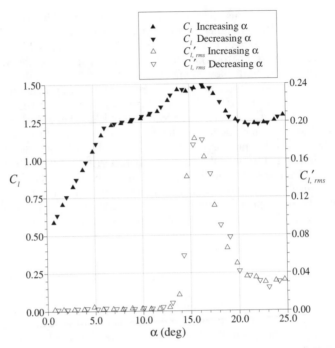

Fig. 10   Mean and fluctuating lift coefficient variation with angle of attack for the (combination thin-airfoil/trailing-edge stall) LRN-1007 airfoil at *Re* = 300,000.

The behavior of the mean lift curve for the LRN-1007 airfoil, Fig. 10, was very unique and obviously resulted from its unconventional geometry. The interpretation of lift variation with angle of attack is aided by flow visualization results from Heinrich[12] and Bragg et al.[14] The reduction in the lift-curve slope that occurred near $\alpha = 6.5$ deg was likely the result of a leading-edge separation bubble that first appeared on the upper surface near this angle of attack. The reason for the increase in the mean lift at $\alpha = 13$ deg is not clear but may have resulted from the boundary-layer separation location moving slightly downstream from $x/c = 0.60$ to 0.70. The decrease in lift beyond $C_{l,\max}$ and the subsequent gradual increase are characteristics reminiscent of the thin-airfoil stall type. In spite of the unique mean lift variation, the fluctuating lift behavior was analogous to the E374 airfoil data (Fig. 9) and to that of the E387 airfoil (not shown), except that the peak values of $C'_{l,\mathrm{rms}}$ (≈ 0.180) were higher than in these two cases.

The fluctuating lift data for the thin-airfoil stall and combination thin-airfoil/trailing-edge stall airfoils are summarized in Fig. 11. The RMS lift coefficient is plotted against angle of attack normalized to the stalling angle of attack (α_{stall}), as given in Table 1. The trends in Fig. 11 illustrate the difference in unsteady behavior for these two stall types. Airfoils of both stall types had large leading-edge separation bubbles, but the combination stall type (thin-airfoil/trailing-edge) also had large extents of separated flow from the trailing edge. This suggests that the extent of boundary-layer separation downstream of the leading-edge bubble may amplify or attenuate the intensity of the unsteady flow. This idea is further discussed in Section III.C.

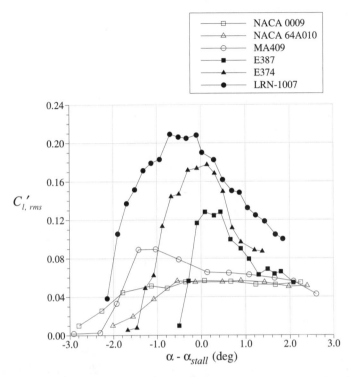

Fig. 11   Comparison of fluctuating lift coefficient variation with angle of attack normalized by α_{stall}.

B.   Frequency Content of the Fluctuating Lift

The previous discussion has shown that there is a distinct relationship between airfoil stall type and elevated $C'_{l,\text{rms}}$ levels near stall. That is, the thin-airfoil and combination thin-airfoil/trailing-edge stall cases had much higher RMS lift than did the other stall types. It was noted that these fluctuations were contained within very low frequency bands. This is illustrated in Fig. 12, which shows sample spectra of both the fluctuating lift (solid line) and wake-velocity (dotted line) signals for the thin-airfoil stall airfoils. The selected spectra shown are for angles of attack in the middle of the low-frequency oscillation range. The low-frequency energy was concentrated in the 2 to 5 Hz range and was much more apparent in the lift signal vs the wake velocity. The spectra for the NACA 64A010 and NACA 0009 airfoils were similar in that the low-frequency peaks had relatively large bandwidth, thus indicating that the lift fluctuations were distributed over a range of frequencies. In contrast, the low-frequency peak in the MA409 spectra was much narrower in bandwidth. The power spectrum magnitude was also larger than for the other two airfoils, which was consistent with the higher RMS values shown in Fig. 11. The lift spectrum for the MA409 airfoil also showed evidence of a first harmonic at approximately 7 Hz. The lift spectra for all three of these airfoils contained various broadband energy peaks for frequencies greater than 20 Hz, which did not occur in the wake-velocity spectra. An extensive study was

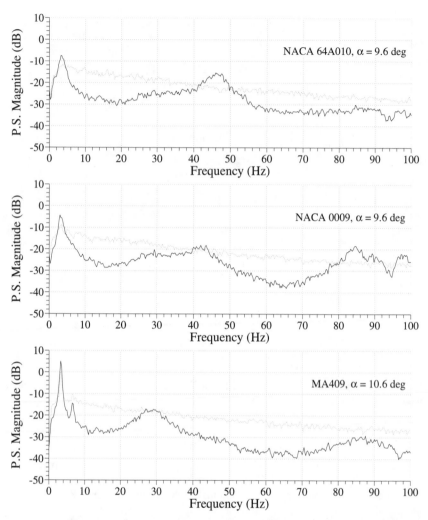

Fig. 12 Comparison of fluctuating lift (solid line) and wake-velocity (dotted line) power spectra for the thin-airfoil stall airfoils at *Re* = 300,000.

performed to determine that these peaks were attributed to structural contamination from the lift-balance and airfoil model and were not directly flow related.[16]

It has already been shown that the lift fluctuations were the most intense for the combination thin-airfoil/trailing-edge stall airfoils. For these airfoils, the low-frequency oscillation was clearly discernable in the wake-velocity signal as well as in the fluctuating lift signal. This is illustrated in Fig. 13 for angles of attack near the middle of the low-frequency oscillation range. The fundamental oscillation frequency and the first harmonic are clearly seen in the lift spectra for each airfoil. The power spectrum magnitude of the fundamental was greatest for the LRN-1007 airfoil, followed by the E374 and E387 airfoils. This is consistent with the RMS data shown in Fig. 11. Similar to the previous figure, the fluctuating lift spectra

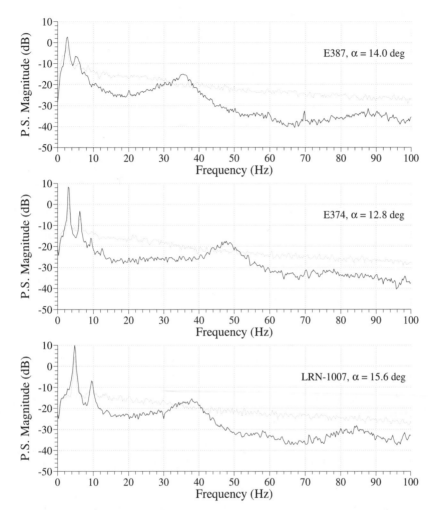

Fig. 13 Comparison of fluctuating lift (solid line) and wake-velocity (dotted line) power spectra for the combination thin-airfoil/trailing-edge stall airfoils at *Re* = 300,000.

also contained broadband peaks centered at frequencies greater than 20 Hz that do not occur in the wake-velocity spectra and were structural in origin.

The frequency corresponding to the fluctuating lift spectral peak was used to calculate the Strouhal number at several angles of attack for each airfoil. These data are summarized in Fig. 14, which shows the Strouhal number variation with angle of attack, normalized by α_{stall}. There was little difference in the nondimensional frequencies between the thin-airfoil stall and combination thin-airfoil/trailing-edge stall airfoils. The Strouhal number dependence on angle of attack was less for the thin-airfoil stall airfoils, but the general trends were otherwise the same. The one anomaly was for the LRN-1007 airfoil, where the Strouhal numbers tended to be higher. The dependence of Strouhal number on angle of attack is mostly due

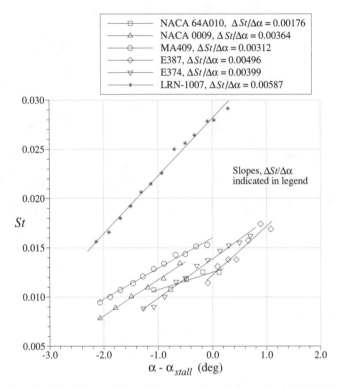

Fig. 14   Comparison of Strouhal number variation with angle of attack normalized by α_{stall}.

to an increase in the dimensional frequency and not the projected airfoil height. In the case of the LRN-1007 airfoil, the Strouhal number increases nearly 100% (from $\approx$0.015 to $\approx$0.030), while the projected airfoil height increases only about 17% [from sin(14.0 deg) to sin(16.5 deg)]. Despite the common trends shown in this figure, the nature of the Strouhal number behavior is yet unknown, but it has been considered in previous studies for the LRN-1007 airfoil.[8,11,12]

C.   The Role of the Separation Bubble

A common feature of the thin-airfoil and combination thin-airfoil/trailing-edge stall airfoils was the leading-edge separation bubble that formed prior to the onset of stall and the accompanying large-scale unsteady flow. The surface-oil flow visualization method proved useful for identifying these boundary-layer characteristics for each angle of attack leading up to the stall. The first evidence of a separation bubble on the NACA 64A010 airfoil was the reduction in the lift-curve slope near $\alpha = 4$ deg in Fig. 7. It is likely that there was already a very small bubble that "burst" into a "long" bubble as described by Tani.[26] This also resulted in an attendant increase in drag.[17] Figure 15 presents a summary of the flow visualization results for the NACA 64A010 airfoil. The plot reveals how the separation bubble reattachment progressed downstream in large increments as the angle of attack was

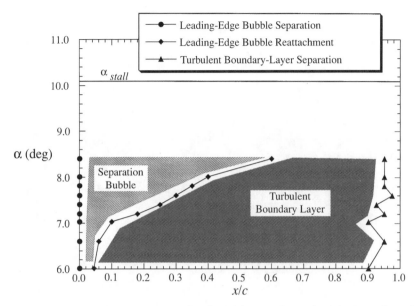

Fig. 15   Summary of upper surface flowfield features determined from surface-oil flow visualization for the NACA 64A010 airfoil at Re = 300,000.

increased, reaching nearly $x/c = 0.60$ at $\alpha = 8.4$ deg. In contrast, the turbulent boundary-layer separation, or trailing-edge separation, remained approximately in the same location ($x/c \approx 0.95$) as angle of attack was increased. These features fit the description of thin-airfoil stall type very well. It is important to note here that the flow visualization was performed over virtually the entire span of the airfoil model and that these features were observed to be uniform across the span. This was true for nearly all of the airfoils tested. Referring to Fig. 15, the separation bubble reattachment region in the oil-flow pattern became somewhat ambiguous ($\pm 0.05\,x/c$), for $\alpha > 7$ deg, and likely resulted from this region being unsteady. This is not unexpected, as Mabey[27] suggests that this region is the most unsteady part of the bubble, in terms of fluctuating pressure. In fact, the oil-flow visualization data only go up to $\alpha = 8.4$ deg, because the patterns at higher angles were very difficult to interpret. A comparison to the fluctuating lift data in Fig. 7 shows that the RMS values began to increase sharply near this angle of attack, thus confirming the onset of the unsteady flow.

The flow visualization data for the E374 airfoil illustrates the combination stall type classification. The summary plot, Fig. 16, shows that there was a separation bubble on the upper surface whose reattachment progressed downstream from $x/c = 0.05$ to $x/c = 0.24$ as the angle of attack was increased from 10 to 11.6 deg. Although this separation bubble was not nearly as large as that observed on the NACA 64A010 airfoil, it does fit the description of thin-airfoil stall. Also, given the change in lift-curve slope, at about $\alpha = 6$ deg, it is likely that a small leading-edge bubble had burst into a long bubble. This is further supported by the drag data.[17] Similarly, the oil-flow patterns on the E374 became somewhat ambiguous for angles of attack above 11.6 deg, which coincided with the rise in the fluctuating lift. The bubble reattachment region likewise became more ambiguous at these

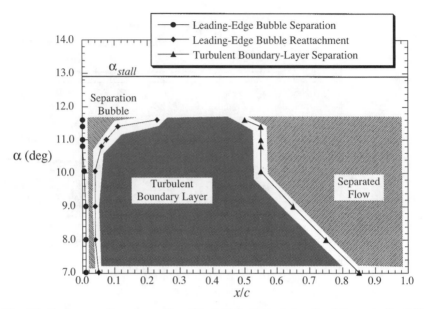

Fig. 16 Summary of upper surface flowfield features determined from surface-oil flow visualization for the E374 airfoil at _Re_ = 300,000.

higher angles. The flow visualization data also show that the there was substantial turbulent boundary-layer separation for the E374, which generally increased with angle of attack. This is a characteristic of trailing-edge stall. Therefore, these data clearly illustrate the combination thin-airfoil/trailing-edge stall type designation.

Previous research has shown that the separation bubble illustrated in Figs. 15 and 16 may play a signal role in the large-amplitude and low-frequency lift fluctuations for airfoils of this stall type. Bragg et al.[14] showed that the low-frequency oscillation on the LRN-1007 airfoil could be eliminated using zigzag tape as a boundary-layer trip. Flow visualization revealed that the zigzag tape eliminated the leading-edge separation bubble and effectively changed the stall type from a combination thin-airfoil/trailing-edge stall to a basic trailing-edge stall type.[13] This is further supported by Fig. 17, which shows the variation in the mean and fluctuating lift with angle of attack using the same zigzag tape. The leading edge of the tape was placed at $x/c = 0.01$ and the tape thickness was 0.036 in. The $C_{l,\max}$ was much lower (1.24 vs 1.50) and occurred at a lower angle of attack (13 vs 16.1 deg) than for the clean, or untripped, case (cf. Fig. 10). Accordingly, the $C'_{l,\mathrm{rms}}$ levels were also much lower in this region. What is peculiar about the tripped case was the increase in $C'_{l,\mathrm{rms}}$ beginning at $\alpha \approx 17$ deg, which was also accompanied by an increase in the mean lift. The magnitude of the fluctuating lift in this region was indicative of low-frequency unsteadiness. The power spectra showed low-frequency unsteadiness at 11 Hz, which converted to a Strouhal number of 0.07. This was considerably higher than the clean, or untripped, value of 0.02 (for $\alpha = 15$ deg) but was of the same order of magnitude. It was not clear if this flow regime was analogous to the low-frequency oscillation in the clean case, or if it was some sort of hybrid unsteadiness near the onset of bluff-body shedding. Unfortunately, the surface-oil flow visualization method did not work well at these high angles and was not

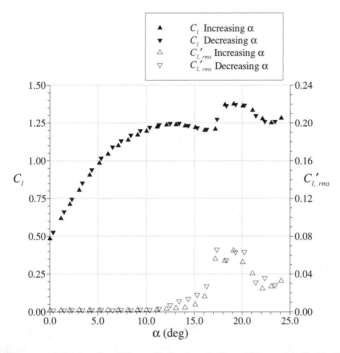

Fig. 17 Mean and fluctuating lift coefficient variation with angle of attack for the LRN-1007 airfoil with boundary-layer trip at $Re = 300,000$.

useful in determining the flowfield characteristics leading up to the elevated RMS lift values. It is clear that in the original low-frequency oscillation angle of attack range ($14 < \alpha < 16.5$ deg), the unsteadiness was eliminated along with the separation bubble.

These results reaffirm the conclusion of Bragg et al.[14] that the source of the low-frequency unsteadiness may be linked to the leading-edge separation bubble. Low-frequency unsteady behavior associated with separation bubble flowfields has been well documented in the literature. For bubbles where the separation location was fixed by geometry, such as for a backward-facing step or the leading-edge of a blunt flat plate, the low-frequency component was intermittent, occurring sporadically and not in a regular, oscillatory manner.[28–30] This was referred to as shear-layer flapping.[29] The bubbles that formed on these airfoils resulted from adverse pressure gradients. Gaster[31] noted a similar low-frequency perturbation for "long" bubbles formed on a flat plate with an imposed adverse pressure gradient. It is likely that the leading-edge bubbles on the thin-airfoil and combination thin-airfoil/trailing-edge stall airfoils were also of this type. The difference, of course, lies in the fact that the low-frequency oscillation on these airfoils was much more periodic than for the shear-layer flapping cases.

The low-frequency unsteady flow is further illustrated in the time series data of Fig. 18. Each plot shows the instantaneous lift coefficient and wake-velocity voltage. The angles of attack at which the data were acquired correspond to the power spectra of Figs. 12 and 13. For the E374 airfoil, the time traces show approximately three periodic cycles (both in the lift and wake-velocity voltage)

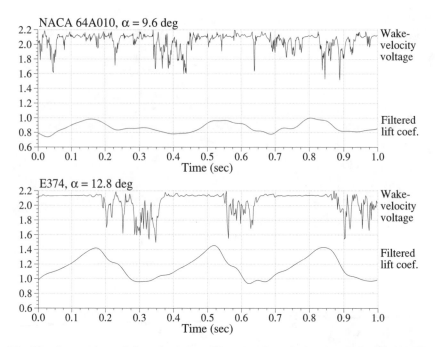

Fig. 18   Comparison of time-dependent lift and wake-velocity signals for the NACA 64A010 and E374 airfoils at *Re* = 300,000.

in the 1-s time trace. This agrees very well with the low-frequency peak in the spectra shown in Fig. 13 for this airfoil. The variation in the lift coefficient was nearly 40% of the mean and was due to the stalling and unstalling process. The smoke flow visualization revealed that the flowfield variation was analogous to the laser Doppler velocimetry data acquired for the LRN-1007 data in an earlier study.[15] During the oscillation, the upper surface flowfield became completely separated from the airfoil, as evidenced by the loss of lift. An attached boundary layer began to form near the leading edge and advanced downstream. Through this process the lift increased as the airfoil "unstalled," ultimately reaching a peak, before the separation process began again. The wake-velocity signal also shows this information, albeit not as clearly. The periodic defects in the time trace correspond to the low-lift, or stalled, condition. Since the flow was separated, the wake was very large and hence the hot-film sensor was immersed in a nonuniform retarded flow. As the flow reattached on the airfoil, the wake decreased in size and the sensor saw a faster, more uniform flow. The large variations in the time-dependent lift and wake-velocity signals illustrate the magnitude of the stall.

This is contrasted with the data for the thin-airfoil stall NACA 64A010 airfoil, also given in Fig. 18. The time-dependent lift trace is not nearly as periodic, nor does it contain large differences in the peak-to-peak variation. The same comments apply to the wake-velocity signal. The smoke flow visualization information for this and the other two pure thin-airfoil stall airfoils showed that the unsteady stalling process was different from that described above for the combination thin-airfoil/trailing-edge stall airfoils. For the thin-airfoil stall airfoils, the boundary-layer development (during the flow oscillation) was characterized by the formation and growth of a very large separation bubble. The stall occurred whenever the bubble failed to reattach, leaving a separated shear layer above the airfoil surface. In contrast, the unsteady flow for the combination thin-airfoil/leading-edge stall cases was characterized by boundary-layer development that led to the formation of a small leading-edge bubble. At least in the case of the LRN-1007 airfoil, it was clear (from the laser Doppler velocimetry data) that fully attached flow developed with the boundary-layer separation location moving downstream before the formation of the leading-edge bubble. The bubble grew in size until it merged with the trailing-edge separation and that led to complete separation over the entire upper surface.[15] For the pure thin-airfoil stall airfoils, it appeared that the flow always remained separated from the leading edge and that the lift fluctuations resulted from the formation of a separation bubble with reattachment, the movement of the reattachment location, and subsequent failure of the bubble to reattach. This process did not cause the catastrophic separation as observed for the combination stall airfoils. Nor did it result in as large lift fluctuations because a higher lift, fully unstalled condition did not develop.

These descriptions of the unsteady flow continue to show that the laminar-separation bubble was a key characteristic. Further, it is speculated that the trailing-edge separation leading up to the stall associated with the combination thin-airfoil/trailing-edge stall type contributed to this more severe case, thus influencing the higher magnitude of the lift fluctuations. The pure thin-airfoil stall airfoils, not exhibiting extensive trailing-edge separation leading up to the stall, did not stall as severely in the unsteady case, thus leading to lower levels of lift fluctuation. Therefore, the amount of trailing-edge separation may enhance or attenuate the level of unsteadiness. Whereas the extent of trailing-edge separation delineating these two stall types may cause the difference in the intensity of the unsteadiness, the frequencies tend to be similar.

IV. Summary and Conclusions

This study has considered the unsteady stalling characteristics of thin airfoils that may be suitable for micro air vehicle applications. Time-dependent lift measurements have been carried out on 12 airfoils with different stalling characteristics to determine the influence of stall type on low-frequency unsteadiness at stall. In addition to the lift data, wake-velocity measurements, spectral analysis, and flow visualization were also performed at a Reynolds number of 300,000. The lift-force fluctuations were determined from the root-mean-square value of the fluctuating lift after low-pass filtering the data to remove unwanted contributions from structural resonances.

The data presented in this chapter show a distinct relationship between stall type and low-frequency/large-scale unsteady flow. Airfoils having trailing-edge stall, leading-edge stall, and their combination stall type exhibited very low levels of the fluctuating lift near stall ($C'_{l,\mathrm{rms}} < 0.030$). However, airfoils with the thin-airfoil stall type were found to have highly unsteady stall characteristics with $C'_{l,\mathrm{rms}}$ values on the order of 0.060 to 0.080 near stall. The combination thin-airfoil/trailing-edge stall type had the highest levels of lift fluctuations at stall, nearly double that of the pure thin-airfoil stall cases. For both the thin-airfoil stall and combination thin-airfoil/trailing-edge stall types, the lift fluctuations occurred at very low frequency, having Strouhal numbers on the order of 0.02. A laminar-separation bubble was a common feature in the airfoil flowfields for both stall types. The bubble played a key role in the unsteady flowfields as its elimination caused a significant reduction in the intensity of the unsteady behavior. The presence of trailing-edge separation, in the case of the combination stall type, may amplify the unsteady effects because the lift fluctuations are much larger in magnitude for this case. The results showed that a classification of airfoils by stall type may offer an effective way of explaining and predicting observed or potential airfoil stall unsteadiness.

Acknowledgments

This work was funded, in part, through a NASA Graduate Student Researchers Program Fellowship. The authors wish to acknowledge K. B. M. Q. Zaman of the NASA Glenn Research Center for his contributions to this research. The authors also wish to acknowledge M. S. Selig of the University of Illinois for providing the wind-tunnel models and part of the apparatus used in this work.

References

[1]Shyy, W., Klevebring, F., Nilsson, M., Sloan, J., Carroll, B., and Fuentes, C., "Rigid and Flexible Low Reynolds Number Airfoils," *Journal of Aircraft*, Vol. 36, No. 3, May–June 1999, pp. 523–529.

[2]Foch, R. J., "Aero, Propulsion & Planform Considerations for Micro-UAVs," presentation at the Micro-UAV Workshop, 9 Nov. 1995.

[3]Jones, B. M. "An Experimental Study of the Stalling of Wings," *Aeronautical Research Council Reports and Memoranda*, No. 1588, Dec. 1933, p. 8.

[4]Farren, W. S., "The Reaction on a Wing Whose Angle of Incidence Is Changing Rapidly— Wind-Tunnel Experiments with a Short-Period Recording Balance," *Aeronautical Research Council Reports and Memoranda*, No. 1648, Jan. 1935.

[5]McCullough, G. B., and Gault, D. E., "Examples of Three Representative Types of Airfoil-Section Stall at Low-Speed," NACA TN 2502, Sept. 1951.

[6]Gault, D. E., "Boundary-Layer and Stalling Characteristics of the NACA 63-009 Airfoil Section," NACA TN 1894, June 1949.

[7]McCullough, G. B., and Gault, D. E., "Boundary-Layer and Stalling Characteristics of the NACA 64A006 Airfoil Section," NACA TN 1923, Aug. 1949.

[8]Zaman, K. B. M. Q., McKinzie, D. J., and Rumsey, C. L., "A Natural Low-Frequency Oscillation over Airfoils near Stalling Conditions," *Journal of Fluid Mechanics*, Vol. 202, 1989, pp. 403–442.

[9]Pfenninger, W., and Vermuru, C. S., "Design of Low-Reynolds Number Airfoils—I," AIAA Paper 88-2572-CP, 1988.

[10]Pfenninger, W., Vermuru, C. S., Mangalam, S. M., and Evangelista, R., "Design of Low-Reynolds Number Airfoils—II," AIAA Paper 88-3764-CP, 1988.

[11]Bragg, M. B., Heinrich, D. C., and Khodadoust, A., "Low-Frequency Flow Oscillation over Airfoils near Stall," *AIAA Journal*, Vol. 31, No. 7, July 1993, pp. 1341–1343.

[12]Heinrich, D. C., "An Experimental Investigation of a Low Frequency Flow Oscillation over a Low Reynolds Number Airfoil near Stall," M. S. Thesis, Dept. of Aeronautical and Astronautical Engineering, Univ. of Illinois, Urbana, IL, 1994.

[13]Balow, F. A., "Effect of an Unsteady Laminar Separation Bubble on the Flowfield over an Airfoil near Stall," M.S. Thesis, Dept. of Aeronautical and Astronautical Engineering, Univ. of Illinois, Urbana, IL, 1994.

[14]Bragg, M. B., Heinrich, D. C., Balow, F. A., and Zaman, K. B. M. Q., "Flow Oscillation over an Airfoil near Stall," *AIAA Journal*, Vol. 34, No. 1, Jan. 1996, pp. 199–201.

[15]Broeren, A. P., and Bragg, M. B., "Flowfield Measurements over an Airfoil During Natural Low-Frequency Oscillations near Stall," *AIAA Journal*, Vol. 37, No. 1, Jan. 1999, pp. 130–132.

[16]Broeren, A. P., "An Experimental Study of Unsteady Flow over Airfoils near Stall," Ph.D. Dissertation, Dept. of Mechanical and Industrial Engineering, Univ. of Illinois, Urbana, IL, 2000.

[17]Selig, M. S., Guglielmo, J. J., Broeren, A. P., and Giguere, P., *Summary of Low-Speed Airfoil Data—Volume 1*, SoarTech Publications, Virginia Beach, VA, 1995.

[18]Selig, M. S., Lyon, C. A., Giguere, P., Ninham, C. P., and Guglielmo, J. J., *Summary of Low-Speed Airfoil Data—Volume 2*, SoarTech Publications, Virginia Beach, VA, 1996.

[19]Lyon, C. A., Broeren, A. P., Giguere, P., Gopalarathanam, A., and Selig, M. S., *Summary of Low-Speed Airfoil Data—Volume 3*, SoarTech Publications, Virginia Beach, VA, 1997.

[20]Giguere, P., and Selig, M. S., "Freestream Velocity Corrections for Two-Dimensional Splitter Plates," *AIAA Journal*, Vol. 35, No. 7, July 1997, pp. 1195–1200.

[21]Broeren, A. P., and Bragg, M. B., "Low-Frequency Flowfield Unsteadiness during Airfoil Stall and the Influence of Stall Type," AIAA Paper 98-2517-CP, 1998.

[22]Coleman, H. W., and Steele, W. G., *Experimentation and Uncertainty Analysis for Engineers*, Wiley, New York, 1989, pp. 40–118.

[23]McGhee, R. J., Walker, B. S., and Millard, B. F., "Experimental Results for the Eppler 387 Airfoil at Low-Reynolds Numbers in the Langley Low-Turbulence Pressure Tunnel," NASA TM 4062, Oct. 1988.

[24]Selig, M. S., Giguere, P., Guglielmo, J. J., and Broeren, A. P., "Wind Tunnel Test Results of Five Free Flight Airfoils," The 28th Annual National Free Flight Society Symposium Report, June 1995, pp. R1–R14.

[25]Selig, M. S., Guglielmo, J. J., Broeren, A. P., and Giguere, P., "Experiments on Airfoils at Low-Reynolds Numbers," AIAA Paper 96-0062, Jan. 1996.

[26]Tani, I., "Low-Speed Flows Involving Separation Bubbles," *Progress in Aeronautical Sciences*, Vol. 5, 1964, pp. 70–103.

[27]Mabey, D. G., "Analysis and Correlation of Data on Pressure Fluctuations in Separated Flow," *Journal of Aircraft*, Vol. 9, No. 9, Sept. 1972, pp. 642–645.

[28]Simpson, R. L., "Turbulent Boundary-Layer Separation," *Annual Review of Fluid Mechanics*, Vol. 21, 1989, pp. 205–234.

[29]Driver, D. M., Seegmiller, H. L., and Marvin, J., "Time-Dependent Behavior of Reattaching Shear Layers," *AIAA Journal*, Vol. 25, No. 7, July 1987, pp. 914–919.

[30]Kiya, M., and Sasaki, K., "Structure of a Turbulent Separation Bubble," *Journal of Fluid Mechanics*, Vol. 137, 1983, pp. 83–113.

[31]Gaster, M., "The Structure and Behaviour of Separation Bubbles," *Aeronautical Research Council Reports and Memoranda*, No. 3595, March 1967.

Part II.   Flapping and Rotary Wing Aerodynamics

Chapter 11

Thrust and Drag in Flying Birds: Applications to Birdlike Micro Air Vehicles

Jeremy M. V. Rayner*
University of Leeds, Leeds, United Kingdom

Nomenclature

B	=	wingspan (m)
M	=	total mass (kg)
S	=	wing area (m^2)
P_{ind}	=	induced power in flapping flight
$P_{ind,r}$	=	induced power in gliding or rigid-wing flight
P_{met}	=	metabolic power (i.e., total power including heat loss) in flight (W)
P_{mr}	=	mechanical flight power at maximum range flight speed (W)
P_{par}	=	parasite power (to overcome body drag) in flapping flight
P_{pro}	=	profile power (to overcome wing drag) in flapping flight
$P_{pro,r}$	=	profile power in gliding or rigid-wing flight
V_{mr}	=	maximum range flight speed (m/s)

I. Introduction

THE majority of development work on animal-based micro air vehicles (MAVs) has so far concentrated on the insect model.[1-4] Despite the challenging problems in mimicking the complex unsteady aerodynamic mechanisms of flapping flight at low Reynolds number and small scales, insects represent a valuable model for MAVs because of the advantages of small size and low power to weight ratio.[4] The relative simplicity of the insect design has prompted Michelson and Reece[2] to propose relatively large insectlike MAVs of size ($\sim$50 g) comparable to that of small birds. By contrast, bird-scale or birdlike MAVs have received less attention,[5-7] despite the considerable and enduring interest in model flying vertebrates[8-11] and in ornithopters.[6]

Birds differ from insects in three major respects. First, they are larger, ranging in mass from about 2 g to about 15 kg (for flying species), or in wingspan from 0.15

---

*Alexander Professor of Zoology, School of Biology.

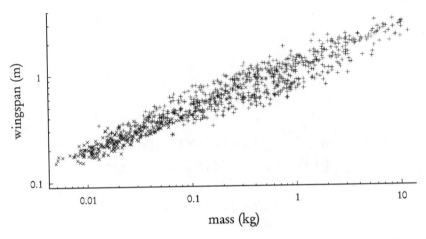

Fig. 1   One aspect of the design space of flying birds. The graph shows the range of body mass and wingspan occupied by extant flying birds, omitting hummingbirds, on log–log axes. Note the relatively narrow band within which flight morphology in birds has evolved. Bats are generally similar, although they do not reach large overall sizes. Extinct pterosaurs tended to have rather larger wingspans than comparably sized birds. For more details see Rayner.[12,18]

to 3 m (Fig. 1). The lower end of the size range overlaps with a few of the largest insects, but the majority of birds are much larger;[12] one consequence of this is that most birds can achieve normal flight with quasi-steady aerodynamic mechanisms, without resource to novel mechanisms of lift generation. Second, birds flap with actively deformed wings with muscles and joints within the wing surface, rather than with deformable insect wings controlled from the wingbase. Third, birds are dominated by mechanisms to reduce drag and improve aerodynamic performance, because they are more constrained by available power requirements than are the much smaller insects.

One reason for the relatively limited aerodynamic application of the bird model to MAVs may be the perceived difficulty of implementing a copy of the deformable, feathered avian wing. Another—although one which with modern developments in actuator technology is probably tractable—is provision of an effective power source. A third reason, which has been the subject of less attention, is that flapping bird or ornithopter models may have failed to be sufficiently "birdlike" to have captured all of the facets of performance optimization that have evolved in birds. In this chapter I explore three aspects of avian adaptation that are probably essential in the design of an efficient flapping MAV with birdlike morphology, capable of achieving birdlike levels of flight performance. These are: 1) mechanisms of generating thrust, 2) mechanisms of reducing drag, and 3) design of the wingtip.

Flapping birdlike MAVs could potentially have a number of advantages over alternatives with different mechanisms of propulsion. Compared to insect-scale devices they will enjoy greater range or endurance, will be less affected by weather and climate, and will be able to fly in open air. With appropriate design the thrust-to-weight ratio is also lower, and flight speeds are higher, but the device is small enough to benefit from aerodynamic advantages of flapping lift and thrust generation, without compromising the application by excessive size or weight. It is generally accepted that at these scales flapping wings are more effective than rotating propellers or autogyros.[4]

Here I concentrate on steady, level flapping flight, at "intermediate flight speeds." (See Section II.A.) This is probably the primitive flight mode in evolution,[13] and for much the same reasons it is likely to be the easiest flight mode to emulate. In normal flight of a relatively long-winged bird the wingbeats are of modest amplitude, there is only limited pro- and supination (i.e., variation in pitch) of the wing about its leading edge, and mechanical power requirements are at their lowest. I do not suggest that this level of performance is all that might be required of an effective avian MAV. However, this is the easiest flight mode, and accordingly this is the problem that should be solved first. As in birds, a sophisticated control system tolerant of the environment (e.g., of gusts) is essential, as also are the abilities to take off and land at slow flight speeds, possibly in uncertain or highly variable conditions, and (for some applications) also to hover. Slow flight and hovering demand more complex wing movements with greater deformation of the wing surface to control the wake vortices than are necessary in normal cruising flight. For a flapping machine based on birds, these flight patterns pose significantly more challenging problems than those considered here.

II. Avian Flight Performance

A. The Power Curve

Extensive theoretical models, and a limited number of experiments, have addressed the question of power consumption in flying birds (for a recent review see Rayner[14]). Because the energy demands of flight are major components of ecological and physiological energetics, flight power has become the key currency for quantification of flight performance and for the understanding of some of the life history decisions made by birds. This has perhaps been to the detriment of consideration of other important quantities that may constrain flight, such as the force available from the flight muscles or the breaking stress of muscles, tendons, and components of the wings.

It is generally agreed that under optimum conditions the mechanical power output required for steady level flight follows a U-shaped curve, which has much in common with comparable curves for low-speed aircraft[14] (Fig. 2). The mechanical power—which is equivalent to the rate of increase of the kinetic energy of the air caused by the passage of the bird—differs from aircraft practice in incorporating explicitly the energy cost of generating thrust. This has the effect of flattening the minimum of the curve and increasing the power required at higher speeds. By comparing this additional cost of thrust to the cost of weight support a speed-dependent aerodynamic efficiency of flight, comparable to the Froude efficiency of a swimming animal,[15] can be defined. Birds are most efficient around the minimum of the power curve (Fig. 2). However, this efficiency has little real biological or physical meaning because any bird requires both thrust and weight support to be able to fly, and both weight support and thrust are obtained from the aerodynamic lift forces on the wings.

The shape of the power curve is one of the reasons for the suggestion cited above that flight evolved first at cruising speeds (near or slightly above the minimum of the curve) and also that the first problem to tackle for an MAV is flight at speeds around the minimum of the curve, where power for level flight is least.

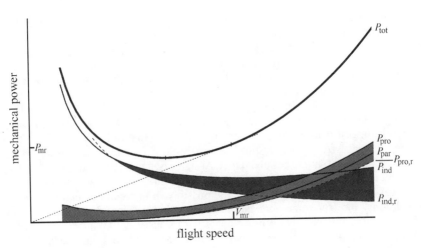

Fig. 2   A typical curve of mechanical power against speed for a bird flying with the continuous vortex gait. The curve is U-shaped, because induced power (P_{ind}) falls as speed increases, but parasite (body; P_{par}) and profile (wing; P_{pro}) drags and powers rise. The model calculations include the cost of generating thrust, which is indicated by the shaded regions, in comparison with the induced and profile powers ($P_{ind,r}$ and $P_{pro,r}$) for a comparable fixed and rigid wing. For further discussion of power curves see Rayner.[14] Maximum range speed V_{mr} and power P_{mr} [Eqs. (1) and (2)] are defined as the speed where the tangent from the origin meets the total mechanical power curve.

In birds, all the work required for flight is produced by the flight muscles that move the wings. Muscle has relatively low efficiency (at best around 25%), and substantial additional energy is released from the bird as heat. Considerable debate surrounds models attempting to estimate heat radiation; it is generally assumed implicitly that in any bird the physiological efficiency is constant, independent of flight speed, and therefore the total power required for flight also follows a U-shape (e.g., Pennycuick[16]). However, results from the few available experiments do not always confirm this hypothesis, although they suggest that it may be appropriate for some birds. Given that a bird controls speed by variation of wingbeat kinematics, that wingbeat frequency and amplitude are directly related to muscle contraction rate, and that muscle efficiency is related to muscle strain rate, it is to be expected that physiological efficiency will vary with flight speed in a manner that is difficult to predict. It is possible that in some birds this efficiency represents a significant constraint on the flight speeds that can be selected.[17]

These constraints on efficiency may not be important for an MAV if the actuator(s) responsible for wing movement can maintain efficiency while varying wingbeat kinematics appropriately. Mechanical efficiency of the power source may impose important design constraints if the speed envelope is not to be too tightly restricted.

B.   Dependence of Performance on Size and Design

Modern actuators may well be able to achieve optimum efficiencies appreciably better than the 10–25% values of birds, but losses from drive systems, joints, etc.

could be greater than for highly effective natural systems. For these reasons direct comparisons with bird performance may not be very helpful. However, values for birds are given here as initial design paradigms for optimal MAV performance. Calculations by the continuous vortex wake model of Rayner[18] give estimates for maximum range speed V_{mr} (Fig. 2) and mechanical power P_{mr} at that speed, in m/s and W respectively, of

$$V_{mr}(m/s) = 12.49 M^{0.190}$$

$$P_{mr}(W) = 14.95 M^{1.161} \tag{1}$$

in terms of total mass M (kg) alone, or

$$V_{mr}(m/s) = 10.00 M^{0.413} B^{-0.553} S^{-0.095}$$

$$P_{mr}(W) = 27.21 M^{1.590} B^{-1.818} S^{0.275} \tag{2}$$

in terms of total mass, wingspan B (m), and wing planform area S (m^2).[19] This speed V_{mr} should be interpreted as characterizing the power curve, and not necessarily as the speed at which a bird (or an MAV) should fly. Provided that an MAV can achieve levels of drag as low as those of birds, these equations should form realistic initial estimates of performance. It is interesting to note that the scaling indices with body mass are very close to predictions based on isometric scaling for general flying bodies in air.[20]

Measurements of *total* power for flight in birds, measured as the total rate of metabolic energy uptake P_{met}, and therefore including energy radiated as heat, are more varied and are sensitive to flight pattern and to phylogeny and also (worryingly) to measurement technique;[19] accordingly these are difficult to interpret. An estimate of total power from measurements of 64 birds in cruising flight[19] is

$$P_{met}(W) = 72.05 M^{0.912} \tag{3}$$

or

$$P_{met}(W) = 114.61 M^{1.145} B^{-1.225} S^{0.523} \tag{4}$$

Total power is at least five times higher than the mechanical power component. Efficiency appears to increase with size, ranging from around 10% in small passerine birds to around 22% in larger flapping species; these estimates are consistent with recent direct measurements of efficiency in starlings.[14,21]

These formulas make it possible to estimate the power-to-mass ratio for birds or for machines that can attain performance comparable to birds. For *mechanical* power output this ranges from about 7 W/kg in small ($\sim$0.01 kg) birds to about 15 W/kg in larger ($\sim$1 kg) birds. To achieve this performance requires birdlike morphology and birdlike wingbeat frequencies; longer wings with higher aspect ratio will result in lower power and better endurance and can generate thrust with lower frequencies, but at the expense of lower speed.[18] If the power source represents an average 15% of total mass, as in birds, then the required mechanical power-to-mass ratio for sustained flight is typically in the range 100–200 W/kg (actuator, but not fuel, mass). Much higher values—up to two to three times this—may apply temporarily in takeoff flight. These levels of power output are relatively high but are not beyond the range that can be foreseen for small actuators.[2] *Metabolic* power-to-total body mass ratio for birds, which includes energy released as heat, ranges from about 100 to 70 W/kg, *decreasing* with size. Technical actuators and modern power sources may well be able to achieve markedly higher efficiencies than birds; if so, P_{met} would overestimate total power and underestimate endurance.

III.   Thrust Generation

The purpose of flapping the wings in natural flight is to generate a mean forward thrust while simultaneously supporting the weight. Flying animals do not have separate engines, and with flapping wings the force vectors for propulsion and weight support cannot be separated.

It is not difficult for a flapping wing to generate thrust, but it is important that it does so economically. It must generate thrust with a low drag penalty (therefore minimizing power) and with simple low-loss movement mechanisms. Emulating the latter is probably the most challenging problem in design of a flapping MAV. In addition to being the sole source of useful force in flight, the wings are also the prime control mechanism in flight. Flapping should not impair control performance.

Selective pressures for efficient thrust generation have led in birds and bats to the evolution of two distinct flapping gaits.[18] These patterns of wing movement and of the associated force generation represent two ways in which an animal can maximize the sum of the mean horizontal component of lift with the (negative) induced drag while maintaining the vertical component of lift. Both represent minimum energy states for the wake vortices. In the *vortex ring* gait, characteristic of slow flight, the wake is composed of ring vortices, whereas in the *continuous vortex* gait, characteristic of cruising flight in long-winged birds, the wake is a pair of sublinear trailing vortices that approaches the line vortices of a fixed wing (Fig. 3). Both are conditions in which the least induced wake energy is required for momentum transport. It is therefore reasonable that flow visualization experiments have so far revealed no other wake geometries in sustained steady flight.

The most important characteristic of the two gaits is that the bound circulation on the wings is held as close to piecewise constant as possible. In the vortex ring gait circulation is constant during the downstroke and falls to zero during the upstroke during which no lift is produced on the wings. In the continuous vortex gait circulation is constant throughout, and a net thrust is obtained by flexing the wrist in the upstroke (Fig. 4). Piecewise constant circulation is a key adaptation as it either eliminates spanwise shed vorticity or confines it to short time periods, and it therefore prevents potentially high induced drag from unfavorable interactions of spanwise vortices behind the trailing edge with bound vorticity on the wing. The emphasis is on the three-dimensional nature of the flow, dominated by wingtip trailing vortices, rather than on a predominantly two-dimensional flapping or heaving wing, for which the sole source of thrust is time variation in circulation. Neither of the two gaits corresponds to conventional models of flapping for which sinusoidal variation of circulation is a key component[6,9,22] (see also Rayner[23]). Fortuitously, piecewise constant circulation enables modeling of induced drag and therefore performance by free vortex theory[23] in a way that is not possible with periodically varying circulation. If the hypothesis is correct then flight with constant circulation and flexed wrist in the upstroke is advantageous. I advance this as a design goal for a birdlike MAV.

The key problem with technological application of this model is that we have yet to determine how birds achieve constant bound circulation (or how close they

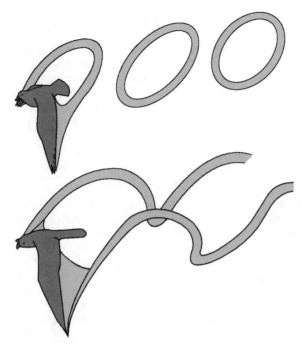

Fig. 3 Sketches showing wake vortex structure in the two gaits typical of birds in steady, level forward flight (after Rayner[18]). (Top) Vortex ring gait, used by all birds in slow flight, and by shorter-winged species at all speeds. (Bottom) Continuous vortex gait, characteristic of longer-winged birds in forward flight. The continuous vortex gait is simpler, requiring only flexure of the wrist in addition to flapping of the whole wing (see Fig. 4). It is this pattern that is proposed to be appropriate for birdlike MAVs.

Fig. 4 Wing flexure during the upstroke in flight of long-winged birds using the constant circulation gait.[18] The wing is flexed at the wrist so that the wingtip moves back relative to the proximal leading edge, and therefore the maximum upstroke wingspan is reduced. Because bound vortex circulation remains constant, the bird achieves a positive mean thrust. (After Steinbacher.[41])

are able to approach this state). Sophisticated control of angle of attack is essential, although this might be achieved passively by elastic deformation of the wing feathers and/or by semi-automatic articulation mechanisms in the shoulder, elbow, and wrist.[24] The magnitude of the potential performance savings from achieving the optimum thrusting wingbeat is also unclear: How closely does an MAV need to approach the condition of constant bound circulation in order to minimize induced drag?

IV. Drag Reduction

The counterpart to optimization of thrust is reduction of drag. Birds have a complex, feathered body shape that is not intuitively a low-drag outline. Although bodies are generally smooth in profile and streamlined, they do not have the slender (fusiform) axisymmetrical shape expected of a low-drag technical body with maximum included volume for minimum total body profile and parasite drag.[25]

Regardless of this, flying birds seem to be able to obtain very low drag values, significantly lower than would be expected from technical comparisons. Recent measurements on frozen bird bodies mounted in a wind tunnel[26,27] show that body drag is of the order of 75% of that normally expected for a body of comparable size and shape, but that this is not the result of the feathered body surface. Test model bird bodies of the same shape but with smooth surface texture have lower drag than natural feathered bodies. Similar results have been found for penguin bodies in water[28] and for starling models.[29-31] Body contour feathers have often been claimed to be a means of reducing drag, probably by inducing a laminar–turbulent transition, and so the result that feathers appear to *increase* drag is initially surprising. A partial explanation is that the bodies are frozen, and mechanisms for active drag reduction by control of feathers[32] are eliminated. Also, it is difficult in a mounted frozen bird to eliminate all irregularities in the integument and to ensure that airflow throughout the body boundary layer is optimal. However, equivalent irregularities, as well as irregularities in the ambient air, are present in natural conditions with real birds. The explanation probably lies in other factors: Birds have feathers for reasons other than simply for flight, the most important of which is thermoregulation. In flight, little heat is dissipated through the body feathers[21] owing to the low conductivity of the contour feathers. A first mechanism is that control of heat flow may be more significant than drag reduction. A second mechanism is that the ability of feathers to flex elastically passively or actively ensures a near-smooth body outline at all times, and in particular it provides a smooth fairing around the base of the wing regardless of wing posture. This may eliminate potentially large and dynamic drag forces resulting from wing–body interactions. Accordingly, a birdlike MAV may not require a "feathered" body surface, but it is probably important to design flexible fairings at the wing base.

I hypothesize that the principal mechanism for low drag in birds is two particular aspects of body shape. The first is that body profile is laterally symmetric, not axisymmetric, and that in typical species there are depressions behind the head and a rapid narrowing of the body at the base of the tail. In both of these regions trapped separations appear (Fig. 5). The second is the role of the tail, which provides a lifting and/or control surface in maneuvers and slow flight but is probably more important in normal flight in controlling flow over the proximal part of the body (Fig. 5). These shape features appear to function by manipulating the boundary layer of the

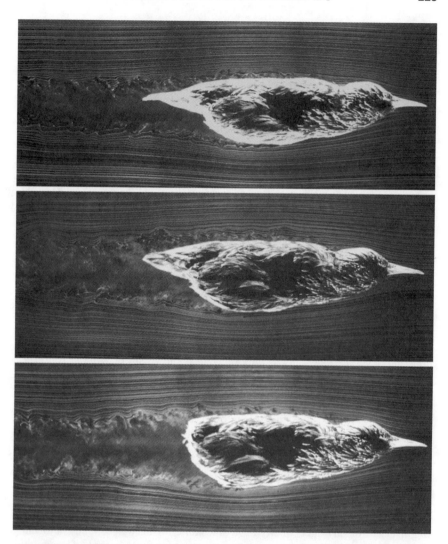

Fig. 5 Flow visualization around a mounted, wingless starling body in a wind tunnel at 9 m/s ($Re = 3.5 \times 10^5$). The boundary layer over most of the body is turbulent, and the turbulence transition is triggered by a laminar-separation bubble in the neck, which is part of a scarf vortex that also extends back as a vortex pair alongside the body. Details of wake are not visible owing to time of exposure and speed of flow. (Top) Intact body; there are shallow regions of reattached separation immediately posterior to the head on the dorsal surface, more posteriorly on the ventral surface, and at the base of the tail. (Middle) Body with main tail feathers removed, but leaving feathers forming a fairing or wedge at the base of the tail; the separation region ventral to the tail no longer reattaches (confirmed by pressure measurements). (Bottom) Body with all tail feathers removed. From top to middle to bottom the height of the wake rises, and the total body drag increases by $\sim 45\%$. The tail feathers act as a wedge/splitter plate,[42] controlling flow over the entire posterior portion of the body and reducing total drag. (From Maybury and Rayner.[27])

body, controlling separation and laminar–turbulent transitions with a sensitivity that would be impossible on a smooth axisymmetric spindle-shaped body.[27] The depression behind the head is responsible for development of a scarf vortex, which persists over the body and tail, and is responsible for tail lift forces.[33] It seems that the remarkable geometry of the body, which lacks either axial symmetry or a continuous convex surface, is responsible for stabilizing the development and location of trapped vortices around the body, serving both to control separation and to reduce total body drag to extremely low values. If this mechanism can be evaluated and copied, it may have far-reaching applications, not only in reducing drag in birdlike MAVs but also in the design of larger, low-drag bodies for technical applications.

Bird body shape is not solely the result of aerodynamic optimization; it also reflects biological factors. Birds have evolved from nonflying quadrupedal animals (Tetrapoda), and bird morphological evolution is constrained by Bauplan constraints resulting from this history. The wings are formed from the forelegs, and the hind legs and their articulation form the rear end of the body. The body cannot simply taper to a streamlined end because of the volume and geometry of the leg musculature and the pelvis. Equally, space is required for the guts and other internal organs within the body cavity. The geometry of the body, wings and tail, and internal organs is also dictated by considerations of stability in pitch during flight.[33] MAVs may be free of some of these constraints.

These mechanisms appear to be responsible for the low drag in birds, but they can be only a part of the full story. As yet my colleagues and I have measured drag and developed these ideas only on rigid bodies, with or without rigid (i.e., gliding) wings. In flapping flight, wing–body interactions associated with wing movements and highly dynamic lift and induced drag forces on the proximal part of the wings may have significant effects on the body boundary layer and therefore on body drag. It would be remarkable if body shape and aerodynamics had not evolved to capitalize on any potential for further drag reduction associated with these little-understood aerodynamic processes.

V. Wing Shape

Flight feathers are a key adaptation of birds and are the main feature that distinguishes birds phylogenetically from dinosaurs.[13] The contour feathers forming the aerodynamic surface of the wing are essential for flight in birds. Despite this, I want to make the striking proposal that on the wing, as on the body, feathers are not essential to birdlike flight. Mimicking the individual feathers with the same combination of strength, elasticity, controlled deformability, and lightness on birdlike scales would be a considerable technological challenge. Flight feathers allow a wing to change shape both passively and actively while still generating aerodynamic force. The most significant function in flapping MAV applications is to allow birds to achieve flapping flight with the constant circulation, low-drag wake that optimizes thrust production. A feathered wing surface is not essential for thrust by these mechanisms, if a suitable flexible alternative wing design can be established. Other aspects of avian morphology, such as a scalloped trailing edge,[34] significant chordwise camber with a concave ventral surface, separation control on the upper wing surface,[35] a notched trailing edge distal to the wrist,[36] and leading-edge structures contributing to lift at high incidence and/or to separation control[37] are likely to have considerable practical benefit if they can be copied, but none of these has to be constructed with feathers. Feathers are the structural elements of the wing for evolutionary as much as aerodynamic reasons, and that viable wings for flapping can be designed in other ways is shown by the membranous wings of bats and the stiffened membranous wings of pterosaurs.[38]

Wing shape varies widely in birds (Fig. 1), as a balance between conflicting performance benefits (e.g., speed, economy, load-carrying, maneuverability, take-off, ability to exploit cluttered habitats[12,18]). Wings differ in shape from normal fixed wing optima (e.g., elliptic loading for minimum induced drag) because of constraints on wing weight and pressures to maximize thrust in flapping flight.[39] Man-made model aircraft and gliders often have much longer wings than birds. Two main constraints appear to act to preclude long wings in birds, namely the structural strength of the loaded wing (and the accompanying pressures on wing weight) and the increased exposure to predators associated with low flight speeds. Neither of these need be a problem for an MAV, but broadly similar tradeoffs between design pressures apply to MAV design. To exploit the constant circulation gait a long, high-aspect-ratio wing is desirable but not essential. MAVs may usefully have rather longer wings than birds (see Fig. 1), benefiting from lower speeds, lower wingbeat frequencies, and lower power but greater endurance.

VI. Conclusions

The performance of an MAV at bird scales is critical because of constraints on power and fuel supply, as previously described, and also because of considerations such as structural strength. It is slavishly impractical to copy the aerodynamic features of birds, and it is unlikely that this will result in the "best" MAV designs, but birds offer a range of aerodynamic adaptations that could usefully be applied to improve MAV performance. No single aerodynamic feature is responsible for the effective flight capacities of birds; rather, birds benefit from a combination of different adaptations, each of which contributes separately to lift maximization or to friction, profile, or induced drag reduction. By taking inspiration from birds comparable reductions in power requirement might be achieved in an MAV. Among factors that are likely to be particularly important are: control of wingbeat kinematics to ensure lift and thrust generation by the continuous vortex gait; division of wing movements into independent dorso-ventral flapping and "wrist" flexure; streamlined but nonaxi-symmetric body design with a tail acting as a splitter plate, so that body boundary layer airflow is constrained; and wings designed to match performance (speed, power) to design constraints and to minimize wing drag and maximize lift.

Two significant issues remain: 1) the flight envelope and 2) considerations of control and stability. Some degree of flexibility of performance is desirable for an MAV as much as for a bird; this may mean the capacity to fly at a range of flight speeds, to take off and land, to maneuver, and to cope with an erratic aerial environment. This may be difficult to emulate, but it need not be a crucial concern for an initial flapping MAV. Control is more significant. Birds are typical of flying animals, but they are unusual compared to technical analogs, in that the same aerodynamic surfaces are used for weight support, for thrust, *and* for control. Contrary to much debate, the tail plays a relatively minor role in maneuverability in many species. The subtle avian control system is poorly understood, and replicating it with technical devices able to sense a range of air flows[40] and to make subtle control movements with rapid response rates is a major challenge. This has proved to be perhaps the greatest design limitation for existing flapping wing models. Control movements must not impair lift production, while at the same time the machine must have the capacity to accommodate any associated distortion of the flight path associated with variation of wing lift to give control forces.

Acknowledgments

This research into flapping wing aerodynamics has been funded by BBSRC Grant S30843. I am grateful to Will Maybury and Laurence Couldrick for permission to cite results from work in progress.

References

[1]Smith, M. J. C., "Simulating Moth Wing Aerodynamics: Towards the Development of Flapping Wing Technology," *AIAA Journal*, Vol. 34, No. 7, July 1996, pp. 1348–1355.

[2]Michelson, R. C., and Reece, S., "Update on Flapping Wing Micro Air Vehicle Research: Ongoing Work to Develop a Flapping Wing, Crawling 'Entomopter'," *Proceedings of the 13th International Conference on Remotely Piloted Vehicles*, Royal Aeronautical Society, London, 1998, Paper 30, p. 11.

[3]Spedding, G. R., and Lissaman, P. B. S., "Technical Aspects of Microscale Flight Systems," *Journal of Avian Biology*, Vol. 29, No. 4, Dec. 1998, pp. 458–468.

[4]Ellington, C. P., "The Novel Aerodynamics of Insect Flight: Applications to Micro-Air Vehicles," *Journal of Experimental Biology*, Vol. 202, No. 23, Dec. 1999, pp. 3439–3448.

[5]DeLaurier, J. D., "An Aerodynamic Model for Flapping Wing Flight," *Aeronautical Journal*, Vol. 97, No. 964, April 1993, pp. 125–130.

[6]DeLaurier, J. D., "The Development of an Efficient Ornithopter Wing," *Aeronautical Journal*, Vol. 97, No. 965, May 1993, pp. 153–162.

[7]Spedding, G. R., and DeLaurier, J. D., "Animal and Ornithopter Flight," *Handbook of Fluid Dynamics and Fluid Machinery*, edited by J. A. Schetz, and A. E. Fuhs, Wiley, New York, 1995.

[8]von Holst, E., "Über 'künstliche Vögel' als Mittel zum Studium des Vogelflugs," *Journal für Ornithologie, Leipzig*, Vol. 91, 1943, pp. 406–447.

[9]von Holst, E., and Küchemann, D., "Biologische und aerodynamische Probleme des Tierfluges," *Naturwissenschaften*, Vol. 29, 1941, pp. 348–362; translated as "Biological and Aerodynamical Problems of Animal Flight," *Journal of the Royal Aeronautical Society*, Vol. 46, 1942, pp. 39–56 (abridged) and NASA-TM 75337, 1980.

[10]Herzog, K., *Anatomie und Flugbiologie der Vögel*, Gustav Fischer Verlag, Stuttgart, and VEB Gustav Fischer, Jena, Germany, 1968.

[11]Brooks, A. N., MacCready, P. B., Lissaman, P. B. S., and Morgan, W. R., "Development of a Wing Flapping Flying Replica of the Largest Pterosaur," AIAA Paper 85-1446, 1985.

[12]Rayner, J. M. V., "Biomechanical Constraints on Size in Flying Vertebrates," *Miniature Vertebrates*, edited by P. J. Miller, Vol. 69, Symposia of the Zoological Society of London, Oxford University Press, Oxford, 1996, pp. 83–109.

[13]Rayner, J. M. V., "On the Origin and Evolution of Flapping Flight Aerodynamics in Birds," *New Perspectives on the Origin and Early Evolution of Birds*, edited by J. Gauthier, Special Publications of the Yale Peabody Museum, (to be published).

[14]Rayner, J. M. V., "Estimating Power Curves for Flying Vertebrates," *Journal of Experimental Biology*, Vol. 202, No. 23, Dec. 1999, pp. 3449–3461.

[15]Lighthill, M. J., *Mathematical Biofluiddynamics*, Society for Industrial and Applied Mathematics, Philadelphia, 1975.

[16]Pennycuick, C. J., *Bird Flight Performance: a Practical Calculation Manual*, Oxford Univ. Press, Oxford, 1989.

[17]Rayner, J. M. V., and Ward, S., "On the Power Curves of Flying Birds," *Proceedings of the XXII International Ornithological Congress*, edited by N. J. Adams and R. H. Slotow, BirdLife South Africa, Johannesburg, 1999, pp. 1786–1809.

[18]Rayner, J. M. V., "Form and Function in Avian Flight," *Current Ornithology*, Vol. 5, 1988, pp. 1–77.

[19]Rayner, J. M. V., "The Mechanics of Flight and Bird Migration Performance," *Bird Migration*, edited by E. Gwinner, Springer-Verlag, Heidelberg, 1990, pp. 283–299.

[20]von Helmholtz, H., "Über ein Theorem, geometrisch ähnliche Bewegungen flüssiger Körper betreffend, nebst Anwendung auf das Problem, Luftballons zu lenken," *Monatsbericht der königliche Akademie der Wissenschaften zu Berlin*, Vol. 1873, 1874, pp. 501–514.

[21]Ward, S., Rayner, J. M. V., Möller, U., Jackson, D. M., Nachtigall, W., and Speakman, J. R., "Heat Transfer from Starlings *Sturnus vulgaris* during Flight," *Journal of Experimental Biology*, Vol. 202, No. 12, June 1999, pp. 1589–1602.

[22]Betteridge, D. S., and Archer, R. D., "A Study of the Mechanics of Flapping Wings," *Aeronautical Quarterly*, Vol. 25, 1974, pp. 129–142.

[23]Rayner, J. M. V., "On Aerodynamics and the Energetics of Vertebrate Flapping Flight," *Fluid Dynamics in Biology*, edited by A. Y. Cheer and C. P. van Dam, Vol. 141, Contemporary Mathematics, American Mathematical Society, Providence, RI, 1993, pp. 351–400.

[24]Vazquez, R. J., "The Automating Skeletal and Muscular Mechanisms of the Avian Wing," *Zoomorphology*, Vol. 114, No. 1, March 1994, pp. 59–71.

[25]Hertel, H., *Struktur, Form, Bewegung*, Krauskopf, Mainz, 1963; translated as *Structure, Form, Movement*, Rheinhold, New York, 1966.

[26]Maybury, W. J., "The Aerodynamics of Bird Bodies," PhD Thesis, School of Biological Sciences, Univ. of Bristol, Bristol U.K., 2000.

[27]Maybury, W. J., and Rayner, J. M. V., "The Avian Tail Reduces Body Parasite Drag by Controlling Flow Separation and Vortex Shedding," *Proceedings of the Royal Society of London B*, Vol. 268, 2001, pp. 1405–1410.

[28]Bannasch, R., "Hydrodynamics of Penguins — An Experimental Approach," *The Penguins: Ecology and Management*, edited by P. Dann, I. Norman, and P. Reilly, Surrey Beatty, Chipping Norton, NSW, Australia, 1995, pp. 141–176.

[29]Gesser, R., Wedekind, F., Kockler, R., and Nachtigall, W., "Aerodynamische Untersuchungen an naturnahen Starenmodellen: 1. Grundlegende Ergebnisse," *Motion Systems*, edited by R. Blickhan, A. Wisser, and W. Nachtigall, Vol. 13, Biona Report, Gustav Fischer Verlag, Jena, Germany, 1998, pp. 229–230.

[30]Gesser, R., Wedekind, F., Kockler, R., and Nachtigall, W., "Aerodynamische Untersuchungen an naturnahen Starenmodellen: 2. Flügel-Rumpf Interferenzen," *Motion Systems*, edited by R. Blickhan, A. Wisser, and W. Nachtigall, Vol. 13, Biona Report, Gustav Fischer Verlag, Jena, Germany, 1998, pp. 257–258.

[31]Nachtigall, W., "Starlings and Starling Models in Wind Tunnels," *Journal of Avian Biology*, Vol. 29, No. 4, Dec. 1998, pp. 478–484.

[32]Homberger, D. G., "The Mechanism of Feather Movements: Implications for the Evolution of Birds and Avian Flight," *Acta Ornithologica*, Vol. 34, 1999, pp. 135–140.

[33]Rayner, J. M. V., Maybury, W. J., and Couldrick, L. B., "Aerodynamic Control by the Avian Tail," *American Zoologist*, (to be published).

[34]Kockler, R., Wedekind, F., Gesser, R., and Nachitgall, W., "Aspekte zur bionischen Übertragung einiger aerodynamischer Strukturen des Vogelflügels," *Motion Systems*, edited by R. Blickhan, A. Wisser, A., and W. Nachtigall, Vol. 13, Biona Report, Gustav Fischer Verlag, Jena, Germany, 1998, pp. 201–202.

[35]Müller, W., and Patone, G., "Air Transmissivity of Feathers," *Journal of Experimental Biology*, Vol. 201, No. 18, Sept. 1998, pp. 2591–2599.

[36]Drovetski, S. V., "Influence of the Trailing Edge Notch on Flight Performance of Galliforms," *Auk*, Vol. 113, No. 4, Oct. 1996, pp. 802–810.

[37]Nachtigall, W., and Kempf, B., "Vergleichende Untersuchungen zur flugbiologischen Funktion des Daumenfittichs (*Alula spuria*) bei Vögeln. I. Der Daumenfittich als Hochauftriebserzeuger," *Zeitschrift für vergleichende Physiologie*, Vol. 71, 1971, pp. 326–341.

[38]Padian, K., and Rayner, J. M. V., "The Wings of Pterosaurs," *American Journal of Science*, Vol. 293-A, 1993, pp. 91–166.

[39]Rayner, J. M. V., and Gordon, R., "Visualization and Modelling of the Wakes of Flying Birds," *Motion Systems*, edited by R. Blickhan, A. Wisser, and W. Nachtigall, Vol. 13, Biona Report, Gustav Fischer Verlag, Jena, Germany, 1998, pp. 165–173.

[40]Brown, R. E., and Fedde, M. R., "Airflow Sensors in the Avian Wing," *Journal of Experimental Biology*, Vol. 179, June 1993, pp. 13–30.

[41]Steinbacher, J., "Der Flug der Vögel," *Der Flug der Tiere*, edited by H. Schmidt, Waldemar Krämer, Frankfurt, 1960, pp. 77–112; also *Natur und Volk*, Vol. 89, 1959, pp. 309–325.

[42]Anderson, E. A., and Szewczyk, A. A., "Effects of a Splitter Plate on the Near Wake of a Circular Cylinder in 2- and 3-Dimensional Flow Configurations," *Experiments in Fluids*, Vol. 23, No. 2, June 1997, pp. 161–174.

Chapter 12

Lift and Drag Characteristics of Rotary and Flapping Wings

C. P. Ellington* and J. R. Usherwood†

University of Cambridge, Cambridge, England

Nomenclature

AR = aspect ratio $(4R^2/S)$
C_D = wing drag coefficient (profile + induced drag coefficients)
$C_{D,\text{pro}}$ = profile drag coefficient
C_L = lift coefficient
c = wing chord
$\bar{c}$ = mean wing chord $(S/2R)$
D'_{pro} = profile drag force per unit span
F' = resultant aerodynamic force per unit span
L' = lift force per unit span
n = wingbeat frequency
Q = propeller torque
R = wing length
Re = mean chord-based Reynolds number $(\bar{c}U_t/\nu)$
r = radial distance from the wing base
$\hat{r}_2$ = nondimensional radius such that $S_2 = \hat{r}_2^2 R^2 S$
$\hat{r}_3$ = nondimensional radius such that $S_3 = \hat{r}_3^3 R^3 S$
S = planform area of both wings
S_2 = second moment of area of both wings about the origin $r = 0$
S_3 = third moment of area of both wings
T = thrust of a two-bladed propeller
U = horizontal flapping velocity in hovering flight
U_r = relative velocity component perpendicular to the longitudinal wing axis
U_t = wingtip velocity (ΩR)

Copyright © 2001 by QinetiQ Limited. Published by the American Institute of Aeronautics and Astronautics, Inc., with permission.
*Professor of Animal Mechanics, Department of Zoology.
†Graduate Student, Department of Zoology.

α = angle of attack in degrees
α_r = effective angle of attack
ε = downwash angle in degrees
ν = kinematic viscosity of air
ρ = mass density of air
Ω = angular velocity of propeller

I. Introduction

ANIMAL flight has always fascinated mankind, particularly the readily observed flight of birds. Da Vinci, Lilienthal, and many others were inspired by birds in their own attempts at flight, and the science of aerodynamics grew from these humble beginnings. Insects have aroused less public attention, perhaps because their wings beat too quickly for casual observation. Nevertheless, their flapping flight performance is matched only by the hummingbirds. Unlike other birds, hummingbirds and insects are capable of sustained hovering, slow flight, and precise maneuvering.

The tables have turned, and the insects can now provide biological inspiration for flight developments. The need for small autonomous flying robots, largely for aerial reconnaissance inside buildings and confined spaces, and often within D^3 environments (dull, dirty, and dangerous), has fostered the recent interest in micro air vehicles (MAVs). Research on MAVs is primarily conducted by aerodynamic and robotic engineers who are attempting to improve the small-size performance of conventional fixed and rotary wings. However, a very successful design already exists for intelligent MAVs with much better lifting performance than offered by conventional wings—the insects. Through natural selection the insects have been experimenting with wings, kinematics, aerodynamics, control, and sensory systems for about 350 million years. Perhaps we can learn a few tricks from them.

Our understanding of the aerodynamics of insect flight has undergone a revolution in the past two decades. The most important testbed for such studies has been hovering flight. Many insects will hover inside filming chambers, and so their wing motion—or *kinematics*—can be recorded without resorting to tethered flight and its potential artifacts. The force and moment balance during flight is simplified by the lack of a net horizontal thrust, and the effects of body lift and body drag can be ignored. The absence of a forward flight velocity also reduces the complexity of the aerodynamic analysis. Not only is hovering relatively simple to analyze, but it is also a most demanding type of flight: Maximum lift and power coefficients are required,[1,2] and unsteady aerodynamic effects are at their most extreme.[3,4] Finally, when designing mechanical models for flow visualization and force measurement experiments, it is much easier to "hover" the models in a stationary fluid.

In this chapter we will review progress on the aerodynamics of hovering insect flight and present new results from propeller experiments on insect-based wings at the higher Reynolds number *Re* range appropriate to MAVs.

II. Aerodynamics of Hovering Insect Flight

Most insects hover with an approximately horizontal *stroke plane*,[3,5] which is the plane defined by the flapping motion of the wings. Variation in the wing elevation angles perpendicular to that plane are relatively small: typically less than 5 deg, and at worst less than 10 deg.[5–7] Such small departures from the stroke plane are usually ignored, and we consider that the downstroke and upstroke are simply confined to a horizontal plane.

High-speed films of hovering insects (e.g., Refs. 3, 5–7) indicate that the wings normally generate lift on both the downstroke and upstroke: Angles of attack are typically about 35 deg for outer wing regions. The wings are usually twisted along their length, like propeller blades, with angles of attack at the wing base some 10–20 deg greater than at the tip. Fruit-fly (Drosophila) wings, however, show a negligible twist; these small wings appear relatively stiffer than those of larger insects and hence more resistant to torsional twisting.[5,8,9] The flapping amplitude is typically about 120 deg, but ranges from 95 to 180 deg. At either end of the wingbeat, the wings rotate about a longitudinal axis to set the angle of attack for the next half-stroke; that is, the leading edge always leads. Thus the morphological upper wing surface is the aerodynamic upper surface on the downstroke, but the morphological lower wing surface is the aerodynamic upper surface on the upstroke. The left and right wings therefore move like contrarotating propellers on the downstroke, flip over and reverse direction for the upstroke, and then flip over to begin the cycle again.

Figure 1a shows the simplified path of the wingtip through the air for a hovering bumblebee. The wingtip path is determined by the vector sum of the flight, flapping, and downwash velocities, although the flight velocity is zero for the case of hovering. The all-important velocity of the wing *relative* to the air is given by the tangent to the path. Lift and profile drags are perpendicular and parallel to that path, respectively, and typical resultant force vectors are drawn for each half-stroke. The downstroke and upstroke are nearly symmetrical, providing comparable weight support but no net horizontal thrust.

A. Blade-Element Analysis

The usual aerodynamic treatment of insect flight is based on the *blade-element theory* of propellers, modified by Osborne[10] for flapping flight. The fundamental unit of analysis is the blade element, or wing element, which is that portion of a wing between the radial distances r and $r + dr$ from the wing base. The aerodynamic force F' per unit span on a wing element is resolved into a lift component L' normal

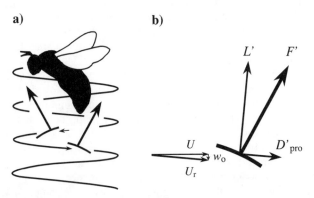

Fig. 1 a) Two-dimensional view of the wingtip path for a hovering bumblebee. b) Force and velocity components in hovering.

to the relative velocity and a profile drag component D'_{pro} parallel to it (Fig. 1b); profile drag consists of the skin friction and form drag on the wing. The force components for a wing section per unit span are

$$L' = \tfrac{1}{2}\rho c U_r^2 C_L \tag{1}$$

$$D'_{\text{pro}} = \tfrac{1}{2}\rho c U_r^2 C_{D,\,\text{pro}} \tag{2}$$

where ρ is the mass density of air, c is the local chord, and U_r is the local relative velocity component perpendicular to the longitudinal wing axis. As with propellers, any spanwise component of the relative velocity is assumed to have no effect on the forces. C_L and $C_{D,\text{pro}}$ are lift and profile drag coefficients. They can be taken from measurements on wings in the steady flow of a wind tunnel, in which case the analysis is quasi-steady and ignores the past history of the wing motion: The coefficients are then functions of the instantaneous Re, profile characteristics, and effective angle of attack α_r relative to U_r. Alternatively, C_L and $C_{D,\text{pro}}$ can be derived from experiments on flapping wings and will thus include unsteady effects as well.

An induced drag coefficient is not calculated explicitly. Instead, the induced power needed to create the downwash corresponding to the vertical force is estimated independently from the ideal momentum jet of propellers, corrected for spatial variation in loading using the differential momentum jet theory and for tip effects using a vortex theory.[11] A new method of calculating induced power, based on the added mass of vortex sheets in the wake,[12] gives good agreement with the more traditional approach. Both methods provide an estimate of the downwash velocity w_0 in the vicinity of the wings, and this is used to calculate the downwash angle ε of the relative velocity (Fig. 1b).

Equations (1) and (2) are resolved into vertical and horizontal components, integrated along the wing length, and averaged over a cycle. The net force balance in hovering then requires the mean vertical force to equal the weight and the mean horizontal force to be zero. One method of applying the blade-element analysis rests on successive stepwise solutions of Eqs. (1) and (2). Complete kinematic data are necessary for this approach: The motion of the longitudinal wing axis, the geometric angle of attack, and the section profile must all be known as functions of time and radial position. Combined with an estimate of the downwash, the effective angle of attack can then be calculated, and appropriate values of C_L and $C_{D,\text{pro}}$ can be selected from experimental results. However, this approach is very prone to error because angles of attack and profile sections cannot be measured with great accuracy and the downwash estimate is crude at best.

The *mean coefficients method* is normally used instead. By treating the force coefficients as constants over a half-stroke, they can be removed from the double integrals resulting from manipulations of the equations, and mean values can be found that satisfy the net force balance. The kinematic detail required for this method is greatly reduced since only the motion of the longitudinal wing axis is needed. The mean lift coefficient is particularly interesting because it is also the minimum value compatible with flight; if C_L varies during the wingbeat, as it surely must, then some instantaneous values must exceed the mean. A typical application of the mean coefficients method also ignores the downwash angle ε. This angle is typically less than 10 deg for hovering insects,[13] which is comparable to the wing elevation angles, which are also neglected. U_r can then be approximated by the horizontal flapping velocity U in Eqs. (1) and (2), and the vertical and horizontal forces are close to the lift and profile drag forces.

Maximum lift coefficients for insect wings in steady airflow in a wind tunnel are typically 0.6–0.9 (reviewed in Ref. 2), although dragonflies show values up to 1.15.[14] Most modern applications of the mean coefficients method have concluded than the mean C_L required for weight support exceeds the maximum for steady flow, sometimes by factors of 2 or 3 (for reviews, see Refs. 2, 15, and 16). These results show that insect wings produce more lift than expected from conventional two-dimensional considerations and that high-lift mechanisms are employed instead.

B. High-Lift Mechanisms

Recently, flow visualization studies on the hawkmoth *Manduca sexta* and a 10× scale mechanical model—the "flapper"—have identified *dynamic stall* as the high-lift mechanism used by most insects.[17–20] During the downstroke, air swirls around the leading edge and rolls up into an intense leading-edge vortex (LEV) as shown in Fig. 2. A laminar LEV is to be expected at these chord-based Re (less than 5000) for thin wings with sharp leading edges operating at high angles of attack. The circulation of the LEV augments the bound vortex and hence the lift. This is an example of dynamic stall, whereby a wing can travel at high angles of attack for a brief period, generating extra lift before it stalls.

Dynamic stall has long been a candidate for explaining the extra lift of insect wings, but two-dimensional aerodynamic studies showed that the lift enhancement is limited to about 3–4 chord lengths of travel;[21] the LEV grows until it becomes unstable at that distance and breaks away from the wing, causing a deep stall. The wings of hawkmoths and other insects travel twice that far during the downstroke at high flight speeds, which should rule out dynamic stall as the high-lift mechanism. However, a strong axial (spanwise) flow in the LEV was discovered for the hawkmoth and the flapper;[17–20] when combined with the swirl motion, it results in a *spiral* LEV with a pitch angle of 46 deg. The axial flow convects vorticity out toward the wing tip, where it joins with the tip vortex and prevents the LEV from growing so large that it breaks away. The spanwise flow therefore stabilizes the

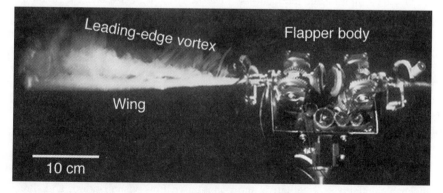

Fig. 2   Flow visualization of the leading-edge vortex over a wing of the flapper during the middle of the downstroke. Smoke is released from the leading edge. The camera view is parallel to the wing surface. (After van den Berg and Ellington.[18])

LEV, prolonging the benefits of dynamic stall for the entire downstroke and generating sufficient lift for weight support. A three-dimensional computational fluid dynamics study of the airflow for a hovering hawkmoth gave excellent agreement with the experimental flow visualization results.[22]

Lift enhancement by the spiral leading-edge vortex bears several similarities to the high-lift devices employed on certain man-made wings. The potential of "attached" vortices to augment lift has long been recognized in aerodynamics, but an axial flow component is essential for the stability of such vortices.[23] The axial flow can be induced by active spanwise suction or blowing over the upper wing surface; for delta wings, it is created instead by the flow component parallel to the swept leading edge. The conical, spiral vortex of the flapper is, in fact, remarkably similar in form to that over delta wings.[19] However, the axial flow for swept delta wings is quite different in origin from that for hovering insects with flapping, nonswept wings. The axial flow over insect wings is due to the dynamic pressure gradient associated with the spanwise flapping velocity gradient.

Axial flow in the leading-edge vortex has not been observed in previous two-dimensional experiments relevant to insect flight, but the spanwise pressure gradient necessary to drive the axial flow is, by definition, absent in two-dimensional studies. Maxworthy's[24] three-dimensional mechanical model of the specialized fling motion used by some insects[3] is the only other case where axial flow has been reported, and it now seems likely that this three-dimensional flow pattern is a common feature of insect flight. Helicopter rotors and wind turbine blades also experience spanwise pressure gradients, but large-scale axial flows have not been observed;[25] however, there is some evidence that a spanwise flow influences their stall characteristics.[26,27] It may be that the Reynolds numbers are too high for a large leading-edge vortex to persist, or that the spanwise flow component is reduced for high aspect ratio airfoils like helicopter blades. The exact conditions for establishing axial flow in a leading-edge vortex for rotary wings are not yet understood.

Since the hawkmoth experiments, Dickinson et al.[28] published an excellent study using a robotic fruit fly operating at $Re = 136$. Direct force measurements and flow visualization confirmed that dynamic stall is responsible for most of the lift, but it fell short of the required force by about 35%. The extra lift was provided by two mechanisms that operated as the wings flipped over between half-strokes: rotational circulation and interaction with vorticity in the near wake. The rotational circulation is that needed to satisfy the Kutta condition at the trailing edge as the angle of attack changes quickly. This rotational motion effectively alters the angle of attack at the $\frac{3}{4}$ chord point, and their results agree well with an earlier prediction of the rotational circulation.[4] The interaction with wake vorticity was complicated and depended on kinematic details, particularly the phase relationship between rotation and flapping; small shifts in the timing of rotation produced unexpectedly large changes in the mean lift coefficient. Drosophila are known to alter the phase of rotation during steering maneuvers,[28] and hoverflies change it during different hovering attitudes.[5] Thus the phasing of rotation provides an exquisite control over the mean force vector as well as a significant contribution to the total lift.

Current research on the aerodynamics of insect flight is proceeding rapidly, using the methods that have proved so successful for the simple case of hovering: mechanical flappers, qualitative and quantitative flow visualization, and direct measurement of wing forces. Characteristics of the spiral LEV and near-wake effects are being investigated, and the next obvious step—a study of forward flapping flight—has been initiated in our laboratory.

C. Applications of Insect Flight to Micro Air Vehicles

The high-lift performance of flapping insect wings offers a tantalizing prospect for the development of MAVs, where the production of a useful payload capacity is a major problem. Can the aerodynamic principles of insect flight be applied to MAVs? The wingspan of a flapping MAV will not be much larger than the hawkmoth, about 10 cm, but for a sufficient payload capacity the lift will have to be increased greatly from the 1–2 g mass of the hawkmoth. The lift in hovering is proportional to $n^2 R^4$, where n is the frequency and R is the wing length.[13] For a target all-up-weight of say, 50 g, the frequency has to increase from 25 Hz to about 150 Hz for hawkmoth-sized wings. Even if the wingspan is increased to 15 cm to take advantage of the fourth power of wing length, the frequency will still be quite high—about 65 Hz. The construction of a controlled, efficient flapping mechanism at such frequencies cannot be a short-term goal, judging by the decades spent developing helicopter rotor heads, which is a comparable problem.

What about a microhelicopter design based on insect wings to take advantage of their high-lift spiral LEV? Such a design could be implemented relatively quickly, if the LEV remains stable during constant rotation and if the wing performance is not degraded at the higher Re of MAVs. We have constructed a rotary wing rig for testing at Re appropriate to insect flight and have discovered that the LEV is indeed stable during rotation at a constant angular velocity (Usherwood and Ellington, submitted for publication to the *Journal of Experimental Biology*). Dickinson et al.[28] also found that their robotic fly wings created constant forces at constant flapping velocities. Thus it appears that the spiral LEV is a robust phenomenon, with the axial flow stabilizing the LEV even during steady rotation. For application to MAVs, however, Re will need to increase up to about 50,000, and the separated shear layer of the LEV will almost certainly become turbulent. Will the spiral LEV mechanism still work?

III. Propeller Experiments at High Re

A propeller rig was constructed to cover the chord-based Re range 10,000 to 50,000. For compatibility with the Re definition for flapping flight, $Re = \bar{c} U_t / \nu$, where $\bar{c}$ is the mean chord, U_t is the tip velocity ($= \Omega R$, where Ω is the angular velocity and R is wing length), and ν is the kinematic viscosity.

A. Propeller Rig

The base of the rig contained a 12-V 15-W dc motor mounted in two ball-races so that it was free to rotate in response to the motor reaction torque. The motor was connected to the propeller via a bellows coupling and drive shaft held between two ball-races within an outer housing. The diameter of the housing and the rotor head was 25 mm (i.e., 10% of the span). A shaft encoder on the motor provided tacho feedback for a custom-built speed control servo. Motor speed readings were averaged over a 100-s acquisition period with a Thandar TF200 frequency meter. The propeller rig (Fig. 3) was stuck to the pan of a Mettler PG2002-S balance with self-adhesive pads for direct measurement of thrust forces; three readings, each averaged over 20 s, were taken for each measurement. Reaction torque was measured on a Mettler BB240 balance; 10 readings, each averaged over 6 s, were taken for each measurement. Torque was transferred via an arm on the motor case

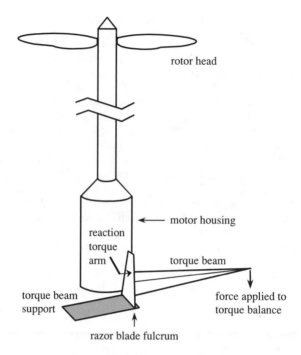

rotor head

motor housing

reaction
torque
arm

torque beam

torque beam
support

force applied to
torque balance

razor blade fulcrum

Fig. 3   The high *Re* propeller rig.

to a beam, pivoted on a razor blade, which converted the horizontal force to a vertical one on the balance pan. Balance readings were calibrated against a known torque; a fine wire pulled horizontally on the arm at a known position, passed around a ball-race, and supported a known weight. Cross talk between the two force measurements was always less than 0.5%. Torque measurements without wings were subtracted from those with wings to account for bearing friction, rotor head drag, etc. The balance pans were shielded from the downwash by a large card, 2.3 rotor diameters below the wings.

The rotor head contained a worm and wheel drive for each wing to adjust angles of attack. The worms were connected by spur gears so that both wings were adjusted together. The worms were fine-threaded screws, and the gear reduction was 72:1. One turn of the screw thus changed the angle of attack of both wings by 5 deg, and it could be set repeatedly to better than ±0.25 deg.

B.   Force Coefficients

It is common practice to reduce thrust and torque measurements of propellers to nondimensional coefficients, and there are two conventions: coefficients based on the disk area $(=\pi R^2)$ or those based on the total blade area. Insect wings show a diversity of planform shapes, however, and for a study of insect-based wings it would be more appropriate to take the shape into account. If we assume that the force coefficients are constant along the span, as in the mean coefficients method, and that the small downwash angle can be neglected, then the integrated blade-element equation for the thrust T of a two-bladed propeller is

$$T \approx \frac{1}{2}\rho\Omega^2 C_L \left[2 \int_0^R cr^2 dr \right] \tag{3}$$

where use of the lift coefficient C_L avoids conflict with traditional thrust coefficient definitions. The bracketed term is the second moment of area S_2 of both wings about the origin $r = 0$. For the analysis of insect wing shapes,[29] we define a nondimensional radius $\hat{r}_2$ such that $S_2 = \hat{r}_2^2 R^2 S$; that is, S_2 is given when the area S of both wings is considered to be concentrated at a radial position $\hat{r}_2 R$. The nondimensional radius therefore provides a measure of the *shape* of the wings, or the distribution of wing area, independent of the area itself. It is also more useful to express the wing area in terms of the wing length and aspect ratio AR, which is equal to the span $(2R)$ divided by the mean chord $(S/2R)$: $AR = 4R^2/S$. Equation (3) then reduces to

$$T \approx 2\rho\Omega^2 R^4 C_L \hat{r}_2^2 / AR \tag{4}$$

and this conveniently separates out the effects of aspect ratio and the normalized wing area distribution.

We treat the torque Q in a similar manner and derive

$$Q \approx 2\rho\Omega^2 R^5 C_D \hat{r}_3^3 / AR \tag{5}$$

where C_D is the drag coefficient, comprised of profile and induced drag, and $\hat{r}_3$ is the nondimensional radius corresponding to the third moment of wing area: $S_3 = \hat{r}_3^3 R^3 S$.

C.   Wing Designs

Because of our extensive work on the hawkmoth *Manduca sexta*, its wings were chosen for the initial design. Of the seven *Manduca* wings whose shape has been analyzed,[2,29] F1 from Willmott and Ellington[2] was selected as the most representative; its aspect ratio and radii for moments of area were closest to the average for all wings. The planform of F1 is shown in Fig. 4, and its characteristics are as follows: $AR = 5.66$, $\hat{r}_2 = 0.511$, and $\hat{r}_3 = 0.560$.

There are two features of *Manduca* wings that might be important for the stability of the spiral LEV. First is the notch in the trailing edge, where the forewing and hindwing overlap. This notch is close to the reattachment line for the flow around the LEV, and it also corresponds to the radial position where the vortex core begins to break down and become unstable.[19] To determine whether this association is causal, the effect of the notch was investigated using a *Smooth* wing design, with the same leading edge shape as *Manduca* but with a smooth trailing edge. Ellington[29] showed that the shape of insect wings can be closely approximated by the Beta distribution, one of the family of Pearson distribution functions in statistics. An analytical expression for the planform is obtained by matching the first two moments of the beta distribution to the first two moments of wing area, and the resulting chords are typically within 5% of the actual values for

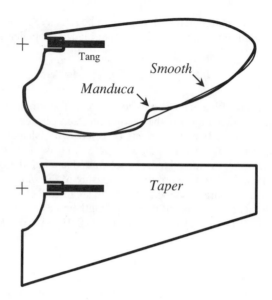

Fig. 4   Three wing designs tested on the propeller rig.

a variety of insect wing designs. The beta distribution is smooth, with no notches, and so it seemed appropriate for the *Smooth* wing design. The distribution was fitted to F1 by matching moments, and the resulting chords were taken from the actual leading edge coordinates of F1 (Fig. 4). Thus the leading edge of *Smooth* is identical to *Manduca*, but the trailing edge is smoothed. Both wing designs have the same second and third moments of wing area, and they should produce identical results if the notch has no effect.

Sweep-back of the leading edge of the *Manduca* wing near its tip is the other feature that might influence the stability of the spiral LEV. The LEV on the flapper curved away from the leading edge as it moved toward the wingtip.[19] A straight leading edge would increase the distance from that edge to the vortex and might cause the flow to break away from the upper wing surface. A *Taper* wing was therefore designed with a straight leading edge (Fig. 4), which again matched the second moment of area of the *Manduca* wing. The third moment of area of a tapered wing differs slightly from that of *Manduca*, however: $\hat{r}_3 = 0.570$.

The wing designs were cut with a wing length of 125 mm from 1.5-mm-thick clear Perspex. The thickness to mean chord ratio was 0.034. All edges were beveled at 45 deg to give a sharp edge at the upper wing surface. An elliptical section was cut out of the base to give clearance of the rotor head at angles of attack up to 50 deg. A flat supporting tang was epoxied into the wing base, and a round stub on the tang was located in the rotor head by a taper pin.

In our propeller experiments at the *Re* appropriate to *Manduca*, camber and twist have very little effect on the lift and drag coefficients. Given the gross flow separation at the leading edge, the robust reattachment, and the stabilizing axial flow, this is not surprising. It was therefore decided to leave the wings flat and untwisted for the high *Re* experiments, to provide an acid test of whether the spiral LEV mechanism still works.

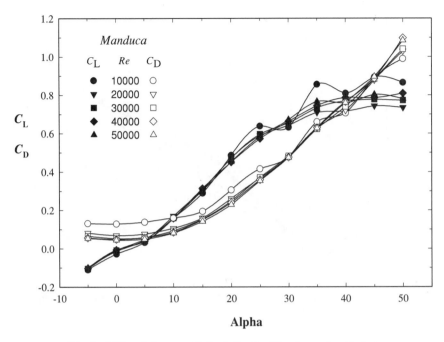

Fig. 5   Lift and drag coefficients for the *Manduca* wing design.

IV.   Results and Discussion

Force coefficients for the *Manduca, Smooth*, and *Taper* wing designs are given in Figs. 5–7, respectively, for a range of *Re* from 10,000 to 50,000. The results for 10,000 are somewhat variable, showing higher C_D for *Manduca* and *Smooth* at small to moderate angles of attack α. C_D for *Taper* at 10,000 is similar at all angles to the values at higher *Re*, but C_L is significantly higher at moderate to large α until it reaches 45 deg. C_L for *Manduca* at 10,000 shows some high values at moderate to large α, but the pattern is irregular. C_L for *Smooth* at 10,000 is similar to that at higher *Re*. Error bars are omitted from Figs. 5–7 for clarity, but the results for $Re = 10,000$ are considerably more variable than for higher values. Coefficients of variation [standard deviation (s.d.), mean] for the three wing designs are generally larger for angles of attack in the range of 15–40 deg than at higher *Re*, by factors of 1.5 to 3, but there is no consistent pattern in the variation.

The erratic results for $Re = 10,000$ suggest that this is the critical *Re* range for these wings. Flow separation must occur at the sharp leading edge. At low *Re*, reattachment in a laminar form is promoted by an enhanced diffusion of vorticity in the separated shear layer, but momentum losses in the boundary layer make the flow prone to further separation downstream. Above the critical *Re*, the separated shear layer mixes with the surrounding air and gains enough kinetic energy to become turbulent, and this energized layer then reattaches to the upper surface. The resulting turbulent boundary layer has a higher skin friction, but flow separation further downstream is delayed and the pressure drag decreases. The net result when the boundary layer is tripped into turbulence is usually a drop in the total drag, as suggested at $Re = 10,000$ for the *Taper* wing at all α, and for the *Manduca* and

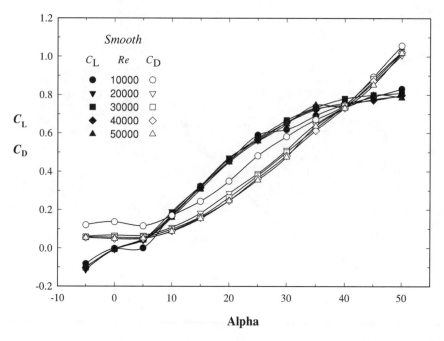

Fig. 6   Lift and drag coefficients for the *Smooth* wing design.

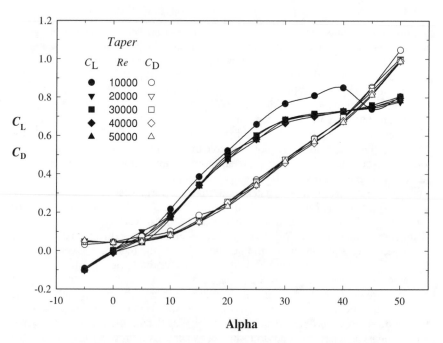

Fig. 7   Lift and drag coefficients for the *Taper* wing design.

Smooth wings at moderate to large α. It seems most likely that the sharp leading edge is working very effectively as a turbulator, bringing the critical Re down to about 10,000, and that the *Taper* wing performs best at this Re. As aeromodelers know all too well, flight in the critical Re range is difficult because of flow hysteresis and instability. However, this need not concern us unduly, because MAVs will need to operate at higher Re for a significant payload capacity.

The force coefficients for all wings over the remaining Re are very similar, indicating that the critical range does not even extend up to 20,000. With increasing Re the minimum of C_D tends to decrease slightly, presumably because of a drop in the skin friction coefficient with increasing Re. There are also small differences between wings, for example, C_L and C_D for *Taper* are often up to 0.05 lower than for *Manduca* and *Smooth* at large angles of attack. However, these effects are minor. There are certainly no major differences between the wing designs. Filling in the trailing-edge notch with the *Smooth* wing, and straightening the leading edge with the *Taper* wing, have little effect on the performance of the original *Manduca* wing. Because the differences are so small, the means for all three wings averaged over the Re range 20,000–50,000 is presented in Fig. 8; s.d. from individual plots were averaged in the same manner, and error bars of ± 2 s.d. are shown when they exceed the symbol size. The small differences mentioned above may be systematic but because they fall within the error bars, we will concentrate only on the means.

The performance at these higher Re is disappointing. C_D at small angles of attack should be very close to the minimum profile drag coefficient, but the observed value of 0.05 is about twice that of conventional thin profiles at $Re = 60,000$, as reviewed

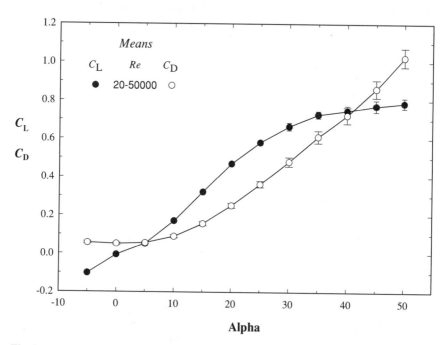

Fig. 8   Mean lift and drag coefficients for all three wing designs, averaged over $Re = 20,000–50,000$. Error bars are ± 2 s.d.

by Lowson.[30] The thickness-to-chord ratios are similar, and so the difference is probably attributable to the beveled leading edge of our models. The bevel is also likely to explain the nonlinearity in C_L at small α. The maximum of C_L, about 0.8, is substantially less than the 1.5 that we have measured in propeller experiments at the Re of real *Manduca* wings (Usherwood and Ellington, submitted for publication to the *Journal of Experimental Biology*). The absence of any abrupt stall provides some consolation, however, and would make for excellent flight characteristics. Lift continues to increase smoothly even up to angles of attack of 50 deg.

A. Evidence for a Leading-Edge Vortex

In his insightful analysis of the LEV over delta wings, Polhamus[31] showed that the drag due to lift divided by the total lift is simply equal to $\tan \alpha$. In other words, the LEV creates a large suction force perpendicular to the chord, and the lift and drag components are given by the cosine and sine of α, respectively. To test whether a LEV does exist over our wings, and whether it does dominate the wing forces, we can compare C_L/C_D against $\cot \alpha$. This ratio for the means is plotted against α in Fig. 9 along with $\cot \alpha$. The fit at large α is reasonable, suggesting that the forces are indeed dominated by LEV suction. We should not expect good agreement at small α because the LEV would make but a small contribution to the total force. At moderate α, however, the high minimum value of C_D should be taken into account for a fair comparison with Polhamus's prediction, which only applies to the drag associated with vortex lift. $C_L/(C_D - C_{D,\min})$ is therefore plotted in Fig. 9 as well. The fit to $\cot \alpha$ is now very good, confirming that the suction from a LEV explains most of the vertical and horizontal force components even at moderate α once the skin friction and form drag are taken into account. If

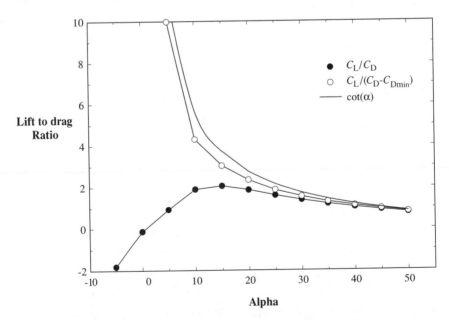

Fig. 9   Lift-to-drag ratios compared with the cot α prediction.

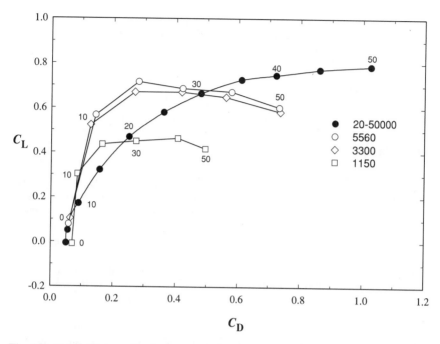

Fig. 10   Polar diagram for the mean propeller forces compared with those for real *Manduca* **wings measured in a wind tunnel at lower** *Re*.

$C_{D,\min}$ is too high, for the reasons given above, then a wing design with a lower value would give quite respectable ratios of C_L/C_D.

Is there axial flow in the LEV, causing it to spiral along the span? At the laminar Re of insect flight the axial flow stabilizes the LEV even at large α, indefinitely delaying dynamic stall in our propeller experiments. C_L reaches values of 1.5, which is about twice that for *Manduca* wings in the steady flow of a wind tunnel.[2] From two-dimensional modeling experiments[21] we expect the lift coefficient in a wind tunnel to oscillate as the LEV periodically grows and breaks away from the wing, giving a lower mean value than when axial flow stabilizes the LEV in flapping flight or propeller experiments. The results we have obtained here are similar to the measurements on *Manduca* wings in a wind tunnel, allowing for a difference in C_D due to Re effects. Figure 10 shows the polar diagram for all wings averaged over $Re = 20{,}000$–$50{,}000$, compared with the measurements on *Manduca* wings in a wind tunnel at lower Re. From the plateau of the lift coefficients it is clear that the mean lift in our propeller experiments is no better than might be expected for a wing that periodically grows and sheds a LEV at high angles of attack.

Our tentative conclusion is therefore that a LEV exists over the propeller wings at high Re, but it is lacking the axial flow that stabilizes the LEV and enables high lift coefficients to be achieved. Instead, the LEV periodically grows and breaks away from the wings, as in two-dimensional flow, limiting the mean value of the lift coefficient. Axial flow, and the resulting spiral LEV, is a robust phenomenon at laminar Re, but it does not survive the transition to turbulence. This is perhaps not surprising in hindsight, since even the laminar LEV eventually exhibits breakdown of the spiraling vortex core and loss of lift along the span.[17,19,22] The introduction of turbulent mixing at higher Re probably destroys the vortex integrity, reducing it to the usual two-dimensional unstable flow.

B. Micro-Helicopter MAVs

Nevertheless, a microhelicopter design still proves feasible for MAVs. Even if we use the C_L/C_D ratio determined here for higher Re, which is poor because of the high $C_{D,min}$ value, the prospects are encouraging. From Eqs. (4) and (5), the thrust to total power ratio is given by

$$T/P = C_L \hat{r}_2^2 / \Omega R C_D \hat{r}_3^3 \tag{6}$$

For a given wing design and operating frequency, the minimum power required to support a given weight is given by the best C_L/C_D ratio: 2.06 at $\alpha = 15$ deg. A target MAV design with a 15-cm wingspan and all-up-weight of 50 g would be reached at a frequency of about 100 Hz, or 6000 rpm, with a power requirement of 7 W. The Re would be almost 85,000, but we assume that the results up to 50,000 can be extrapolated that far. Power outputs of 1 W/g (engine mass) are readily achieved by small internal combustion engines.[32] Even allowing for the mass of fuel, a significant payload capacity would remain for a micro-helicopter based on the above design.

We have shown, however, that the high-lift spiral LEV mechanism fails at the turbulent Re of MAVs. A more conventional helicopter design would be better, based on traditional thin profiles and two-dimensional attached flow. Such a design would give much higher C_L/C_D ratios, at relatively small angles of attack, and the power requirement would be much less than that estimated above.

Acknowledgments

This work has been supported by grants from DERA (MOD CRP TG04 Materials and Structures Programme, Programme Manager Dr. E. S. O'Keefe), BBSRC, and DARPA.

References

[1] Dudley, R., and Ellington, C. P., "Mechanics of Forward Flight in Bumblebees. II. Quasi-Steady Lift and Power Requirements," *Journal of Experimental Biology*, Vol. 148, 1990, pp. 53–88.

[2] Willmott, A. P., and Ellington, C. P., "The Mechanics of Flight in the Hawkmoth *Manduca sexta*. II. Aerodynamic Consequences of Kinematic and Morphological Variation," *Journal of Experimental Biology*, Vol. 200, 1997, pp. 2723–2745.

[3] Weis-Fogh, T., "Quick Estimates of Flight Fitness in Hovering Animals, Including Novel Mechanisms for Lift Production," *Journal of Experimental Biology*, Vol. 59, 1973, pp. 169–230.

[4] Ellington, C. P., "The Aerodynamics of Hovering Insect Flight. IV. Aerodynamic Mechanisms," *Philosophical Transactions of the Royal Society B*, Vol. 305, 1984, pp. 79–113.

[5] Ellington, C. P., "The Aerodynamics of Hovering Insect Flight. III. Kinematics," *Philosophical Transactions of the Royal Society B*, Vol. 305, 1984, pp. 41–78.

[6]Dudley, R., and Ellington, C. P., "Mechanics of Forward Flight in Bumblebees. I. Kinematics and Morphology," *Journal of Experimental Biology*, Vol. 148, 1990, pp. 19–52.

[7]Willmott, A. P., and Ellington, C. P., "The Mechanics of Flight in the Hawkmoth *Manduca sexta*. I. Kinematics of Hovering and Forward Flight," *Journal of Experimental Biology*, Vol. 200, 1997, pp. 2705–2722.

[8]Vogel, S., "Flight in *Drosophila* II. Variations in Stroke Parameters and Wing Contour," *Journal of Experimental Biology*, Vol. 46, 1967, pp. 383–392.

[9]Weis-Fogh, T., "Energetics of Hovering Flight in Hummingbirds and in Drosophila," *Journal of Experimental Biology*, Vol. 56, 1972, pp. 79–104.

[10]Osborne, M. F. M., "Aerodynamics of Flapping Flight with Application to Insects," *Journal of Experimental Biology*, Vol. 28, 1951, pp. 221–245.

[11]Ellington, C. P., "The Aerodynamics of Hovering Insect Flight. V. A Vortex Theory," *Philosophical Transactions of the Royal Society B*, Vol. 305, 1984, pp. 115–144.

[12]Sunada, S., and Ellington, C. P., "An Approximate Added Mass Method for Estimating Induced Power for Flapping Flight," *AIAA Journal*, Vol. 38, 2000, pp. 1313–1321.

[13]Ellington, C. P., "The Aerodynamics of Hovering Insect Flight. VI. Lift and Power Requirements," *Philosophical Transactions of the Royal Society B*, Vol. 305, 1984, pp. 145–181.

[14]Wakeling, J. M., and Ellington, C. P., "Dragonfly Flight I. Gliding Flight and Steady-State Aerodynamic Forces," *Journal of Experimental Biology*, Vol. 200, 1997, pp. 543–556.

[15]Ellington, C. P., "Unsteady Aerodynamics of Insect Flight," *Biological Fluid Dynamics*, edited by C. P. Ellington and T. J. Pedley, Vol. 49, Symposia of the Society for Experimental Biology, Cambridge, U.K. 1995, pp. 109–129.

[16]Wakeling, J. M., and Ellington, C. P., "Dragonfly Flight III. Lift and Power Requirements," *Journal of Experimental Biology*, Vol. 200, 1997, pp. 583–600.

[17]Ellington, C. P., van den Berg, C., Willmott, A. P., and Thomas, A. L. R., "Leading-Edge Vortices in Insect Flight," *Nature*, Vol. 384, 1996, pp. 626–630.

[18]van den Berg, C., and Ellington, C. P., "The Vortex Wake of a 'Hovering' Model Hawkmoth," *Philosophical Transactions of the Royal Society B*, Vol. 352, 1997, pp. 317–328.

[19]van den Berg, C., and Ellington, C. P., "The Three-Dimensional Leading-Edge Vortex of a 'Hovering' Model Hawkmoth," *Philosophical Transactions of the Royal Society B*, Vol. 352, 1997, pp. 329–340.

[20]Willmott, A. P., Ellington, C. P., and Thomas, A. L. R., "Flow Visualization and Unsteady Aerodynamics in the Flight of the Hawkmoth *Manduca sexta*," *Philosophical Transactions of the Royal Society B*, Vol. 352, 1997, pp. 303–316.

[21]Dickinson, M. H., and Götz, K. G., "Unsteady Aerodynamic Performance of Model Wings at Low Reynolds Numbers," *Journal of Experimental Biology*, Vol. 174, 1993, pp. 45–64.

[22]Liu, H., Ellington, C. P., Kawachi, K., van den Berg, C., and Willmott, A. P., "A Computational Fluid Dynamic Study of Hawkmoth Hovering," *Journal of Experimental Biology*, Vol. 201, 1998, pp. 461–477.

[23]Wu, J. Z., Vakili, A. D., and Wu, J. M., "Review of the Physics of Enhancing Vortex Lift by Unsteady Excitation," *Progress in Aerospace Science*, Vol. 28, 1991, pp. 73–131.

[24]Maxworthy, T., "Experiments on the Weis-Fogh Mechanism of Lift Generation by Insects in Hovering Flight. Part 1. Dynamics of the 'Fling'," *Journal of Fluid Mechanics*, Vol. 93, 1979, pp. 47–63.

[25]De Vries, O., "On the Theory of the Horizontal-Axis Wind Turbine," *Annual Review of Fluid Mechanics*, Vol. 15, 1983, pp. 77–96.

[26]Harris, F. D., *Journal of the American Helicopter Society*, Vol. 11, 1966, pp. 1–21.

[27]Clausen, P. D., Piddington, D. M., and Wood, D. H., "An Experimental Investigation of Blade Element Theory for Wind Turbines. Part 1. Mean Flow Results," *Journal of Wind Engineering and Industrial Aerodynamics*, Vol. 25, 1987, pp. 189–206.

[28]Dickinson, M. H., Lehmann, F.-O., and Sane, S., "Wing Rotation and the Aerodynamic Basis of Insect Flight," *Science*, Vol. 284, 1999, pp. 1881–2044.

[29]Ellington, C. P., "The Aerodynamics of Hovering Insect Flight. II. Morphological Parameters," *Philosophical Transactions of the Royal Society B*, Vol. 305, 1984, pp. 17–40.

[30]Lowson, M. V., "Aerodynamics of Aerofoils at Low Reynolds Numbers," *Proceedings of the UAVs Fourteenth International Conference*, Univ. of Bristol, U.K., 1999, pp. 35.1–35.16.

[31]Polhamus, E. C., "Predictions of Vortex-Lift Characteristics by a Leading-Edge Suction Analogy," *Journal of Aircraft*, Vol. 8, 1971, pp. 193–199.

[32]Baxter, D. R. J., and East, R. A., "A Survey of Some Fundamental Issues in Micro-Air-Vehicle Design," *Proceedings of the UAVs Fourteenth International Conference*, Univ. of Bristol, U.K., 1999, pp. 34.1–34.13.

Chapter 13

A Rational Engineering Analysis of the Efficiency of Flapping Flight

Kenneth C. Hall[*]
Duke University, Durham, North Carolina
and
Steven R. Hall[†]
Massachusetts Institute of Technology, Cambridge, Massachusetts

Nomenclature

$\mathcal{A}$	=	surface bounding Trefftz volume $\mathcal{V}$
b	=	wingspan
C_d	=	coefficient of profile drag
C_l	=	coefficient of profile lift
C_{d0}, C_{d2}	=	parameters of drag coefficient curve fit
C_{l0}	=	sectional lift coefficent at minimum sectional drag
$C_{\mathcal{L}_1}$	=	coefficient of time-averaged lift
$C_{\mathcal{P}}$	=	coefficient of time-averaged power
$C_{\mathcal{P}0}$	=	coefficient of power required for steady gliding flight
$C_{\mathcal{T}_1}$	=	coefficient of time-averaged thrust
c	=	aerodynamic chord
F, G	=	real and imaginary parts of unsteady lift
f	=	frequency of flapping motion in hertz
H	=	amplitude of flapping motion transverse to the direction of flight
$\boldsymbol{i}, \boldsymbol{j}, \boldsymbol{k}$	=	unit vectors in x, y, and z directions
j	=	$\sqrt{-1}$
k	=	reduced frequency $= \Omega b / U$
$\mathcal{L}_1$	=	time-averaged aerodynamic lift
M	=	number of discrete vortices on airfoil in two-dimensional vortex-lattice model
N	=	total number of discrete vortices in two-dimensional vortex-lattice model

[*]Professor and Chair, Department of Mechanical Engineering and Materials Science. Associate Fellow AIAA.
[†]Professor, Department of Aeronautics and Astronautics. Associate Fellow AIAA.

n	=	unit vector normal to wake
p	=	pressure
$\mathcal{P}$	=	time-averaged power
$\mathcal{P}_i$	=	time-averaged induced power
$\mathcal{P}_v$	=	time-averaged viscous power
$\mathcal{P}_{v0}$	=	time-averaged viscous power with zero circulation
Re	=	Reynolds number
S_1	=	time-averaged aerodynamic lateral force
St_H	=	Strouhal number based on amplitude of flapping motion
s	=	distance along airfoil path
T	=	period of flapping motion $(1/f)$
T_1	=	time-averaged force tangent to the direction of flight
t	=	time
U	=	flight velocity
V	=	relative velocity of airfoil section through fluid
$\mathcal{V}$	=	trefftz volume
$\mathbf{V}$	=	vector velocity of fluid
$\mathcal{W}$	=	wake surface
w	=	component of induced wash normal to surface
$\mathbf{w}$	=	induced wash
x, y, z	=	Cartesian coordinates
γ	=	vorticity
Γ	=	circulation (jump in potential across wake)
Γ_0	=	circulation corresponding to the minimum sectional coefficient of drag
ϵ	=	size of vortex core
ζ	=	coordinate along wingspan
η	=	propulsive efficiency
θ	=	amplitude of flapping motion
λ	=	vector of equality constraint Lagrange multipliers
μ	=	advance ratio; also, fluid viscosity
ν	=	inequality constraint Lagrange multiplier
ξ	=	Kelvin linear impulse
Π	=	Lagrangian power
ρ	=	fluid density
σ	=	dummy variable of integration
ϕ	=	velocity potential
Ω	=	flapping frequency $(2\pi f)$

Subscripts

$(\cdot)_a$	=	airfoil
$(\cdot)_i$	=	inviscid
$(\cdot)_{max}$	=	value corresponding to maximum lift condition
$(\cdot)_R$	=	required value
$(\cdot)_v$	=	viscous

I. Introduction

A NUMBER of animals and man-made devices use flapping as a means of producing aerodynamic or hydrodynamic thrust. For example, aircraft and ship propellers "flap" by rotating, producing a helical wake of trailing vorticity.

Helicopter rotors also rotate, but in forward flight they produce a skewed wake containing both shed and trailing vorticity. Similarly, birds, bats, and man-made ornithopters generate an undulating wake with both shed and trailing vorticity. For a discussion of the structure of the wakes of birds, bats, and flapping airfoils, see Refs. 1–8. In all these cases, the wing of the device is made to move with a component of velocity normal to the direction of flight, a condition required to generate thrust. In this chapter, we seek to find optimal solutions to the flapping problem. In particular, we present an approach for computing the circulation distribution along a flapping wing that produces a desired aerodynamic force (lift and thrust) with the minimum required power.

A number of investigators have studied the problem of flapping flight by modeling the flowfield induced near the flapping wings caused by the system of trailing and shed vorticity in the wake. Wilmott[9] developed an unsteady lifting-line theory using the method of matched asymptotic expansions for the general motion of a wing with high aspect ratio. Phlips et al.[10] modeled flapping using an unsteady lifting-line theory in which the shed or transverse vorticity in the wake was lumped at the start of each stroke. Ahmadi and Widnall[11] developed an unsteady lifting-line theory using matched asymptotic expansions, with the inverse of the aspect ratio being the small parameter. Lan[12] developed an unsteady quasi-vortex-lattice method, which he then applied to predict the flapping efficiency of various planforms and flapping motions. All of these theories, however, are restricted to low-frequency flapping, that is, cases in which the reduced frequency $k \equiv \Omega b/U \ll 1$, where Ω is the flapping frequency (in rad/s), b is the wingspan, and U is the flight velocity.

A number of investigators have studied flapping propulsion in animals. For example, Lighthill[13] developed a two-dimensional small-amplitude model of the hydromechanics of flapping airfoils using an acceleration potential method. Based on this theory, Lighthill suggested that high thrust and hydromechanical efficiency are achieved when the heaving motion of the airfoil lags the pitching motion by 90 deg and the pitching axis is located near the trailing edge. Wu[14] has studied the thrust and efficiency characteristics of a two-dimensional waving plate of finite chord. Chopra[15] computed the thrust and power generated by a rectangular flat plate undergoing small-amplitude pitching and heaving motions in an inviscid fluid. Later, Chopra[16] developed a two-dimensional theory of carangiform propulsion; this model removed the earlier restriction of small-amplitude motion. However, because of its two-dimensional nature, the model does not include losses associated with streamwise vorticity. Chopra and Kambe[17] generalized Chopra's[15] model allowing flat-plate wings of arbitrary planform to be analyzed. Cheng et al.[18] have analyzed propulsive forces generated by three-dimensional waving plates. Lan[12] has developed a three-dimensional quasi-vortex-lattice method for analyzing animal propulsion and has applied the method to swept wings and tandem wing configurations. All of the above analyses have provided great insight into the fluid mechanics of swimming. Nevertheless, all are somewhat deficient in that viscous effects are not included, and most are restricted to small-amplitude motions. Furthermore, wake roll up is not considered.

None of the above analyses address the issue of minimum power flapping flight. The problem of finding the minimum induced loss (MIL) circulation distribution on a flapping wing is conceptually the same as finding the MIL solution for a propeller. The Betz[19] criterion describes the conditions that must be met to obtain an optimum circulation distribution for propellers operating in an inviscid fluid. Goldstein's[20] classic solution for MIL propellers was based on the Betz criterion. Later, Jones[21] applied a similar criterion to solve the problem of thrust generation using low-frequency flapping ($k \ll 1$) of high aspect ratio wings for the case where the left and right wings flap symmetrically about a common hinge point on the longitudinal axis. Jones was able to find a closed-form solution for the optimum circulation distribution. However, his theory is restricted to low-frequency small-amplitude flapping motions. Notwithstanding these assumptions, his results predict that high induced propulsive efficiencies are achieved using flapping motions with large amplitudes and/or high frequencies.

In an unpublished work, D. Munro (private communication, MIT, Cambridge, MA, 1979) showed that Betz's optimality condition could be generalize to devices other than propellers, such as helicopter rotors. Potze and Sparenberg[22] developed a similar optimality criterion for MIL "sculling" propulsion of ships. Hall et al.,[23] also using a generalized Betz criterion, predicted the optimum circulation distribution for helicopters in forward flight for the case where the rotor must simultaneously generate a prescribed lift and rolling and pitching moments. More recently, Hall and Hall[24] used this approach to compute the optimum circulation distribution along the span of wings undergoing large-amplitude high-frequency flapping motion that generate both lift and thrust. Sparenberg,[25] who has written extensively on optimal hydrodynamic propulsion, also observed that the MIL flapping flight problem could be solved using this approach, although no numerical examples were presented. De Jong [26,27] considered the problem of optimizing the performance of screw propellers, including the effects of viscosity in his analysis. He considered the effect of varying Reynolds number on the viscous drag, but he neglected the potential influence of loading (i.e., the dependence of the viscous drag on the lift coefficient as described by the sectional drag polar). De Jong did note that this dependence could be included, if necessary.

In this chapter, a method for predicting the minimum power circulation distribution required to generate thrust and lift via the flapping motion of wings is presented. The model extends the method of Hall and Hall[23] by including profile, or viscous, losses. In particular, we allow the sectional coefficient of drag to depend on the sectional lift coefficient. This is especially important for flapping flight because the sectional coefficients of lift and drag vary significantly over the flapping cycle. A variational principle, which is similar in form to the Betz[19] criterion for minimum induced loss propellers, describes the optimum circulation distribution in the wake of the flapping wing. A numerical method for solving for the optimality conditions is described. Finally, numerical examples of minimum power circulation distributions for propellers and flapping wings are presented, and the implications of viscous forces on flapping flight are discussed.

In Section II of this chapter, a two-dimensional flapping airfoil is analyzed using a vortex-lattice model of the airfoil and wake. It is shown by numerical example that the influence of wake roll up on the flow near the airfoil is negligible at flapping frequencies of practical interest, even for large-amplitude flapping motions with large hydrodynamic loadings. In Section III, a variational principle for determining the optimum distribution of circulation about a three-dimensional flapping wing is developed. The optimum circulation is defined as that circulation which produces a desired thrust with minimum induced power. This variational principle is the unsteady equivalent of the Betz[19] criterion for minimum induced loss. Included in the variational principle is a quasi-steady viscous correction based on measured or computed sectional drag polars. In Section IV, we analyze both high and low Reynolds number flapping flight using a numerical implementation of the variational principle.

II.   The Influence of Wake Roll Up on Flapping Flight

A.   A Vortex-Lattice Model of Flapping

Consider a two-dimensional airfoil moving with speed U through an inviscid and incompressible fluid. To begin, the flow is assumed to be irrotational everywhere except in the wake shed behind the airfoil. Under these assumptions, the flow outside of the wake is described by a single scalar velocity potential $\phi(x, y)$. The velocity potential is defined here such that the velocity vector V is given by

$$V = Ui + \nabla\phi \tag{1}$$

The conservation of mass is then just Laplace's equation,

$$\nabla^2\phi = 0 \tag{2}$$

For the moment, the coordinates x, y are referenced to the mean position of the airfoil, which moves through the fluid (to the left) at speed U. Therefore, in the x, y coordinate system, the flow in the far field moves in the positive x direction (to the right) with speed U.

Consider the case where the airfoil flaps in a pure heaving motion such that the y position of the airfoil is given by

$$y_a(t) = \frac{H}{2} \cos(\Omega t) \tag{3}$$

where Ω is the circular frequency of oscillation ($\Omega = 2\pi f$). Because the flow is described by Laplace's equation, the airfoil may be represented by a sheet of bound vorticity $\gamma_a(s)$ distributed from $x = 0$ to $x = c$. Similarly, the wake is represented by a sheet of shed vorticity $\gamma_w(s)$. Here, s is a coordinate that measures arc length along the airfoil and curved wake. Thus, the normal wash w on the surface of the airfoil (for the case of pure heaving) is given by

$$w(x, t) = \frac{dy_a}{dt} = \frac{1}{2\pi} \int_0^c \frac{\gamma_a(\sigma)}{x - \xi} d\sigma + \frac{1}{2\pi} \int_c^\infty \frac{\gamma_w(\sigma)(x - \xi)}{(x - \xi)^2 + (y_a - \eta)^2} d\sigma \tag{4}$$

where x, y_a, are the coordinates of the point on the airfoil at which the normal wash is measured and ξ, η are the coordinates of the vorticity located at arc length σ.

The strength of the shed vorticity in the wake is governed by the condition that the vorticity surrounding a fixed set of fluid particles must be constant (Kelvin's theorem). Thus, the vorticity at the trailing edge is given by

$$\gamma_w(c) = -\frac{1}{V}\frac{d\Gamma}{dt} \tag{5}$$

where $\Gamma(t)$ is the circulation around the airfoil and V is the magnitude of the velocity of the flow at the trailing edge (more precisely, the average of the velocity on either side of the trailing edge). Furthermore, once vorticity is shed into the wake, it is simply convected with the flow, so that

$$\frac{D\gamma_w}{Dt} = 0 \tag{6}$$

Fig. 1   Vortex-lattice model of flapping airfoil.

Finally, the well-known Kutta condition is required; that is, the pressure jump across the trailing edge of the airfoil is zero. For steady flow, this is equivalent to requiring that the vorticity go to zero at the trailing edge. For unsteady flows, the pressure jump across the trailing edge, given by the unsteady Bernoulli equation

$$\Delta p = \rho \left(V \gamma_a + \frac{d\Gamma}{dt} \right) \qquad (7)$$

must go to zero. Combining this requirement with Eqs. (4) and (6), one concludes that $\gamma_a = \gamma_w$ at the trailing edge; thus the Kutta condition requires that the vorticity distribution γ must be continuous at the trailing edge.

To solve Eqs. (4–7), a time-marching vortex-lattice model was constructed (see Fig. 1). Using this approach, the airfoil is divided into M elements of equal size Δx. At the quarter chord of the ith panel on the airfoil a point vortex of strength Γ_i is placed. The wake is modeled with additional $N - M$ discrete vortices. At each time step, a point vortex is shed into the wake a distance $\Delta x/4$ behind the trailing edge. The shed vortex is then convected downstream at each successive step. The strength of the point vortices on the airfoil is chosen at each time step so that the upwash condition, Eq. (4), is satisfied at M collocation points located at the three-quarter chord of each airfoil panel.

A few comments on some details of the numerical model are in order. With regard to the placement of the vortices at the quarter chord point and the collocation point at the three-quarter chord point of each panel, this is the traditional placement. The placement of the collocation points halfway between the vortices is in some respect equivalent to the requirement that the first integral in Eq. (4) be evaluated as a Cauchy principal value integral. Consider an infinite line of vortices of constant strength placed on the x axis. In the continuous case, the normal wash on the x axis would be constant. In the discrete case, however, the induced normal wash will vary with x and is in fact singular at the vortex points. However, the average of this wash will be very close to the exact continuous solution. Furthermore, the average wash will just be equal to the wash at a point midway between two vortices. In other words, picking the vortex points at the quarter chord and the collocation points at the three-quarter chord point measures the smooth part of the normal wash distribution, in effect discarding the nonsmooth part arising from the discrete nature of the model.

Lifanov and Polonskii[28] have provided two mathematical theorems relevant to this discussion, although both theorems are restricted to steady flows. The first theorem states that for the Cauchy singular integral problem given by

$$w(x) = \frac{1}{2\pi} \int_0^c \frac{\gamma_a(\sigma)}{x - \xi} d\sigma \tag{8}$$

the requirement that $\gamma(c)$ be bounded implies that $\gamma(c)$ vanish. The second theorem states that if Eq. (8) is solved using a discrete vortex model of the sort proposed in this chapter with the vortex and collocation points located at the quarter chord and three-quarter chord of the panels, respectively, then the numerical solution will converge uniformly to the exact solution of the Cauchy singular integral problem as the number of panels goes to infinity.

For unsteady flows, the theorems of Lifanov and Polonskii do not apply without modification because of the presence of the vorticity in the wake. However, a number of investigators, including Albano and Rodden,[29] have found by numerical experiment that the Kutta condition is satisfied in the limit when the quarter-chord, three-quarter-chord rule is used. Bliss and Epstein[30] have shown by numerical experiment that the computed lift is in fact quite sensitive to the position of the collocation points and that the lift will not be correctly predicted if the collocation points are not placed midway between point vortices.

B. Numerical Results

In the previous section, a numerical free-wake model for calculating the unsteady flow about a flapping airfoil was described. In this section, a number of numerical calculations used to validate the model are presented. To begin, consider the case where an airfoil with chord c vibrates with a small-amplitude heaving motion, say $y_a = H \times \exp(j\Omega t)/2$ with $H/c = 0.04$ and $j = \sqrt{-1}$. Using the free-wake model, the unsteady lift was calculated for a range of reduced frequencies $k = \Omega c/(2U)$. The computed results are shown in Fig. 2. Here, the nondimensional lift is given by $F + jG$, where F is the component of lift in phase with the heaving

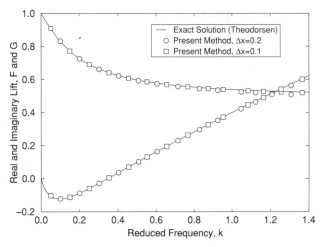

Fig. 2 Comparison of present vortex-lattice method to Theodorsen's [31] exact solution for small-amplitude heaving. The ——— represents Theordorsen's exact solution □, $\Delta x/c = 0.1$; with $\epsilon = 2.0$; and o, $\Delta x/c = 0.1$, $\epsilon = 2.0$.

velocity and G is the component out-of-phase with the heaving velocity, that is,

$$F + jG = \frac{2L}{j\Omega\pi\rho UHc} \qquad (9)$$

Note that the lift is nondimensionalized such that the quasi-steady lift is unity. To test the sensitivity of the present method to nonphysical model parameters, the results were computed for two different vortex panel sizes, $\Delta x/c = 0.1$ and 0.2. In both cases, the vortex core size ϵ was taken to be 2.0, and the length of the computational wake was 39 times the airfoil chord.

Also shown for comparison is the exact solution obtained by Theodorsen.[31] The present method is seen to be in very good agreement with the exact solution for reduced frequency k at least as large as 1.4—a reduced frequency significantly larger than is observed in fish, cetaceans, and birds. Furthermore, the numerical solution is insensitive to the size of the vortex panels, at least for panel sizes $\Delta x/c$ smaller than 0.2.

Having validated the vortex-lattice model for small-amplitude flapping motion, we next consider the case where an airfoil with chord c begins large-amplitude flapping at time $t = 0$ with motion given by Eq. (3). For the case considered here, the double amplitude of flapping motion H/c is 2.0 and the Strouhal number $St_H = \omega H/\pi U = 0.3$. This corresponds to a reduced frequency k based on a semichord of 0.471. The unsteady vorticity distribution and wake position was computed using the vortex-lattice method described above for both the prescribed-wake and free-wake models. Ten vortex lattice panels were used to model the airfoil ($\Delta x/c = 0.1$), and an additional 390 point vortices were used to model the wake. The free-wake calculations were performed using two different core sizes, $\epsilon = 1.0$ and 5.0. Figure 3 shows the computed wake position at time $Ut/c = 100$ for the two models. Note that the position of the wake for the prescribed-wake model is just the trace of the trailing edge of the airfoil. The experimentally observed behavior of a real wake is, of course, more nearly predicted by the free-wake model. Note that the wake rapidly rolls up so that the vorticity is concentrated in discrete vortex clouds of alternating direction. This wake roll-up behavior has been observed experimentally (Triantafyllou et al.[32]) and has also been predicted numerically by numerous investigators using models similar to the one presented here (see Katz and Weihs,[33,34] McCune et al.,[35] and Jones et al.[36]). The position of the vortex clouds and the vortex sheet that connects them is seen to be insensitive to the vortex core size.

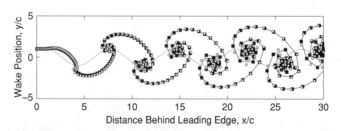

Fig. 3 Instantaneous wake position at time $Ut/c = 100$ with $H/c = 2.0$ and $St_H = 0.3$. The ——o—— indicates free-wake model with $\epsilon = 1.0$. The ——□—— indicates free wake model with $\epsilon = 5.0$; ------, prescribed-wake model.

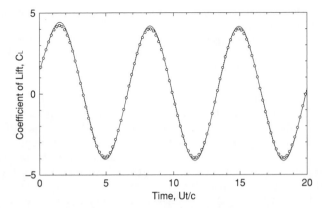

Fig. 4   Time history of unsteady lift resulting from heaving motion with $H/c = 2.0$ and $St_H = 0.3$. The —— indicates prescribed-wake model with $\Delta x = 0.1$. The ○ indicates free-wake model with $\Delta x/c = 0.2$. The – – – – indicates free-wake model with $\Delta x/c = 0.1$.

Figure 4 shows the computed unsteady lift (force in the y direction) acting on the airfoil as a function of time. For the moment, the coefficient of lift is defined as $C_L = L/(\frac{1}{2}\rho U^2 c)$. Shown are three calculations, one with a prescribed wake ($\Delta x/c = 0.1$), and two with a free wake, but with different vortex element sizes ($\Delta x/c = 0.2$ and 0.1, and in both cases $\epsilon = 1.0$). Given the vigorous roll up of the wake predicted by the free-wake model, it is indeed remarkable that the unsteady lift values predicted by the free-wake model and the prescribed-wake model are nearly identical, with only very minor differences seen in the peak values. Also note that the free-wake solutions appear to be insensitive to Δx, the size of the vortex panels.

Figure 5 shows the instantaneous vorticity distribution along the airfoil at time $Ut/c = 98$. Again, the vorticity distributions (and by implication the lift and

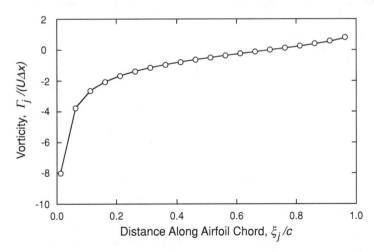

Fig. 5   Vorticity distribution along airfoil at time $Ut/c = 98$ with $H/c = 2.0$, $St_H = 0.3$, $\Delta x/c = 0.05$, and $\epsilon = 1.0$. The ○ indicates free-wake model and the —— indicates prescribed-wake model.

thrust) predicted by the prescribed- and free-wake models are nearly identical. These results, and those shown in Fig. 4, indicate that the effect of wake roll up can be neglected for the purposes of computing the hydrodynamic forces acting on flapping airfoils—at least for frequencies and flapping amplitudes not too different from those used in these examples.

The observant reader will have noticed that the unsteady lift coefficients for the present example (see Fig. 4) are unrealistically large (on the order of 4). Clearly, such large lift coefficients cannot be achieved, because airfoil sections typically stall for lift coefficients slightly larger than unity. The values of lift coefficient used here are artificially large, but nevertheless they demonstrate the unimportance of wake roll up, which should be most pronounced at large lift coefficients. In fact, wake roll up has only a minor effect on the unsteady lift at these large lift coefficients and is negligible at more realistic lift coefficients. Thus, throughout the rest of this chapter, wake roll up is neglected in favor of the simpler prescribed-wake model.

III.   Minimum Loss Flapping Theory

A.   Lift and Thrust

Minimum induced loss propeller theory is based on the observation that thrust is a consequence of momentum in the wake and induced power loss is a consequence of excess kinetic energy in the wake. Therefore, the thrust and power loss can be deduced directly from the structure of the wake, without any reference to how the wake was generated. Likewise, the lift and thrust forces produced in flapping flight are due to momentum in the wake, and the induced power loss is due to excess energy deposited in the wake.

The key assumption required in the following development is the *light loading* assumption. For lightly loaded flapping wings, the induced velocities in the wake are small compared to the velocity of the wings. This implies that the wake sheet left behind the wing will be undistorted by the induced flows, at least for a considerable distance behind the wing. As a result, the *far wake* will be undistorted and periodic in the flight direction. Hence, the light loading assumption is equivalent to the rigid wake assumption often used in vortex-lattice methods. However, as shown in the previous section, wake roll up actually has negligible effect on the unsteady aerodynamic response of a flapping airfoil. Therefore, we may expect the theory presented here to be valid even for moderately loaded wings.

For the purposes of computing the induced losses due to wing flapping, we assume that the flow resulting from flapping is inviscid, incompressible, and irrotational (except for the trailing and shed vorticity in the wake). Thus, the three-dimensional flow about the wings and wake is governed by Laplace's equation,

$$\nabla^2 \phi = 0 \qquad\qquad (10)$$

where the velocity $V = \nabla\phi$. The Cartesian coordinates x, y, and z are taken to be along the longitudinal, lateral, and vertical axes, as shown in Fig. 6. Furthermore, the coordinates are fixed to the fluid frame of reference, so that the velocity of the fluid goes to zero at infinity.

The aerodynamic forces (thrust, side force, and lift) acting on the flapping wing arise when the wing imparts linear momentum to the surrounding fluid. The force averaged over one period of flapping motion is equal and opposite to the time-averaged rate of change in momentum of the flowfield. The linear momentum

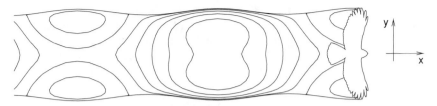

Fig. 6 Top view of bird in flight showing coordinate system and vortex filaments (trailing and shed vorticity) in the wake.

ξ deposited in the wake per temporal period T of wing flapping is given by

$$\xi = \rho \iiint_{\mathcal{V}} \nabla\phi \, d\mathcal{V} \qquad (11)$$

where $T = 2\pi/\Omega$. In Eq. (11), $\mathcal{V}$ is a control volume enclosing one period of the far downstream wake and extending to infinity in the y and z directions (see Fig.7).

Using Gauss's theorem, the volume integral in Eq. (11) may be converted to a surface integral over the surface $\mathcal{A}$ bounding the volume $\mathcal{V}$, so that

$$\xi = -\rho \iint_{\mathcal{A}} \phi \boldsymbol{n} \, d\mathcal{A} \qquad (12)$$

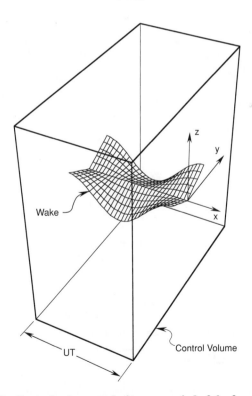

Fig. 7 Control volume enclosing one period of the far wake.

Note that ϕ is periodic in the x direction, and ϕ is continuous, except across the wake. Therefore, the potential ϕ is equal on the upstream and downstream sides of $\mathcal{V}$. Also, the unit normals point in opposite directions on opposite sides of $\mathcal{V}$, and thus the portion of the integral over the upstream and downstream sides cancel. Finally, ϕ goes to zero exponentially as $|y|$ or $|z|$ goes to infinity. Thus, Eq. (12) reduces to

$$\xi = -\rho \iint_S \phi n \, dA \tag{13}$$

where the surface S includes just the upper and lower sides of the wake.

On either side of the wake, the unit normals n point in opposite directions. Furthermore, the difference in the potentials across the wake is just the bound circulation Γ around the wing at the time the trailing edge of the wing passed by that point in space. Therefore, Eq. (13) can be rewritten as

$$\xi = -\rho \iint_W \Delta\phi n \, dA = -\rho \iint_W \Gamma n \, dA \tag{14}$$

where W is the upper surface of the wake.

Equation (14) may be recognized as the Kelvin linear impulse[37] generated by one period of the wake sheet. The time-averaged force F on the flapping wing is just the negative of the Kelvin linear impulse generated in one wake period, divided by the temporal period T. Therefore,

$$F = -\frac{\xi}{T} = \frac{\rho}{T} \iint_W \Gamma n \, dA \tag{15}$$

(For a detailed derivation demonstrating the validity of Eq. (15), see Hall and Hall.[24]) Dotting F with the unit normals i, j, and k gives the thrust, side force, and lift, respectively, as

$$T_1 = \frac{\rho}{T} \iint_W \Gamma i \cdot n \, dA \tag{16}$$

$$S_1 = \frac{\rho}{T} \iint_W \Gamma j \cdot n \, dA \tag{17}$$

$$\mathcal{L}_1 = \frac{\rho}{T} \iint_W \Gamma k \cdot n \, dA \tag{18}$$

The subscript "1" denotes that these are first-order forces only; induced forces are bookkept as induced power losses and are considered in the following section.

B. Inviscid Induced Power

The induced power losses resulting from lift and thrust of a flapping wing arise from the deposition of kinetic energy into the wake. Hence, the time-averaged induced power $\mathcal{P}_i$ is equal to the kinetic energy contained in one period of the wake, divided by the wing-beat period T, so that

$$\mathcal{P}_i = \frac{1}{T} \iiint_V \frac{1}{2}\rho |V|^2 \, dV = \frac{\rho}{2T} \iiint_V |\nabla\phi|^2 \, dV \tag{19}$$

Using the first form of Green's theorem, we have

$$\mathcal{P}_i = \frac{\rho}{2T}\left(-\iint_A \phi\nabla\phi\cdot\boldsymbol{n}\,\mathrm{d}\mathcal{A} - \iiint_V \phi\nabla^2\phi\,\mathrm{d}\mathcal{V}\right) \tag{20}$$

The second integral above is identically zero, because $\nabla^2\phi = 0$ in the interior of the volume. The portion of the first integral over the outer surface of $\mathcal{A}$ is zero as well, by an argument similar to the argument used earlier to compute the Kelvin linear impulse. The remainder of the integral is over the upper and lower surface of the wake, so that

$$\mathcal{P}_i = -\frac{\rho}{2T}\iint_W \Delta\phi\nabla\phi\cdot\boldsymbol{n}\,\mathrm{d}\mathcal{A} \tag{21}$$

But $\nabla\phi\cdot\boldsymbol{n}$ is just equal to $\boldsymbol{w}\cdot\boldsymbol{n}$, the normal wash induced at the surface of the wake, which by continuity must be equal on the upper and lower surface of the wake. Therefore, Eq. (21) may be expressed as

$$\mathcal{P}_i = -\frac{\rho}{2T}\iint_W \Gamma\boldsymbol{w}\cdot\boldsymbol{n}\,\mathrm{d}\mathcal{A} \tag{22}$$

Note that the induced wash $\boldsymbol{w}$ is linearly related to the circulation Γ through the Biot–Savart law. Furthermore, the lift and thrust are proportional to the circulation Γ. Thus, the induced power is quadratic in the lift and thrust.

C. Viscous Profile Power

Because of the difficulty in computing complex unsteady viscous flows about oscillating airfoils, the simplifying assumption is made that the wing has a large aspect ratio, that is, $c/b \ll 1$. We further assume that the reduced frequency of flapping based on the airfoil chord is small, so that

$$\Omega c/V \ll 1 \tag{23}$$

where V is the magnitude of the relative velocity of the airfoil through the fluid, given by

$$V = U\frac{\mathrm{d}s}{\mathrm{d}x} \tag{24}$$

and $\mathrm{d}s/\mathrm{d}x$ is the distance traveled by the wing per unit distance traveled by the wing in the x direction. Under these conditions, the viscous forces may be accurately modeled using quasi-steady lift–drag correlations.

Consider, for example, the drag polar shown in Fig. 8 for a NACA 4412 airfoil with standard surface roughness operating at a Reynolds number Re of 6×10^6, where $Re \equiv \rho Vc/\mu$. Shown is the experimentally measured coefficient of sectional drag C_d as a function of the coefficient of sectional lift C_l for steady flow as reported by Abbott et al.,[38] where

$$C_l \equiv \frac{l}{\frac{1}{2}\rho V^2 c} \quad\text{and}\quad C_d \equiv \frac{d}{\frac{1}{2}\rho V^2 c} \tag{25}$$

For moderate to large Reynolds numbers, the sectional drag is predominantly a function of the sectional lift and is only weakly dependent on the Reynolds number.

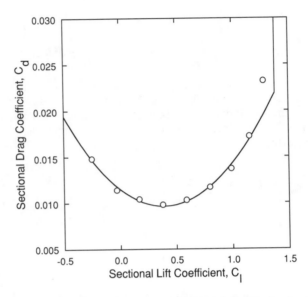

Fig. 8   Two-dimensional drag polar for NACA 4412 airfoil with standard roughness, $Re = 6 \times 10^6$. The ○ indicates experimental data (Abbott et al.[38]) and the ——— indicates curve fit ($C_{l0} = 0.3863$, $C_{d0} = 0.00964$, $C_{d2} = 0.01242$, $C_{l\max} = 1.38$).

Also, for angles of attack away from stall, the drag can generally be approximated as a quadratic function of the lift, so that

$$C_d \approx C_{d0} + C_{d2}(C_l - C_{l0})^2 \tag{26}$$

where C_{d0}, C_{d2}, and C_{l0} are constants that depend on the particular airfoil profile and Reynolds number. For the present example of a NACA 4412 airfoil, $C_{l0} = 0.3863$, $C_{d0} = 0.00964$, and $C_{d2} = 0.01242$. This correlation is plotted in Fig. 8 and is seen to be in good agreement with the experimental data over a wide range of sectional lift coefficients.

For large angles of attack, the airfoil will stall, and the correlation given by Eq. (26) will fail. Physically, such situations are to be avoided since they result in large viscous power losses with little or no additional thrust production. Thus, a constraint is imposed that the flapping motion be such that

$$\Gamma \leq \Gamma_{\max} \tag{27}$$

The coefficient of lift c_l may be expressed in terms of the circulation as

$$C_l = \frac{l}{\frac{1}{2}\rho V^2 c} = \frac{2\Gamma}{Vc} \tag{28}$$

Hence, $\Gamma_{\max} = \frac{1}{2}VcC_{l\max}$, with $C_{l\max}$ taken here to be 1.38.

The profile power $\mathcal{P}_v$ resulting from flapping may be expressed as the work per cycle divided by the period, so that

$$\mathcal{P}_v = \frac{1}{T}\int\int_W \frac{1}{2}\rho V^2 c C_d \, d\mathcal{A} \tag{29}$$

Substitution of Eqs. (24–26) and Eq. (28) into Eq. (29) gives

$$P_v = \frac{\rho}{2T} \iint_W (U^2 c C_{d0}) \cdot \left(\frac{ds}{dx}\right)^2 dA + \frac{\rho}{2T} \iint_W \left(\frac{4C_{d2}}{c}\right) \cdot (\Gamma - \Gamma_0)^2 \, dA \tag{30}$$

Finally, the total power loss P, which is the sum of the induced and profile powers, is given by

$$P = P_i + P_v$$

$$= -\frac{\rho}{2T} \iint_W \Gamma w \cdot n \, dA + \frac{\rho}{2T} \iint_W \left(\frac{4C_{d2}}{c}\right)(\Gamma - \Gamma_0)^2 \, dA$$

$$+ \frac{\rho}{2T} \iint_W (U^2 c C_{d0}) \cdot \left(\frac{ds}{dx}\right)^2 dA \tag{31}$$

Note that the total power loss may be expressed in terms of the shape of the wake, the planform of the wing, the quasi-steady sectional drag polar, and the circulation distribution in the wake. For a given wing undergoing a prescribed flapping motion, only the circulation is unknown. Furthermore, the profile power is quadratic in the circulation Γ, a fact that will prove useful when computing the optimum circulation distribution.

D. Optimal Solution to the Large-Amplitude Flapping Problem

To find the circulation that minimizes the total power loss in the prescribed flight condition, the constraints are adjoined to the power using Lagrange multipliers λ and ν to form the Lagrangian power

$$\Pi = P + \lambda \cdot (F - F_R) + \frac{\rho}{T} \iint_W \nu \, (\Gamma - \Gamma_{max}) \, dA \tag{32}$$

The inequality constraint Lagrange multiplier ν is zero if the inequality constraint [Eq. (27)] is inactive and is positive if it is active. Taking the variation of Eq. (32), one obtains

$$\delta\Pi = \delta\lambda \cdot (F - F_R) + \frac{\rho}{T} \iint_W \delta\nu \, (\Gamma - \Gamma_{max}) \, dA$$

$$+ \frac{\rho}{T} \iint_W \left[\lambda \cdot n - w \cdot n + \left(\frac{4C_{d2}}{c}\right)(\Gamma - \Gamma_0) + \nu\right] \delta\Gamma \, dA \tag{33}$$

where the potential flow identity

$$\iint_W w \cdot n\delta\Gamma \, dA = \iint_W \delta w \cdot n\Gamma \, dA \tag{34}$$

is used to eliminate δw from the integrand of the second integral in Eq. (33). At the constrained optimum, Π is stationary ($\delta\Pi = 0$). Therefore, the necessary conditions for minimum power are

$$T_1 = T_R \tag{35}$$

$$S_1 = S_R \tag{36}$$

$$\mathcal{L}_1 = \mathcal{L}_R \tag{37}$$

and

$$\begin{aligned} \nu \geq 0 &\quad \text{if } \Gamma = \Gamma_{\max} \\ \nu = 0 &\quad \text{if } \Gamma < \Gamma_{\max} \end{aligned} \tag{38}$$

and finally,

$$\boldsymbol{w} \cdot \boldsymbol{n} = \boldsymbol{\lambda} \cdot \boldsymbol{n} + \frac{4C_{d2}}{c}(\Gamma - \Gamma_0) + \nu \tag{39}$$

Equation (39) is the viscous equivalent of the Betz[19] criterion. The physical interpretation of Eq. (39), at least for cases when the inequality constraint is inactive ($\nu = 0$), is that the optimum induced normal wash on the wake is equal to the normal wash induced at the surface of the wake by an impermeable surface (the wake sheet) translating with velocity $\boldsymbol{\lambda}$, plus a term proportional to the coefficient C_{d2} times the deviation of the circulation Γ away from the minimum viscous drag circulation Γ_0. Note in particular that because C_{d0} has no effect on the first-order thrust or lift, and is not a function of Γ, it has no influence on the optimum distribution of circulation along the wake. Furthermore, the coefficient C_{d2} is usually small, typically $\mathcal{O}(10^{-2})$. Thus, at least for lightly loaded cases, the optimum viscous circulation distribution is nearly equal to the optimum inviscid distribution. Of course, the total power loss $\mathcal{P}$ is increased by the presence of viscous forces, even if the optimum circulation distribution is not significantly different from the inviscid case.

If the inequality constraint is active, then the above interpretation of the optimality conditions is slightly modified. In regions of the wake where $\Gamma < \Gamma_{\max}$, the optimal induced wash at the wake is as described above. In regions where $\Gamma = \Gamma_{\max}$, however, the induced normal wash (which is usually negative) is increased by the inequality constraint Lagrange multiplier ν (i.e., the magnitude of the normal wash is usually reduced).

The numerical solution technique used to find the optimal circulation distribution is described in detail by Hall et al.[39] The variational principle is discretized using a vortex-lattice model of the far-wake region, and the optimum solution is computed numerically.

IV. Results

A. Flapping of a Rigid Wing at High Reynolds Number

In this section, we compute the optimal circulation distributions and corresponding power requirements for a wing in flapping flight. The wing is assumed to have a rectangular planform with span b and an aspect ratio of 6.0. The airfoil section is a NACA 4412 airfoil with drag polar as in Fig. 8 for the case where the Reynolds number is 6×10^6. We consider flapping motions where the wake has the following shape:

$$\left. \begin{aligned} z &= |\zeta| \sin[\theta \cos(\Omega x/U)] \\ y &= |\zeta| \cos[\theta \cos(\Omega x/U)] \end{aligned} \right\} \quad \text{for } |\zeta| \leq b/2 \tag{40}$$

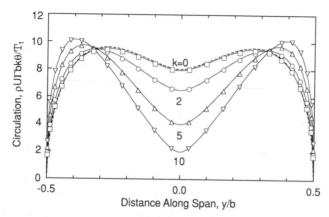

Fig. 9 Minimum induced power circulation distributions for thrust due to flapping. The ——— indicates small-amplitude harmonic theory of Hall and Hall.[25] The — — — indicates Jones's theory.[21] The □, ○, △, and ▽ are present method with $\theta = 1$ deg.

where Ω is the flapping frequency and θ is the angular amplitude of flapping motion. This motion corresponds to a wing with a straight, unswept trailing edge flapping rigidly about a hinge point on the longitudinal axis.

For the first example, the amplitude of flapping is small with $\theta = 1$ deg, and only induced losses are considered. Shown in Fig. 9 is the optimum circulation required to produce thrust with no lift. The circulation is plotted at the point in the downstroke when the wings pass through the horizontal position. These and subsequent results were computed using a vortex-lattice mesh with 24 elements per period in the flight direction and 24 elements in the spanwise direction. Each solution required about 18 s of CPU time to compute on a Silicon Graphics Inc. (SGI) Power Indigo^2 workstation. Also shown is Jones's[21] exact solution to the optimal circulation problem for low-frequency small-amplitude flapping and the small-amplitude harmonic theory developed by Hall and Hall.[24] The present method is seen to be in excellent agreement with the other two theories, at least for this small-amplitude case.

Next, we consider the case of large-amplitude flapping where the flapping wing must simultaneously generate both thrust and lift. Shown in Fig. 10 is the optimal circulation distribution in the wake of a flapping wing with $1/\mu = 2.0$, $\theta = 45$ deg, $C_{\mathcal{L}_1} = 0.1$, and $C_{\mathcal{T}_1} = 0.01$, where

$$k \equiv \frac{1}{\mu} \equiv \frac{\Omega b}{U} \tag{41}$$

and

$$C_{\mathcal{L}_1} \equiv \frac{\mathcal{L}_1}{\frac{1}{2}\rho U^2 b^2}, \qquad C_{\mathcal{T}_1} \equiv \frac{\mathcal{T}_1}{\frac{1}{2}\rho U^2 b^2}, \qquad C_{\mathcal{P}} \equiv \frac{\mathcal{P}}{\frac{1}{2}\rho U^3 b^2} \tag{42}$$

Two cases are shown, one in which the optimal circulation distribution is computed considering only induced losses and the other with profile losses included. The two solutions are very similar, with the largest differences occurring at the middle

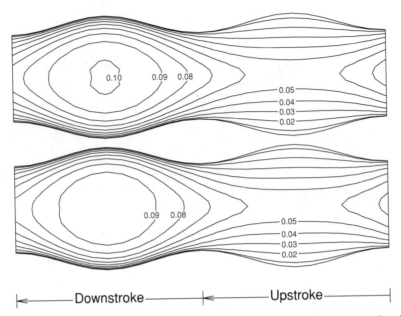

Fig. 10   Top view of optimal circulation distribution in wake of flapping wing for NACA 4412 airfoil with $Re = 6 \times 10^6$, $\theta = 45$ deg, $\mu = 0.5$, $C_{L_1} = 0.1$, and $C_{T_1} = 0.01$. (Top) Inviscid solution; (Bottom) viscous solution. Direction of flight is right to left. Contours of Γ/Ub are plotted.

of the wake during the downstroke. One sees that the inviscid circulation is very slightly larger in this region. In the viscous solution, the circulation is reduced slightly in this region to minimize the profile losses associated with large sectional lift coefficients.

Figure 11 shows the optimal circulation distribution for the same case shown in Fig. 10, but with the thrust coefficient C_{T_1} now increased to 0.2. Over a portion of the downstroke, the maximum coefficient of sectional lift constraint is active over the inboard portion of the wing, reducing the circulation in this region. Also note that over a portion of the upstroke, the circulation near the tips is negative.

Next, the power losses were computed for the flapping wing. Figure 12 shows the power loss for an optimally loaded flapping wing for a range of flapping frequencies $1/\mu$ for the case where $\theta = 45$ deg, $C_{L_1} = 0.1$, and $C_{T_1} = 0.01$. The induced losses generally decrease with increasing flapping frequency, whereas the profile losses generally increase. The induced losses become relatively "flat" at moderate reduced frequencies and may even increase slightly for very large frequencies. This is because a flapping wing must generate both thrust and lift. Whereas the (inviscid) generation of thrust becomes more efficient at large flapping frequencies, approaching the actuator disk limit, the generation of lift is most efficient at low frequencies. For this example, the optimal flapping frequency $1/\mu$ is approximately 2.1.

For a propeller, which has the sole purpose of generating thrust, the propulsive efficiency is simply the thrust power produced divided by the total power (thrust power plus power losses). For a flapping wing, which must produce both thrust

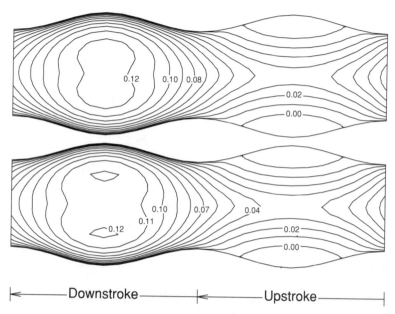

Fig. 11 **Top view of optimal circulation distribution in wake of flapping wing for NACA 4412 airfoil with** $Re = 6 \times 10^6$, $\theta = 45$ **deg,** $\mu = 0.5$, $C_{\mathcal{L}_1} = 0.1$, **and** $C_{\mathcal{T}_1} = 0.02$. **(Top) Inviscid solution; (Bottom) viscous solution.**

and lift, we may define the propulsive efficiency as

$$\eta \equiv \frac{C_{\mathcal{T}_1}}{C_{\mathcal{T}_1} + C_{\mathcal{P}} - C_{\mathcal{P}0}} \tag{43}$$

where $C_{\mathcal{P}0}$ is the power loss with $\theta = 0$, $C_{\mathcal{T}_1} = 0$, and $C_{\mathcal{L}_1} = 0.1$. In other words, only the increase in power required to generate thrust should be counted in the calculation of efficiency. Using this definition, we find that the propulsive

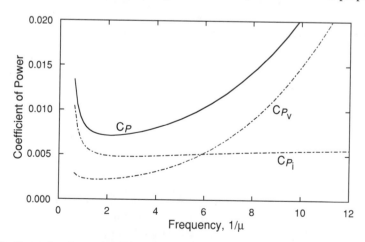

Fig. 12 **Power loss for optimally loaded flapping wing at various flapping frequencies for NACA 4412 airfoil with** $Re = 6 \times 10^6$, $\theta = 45$ **deg,** $C_{\mathcal{L}_1} = 0.1$, **and** $C_{\mathcal{T}_1} = 0.01$.

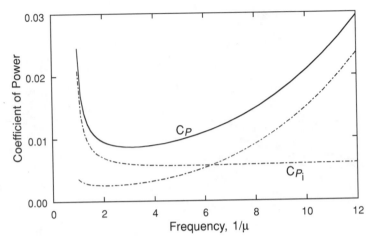

Fig. 13 Power loss for optimally loaded flapping wing at various flapping frequencies for NACA 4412 airfoil with $Re = 6 \times 10^6$, $\theta = 45$ deg, $C_{\mathcal{L}_1} = 0.1$, and $C_{\mathcal{T}_1} = 0.02$.

efficiency of the flapping wing discussed in the previous example is approximately equal to 0.808 at a flapping frequency $1/\mu$ of 2.1.

Figure 13 shows the optimal power loss for the same case presented in Fig. 12, but with $C_{\mathcal{T}_1} = 0.02$. For this more highly loaded case, the optimal flapping frequency $1/\mu$ is about 3.1, and the corresponding propulsive efficiency η is 0.837.

In the previous two examples, the minimum power loss for flapping flight was computed for two different thrust coefficients. The flapping amplitude was fixed ($\theta = 45$ deg) and the flapping frequency $1/\mu$ was varied. Next, we consider the effect of flapping amplitude on power losses. Shown in Fig. 14 are contours of power loss $C_{\mathcal{P}}$ as a function of flapping frequency and amplitude for a thrust coefficient $C_{\mathcal{T}_1} = 0.02$ and lift coefficient $C_{\mathcal{L}_1} = 0.1$. Note that there is an optimum amplitude and frequency that produces the minimum power loss. The minimum power $C_{\mathcal{P}}$ for this case is approximately 0.00825, corresponding to $\theta = 27$ deg and $1/\mu = 5.3$. This flapping frequency equartes to slightly less than one wingbeat per wingspan traveled. The corresponding propulsive efficiency η is about 0.852. For a lower thrust loading case, $C_{\mathcal{T}_1} = 0.01$ at the same lift coefficient (not shown), the minimum power $C_{\mathcal{P}}$ is about 0.00641, corresponding to $\theta = 20$ deg and $1/\mu = 5.3$. Under these conditions, the propulsive efficiency η is approximately 0.859. Thus, flapping flight is seen to be remarkably efficient, with propulsive efficiencies rivaling that of propellers. Note that the optimum flapping frequency is relatively insensitive to thrust loading. The optimum flapping amplitude, however, is sensitive to thrust loading.

B. Flapping of a Rigid Wing at Low Reynolds Number

In this section, we compute the optimal circulation distributions and corresponding power requirements for a wing similar to that presented in the previous section. In this case, however, the Reynolds number is 6×10^4, which is more typical of a micro unmanned air vehicle. The airfoil used here is an AG02, an airfoil designed for use in high-performance hand-launched model airplanes that operate at low Reynolds numbers (M. Drela, personal communication, April 2000). The drag

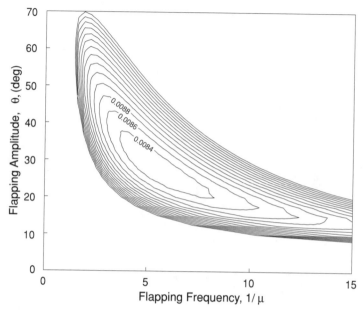

Fig. 14   Power loss $C_\mathcal{P}$ for various flapping amplitudes and frequencies for NACA 4412 airfoil with $Re = 6 \times 10^6$, $C_{\mathcal{L}_1} = 0.1$, and $C_{\mathcal{T}_1} = 0.02$.

polar for this airfoil was computed using the computer code XFOIL[40] and is shown in Fig. 15. Note that the XFOIL code predicts a fairly constant sectional lift coefficient at low coefficients of lift, followed by a rapid rise in drag for coeffcients of lift greater than about 0.5. This rise in drag is associated with transition of the boundary layer from laminar to turbulent flow at high lift coefficients. Here, the drag polar data were curve fit with a quadratic function over a range of C_l from about 0.1 to 0.9, the range of expected coefficients of lift. The result of this curve fit gives $C_{l0} = 0.308$, $C_{d0} = 0.00991$, and $C_{d2} = 0.138$.

Next, using this approximate drag polar, we computed the optimal circulation distribution for low Reynolds number flapping. Shown in Fig. 16 is the optimal circulation distribution in the wake of a flapping wing with $1/\mu = 2.0$, $\theta = 45$ deg, $C_{\mathcal{L}_1} = 0.1$, and $C_{\mathcal{T}_1} = 0.01$. Again, two cases are shown, one in which the optimal circulation distribution is computed considering only induced losses and the other with profile losses included. Even for this low Reynolds number case, the optimal circulation distributions for the inviscid and viscous cases are remarkably similar. As in the high Reynolds number case, the circulation is reduced in the downstroke region to minimize the profile losses associated with large sectional lift coefficients, although here the effect is more pronounced.

Figure 17 shows the power loss for the optimally loaded low Reynolds number flapping wing for a range of flapping frequencies $1/\mu$ for the case where $\theta = 45$ deg, $C_{\mathcal{L}_1} = 0.1$, and $C_{\mathcal{T}_1} = 0.02$. Note that, in this case, the viscous power is a much larger component of the total power than in the high Reynolds number example, so that the minimum power frequency (for this flapping amplitude) is more strongly influenced by the viscous loss.

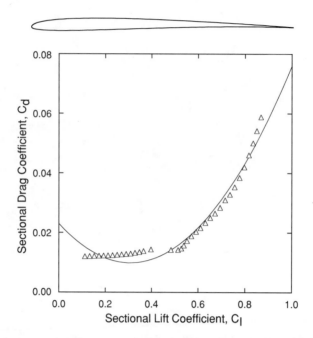

Fig. 15   Two-dimensional drag polar for AG02 airfoil with $Re = 6 \times 10^4$: $\triangle$, XFOIL calculation (Drela[40]); ———, curve fit ($C_{l0} = 0.308$, $C_{d0} = 0.00991$, $C_{d2} = 0.138$).

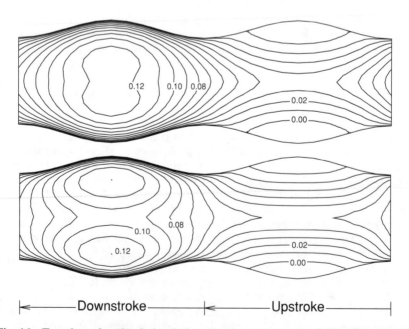

Fig. 16   Top view of optimal circulation distribution in wake of flapping wing for AG02 airfoil with $Re = 6 \times 10^4$, $\theta = 45$ deg, $\mu = 0.5$, $C_{\mathcal{L}_1} = 0.1$, and $C_{\mathcal{T}_1} = 0.02$. (Top) Inviscid solution; (Bottom) viscous solution.

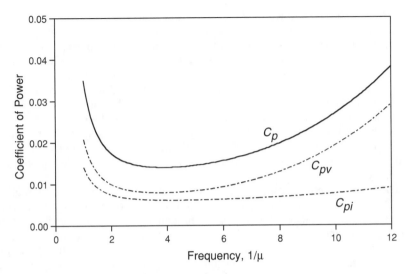

Fig. 17   Power loss for optimally loaded flapping wing at various flapping frequencies for AG02 airfoil with $Re = 6 \times 10^4$, $\theta = 45$ deg, $C_{\mathcal{L}_1} = 0.1$, and $C_{\mathcal{T}_1} = 0.02$.

V.   Summary and Discussion

A variational method has been developed for computing the minimum power circulation distribution along the span of a flapping wing that must simultaneously generate thrust and lift. Using this approach, the sum of the induced (inviscid) power and profile (viscous) power are minimized subject to the constraints that the desired thrust and lift be obtained and that no portion of the wing may operate in a stalled condition. The resulting variational statement is the viscous equivalent of the well-known Betz criterion for minimum induced loss propellers. This variational principle is solved numerically using a vortex-lattice method to compute the lift, thrust, and induced power and a quasi-steady drag polar correlation to model profile drag. Inequality constraints are implemented using an augmented Lagrangian method. The present method is computationally very efficient. For cases in which the maximum sectional coefficient of lift constraint is active, the analysis typically requires less than 2 min of CPU time on an SGI Power Indigo.[2] For lightly loaded cases in which the constraint is inactive, solutions require about 18 s of computer time.

Not included in the variational principle is the effect of wake roll up—a key feature in Triantafyllou et al.'s[32] explanation for the optimum frequency range of carangiform swimming, a problem similar in many ways to the flapping-flight problem. Triantafyllou et al. are correct that the wake rolls up into vortex clouds of alternating sign and that the wake may roll up more vigorously for certain Strouhal numbers. However, the effect of the wake roll up on the propulsive efficiency is exceedingly small. This was demonstrated conclusively, in the section discussing the influence of wake roll up on flapping flight, using time-marching numerical simulations of flapping. In one computation, the wake was convected with the mean flow (a prescribed wake); in another computation, the wake was convected under its own influence (a free wake). Although the resulting wake shapes for these two models differ dramatically, the circulation histories and the instantaneous vorticity distributions on the surface of the airfoil—and hence the thrusts—are virtually identical for these two models. One concludes that the effect of roll up is indeed small and may be ignored in the analysis of carangiform propulsion. In support of this view, it is noted that the present theory is an adaptation of Goldstein's[20] minimum induced loss propeller theory, which predicts the optimum circulation distribution along the span of a propeller blade required to generate a certain thrust with minimum power. Goldstein's theory is based on the assumption that the wake left behind a propeller is helical in shape and does not roll up. Of course, real propeller wakes roll up rapidly forming a chaotic jet of high-momentum fluid behind the propeller. Nevertheless, Goldstein's theory accurately predicts the thrust and induced power and is still used by engineers to design propellers more than six decades after its introduction.

The present method was used to compute minimum power loss circulation distributions for (inviscid) flapping wings. For the case of small-amplitude wing flapping, the present theory recovers the minimum induced circulation distribution predicted by Jones[21] and the small-amplitude flapping theory of Hall and Hall.[24]

The method was also used to compute optimal circulation distributions for flapping wings operating at both high and low Reynolds numbers. One interesting result is that the optimal viscous circulation distribution is very similar to the optimal inviscid circulation distribution. For typical levels of required thrust for the high Reynolds number case, the optimal propulsive efficiency of a flapping wing that must generate both thrust and lift was found to be about 0.85 for two different thrust levels. The optimal flapping frequency $1/\mu$ was found to be about 5.3 and was relatively insensitive to thrust requirements. This flapping frequency means that the wing should flap slightly less than once for every one wingspan the wing flies forward through the air. Again, this level of efficiency is not expected to be significantly different from that of a propeller-driven aircraft.

Although it has been argued that flapping flight is more efficient than propeller-driven flight at low Reynolds number, our analysis (from a previous similar study[41]) does not support that conclusion. The preference for flapping flight in biological systems is probably due to the available evolutionary pathways. Flapping flight in mechanical systems is probably disadvantageous for all but very narrow niche applications, such as stealth-by-mimicry.

Acknowledgments

Portions of this chapter were presented as Paper 97-0827 at the AIAA 35th Aerospace Sciences Meeting in Reno, Nevada, held 6–9 January 1997, by Kenneth C. Hall, Steven A. Pigott, and Steven R. Hall. The authors are endebted to Mr. Pigott for his assistance on the earlier paper. The authors would also like to thank Professor Mark Drela of the Massachusetts Institute of Technology for his selection of the AGO2 airfoil as a suitable low Reynolds number airfoil and acknowledge the use of his computer model XFOIL for the computation of the low Reynolds number drag polar.

References

[1]Rayner, J. M. V., "Wake Structure and Force Generation in Avian Flapping Flight," *Acta XX Congressus Internationalis Ornithologici*, edited by B. D. Bell, Vol. 2, 1991, pp. 702–715.

[2]Rayner, J. M. V., "A Vortex Theory of Animal Flight. Part 2. The Forward Flight of Birds," *Journal of Fluid Mechanics*, Vol. 91, No. 4, 1979, pp. 731–763.

[3]Rayner, J. M. V., "On Aerodynamics and the Energetics of Vertebrate Flapping Flight," *Fluid Dynamics in Biology*, edited by A. Y. Cheer and C. P. van Dam, Vol. 141, Contemporary Mathematics, American Mathematical Society, Providence, RI, 1993, pp. 351–400.

[4]Rayner, J. M. V., Jones, G., and Thomas, A., "Vortex Flow Visualizations Reveal Change in Upstroke Function with Flight Speed in Bats," *Nature*, Vol. 321, 8 May 1986, pp. 162–164.

[5]Spedding, G. R., "The Wake of a Jackdaw (*Corvus monedula*) in Slow Flight," *Journal of Experimental Biology*, Vol. 125, 1986, pp. 287–307.

[6]Spedding, G. R., "The Wake of a Kestrel (*Falco tinnunculus*) in Flapping Flight," *Journal of Experimental Biology*, Vol. 127, 1987, pp. 59–78.

[7]Spedding, G. R., Rayner, J. M. V., and Pennycuick, C. J., "Momentum and Energy in the Wake of a Pigeon (*Columba livia*) in Slow Flight," *Journal of Experimental Biology*, Vol. 111, 1984, pp. 81–102.

[8]Jones, K. D., Dohring, C. M., and Platzer, M. F., "Wake Structures behind Plunging Airfoils: A Comparison of Numerical and Experimental Results," AIAA Paper 96-0078, January 1996.

[9]Wilmott, P., "Unsteady Lifting-Line Theory by the Method of Matched Asymptotic Expansions," *Journal of Fluid Mechanics*, Vol. 186, Jan. 1988, pp. 303–320.

[10]Phlips, P. J., East, R. A., and Pratt, N. H., "An Unsteady Lifting Line Theory of Flapping Wings with Application to the Forward Flight of Birds," *Journal of Fluid Mechanics*, Vol. 112, Nov. 1981, pp. 97–125.

[11]Ahmadi, A. R., and Widnall, S. E., "Unsteady Lifting-Line Theory as a Singular-Perturbation Problem," *Journal of Fluid Mechanics*, Vol. 153, April 1985, pp. 59–81.

[12]Lan, C. E., "The Unsteady Quasi-Vortex-Lattice Method with Applications to Animal Propulsion," *Journal of Fluid Mechanics*, Vol. 93, No. 4, 1979, pp. 747–765.

[13]Lighthill, M. J., "Aquatic Animal Propulsion of High Hydromechanical Efficiency," *Journal of Fluid Mechanics*, Vol. 44, 1970, pp. 265–301.

[14]Wu, T. Y., "Swimming of a Waving Plate," *Journal of Fluid Mechanics*, Vol. 10, No. 3, 1961, pp. 321–344.

[15]Chopra, M. G., "Hydromechanics of Lunate-Tail Swimming Propulsion," *Journal of Fluid Mechanics*, Vol. 64, 1974, pp. 375–391.

[16]Chopra, M. G., "Large Amplitude Lunate Tail Theory of Fish Locomotion," *Journal of Fluid Mechanics*, Vol. 74, 1976, pp. 161–182.

[17]Chopra, M. G., and Kambe, T., "Hydromechanics of Lunate-Tail Swimming Propulsion: Part 2," *Journal of Fluid Mechanics*, Vol. 79, 1977, pp. 49–69.

[18]Cheng, J. Y., Zhuang, L. X., and Tong, B. G., "Analysis of Swimming Three-Dimensional Waving Plates," *Journal of Fluid Mechanics*, Vol. 232, 1991, pp. 341–355.

[19]Betz, A., "Schraubenpropeller mit geringstem Energieverlust," *Nachrichten von der Gesellschaft der Wissenschaften zu Göttingen, Mathematisch-Physikalische Klasse*, Vol. 1919, 1919, pp. 193–217.

[20]Goldstein, S., "On the Vortex Theory of Screw Propellers," *Proceedings of the Royal Society of London, Series A*, Vol. 123, 1929, pp. 440–465.

[21]Jones, R. T., "Wing Flapping with Minimum Energy," *Aeronautical Journal*, Vol. 84, July 1980, pp. 214–217.

[22]Potze, W., and Sparenberg, J. A., "On the Efficiency of Optimum Finite Amplitude Sculling Propulsion," *International Shipbuilding Progress*, Vol. 30, No. 351, 1983, pp. 238–244.

[23]Hall, S. R., Yang, K. Y., and Hall, K. C., "Helicopter Rotor Lift Distributions for Minimum Induced Power Loss," *Journal of Aircraft*, Vol. 31, No. 4, 1994, pp. 837–845.

[24]Hall, K. C., and Hall, S. R., "Minimum Induced Power Requirements for Flapping Flight," *Journal of Fluid Mechanics*, Vol. 323, 25 Sept. 1996, pp. 285–315.

[25]Sparenberg, J. A., *Hydrodynamic Propulsion and Its Optimization: Analytic Theory*, Kluwer Academic, Boston, 1995, pp. 216–217.

[26]de Jong, K., "On the Optimization, Including Viscosity Effects, of Ship Screw Propellers with Optional End Plates (Part I)," *International Shipbuilding Progress*, Vol. 38, No. 414, 1991, pp. 115–156.

[27]de Jong, K., "On the Optimization, Including Viscosity Effects, of Ship Screw Propellers with Optional End Plates (Part II)," *International Shipbuilding Progress*, Vol. 38, No. 415, 1991, pp. 211–252.

[28]Lifanov, I. K., and Polonskii, I. E., "Proof of the Numerical Method of 'Discrete Vortices' for Solving Singular Integral Equations," *Journal of Applied Mathematics and Mechanics*, Vol. 39, No. 4, 1975, pp. 742–746.

[29]Albano, E., and Rodden, W., "A Doublet-Lattice Method for Calculating Lift Distributions on Oscillating Surfaces in Subsonic Flows," *AIAA Journal*, Vol. 7, No. 2, 1969, pp. 279–285.

[30]Bliss, D. B., and Epstein, R. J., "A Novel Approach to Aerodynamic Analysis Using Analytical/Numerical Matching," AIAA Paper 95-1840-CP, 1995.

[31]Theodorsen, T., "General Theory of Terodynamic Instability and the Mechanism of Flutter," NACA Rept. 496, 1935.

[32]Triantafyllou, G. S., Triantafyllou, M. S., and Grosenbaugh, M. A., "Optimal Thrust Development in Oscillating Foils with Application to Fish Propulsion," *Journal of Fluids and Structures*, Vol. 7, No. 2, 1993, pp. 205–224.

[33]Katz, J., and Weihs, D., "Behavior of Vortex Wakes from Oscillating Airfoils," *Journal of Aircraft*, Vol. 15, No. 12, 1978, pp. 861–863.

[34]Katz, J., and Weihs, D., "Wake Rollup and the Kutta Condition for Airfoils Oscillating at High Frequency," *AIAA Journal*, Vol. 19, No. 12, 1981, pp. 1604–1606.

[35]McCune, J. E., Lam, C.-M. G., and Scott, M. T., "Nonlinear Aerodynamics of Two-Dimensional Airfoils in Severe Maneuver," *AIAA Journal*, Vol. 28, No. 3, 1990, pp. 385–393.

[36]Jones, K. D., Dohring, C. M., and Platzer, M. F., "Wake Structures behind Plunging Airfoils: A Comparison of Numerical and Experimental Results," AIAA Paper 96-0078, Jan. 1996.

[37]Ashley, H., and Landahl, M., *Aerodynamics of Wings and Bodies*, Addison–Wesley, Reading, MA, 1965, pp. 21–50.

[38]Abbott, I. H., von Doenhoff, A. E., and Stivers, L. S. Jr., "Summary of Airfoil Data," NACA Rept. 824, 1945.

[39]Hall, K. C., Pigott, S. A., and Hall, S. R., "Power Requirements for Large-Amplitude Flapping Flight," *Journal of Aircraft,* Vol. 35, No. 3, May–June 1998, pp. 352–361.

[40]Drela, M., "XFOIL: An Analysis and Design System for Low Reynolds Number Airfoils," *Low Reynolds Number Aerodynamics*, Springer-Verlag Lecture Notes in Engineering, edited by T. J. Mueller, No. 54, Springer-Verlag, Berlin, 1989.

[41]Hall, S. R., "Micro Air Vehicles: Propulsion and Actuation Issues," *Defense Science Study Group V: Study Reports 1996–1997, Volume I: Papers 1–13*, IDA Paper P-3414, Institute for Defense Analyses, Alexandria, VA, February 1998, pp. 193–206.

Chapter 14

Leading-Edge Vortices of Flapping and Rotary Wings at Low Reynolds Number

Hao Liu*

Institute of Physical and Chemical Research (RIKEN), Saitama, Japan

and

Keiji Kawachi[†]

University of Tokyo, Tokyo, Japan

Nomenclature

C_D = drag coefficient [drag force$/(0.5U_{ref}S\rho)$]
C_L = lift coefficient [lift force$/(0.5U_{ref}S\rho)$]
c_m = reference length, mean wing chord length
f = flapping frequency of the flapping wing
k = reduced frequency [$\omega c_m/(2U_{ref})$]
R = wing length
Re = Reynolds number ($U_{ref}c_m/\nu$)
Ro = Rossby number [$U_{ref}/(\omega c_m)$]
$r_2(s)$ = radius of the second moment of wing area as a fraction of R
S = wing area
U_{ref} = reference velocity, at the radius of the second moment of wing area [$\Omega r_2(s)R$]
u, v, w = velocity components in the wing-fixed coordinate system
x, y, z = wing-fixed Cartesian coordinates
ρ = density
ν = kinematic viscosity
Φ = beat amplitude of the flapping wing
Ω = maximum Angular velocity of the flapping wing ($\pi f \Phi$)
ω = angular velocity of the rotary wing ($2\pi f$)

---

*Senior Research Scientist.
[†]Professor, Research Center for Advanced Science and Technology (RCAST).

I. Introduction

T HERE is a growing recognized need for miniature flight vehicles with multi-functional capabilities, such as micro air vehicles (MAVs), with wingspans of 15 cm or less and flight speed of 30 to 60 km/h for military and civilian applications.[1] Our knowledge of how to design conventional fixed wing aircraft with moderate aspect ratio is very likely not directly applicable to designing such MAVs or smaller flight vehicles because the aerodynamic performance of stationary airfoils drops at such low Reynolds numbers (10^3–10^5).[2–4] Flapping wing flight of insects provides us with a sophisticated example of utilizing unsteady aerodynamics to mechanize the miniature flight structures. However, our understanding of insect flight is that the familiar conventional aerodynamic theories fail to predict sufficient lift to support the insect.[1,5–8] After several studies of visualizing airflow around flapping insect wings in an attempt to identify the high-lift mechanism, it is has been found[7] that a micro-leading-edge vortex (LEV) mechanism during the translational motion of downstroke is of significant importance in flapping wing insect flight.

To establish an overall understanding of the vortex structures at downstroke, upstroke, and during the rotational motion of supination and pronation, and of how this LEV is created and contributes to enhance the lift generation, we developed a computational fluid dynamics (CFD) methodology for modeling the flapping insect wing.[9,10] Our model simulated the complete three-dimensional, unsteady flowfields with large-scale vortices around deformable wings, which are based on real insect wings and can mimic their real kinematics. Successful modeling of the hovering flight of a hawkmoth wing thereby established the importance of accurately predicting the existence and timing of the complex dynamic vortex structure at low Reynolds numbers ranging from 10^3 to 10^4.

The aerodynamic performance of airfoils and wings at low Reynolds numbers ($<10^4$) greatly depends upon the features of the leading-edge bubbles (i.e., the LEVs), which can be largely affected by both three-dimensional geometry and wing dynamics.[2–4] LEVs can also be found in the case of fixed airfoils and wings[2,3] but are quite unstable in general, resulting in large-scale separation and reattachment that is difficult to control merely through optimal selection of the profile. Fortunately, as often seen in insect and bird flight, appropriate active control of the wing motion, or kinematics, can be an efficient way of achieving high and stable lift generation. One successful man-made aircraft that actively utilizes the wing motion is the rotary-blade helicopter. However, whether the rotary wing (blade) still works well at the size of insect and bird wings, or, if not, how its aerodynamic characteristics change, is not yet clear. Motivated by such thinking we started the present study, aiming to clarify the similarities and differences of the LEVs of flapping and rotary wings through an extensive CFD study of flows around a rotary wing as well as their relationship with the aerodynamic characteristics.

In the following, we first give a brief description of the numerical algorithm with particular emphasis on the formulation in a rotating frame of reference and the data set used in the computations associated with a moth. Results are then presented for the flapping and rotary wing. Finally, we address a comprehensive understanding of the characteristics of the leading-edge vortex in terms of the wing rotation.

II. Computational Modeling of a Rotary Wing

Consider the flow around a rotary wing (blade) with a constant angular velocity ω in a wing-fixed system (x, y, z), as depicted in Fig. 1. The governing equations are the three-dimensional, incompressible, unsteady Navier–Stokes equations written in strong conservation form for mass and momentum. In the manner of finite volume discretization, an integrated form of the nondimensionalized governing equations written in a general curvilinear coordinate system is

$$\int_{V(t)} \left(\frac{\partial \mathbf{q}}{\partial \tau} + f_{\text{cent}} + f_{\text{cor}} \right) dV + \int_{V(t)} \left(\frac{\partial \mathbf{F}}{\partial x} + \frac{\partial \mathbf{G}}{\partial y} + \frac{\partial \mathbf{H}}{\partial z} + \frac{\partial \mathbf{F}_v}{\partial x} \right.$$

$$\left. + \frac{\partial \mathbf{G}_v}{\partial y} + \frac{\partial \mathbf{H}_v}{\partial z} \right) dV = 0 \tag{1}$$

where

$$\mathbf{q} = \begin{bmatrix} u \\ v \\ w \\ p \end{bmatrix}, \quad f_{\text{cent}} = \begin{bmatrix} -x/Ro^2 \\ -y/Ro^2 \\ 0 \\ 0 \end{bmatrix}, \quad f_{\text{cor}} = \begin{bmatrix} -2v/Ro \\ 2u/Ro \\ 0 \\ 0 \end{bmatrix}, \tag{2}$$

$$\mathbf{F} = \begin{bmatrix} u^2 + p \\ uv \\ uw \\ \beta u \end{bmatrix}, \quad \mathbf{G} = \begin{bmatrix} vu \\ v^2 + p \\ vw \\ \beta v \end{bmatrix}, \quad \mathbf{H} = \begin{bmatrix} wu \\ wv \\ w^2 + p \\ \beta w \end{bmatrix}$$

$$\mathbf{F}_v = -\left(\frac{1}{Re} + v_t \right) \begin{bmatrix} 2u_x \\ u_y + v_x \\ u_z + w_x \\ 0 \end{bmatrix}, \quad \mathbf{G}_v = -\left(\frac{1}{Re} + v_t \right) \begin{bmatrix} v_x + u_y \\ 2v_y \\ v_z + w_y \\ 0 \end{bmatrix}, \tag{3}$$

$$\mathbf{H}_v = -\left(\frac{1}{Re} + v_t \right) \begin{bmatrix} w_x + u_z \\ w_y + v_z \\ 2w_z \\ 0 \end{bmatrix}$$

In the preceding equations, β is the pseudo-compressibility coefficient; p is pressure; u, v, and w are velocity components in Cartesian coordinate system x, y, and z; t denotes physical time and τ is pseudo time; Re is the Reynolds number; Ro is the Rossby number; and v_t is the eddy viscosity in turbulence simulation. Note that the effect of using a rotating frame of reference (i.e., a noninertial frame) can be taken into account by introducing centrifugal and Coriolis forces, f_{cent} and f_{cor} as defined in the Eq. (2), respectively. The term $\mathbf{q}$ associated with the pseudo time is designed for an inner iteration at each physical time step and will vanish when the divergence of velocity is driven to zero to satisfy the equation of continuity.

Schematic diagrams of the flapping- and rotary-wing models are illustrated in Fig. 1. A complete description of the methodology, including formulation of the governing equations, generation of grids, boundary conditions, details of the computations, validation study of grid resolution and time-step size, features of the vortex flowfields, and comprehensive results can be found in Ref. 9.

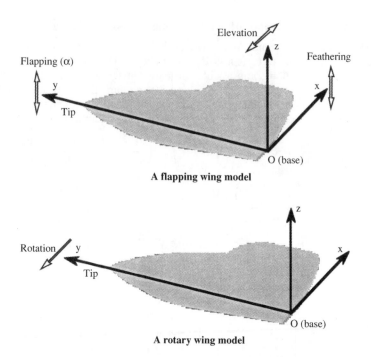

Fig. 1   Schematic diagrams of flapping- and rotary-wing models.

The geometric model of the wing was based on a tracing of the outline of a fore- and hindwing of a hawk moth. The wing has a mean wing chord length c_m of 1.83 cm, a wing length R of 4.83 cm, a thickness of 2.5% of the mean wing chord length, and a radius of the second moment of the wing area expressed as a fraction of R, $r_2(s) = 0.52$. The flapping frequency of the wing f is 26.1 s^{-1}, and the wingbeat amplitude is taken to be approximately 2.0 rad. Hence, in the case of hovering the Reynolds number Re is calculated to be about 5000 and the reduced frequency k is 0.37. Note that the $Re = 4000$ value was actually used in our systematic studies of flapping wings under the assumption that the Reynolds number effect may not be significant because we have experimental results available at this Reynolds number[2,3] for comparison. The three-dimensional movement of the flapping wing was based on the kinematic data of the hovering hawk moth. It consists of three basic motions: 1) flapping in the stroke plane by the positional angle α, 2) elevation of the wing on either side of the stroke plane, and 3) rotation (feathering) of the forewing (leading edge) and hindwing (trailing edge) with respect to the stroke plane by the angle of attack of the forewing and the angle of attack of the hindwing (Fig. 1).

For the rotary wing, the same geometric model is utilized but in a steadily rotating mode with respect to the z axis. The Reynolds number and the reduced frequency are identical to those in the flapping wing model. In the rotating frame of reference, flows around the rotary wing may be solved by computing the flows around a wing, as described in Eqs. (1–3), which starts impulsively from rest, advancing forward with a spanwise gradient of oncoming velocity. The influence of the centrifugal and Coriolis forces is considered by introducing the Rossby number, which is calculated to be 1.37—quite small in this case. To avoid the influence of the induced velocity and wake on the rotary wing, all the computations were conducted only during the first cycle of rotation.

III. Numerical Accuracy

A complete description of the numerical algorithm, including the governing equations, numerical schemes, generation of grids, boundary conditions, details of the computations and comprehensive comparison with reliable experimental results may be found in Ref. 9. In the present steady computations, the same grid systems, time-step size, and other computational parameters used in the previous simulation of the flapping insect wing were also employed; these were confirmed to be able to validate the simulated results even for the rotary wing.

IV. Results

A. Leading-Edge Vortex of Flapping Wing

For a better understanding of the unsteady aerodynamics of flapping wing flight, we first highlight the characteristics of the dynamic vortex flow around the flapping wing of a hovering hawkmoth.[9,10] Figures 2a–d show the vortex structures

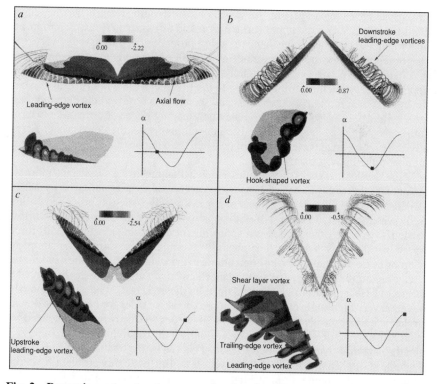

Fig. 2 Dynamic vortex structure around a flapping moth wing during a complete wing beat cycle: a) downstroke; b) supination; c) upstroke; d) pronation.

during a complete flapping cycle, where vortices were visualized by instantaneous streamlines released from the upper wing surface. Negative pressure contours are shown in which red denotes high pressure and blue represents low pressure. During most of the downstroke (Fig. 2a), an intense, conical spiral leading-edge vortex is detected with strong axial flow at the core. This leads to a steadily if increasing negative pressure region on the upper wing surface and hence augments the lift force. During the latter half of the downstroke, another leading-edge vortex is observed at the wing tip (Fig. 2b), which also shows a spiral flow at the core but runs toward the base. Although the combination of these two vortices results in a very large lift force, this second vortex could not be detected in the airflow visualization[6] because smoke was only released from the leading edge over the inner half of the forewing. During the subsequent supination the two vortices change their shapes gradually, and as the wing approaches the bottom of the wing stroke, they combine with a tip vortex, which appeared during early supination to form a hook-shaped vortex (Fig. 2b). During the latter half of supination, the vortices are strongly deformed, quickly collapse beyond the midspan joint of the two leading-edge vortices, and are shed from the trailing edge. This was also observed in the flow visualization experiments of a tethered peacock butterfly.[5] During the early upstroke, no leading-edge vortex is observed, and the flow is quite smooth over both the upper and lower surfaces. When the wing precedes to the latter half of the upstroke, a leading-edge vortex (Fig. 2c) appears with a comparable size to those observed during the first half of the downstroke. During pronation, the upstroke leading-edge vortex (Fig. 2d) is still observed rolling over the leading edge, but it is not located on the upper surface as would be expected if it were produced by a rotational circulation mechanism during pronation. A trailing-edge vortex is also observed below the lower surface as a result of the wing rotation during the first half of the pronation. Additionally, a shear layer vortex is observed on the upper surface, stretching from the base to the tip. When the wing is ready to start the translational motion of downstroke, all the vortices observed during pronation are shed and thus the wing must produce a new leading-edge vortex in order to have some mechanism of enhancing the life-generation during the downstroke. Actually, the leading-edge vortex does initiate when the wing just starts the translational motion of the downstroke.

Estimation of the forces during one complete flapping cycle shows that approximately 80% of the lift force[9,10] (here we mean vertical force against gravity with a calculated inclination of stroke plane of approximately 24 deg) is produced during the downstroke and the latter half of the upstroke. Hence the leading-edge vortex, particularly on the downstroke, is responsible for generating most of the lift force for an insect to stay airborne. The mean lift (vertical) force during one complete flapping cycle is calculated to be comparable to the weight (1.5 g) of the hawkmoth.

Our previous study suggests that the mechanism of augmenting the lift force by means of LEV generation can be explained by dynamic stall: The wing obviously travels at very high angle of attack[10] of approximately 50 deg for most of the downstroke, and this exceeds the stall angle of 20–30 deg in steady flow[3] if reduced by approximately 15 deg by the downwash. But how the rotational motion, in particular the *potential* spanwise velocity gradient caused by the rotation, influences this dynamic stall phenomenon is still not yet clear. To isolate this problem from the complexities of flapping flight, we have therefore performed an extensive study of the aerodynamic characteristics of a rotary wing with the same geometry of the hawkmoth wing.

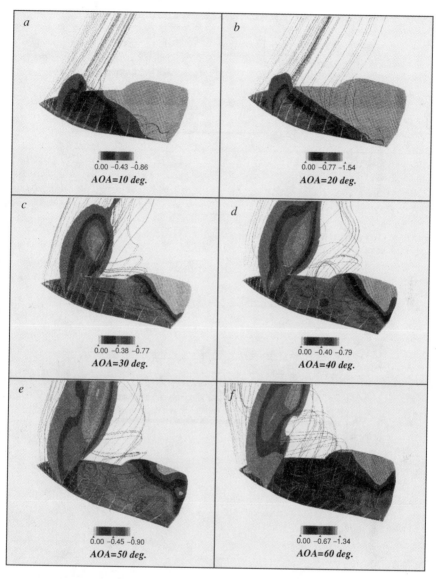

Fig. 3 Flow visualization of leading-edge vortices around a rotary moth wing at angles of attack of a) 10, b) 20, c) 30, d) 40, e) 50, and f) 60 deg.

B. Leading-Edge Vortex of Rotary Wing

The rotary-wing model rotates with respect to the z axis at angles of attack ranging from 10 to 60 deg. The angular velocity of rotation is taken equal to the maximum achieved by the flapping hawkmoth wing. Figure 3 shows the visualized LEV structure, and the lift and drag coefficients are plotted in Fig. 4. Unlike the typical stall of stationary airfoils at high Reynolds number ($>10^6$), the lift coefficient

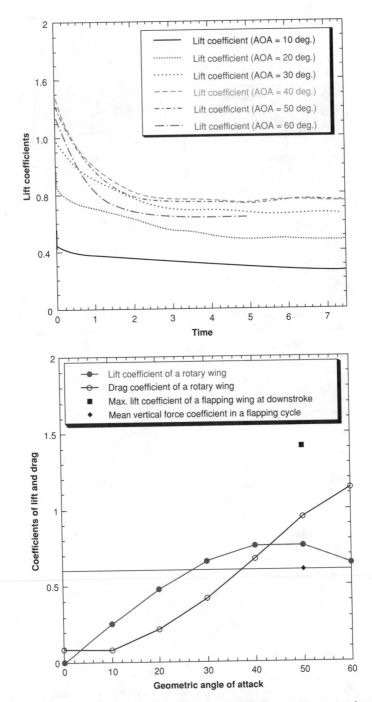

Fig. 4   Comparison of the lift coefficients for a flapping- and rotary-wing moth.
a) Hysterisis of the lift coefficients b) Lift coefficients against the angles of attack.

does not drop even at very large angles of attack (>40 deg) (Fig. 4b). An apparent stall angle where the linear increase of the lift reaches a halt is observed around 30 deg. Figure 3 clearly identifies the leading-edge vortex as the source of lift generation at this moderate Reynolds number. The vortex is constantly created, and it grows with increasing angle of attack but does not collapse; correspondingly, a negative pressure region is induced on the leading edge, which is responsible for the lift generation.[11]

Vortices initiated at the leading edge are quite two-dimensional but become larger as they move toward the wingtip, which is thought to be due to the rotation-based potential velocity gradient in the spanwise direction. The LEV is subject to a combination of centrifugal force, Coriolis force, dynamic pressure gradient corresponding to the spanwise velocity gradient, and pressure gradient through the cores of the LEVs, which originates from the LEV itself. Eventually, the LEVs take on a characteristic spiral three-dimensional shape with apparent axial (radial) flow at their cores (Fig. 3), as observed in the flapping wing (Fig. 2). The LEVs tend to enlarge in size with increasing angle of attack, and apparently they initiate shedding at the wingtip toward the wing base, for larger angles of attack exceeding 30 deg. This indicates that the axial flow at the core of LEV may play a key role in stabilizing the vortices and hence delaying the stall. This result also supports the observation of a stall delay phenomenon associated with a propeller from Himmelskamp's experimental work by way of Schlichtings,[12] which indicated that in moving toward the center of the rotor (i.e., the wing base), the stall delay became more and more pronounced.

Here we ask the question: What force plays the key role in creating this axial flow? We believe that the axial flow is dominated by the spanwise pressure gradient originating from the LEVs. Note that the potential spanwise velocity gradient due to the rotation results in many two-dimensional LEVs with different effective angles of attack, whose sizes and strengths increase in moving toward the wingtip. We believe that this leads to a much steeper pressure gradient than can be obtained by changing the other three factors. This LEV-induced pressure gradient runs through the cores of the LEVs and gives the vortices a characteristic spiral conical three-dimensional shape in toto. Hence, the axial flow is observed, closely corresponding to the LEVs at all angles of attack (Fig. 3). We can see the axial flow as well as the three-dimensional spiral shape even when the LEV is shed from the wing toward the wingtip. However, helicopter rotors and wind turbine blades at high Reynolds numbers and at smaller angles of attack generally will not result in a stable vortex and a radial flow. Within the helicopter community, many studies[13,14] reported that stall delay caused by rotation could be observed in both laminar and turbulence boundary layers, which led to additional lift generation. At high Reynolds number and small Rossby number, a combination of dynamic pressure gradient, centrifugal force, and Coriolis force may become dominant in influencing the radial flow and the characteristics of the boundary layer.

For comparison, consider the axial flow in the case of the flapping wing. As illustrated in Fig. 2a, the axial flow is accelerated by the spanwise pressure gradient up to 60–70% of the wing length, where a lower pressure island is observed and the spanwise pressure gradient reverses. As a result, the axial flow decelerates to a halt at about 75–80% of the wing length, where the vortex core breaks down and eventually separates from the wingtip. Because the axial flow is essential for the stability of the three-dimensional attached vortex,[4] this breakdown is clearly due to the instability of the vortex at the position where the axial flow disappears and the vortex enlarges quickly. Hence, the reverse pressure gradient at the wingtip is a key feature influencing the breakdown of the leading-edge vortex. Additionally, a large angle of attack of the wing can result in a steep velocity gradient at the wingtip and thus produce a reverse pressure gradient. During the latter half of the downstroke the velocity gradient over the wingtip becomes steeper, which increases the reverse pressure gradient. As a result, the point of breakdown is continuously pushed toward the base, and eventually at the end of the downstroke it is located near the middle of the wing (Fig. 2b).

Comparing Fig. 3 with Fig. 2, we see an obvious discrepancy between the LEVs of flapping and rotary wings. Overall, the LEVs in the rotary wing (Fig. 3) are not coherently attaching to the leading edge and hence they are not so intense: The maximum lift coefficient at angles of attack of 40–50 deg is only half of that achieved by the flapping wing in the middle of the downstroke (Fig. 4). This means that the LEV characteristics are strongly affected by the unsteady movements of the wing. So why can the flapping wing create a more intense leading-edge vortex and hence higher lift than the rotary wing? It may be suggested that time-varying inflow with respect to an oscillating airfoil may lead to intense vortex generation, mostly on rotation with low translation velocity, and hence the high lift-production mechanism,[15] with the wing in the middle of downstroke, corresponds temporally to the high velocity of the local inflow with moderate variation.[10] Acceleration (or added mass effect) also cannot explain this high-lift generation since the highest lift occurs around the mid-downstroke when the acceleration of the wing passes through zero. The nature of the intense leading-edge vortex that its strength increases along the wingspan with time (Fig. 2a) on most of the downstroke clearly identifies dynamic stall as the aerodynamic mechanism responsible for the high-lift production. Furthermore, dynamic stall for flapping wings is also characterized by the potential spanwise velocity gradient as well as the wing dynamics. Hence, by means of the sophisticated kinematics of the flapping wing, this dynamic stall mechanism is able to further largely delay the apparent stall angle as shown in Fig. 4 up to very high angles of attack; and with the axial flow convecting the vorticity toward the wingtip, it can significantly strengthen the LEV that is responsible for producing the extra lift. Obtaining more accurate measurements of three-dimensional flapping wing dynamics involving the three basic motions of flapping, elevation, and feathering as well as passive elastic deformation of the wing can be a promising step in deepening our understanding of the dynamic stall mechanism and may help us to sort out some applicable principles for man-made miniature flight vehicles.

V. Conclusions

The leading-edge vortices induced by flapping and rotary wings at low Reynolds number have been compared. We summarize our results as follows:

1) Rotary wings can generate a conical spiral LEV with axial flow that is dominated by the potential spanwise velocity gradient caused by rotation. The axial flow at the core of the LEV can delay stall and hence leads to stable lift generation with large "apparent" stalling angle of approximately 30 deg.

2) Flapping wings can create an intense, conical spiral LEV with strong axial flow, which is linked to a dynamic stall mechanism and is affected by the potential

spanwise velocity gradient as well as the wing dynamics. Hence, much higher lift force is generated, in particular during the downstroke.

3) A spanwise pressure gradient originating from the LEV itself creates the axial flow at the core of the LEV of both flapping and rotary wings.

Acknowledgment

This work was partially supported by the Japan Science and Technology Corporation.

References

[1] Shyy, W., Berg, M., and Ljungqvist, D., "Flapping and Flexible Wings for Biological and Micro Air Vehicles," *Progress in Aerospace Sciences*, Vol. 35, No. 5, 1999, pp. 455–506.

[2] Sunada, S., Sakaguchi, A., and Kawachi, K., "Airfoil Section Characteristics at a Low Reynolds Number," *Journal of Fluid Engineering*, Vol. 119, 1997, pp. 129–135.

[3] Ikehata, M., Inoue, T., Ozawa, M., and Matsumoto, S., "Experimental Investigation on Flow Fields of Viscous Fluid around Two-dimensional Wings: Comparison with Computational Results," *Journal of Marine Science and Technology*, Vol. 2, 1997, pp. 62–76.

[4] Willmott, A. P., and Ellington, C. P., "The Mechanics of Flight in the Hawkmoth *Manduca sexta*. II. Aerodynamic Consequences of Kinematic and Morphological Variation," *Journal of Experimental Biology*, Vol. 200, 1997, pp. 2723–2745.

[5] Brodsky, A. K., "Vortex Formulation in the Tethered Flight of the Peacock Butterfly *Inachis io* L. (Lepidoptera, Nymphalidae) and Some Aspects of Insect Flight Evolution," *Journal of Experimental Biology*, Vol. 166, 1991, pp. 77–95.

[6] Ellington, C. P., "Unsteady Aerodynamics of Insect Flight," *Biological Fluid Dynamics*, edited by C. P. Ellington and T. J. Pedley, *Symposium of the Society for Experimental Biology*, The Company of Biologist Limited, Cambridge, U.K., Vol. 49, 1995, pp. 109–129.

[7] Ellington, C. P., Van den Berg, C., Willmott, A. P., and Thomas, A. L. R., "Leading-Edge Vortex in Insect Flight," *Nature*, Vol. 384, 1996, pp. 626–630.

[8] Ennos, R., "Unconventional Aerodynamics," *Nature*, Vol. 344, No. 5, 1990, pp. 67–69.

[9] Liu, H., and Kawachi, K., "A Numerical Study of Insect Flight," *Journal of Computational Physics*, Vol. 146, 1998, pp. 1–33.

[10] Liu, H., Ellington, C. P., Kawachi, K., Van den Berg, C., and Willmott, A. P., "A Computational Fluid Dynamic Study of Hawkmoth Hovering," *Journal of Experimental Biology*, Vol. 201, No. 4, 1998, pp. 461–477.

[11] Wu, J. E., Vakili, A. D., and Wu, J. M., "Review of the Physics of Enhancing Vortex Lift by Unsteady Excitation," *Progress in Aerospace Sciences*, Vol. 28, 1991, pp. 73–131.

[12] Schlichtings, H., *Boundary Layer Theory*, 7th ed., McGraw-Hill, New York, 1979.

[13] Harris, F. D., "Preliminary Study of Radial Flow Effects on Rotor Blades," *Journal of the American Helicopter Society*, Vol. 11, No. 3, 1966, pp. 1–21.

[14] Corrigan, J. J., and Schlichting, H., "Empirical Model for Stall Delay due to Rotation," *Proceedings of the American Helicopter Society Aeromechanics Specialists Conference*, San Francisco, CA, 19–21 Jan. 1994.

[15] Gursul, I., and Ho, C. M., "High Aerodynamic Loads on an Airfoil Submerged in an Unsteady Stream," *AIAA Journal*, Vol. 30, No. 4, 1992, pp. 1117–1120.

Chapter 15

On the Flowfield and Forces Generated by a Flapping Rectangular Wing at Low Reynolds Number

Richard Ames,* Oliver Wong,* and Narayanan Komerath[†]
Georgia Institute of Technology, Atlanta, Georgia

Nomenclature

C_L = lift coefficient, $L/[(1/2)\rho U_\infty^2 S]$
c = wing chord length
d = distance from wing pivot to center of mass
F_{meas} = total measured force
L = lift
m = mass of wing
S = wing area
T = flapping period
t = time
U = velocity magnitude
U_∞ = freestream velocity
u, v = x, y components of velocity, respectively
α = wing angle of attack
γ = wing angular acceleration, $d\Omega/dt$
θ = flapping angle
Ω = wing angular speed, $d\theta/dt$
ω_z = z component of vorticity
$\dot{}$ = superscript dot denotes differentiation with respect to time
$\bar{}$ = overbar denotes average

I.   Introduction

THIS chapter describes an experimental study of the flowfield and forces produced by a flapping flat plate where the amplitude of the flapping is large,

*Graduate Research Assistant, School of Aerospace Engineering.
[†]Professor, School of Aerospace Engineering. Associate Fellow AIAA.

the reduced frequency is moderate, and the chord Reynolds number is low. The objective of the work is to generate a basic experimental test case that can be used to connect and validate the results on unsteady aerodynamics over a wide range of Reynolds number and unsteadiness.

The aerodynamics of small-amplitude wing motions in the reduced frequency range $0 < k < 0.1$ is well understood for applications where the chord Reynolds number is large. Many applications for autonomous flight vehicles in the atmospheres of Earth and other planets occur at Reynolds number below 100,000, with some as low as 10,000. Several of these applications require very small vehicles, low flight speeds, and high levels of unsteady motion caused by either gusts or quick maneuvers. Vehicle designs in this range may benefit from the use of wings flapping with high amplitudes and frequencies.

Past work in this area has been driven by interest in fixed wing and rotary wing uninhabited aerial vehicles, large-scale ornithopters, and the flight mechanisms of actual birds and insects. The aerodynamics of all but the smallest of birds, and of insects, appears to be predictable using aerodynamic theory developed for larger aircraft. Thus, steady and quasi-steady analyses using panel codes appear to suffice for fixed wing unmanned air vehicles (UAVs) and the larger birds where the reduced frequency of wing motion is small and the flight Reynolds number is on the order of 50,000 or more. Attention has been devoted to airfoil aerodynamics at low Reynolds number and low to moderate angles of attack.

II. Previous Work

There is as yet only a small amount of research that has been directed toward the problem of low Reynolds number, unsteady flows. Most of the work related to the low Reynolds number flight regime has been done with regard to design and optimization at low angles of attack and steady flight conditions. The majority of the work in the field of unsteady aerodynamics has been directed at higher Reynolds number flows. There is a vast amount of research directed toward the measurement and prediction of dynamic stall and of airfoil/wing behavior for small excursions around some nominal angle of attack. This work has focused on developing the ability to predict such phenomena as flutter, aircraft gust response, and pitch rate effects (with particular emphasis on helicopter aerodynamics). Figure 1 gives a graphical description of the existing general aerodynamic knowledge database in terms of reduced frequency and Reynolds number.

A. Numerical Studies

1. General Two-Dimensional Flows

The well-known lift-deficiency theory of Theodorsen is probably the most widely applied theory in the field of unsteady aerodynamics. However, inherent within this theory is the assumption of infinitely thin regions of vorticity that are easily separated into bound circulation on the airfoil and that contained within it a planar wake. Although the theory is appropriate to attached, high Reynolds number flows where the effects of vorticity are generally contained within the thin airfoil boundary layer, the increased (and widespread) effects of viscosity at low Reynolds number brings to question its applicability.

Using a numerical approach based on singularity distributions[1,2] provides a means by which airfoil thickness can be accounted for and the planar wake

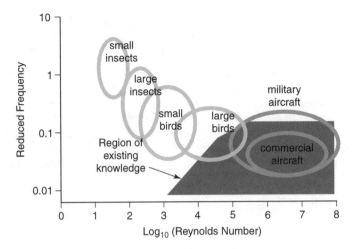

Fig. 1 Existing general aerodynamic knowledge database in terms of reduced frequency and Reynolds number.

assumption can be removed. However, the assumption of a singular vorticity distribution still confines the effects of vorticity to narrow regions around the airfoil and the wake. Full Navier–Stokes calculations have also been performed for low Reynolds number cases,[3–11] removing all assumptions about vorticity confinement. However, the computational time required for these numerical investigations is prohibitive because of the low Re and combined laminar/turbulent nature of the free shear layers generated.

2. General Three-Dimensional Flows

There has been significant work in expanding the results of two-dimensional unsteady aerodynamics into the third dimension,[2,12–14] even at the level of Navier–Stokes computations.[15] Much of this work has been an extension of methods applied to helicopter aerodynamics and, as such, has been limited to high Reynolds number flows. Most of these models (all numerical, none analytical) have fallen into three categories: blade-element-type computations where the three-dimensional nature of the body is represented by a collection of two-dimensional entities, lifting-line computations that replace wings and rotors with filaments of vorticity bound along the span, and singularity distributions over a paneled representation of the body (panel codes).

These unsteady numerical models have been developed both in the time domain and the frequency domain[16,17] and even for separated flows,[18] but all share the same high Reynolds number restrictions. There have been examples of the application of these methods to unsteady, low Reynolds number flight[14,19–21] but the results should be viewed in the light of the mismatch between the assumptions behind the computations and the flight vehicle/conditions modeled.

B. Experimental Studies

The difficulties associated with low Reynolds number aerodynamic testing are well documented, as are the difficulties associated with unsteady measurements.

Briefly, most experimenters have encountered serious problems with repeatability and facility dependence of lift and drag measurements at Reynolds numbers below 300,000. This is usually attributed to the critical dependence of transition lines, separation lines, and separated flow regions on various parameters that are difficult or impossible to control adequately. The vast body of work done by the NACA in measuring force and moment coefficients for various airfoils was limited to Reynolds number greater than 300,000. Much work since then has been done to extend this body of experimental measurements to the low Reynolds number regime.[22,23] However, these measurements have all been directed at steady flows.

The mechanisms of bird and insect flight have also been studied with flow visualization techniques.[24] A limited number of force measurements have been made for low Reynolds number flapping wings.[25,26] Higher Reynolds number force measurements on flapping wings have been done with primary emphasis on thrust development.[27]

III. Scope of Present Work

Referring to Fig. 1 and the previous work, some comments about the terms "high" and "low" in the context of reduced frequency and Reynolds number, respectively, would serve to set the present work in context. Textbooks on unsteady aerodynamics generally define the ranges of unsteadiness in terms of the Theodorsen function and the regimes where various assumptions hold. Thus, $0 < k < 0.03$ is nominally considered "quasi-steady" (wake effects are not very significant), $0.03 < k < 0.1$ can be considered "quasi-unsteady" (wake effects are significant but apparent-mass acceleration effects are negligible), and beyond that, flow is considered "fully unsteady" (all unsteady effects are important). For $k > 1$, acceleration effects begin to dominate. Figure 1 shows that little is really known about flowfields beyond a reduced frequency of 0.5. Insect flight and the wing flapping of hummingbirds are beyond this range. It is apparent that airfoil aerodynamics below a Reynolds number of 300,000 poses problems, and below 50,000, little is known at any significant level of unsteadiness. These help define the parameter range of the present experiments. The Reynolds number is kept well below 50,000. The reduced frequency is kept in the range from 0.1 to 0.6 (the present data are at 0.5). This allows comparisons with theory to begin constructing potential-flow methods with some reasonable expectation of being able to identify similarities and issues. Even here, the flapping rate is sufficiently high to induce instantaneous angles of attack that are well beyond the attached-flow regime.

IV. Experimental Setup

Velocity measurements were carried out using a laser velocimetry system (LV) in the John J. Harper Wind Tunnel at Georgia Tech. The closed-return tunnel has a 7 × 9 ft atmospheric-pressure test section and was run at a freestream speed of 10 ft/s, giving a chord Reynolds number of approximately 20,000.

The LV system used a 6-W argon-ion laser with the light transmitted and received through a fiber-optic probe. The probe was installed on a two-axis traverse with a position accuracy of better than 0.001 in. and was controlled via a serial interface to a PC with an Intel® Pentium® processor. The velocity was measured using a counter/processor interfaced to the PC. The LV counter/processor was configured to use 32 cycles/burst with 1% comparison to define a data point. A once-per-cycle timing pulse from the wing mechanism provided the phase reference to sort the velocity data. The data were sorted into 500 bins per cycle and ensemble-averaged over approximately 120 data points per bin. The root-mean-square velocity fluctuation in each bin was also computed. The velocity measurements were performed one component at a time.

The wing model (see Fig. 2 and Table 1) had a rectangular planform spanning 10 in. with a chord of 4.125 in. The wing was constructed of 1/16-in. sheet balsa wood with 1/32-in. sheet aluminum reinforcement near the root. The wing was covered with Monokote^{TM}. The wing had a circular arc camber of 3%, concave to the x axis. The reference axes were defined as shown in Fig. 2 with the wing in the vertical position. Positive flapping angles are given by the right-hand rule in the x–y plane, measured from the y axis. The freestream is in the $+z$ direction (see Fig. 2).

V.   Wing Motion

A MicroMo DC servomotor with 76:1 gearbox provided the flapping mechanism. The reduced frequency is given by

$$k = \frac{\pi f c}{U_\infty}$$

Using this definition gives $k = 0.5$ for the case where the flapping was at 4.5 Hz. The flapping motion was approximately sinusoidal (as shown in Fig. 3) with a maximum at $+35.2$ deg and a minimum at -13.4 deg. Figure 3 also shows the wing angle of attack, angular velocity, and angular acceleration. The master once-per-cycle trigger fired at $T_0 = 0$ s corresponding to a wing position that is just short of its maximum leftward $(-x)$ position. This position also corresponds to the time just before the wing begins the downstroke.

Wing position data were obtained using a US Digital optical encoder and a Technology 80 TE5312 encoder counter. The encoder uses a 1024 count/rev disk in quadrature mode, giving 4096 counts/rev or a precision of ±0.088 deg. Flexibility in the wing is not accounted for; strobe light visualization showed no evidence of significant motion-induced bending.

Wing angle of attack, angular velocity, and angular acceleration were calculated based on the wing position measurements. The instantaneous angular velocity was computed based on the encoder data and used to compute the motion-induced velocity at 80% span. Defining the angle of attack as

$$\alpha = \tan^{-1}\left(\frac{v_i}{U_\infty}\right)$$

where v_i is the motion-induced velocity gives the angle of attack variation shown in Fig. 3. As shown in the plot, the angle of attack at 80% span ranges ±38 deg.

VI.   Velocity Data Planes

The velocity and flow visualization data were taken in five planes. These planes were in the plane of flapping (x–y plane) at the leading edge, the wing midchord, the trailing edge, one chord length, and two chord lengths downstream of the trailing edge.

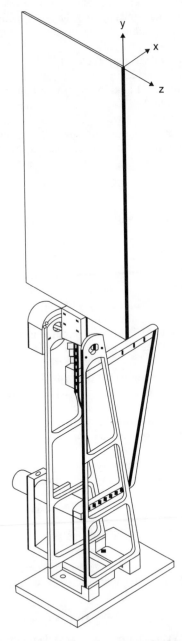

Fig. 2   Wing configuration. Freestream is along the _z_ axis.

Table 1   Flapping wing experimental parameters

Planform: rectangular plate	Freestream speed: 10 ft/s
Chord: 4.125 in.	Flapping frequency: 4.5 Hz
Span: 10.00 in.	Amplitude: +35.2 deg/−13.4 deg
Camber: 3% circular arc	Reduced frequency: 0.5

The measurement points in the first three planes were spaced at $0.048c$ in the x direction and $0.061c$ in the y direction (see Fig. 4). These planes spanned $1.21c$ in the y direction and $2.91c$ in the x direction, giving a total of 1281 measurement points per plane. The (0,0) point was placed at the tip of the wing in the 0 deg position and the measurement plane extended $0.67c$ above, $0.55c$ below, and $1.45c$ to either side. The wake measurement planes used the same zero reference point but extended $1.45c$ above, $0.97c$ below, $2.42c$ in the $-x$ direction, and $1.45c$ in the $+x$ direction, giving 693 measurement points per plane. Grid spacing was $0.061c$ in both the x and y directions in the wake measurement planes.

VII.   Velocity Field Data Analysis

The measured velocity data were sorted into 500 bins per cycle and were used to numerically compute the in-plane component of vorticity, given by

$$\omega_z = \frac{1}{2}\left(\frac{dv}{dx} - \frac{du}{dy}\right)$$

Contours of the in-plane vorticity component were animated to view the dynamic flowfield. The mean flowfield values were computed by averaging over all bins for a particular measurement point. Vortex core positions were plotted by locating the center of rotation from vector plots of the LV data.

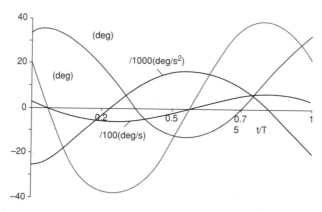

Fig. 3   Wing flapping angle θ, angle of attack α, angular velocity Ω, and angular acceleration γ.

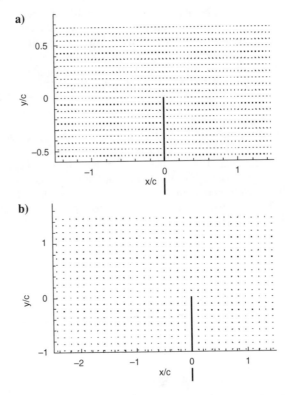

Fig. 4   Grid points for LV measurements: a) leading edge, mid-chord, trailing edge; b) wake. The thick line shows the wing in zero degree position.

The root-mean-square (rms) velocities were also computed for each bin of each LV grid location. The rms velocity is given by

$$U_{\text{rms}} = \sqrt{\frac{1}{n}\sum_{i=1}^{n}(U_i - \bar{U})^2}$$

where n is the total number of data points taken for a given bin, U_i are the velocity measurements, and $\bar{U}$ is the ensemble average for all n data points in a given bin. Note that U_{rms} is the rms velocity for a particular bin at a particular grid location. As a result, for a given grid location, the value of U_{rms} will (in general) vary in time.

VIII.   Force Measurements

Force measurements were obtained by mounting the entire flapping wing assembly onto a two-component force/moment balance. The data were recorded and stored using a National Instruments® analog-to-digital conversion system. The once-per-cycle timing reference was used to trigger the acquisition to provide a phase reference for the force measurements. In this experiment, the flapping rates and freestream speeds were varied over a wide range. The wing motion histories changed from their intended sinusoidal nature at the higher speeds, as the load on the motor varied through the cycle. Actual motion histories detailed in Ref. 28 were used in the panel code computations described later in the chapter.

At the high flapping rates, the inertial loading on the force balance was a significant portion of the total measured force. As a result, the following dynamic model was used to correct the measured force data:

$$L = F_{\text{meas}} - md[\ddot{\theta}\sin(\theta) + \dot{\theta}^2\cos(\theta)]$$

where F_{meas} is the total measured force, L is the aerodynamic force (lift), m is the mass of the wing, d is the distance from the wing pivot to its center of mass, and θ is the measured flapping angle. Inertial loading resulting from unsteady flex in the force balance was not considered.

IX. Results and Discussion

The plots in Figs. 5–9 give the normalized velocity and vorticity field measurements for the leading edge, midchord, trailing edge, one chord downstream, and two chord downstream planes, respectively. Figure 10 gives the normalized rms

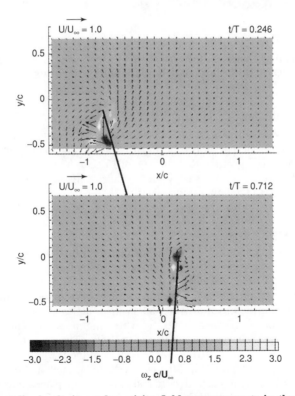

Fig. 5 Normalized velocity and vorticity field measurements in the leading-edge plane. The wing is indicated by the thick dark line. The unit vector is shown in the upper left of each plot; the time stamp is shown in the upper right.

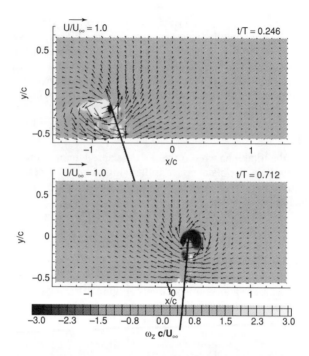

Fig. 6 Normalized velocity and vorticity field measurements in the midchord plane. The wing is indicated by the thick dark line. The unit vector is shown in the upper left of each plot; the time stamp is shown in the upper right.

velocity contours in the two chord downstream plane. The velocity vectors and rms velocities were normalized to U_∞ and the vorticity values were normalized to U_∞/c. This collection of plots is necessarily limited due to the print medium; a complete collection is given in Ref. 28. Figure 15 gives a matrix of plots showing the measured and computed unsteady lift coefficients for various combinations of freestream speeds and flapping frequencies.

A. Velocity Data

There are four characteristics of the velocity field that are of note: First, the nature of the vorticity concentrated within the tip vortex is unlike that of a high Re, steady fixed wing vortex. Rather than seeing concentrated circular tip vortices, we see separated shear/vorticity layers forming well away from the nominal center of rotation, resembling the situation over a delta wing at high angles of attack. Second, the tip vortex becomes very far removed from the wingtip at the trailing-edge plane, moving as far as $0.75c$ away from the tip of the wing. The outward motion of the vortex is attributed to the centrifugal acceleration of the flow as the wing rotates through its flapping motion. Third, the vortex "core" is stretched in the direction of flapping, particularly at the midchord and trailing-edge planes. Finally, the rms velocities in the wake are much higher than what would be expected from a low Re (presumably laminar) flow. This is attributable to the development of discrete vortical structures in the separated shear layers (see Fig. 10).

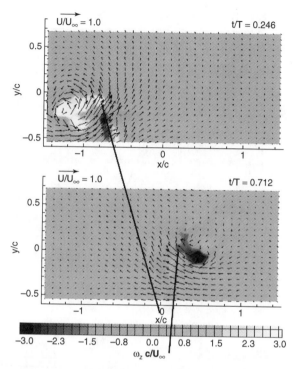

Fig. 7 Normalized velocity and vorticity field measurements in the trailing-edge plane. The wing is indicated by the thick dark line. The unit vector is shown in the upper left of each plot; the time stamp is shown in the upper right.

In the midchord and trailing-edge planes (Figs. 6 and 7), the areas of concentrated vorticity more closely resemble a shed vortex sheet rather than a single, concentrated vortex. The vorticity tends to be positioned mostly to one side of the core, a fact that is manifested in a highly asymmetric velocity distribution across the vortex core (with much higher velocities on the "vorticity" side of the core).

This asymmetry exists far away from the wing, as well. As shown in Fig. 11 (which shows circumferential velocity across the trailing tip vortex $5c$ downstream), the velocities on one side of the vortex core are nearly five times as large as those on the other. Furthermore, the core region (i.e., the region of increasing circumferential velocity) extends from approximately $0.30c$ to $0.25c$, giving a core diameter of more than $0.50c$, substantially larger than the $0.03–0.08c$ typical of higher Re tip vortices.

The velocity field data in Fig. 7 show that the tip vortex becomes very far removed from the wingtip as the wing begins the downstroke. The center of rotation stays in very nearly the same location for $t/T = 0.2$ to 0.4 while the wing is pulling away during the downstroke. Figure 12 shows the normalized distance from the tip of the wing to the center of the tip vortex at the midchord and trailing-edge planes. As shown in the plot for the trailing-edge data, the wing tip moves nearly $0.75c$ away from the tip vortex before it begins to catch up. The same phenomenon is

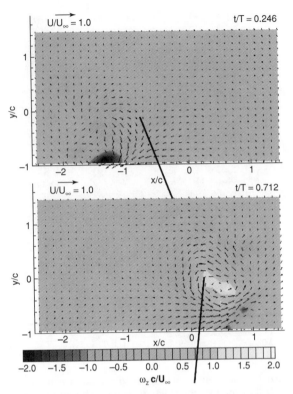

Fig. 8   Normalized velocity and vorticity field measurements in the one chord downstream plane. The wing is indicated by the thick dark line. The unit vector is shown in the upper left of each plot; the time stamp is shown in the upper right.

observed on the upstroke, as well, and in the midchord plane, but the effect is not nearly as pronounced as it is at the trailing edge on the downstroke.

As the tip vortex begins to catch up to the wingtip, the core appears to be "stretched" in the direction of flapping. As shown in Fig. 13, the region that would normally be considered the "inner core region" of the tip vortex is stretched in the direction of flapping from approximately $-0.70c$ to $-0.20c$. This core-stretching phenomenon is not apparent in any of the other data planes and does not appear downstream. It has to be attributed to distortions in the flowfield caused by the unsteady accelerations of the flow, but beyond that, it has to be left to flowfield computations to identify the precise causes and effects.

The normalized rms velocities in the $2c$ downstream plane are shown in Fig. 10. The plots show large regions where the magnitude of the velocity fluctuations approach 40% of the freestream speed. A comparison between Figs. 10 and 9 shows that the areas of high fluctuation almost exactly follow the areas of concentrated vorticity. Because these values are extraordinarily high, they have been validated against hot-film measurements at several locations. Such high values can only result from the roll up of discrete vortical structures in the shear layer. Such structures would have a range of sizes and involve kinetic energy over a broad range of frequencies. There is no reason to believe that these vortical structures

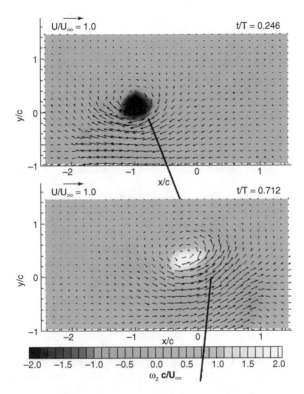

Fig. 9 Normalized velocity and vorticity field measurements in the two chord downstream plane. The wing is indicated by the thick dark line. The unit vector is shown in the upper left of each plot; the time stamp is shown in the upper right.

will bear an integer relationship to the flapping frequency; hence these velocity fluctuations will appear as perturbations in the phase-resolved velocity measurements. The Reynolds number is too low, of course, to attribute these to random turbulence.

B. Force Data

As noted above, the inertial loading resulting from the motion of the wing was a substantial portion of the total measured force. Figure 14 shows a breakdown of the force components for a typical flapping cycle at 4.5 Hz and 10 ft/s freestream speed. The plot shows that the inertial loading is highest at the start of the flapping cycle, peaking at more than 50% of the aerodynamic force shortly after the downstroke has begun. The asymmetry in the inertial loading is a consequence of an asymmetric flapping setup in the wind tunnel and the slightly nonharmonic flapping motion. At the higher tunnel speeds used in these experiments, the motion histories became substantially distorted (see Ref. 28 for details).

The force measurement data were compared against computations from an unsteady panel code. The flapping wing model was simulated in an unsteady vortex lattice code with a free wake generated at the trailing-edge panels. The code used 40 panels to represent the wing geometry and 300 time steps over two flapping

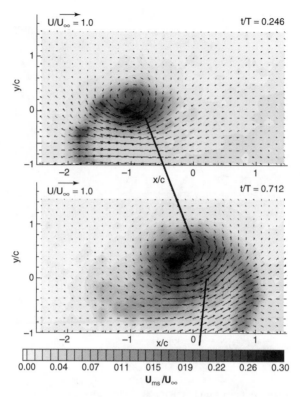

Fig. 10   Normalized velocity and rms velocity field measurements at two chord lengths downstream. The wing is indicated by the thick dark line. The unit vector is shown in the upper left of each plot; the time stamp is shown in the upper right.

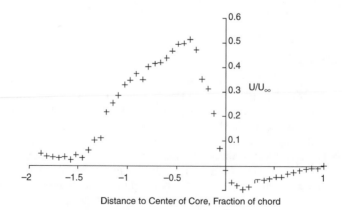

Fig. 11   Normalized circumferential velocity across the vortex core in the 5c downstream plane. Data are taken from the $t/T = 0$ bin.

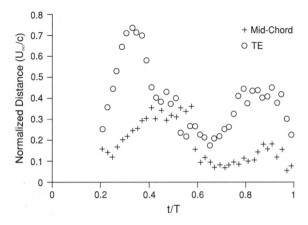

Fig. 12 Tip vortex core distance from wingtip (normalized).

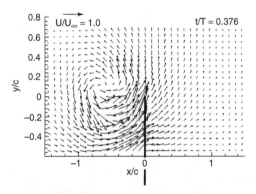

Fig. 13 Stretched vortex core indicated by double-arrowhead line in the trailing-edge plane. Wing motion is left to right at $t/T = 0.376$; instantaneous angular velocity is 578 deg/s. Reference unit vector is shown in the upper left.

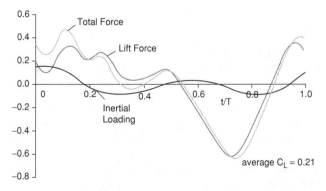

Fig. 14 Breakdown of force components: unsteady lift, inertial loading, and aerodynamic force in lb.

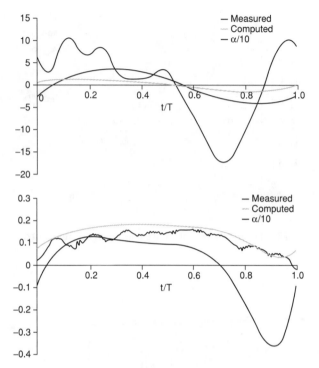

Fig. 15   Measured and computed force data, and a curve fit to the motion history, for (top) 10 ft/s, and 4.5 Hz and (bottom) 60 ft/s and 1.0 Hz.

cycles for the various combinations of freestream speed and flapping frequency. The influence of the free wake on the wing and on itself was calculated for the 75 most recent time steps. Wake panels older than 75 time steps were neglected. The wing motions used in the panel code were based on curve fits[31] of the measured wing motion. Thus, higher-frequency unsteadiness within a cycle would be under-predicted by the code. Results from the second cycle of the wing motion are used in Fig. 15 to reduce the effects of the starting vortex at the implusive start of the computations.

Figure 15 shows cases at two extremes of the reduced frequency/Reynolds number ranges studied. The top graph shows results at a flapping frequency of 4.5 Hz and freestream velocity of 10 ft/s. This combines the highest reduced frequency and lowest Reynolds number studied. Here the wing motion is largely simple harmonic, but the experimental result shows large fluctuations. The experimental results have not been averaged over many cycles, so that fluctuations that are not synchronized with the flapping phase are also captured. The panel code greatly underpredicts the force variation. The matrix of plots in Fig. 15 shows two compet-ing trends: For a given freestream speed, as the flapping frequency increases, the unsteady panel code computations underpredict the force generated by the wing. However, in the bottom graph of Fig. 15, the agreement between calculation and experiment is seen to be quite good. Here the freestream velocity is 60 ft/s, the highest studied, and the flapping frequency is 1 Hz, the lowest reduced-frequency case studied. The wing motion here is not simple harmonic: The cambered wing experiences substantially more force during motion to one side than to the other, and the motor is unable to overcome this asymmetry completely. The panel code overpredicts the force slightly in this case.

There is a substantial amount of higher harmonic content in the force measurements than predicted by the panel code, particularly on the downstroke. This type of fluctuating force hints at some type of shedding of the wing and is also evidenced in the LV data in the two chord downstream plane. Here, discrete patches of negative vorticity are seen circulating around a vortex core from $t/T = 0.19$ to 0.32 (though it should be noted that this vorticity is associated with the upstroke). This aspect is not understood at present. From the above, it appears that the panel code is an adequate representation of the latter test case but performs poorly in the former. At the limit of high reduced frequency and low Reynolds number, the flowfield is dominated by large instantaneous angles of attack, separated flows, smeared distributed regions of vorticity, and shear layers with discrete vortical structures, none of which can be expected to be captured in the panel code.

X. Conclusions

This chapter has presented quantitative velocity field data and panel code computations for a rectangular wing undergoing large-amplitude flapping at moderate reduced frequency and low chord Reynolds number. The observed characteristics of the velocity field include the following:

1) Vorticity is distributed over the tip vortex, unlike what is seen in vortices at higher Reynolds number.

2) The tip vortex cores become stretched in the direction of wing flapping during the downstroke.

3) The velocity distribution across the vortex core is highly asymmetric.

4) Predictions from an unsteady panel code approximate measurements at the test case of highest Reynolds number and lowest reduced frequency.

5) At higher reduced frequency and low Reynolds number, the panel code underpredicts force by an order of magnitude.

6) The lift exhibits substantial higher-frequency spectral content. These are attributed to the generation of discrete vortical structures in shear layers and near the concentrated vortices as the result of unsteady motion.

References

[1]DeLaurier, J., and Winfield, J., "Simple Marching-Vortex Model for Two-Dimensional Unsteady Aerodynamics," *Journal of Aircraft*, Vol. 27, No. 4, April 1990, pp. 376–378.

[2]Katz, J., and Plotkin, A., *Low-Speed Aerodynamics*, McGraw-Hill, New York, 1991.

[3]Muti Lin, J. C., and Pauley, L. L., "Low-Reynolds-Number Separation on an Airfoil," *AIAA Journal*, Vol. 34, No. 8, Aug. 1996, pp. 1570–1577.

[4]Shyy, W., Jenkins, D. A., and Smith, R. W., "Study of Adaptive Shape Airfoils at Low Reynolds Number in Oscillatory Flows," *AIAA Journal*, Vol. 35, No. 9, pp. 1545–1548.

[5]Wu, J. C., Sampath, S., and Sankar, L. N., "A Numerical Study of Unsteady Viscous Flows around Airfoils," *AGARD Conference Proceedings, No. 227: Unsteady Aerodynamics*, AGARD, Neuilly-sur-Seine, France, 1978, pp. 24.1–24.18.

[6]Monttinen, J. T., Shortridge, H. L., and Saric, W. S., "Adaptive, Unstructured Meshes for Solving the Navier–Stokes Equations for Low-Chord-Reynolds-Number Flows,"

Conference on Fixed, Flapping, and Rotary Wing Vehicles at Low Reynolds Number, Univ. of Notre Dame, Notre Dame, IN, 5–7 June 2000, pp. 142–152.

[7]Kunz, P., and Kroo, I., "Analysis, Design, and Testing of Airfoils for Use at Ultra-Low Reynolds Numbers," *Conference on Fixed, Flapping, and Rotary Wing Vehicles at Low Reynolds Number*, Univ. of Notre Dame, Notre Dame, IN, 5–7 June 2000, pp. 349–372.

[8]Drela, M., "Transonic Low-Reynolds Number Airfoils," *Journal of Aircraft*, Vol. 29, No. 6, Nov.–Dec. 1992, pp. 1106–1113.

[9]Tuncer, I. H., and Sankar, L. N., "Unsteady Aerodynamic Characteristics of a Dual-Element Airfoil," *Journal of Aircraft*, Vol. 31, No. 3, May–June 1994, pp. 531–537.

[10]Kantha, L. H., "Empirical Model of Transport and Decay of Aircraft Wake Vortices," *Journal of Aircraft*, Vol. 35, No. 4, July–Aug. 1998, pp. 649–652.

[11]Alsalihi, Z., "Compressible Navier–Stokes Solutions over Low Reynolds Number Airfoils," *Lecture Notes in Engineering*, edited by T. J. Mueller, Vol. 54, Springer-Verlag, Berlin, 1989, pp. 343–357.

[12]Chiu, T. W., Broers, C. A. M., Wood, B. D., and Berney, A. H., "The Application of Modern Panel Method Techniques to Sport Aero/Hydrodynamics," *26th AIAA Fluid Dynamics Conference*, AIAA, Washington, DC, 19–22 June 1995.

[13]Reed, H. L., and Toppel, B. A., "Method to Determine the Performance of Low-Reynolds-Number Airfoils under Off-Design Unsteady Freestream Conditions," *Lecture Notes in Engineering*, edited by T. J. Mueller, Vol. 54, Springer-Verlag, Berlin, 1989, pp. 218–230.

[14]Smith, M. J. C., Wilkin, P. J., and Williams, M. H., "The Advantages of an Unsteady Panel Method in Modeling the Aerodynamic Forces on Rigid Flapping Wings," *Journal of Experimental Biology*, Vol. 199, 1996, pp. 1073–1083.

[15]Liu, H., and Kawachi, K., "A Numerical Study of Insect Flight," *Journal of Computational Physics*, Vol. 146, No. 1, 1998, pp. 124–156.

[16]Chin, S., and Lan, E. C., "Fourier Functional Analysis for Unsteady Aerodynamic Modeling," *AIAA Journal*, Vol. 30, No. 9, Sept. 1992, pp. 2259–2266.

[17]Cho, J., and Williams, M. H., "Propeller–Wing Interaction Using a Frequency-Domain Panel Method," *Journal of Aircraft*, Vol. 27, No. 3, 1990, pp. 196–203.

[18]Chi, R. M., "Separated Flow Unsteady Aerodynamic Theory," *Journal of Aircraft*, Vol. 22, No. 11, Nov. 1985, pp. 956–964.

[19]DeLaurier, J. D., "Aerodynamic Model for Flapping Wing Flight," *Aeronautical Journal*, Vol. 97, No. 964, April 1993, p. 125.

[20]Jones, K. D., Lund, T. C., and Platzer, M. F., "Experimental and Computational Investigation of Flapping Wing Propulsion for Micro-Air Vehicles," *Conference on Fixed, Flapping, and Rotary Wing Vehicles at Low Reynolds Number*, Univ. of Notre Dame, Notre Dame, IN, 5–7 June 2000, pp. 421–445.

[21]Phlips, P. J., East, R. A., and Pratt, N. H., "Unsteady Lifting Line Theory of Flapping Wings with Application to the Forward Flight of Birds," *Journal of Fluid Mechanics*, Vol. 112, Nov. 1981, pp. 97–125.

[22]Gopalarathnam, A., Giguere, P., Lyon, C., and Selig, M., "Systematic Airfoil Design Studies at Low Reynolds Numbers," *Conference on Fixed, Flapping, and Rotary Wing Vehicles at Low Reynolds Number*, Univ. of Notre Dame, Notre Dame, IN, 5–7 June 2000, pp. 101–127.

[23]Donovan, J. F., and Selig, M. S., "Low Reynolds Number Airfoil Design and Wind Tunnel Testing at Princeton University," *Lecture Notes in Engineering*, edited by T. J. Mueller, Vol. 54, Springer-Verlag, Berlin, 1989, pp. 39–57.

[24]Freymuth, P., "Unsteady Model of Animal Hovering," *Lecture Notes in Engineering*, edited by T. J. Mueller, Vol. 54, Springer-Verlag, Berlin, 1989, pp. 231–245.

[25]Ellington, C., "Lift and Drag Characteristics of Rotary and Flapping Wings,"*Conference on Fixed, Flapping, and Rotary Wing Vehicles at Low Reynolds Number*, Univ. of Notre Dame, Notre Dame, IN, 5–7 June 2000.

[26]Latek, B., Mönttinen, J., Reed, H., and Saric, W., "Experiments on Wing–Body and Wing–Winglet Aerodynamics (Force Balance, Hot Wires, and Flow Visualization)," *Conference on Fixed, Flapping, and Rotary Wing Vehicles at Low Reynolds Number*, Univ. of Notre Dame, Notre Dame, IN, 5–7 June 2000, p. 477.

[27]Archer, R. D., Sapuppo, J., and Betteridge, D. S., "Propulsion Characteristics Of Flapping Wings,"*Aeronautical Journal*, Vol. 83, No. 825, Sept. 1979, pp. 355–371.

[28]Ames, R. G., "On the Flowfield and Forces Generated by a Flapping Wing," Ph.D. Thesis, Georgia Institute of Technology, Atlanta, GA.

Chapter 16

Experimental and Computational Investigation of Flapping Wing Propulsion for Micro Air Vehicles

K. D. Jones,[*] T. C. Lund,[†] and M. F. Platzer[‡]
Naval Postgraduate School, Monterey, California

Nomenclature

AR = aspect ratio (b/c)
b = wingspan
C_d = drag coefficient $[D/(q_\infty S)]$
C_l = lift coefficient $[L/(q_\infty S)]$
C_m = moment coefficient $[M/(q_\infty S c)]$
C_p = power coefficient $(-C_l \dot{y} - C_m \dot{\alpha})$
C_t = thrust coefficient $[T/(q_\infty S) = -C_d]$
c = chord length
D = drag
f = frequency in hertz
h = vertical plunge amplitude in terms of c
h_{te} = vertical amplitude of the trailing edge in terms of c
k = reduced frequency $(2\pi f c/U_\infty)$
k_G = Garrick's reduced frequency $(\pi f c/U_\infty)$
L = lift
M = moment
q_∞ = freestream dynamic pressure $(\rho_\infty U_\infty^2/2)$
Re_c = chord Reynolds number $(U_\infty c/\nu_\infty)$
S = wing area
Sr = Strouhal number $(2 f c h_{te}/U_\infty = k h_{te}/\pi)$
T = thrust $(-D)$
t = time
U_∞ = freestream velocity

*Research Assistant Professor, Department of Aeronautics and Astronautics. Senior Member AIAA.
†Graduate Student, Department of Aeronutics and Astronautics. Lt. U.S. Navy.
‡Professor, Department of Aeronautics and Astronautics. Fellow AIAA.

x_p = pivot location from leading edge in terms of c
$y(\tau)$ = vertical displacement in terms of c
α = angle of attack
α_i = induced angle of attack
$\Delta\alpha$ = pitch amplitude in degrees
η = propulsive efficiency (C_t/C_p)
ρ_∞ = freestream density
τ = nondimensional time (tU_∞/c)
ν_∞ = freestream kinematic viscosity
ϕ = phase between pitch and vertical plunge
$(\dot{\ })$ = rate of change w.r.t. τ
$(')$ = per unit span

I. Introduction

$\mathbf{C}$ URRENTLY, the field of low Reynolds number aerodynamics is receiving considerable attention and stimulus because of the recently announced requirement to develop micro air vehicles (MAVs). This raises the question whether the flapping wing principles used by birds and insects for many millenia can be applied to the development of these new flight vehicles.

This chapter reviews the state of the art of flapping wing aerodynamics and describes the experimental and computational flapping wing investigations carried out at the Naval Postgraduate School. The material is organized primarily by the type of motion, with sections dedicated to pure plunging, pure pitching, and pitch/plunge motions of single wings, and concluding with a section about multiple wing configurations that utilize some form of interference to enhance performance. Each of these sections includes a brief historical review of key past work on the topic, followed by a summary of numerical and experimental investigations carried out by the present authors and a number of associates at the Naval Postgraduate School. These sections are preceded by a brief overview of the parametric equations of motion and the configurations investigated, and they are succeeded by a summary with a prospective of ongoing work and future plans.

II. General Kinematics

In this section the flapping wing motions are defined, illustrations of the configurations are included for reference, and the relevant nondimensional variables and performance parameters are defined.

A. Configurations

Several configurations have been considered by the present authors and colleagues in past and present studies. Although all of the experimental studies have included some level of flow three-dimensionality, all of the numerical work performed to date has been two dimensional. Consequently, the additional degrees of freedom allowed for three-dimensional motion are not included here. In the general case, the numerical model provides three-degree-of-freedom harmonic motion. However, only two degrees of freedom are considered here, as shown in Fig. 1. A variation of the code allows for airfoil combinations, with two independently moving wings, as shown in Fig. 2.

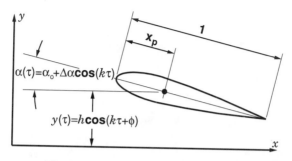

Fig. 1 Two-degree-of-freedom motion.

The single and multiwing configurations investigated in past and present studies are shown in Fig. 3. The first configuration a) is the single flapping wing. The second b) is the two-airfoil combination first suggested by Schmidt, where the leading airfoil flaps, and the trailing airfoil is stationary in the wake of the flapping foil. The third c) is the opposed plunge, or flapping in ground-effect simulation. The fourth d) is an experimental variation of the second, with two leading airfoils performing the opposed plunge motion of c), and two, stationary trailing airfoils in the wake. The last e) is the so-called opposed-pitch configuration, where the two wings are pitched about a point several chord lengths upstream of the leading edge, resulting in a combined pitch and plunge motion of the leading edge. It is a mechanically simple variation of c), developed for the MAV models, which cannot easily incorporate complicated machinery.

B. Equations of Motion

The equations of motion shown in Fig. 1 are given in nondimensional form using the chord length c as the reference length, using the freestream velocity U_∞ as the reference velocity, and using their ratio c/U_∞ as the reference time.

This nondimensionalization results in a nondimensional or reduced frequency of

$$k = \frac{2\pi f c}{U_\infty} \tag{1}$$

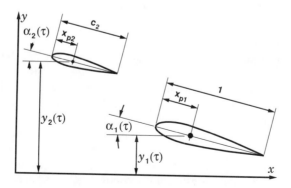

Fig. 2   Schematic of two-airfoil simulations.

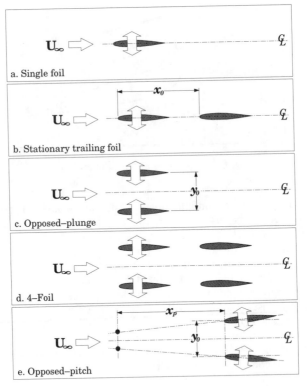

Fig. 3 Numerical and experimental configurations.

Flapping wing flowfields are generally considered to be wake-dominated flows, where the Strouhal number is usually considered to be the definitive parameter. The Strouhal number is a reduced frequency of sorts, usually replacing the chord length by the wake width for the reference length and using the real frequency in place of the circular frequency. It is given by

$$Sr = \frac{2 f c h_{te}}{U_\infty} = \frac{k h_{te}}{\pi} \qquad (2)$$

where h_{te} is the oscillatory amplitude of the trailing edge.

C. Performance Criteria

The propulsive efficiency of the flapping wings is usually defined as the ratio of the *power out* to the *power in*, where the *power out* is the product of the thrust and the free stream velocity, and the *power in* is the time rate of work done to move the wing. In nondimensional form this becomes the ratio of the thrust coefficient to the power coefficient:

$$\eta = \frac{power\ out}{power\ in} = \frac{T U_\infty}{P} = \frac{C_t}{C_p} \qquad (3)$$

Note that throughout this investigation the effects of mechanical and inertial losses of the flapping mechanism have been neglected. In the present experimental mechanisms, by far the largest mechanical loss is the work done to harmonically accelerate the mass of the wings at high frequencies. Although these losses are not small, several means of diminishing these losses have been observed in nature. It is theorized that birds make minor pitch and/or camber changes to use aerodynamic forces to facilitate the acceleration of their wings. There is additional evidence that some insects may actually have a springlike mechanism coupling their wings and that they merely excite the system at the natural frequency, allowing the spring to provide much of the energy for the harmonic acceleration. These methods may be explored in future models but have not been utilized thus far.

III. Plunging Airfoils

Wings undergoing pure plunging motions ($\alpha = 0$) are considered in this section. A few landmark studies of the past are mentioned, and numerical and experimental investigations by the present authors and a number of associates are summarized.

A. Historical Perspective

The earliest scientific theories concerning flapping wing flight pertained to purely plunging airfoils. Knoller[1] and Betz[2] provided the first theoretical explanation of thrust generation in independent papers published in 1909 and 1912, respectively. Both noted that a flapping wing encounters an induced angle of attack, which cants the normal-force vector forward such that it includes both cross-stream (lift) and streamwise (thrust) force components.

Katzmayr[3] provided the first experimental verification of the Knoller–Betz effect in 1922, interestingly, by placing a stationary airfoil in an oscillatory flowfield. A few years later Birnbaum[4,5] identified the conditions that lead to flutter or to thrust generation (really, a pitch/plunge effect) and suggested the use of a flapping (plunging) wing as an alternative to the conventional propeller.

In 1935 von Kármán and Burgers[6] offered the first explanation of drag or thrust production based on the observed location and orientation of the wake vortices, as illustrated in Figs. 4 and 5 for drag-indicative and thrust-indicative wakes,

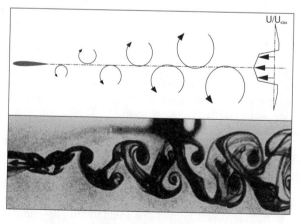

Fig. 4   Drag-indicative vortex street.[10]

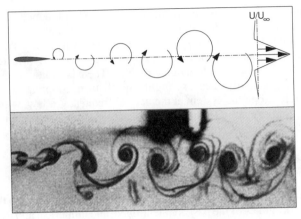

Fig. 5   Thrust-indicative vortex street.[10]

respectively. Note that although the location and orientation of the vortices in the thrust-indicative wake induce the well-known jetlike velocity profile, as shown in Fig. 5, the vortices are not generating the thrust. To the contrary, the vortices are not only lost energy expended by the airfoil, but the proximity of the vortices acts to reduce the propulsive performance of the wing. These effects will be discussed in more detail in later sections.

In 1936 Garrick[7] applied Theodorsen's inviscid, incompressible, oscillatory, flat-plate theory[8] to the determination of the thrust force providing the first notable numerical predictions of thrust. Garrick's formulation for a pure plunge motion is

$$C_t = \pi 4 k_G^2 h^2 (F^2 + G^2) \tag{4}$$

where F and G are the real and imaginary parts of the Theodorsen lift-deficiency function, respectively. Equation (4) suggests that the thrust is related to the frequency squared and the plunge amplitude squared, although, since both F and G are functions of k, there is an additional, nonlinear dependence on k. The linear theory of Garrick treats the wing as a flat-plate and the wake as a vortex sheet in the plane of the airfoil. Use of a planar, nondeforming wake alters the influence of the wake on the airfoil. This will be discussed in more detail in later sections.

In 1950 Bratt[9] performed flow visualization experiments that corroborated von Kármán and Burgers's observations. Of particular interest, Bratt's experimental data included several cases where a nonsymmetrical, deflected wake pattern was recorded, but no comment was made on these deflected wakes, and, in fact, they were never again reported until 1998 when Jones et al.[10] showed them to be reproducible both numerically and experimentally. This will be discussed in more detail in later sections.

Birnbaum's suggestion to regard a flapping foil as an alternative (two-dimensional) propeller generated some interest over the years. Most noteworthy is Küchemann and Weber's book[11] in which they comment on aerodynamic propulsion in nature and observe that the propulsive efficiency of an idealized flapping wing is greater than that of a simplified propeller model because of the disadvantageous trailing vortex system generated by the propeller.

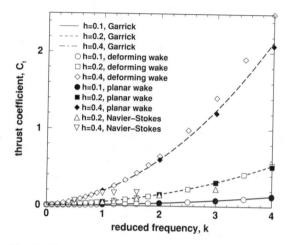

Fig. 6 Thrust as a function of h, k, and wake model.

B. Numerical Simulations

Several flow solvers have been used to simulate the flow about flapping wings and wing combinations. An unsteady, two-dimensional, potential-flow code with a nonlinear, deforming-wake model has been used for the majority of the results presented here, and an unsteady, two-dimensional, Euler/Navier–Stokes solver has been applied in a few cases either to verify results of the panel code or to explore flowfields where the panel code is not applicable. Details of the panel-code algorithm and applications can be found in Refs. 10 and 12–22, and details of the Euler/Navier–Stokes algorithm and applications can be found in Refs. 23–30.

Because of its computational efficiency, the panel code has been used extensively to explore the parameter space of flapping wing flight. It is compared to Garrick's linear theory and to the Navier–Stokes results of Tuncer et al.[29] in Figs. 6 and 7, for

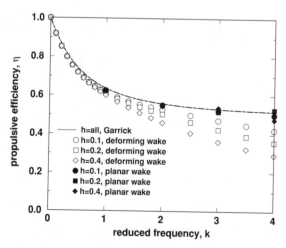

Fig. 7 Propulsive efficiency, η as a function of h, k, and wake model.

variations in plunge amplitude and reduced frequency. As mentioned above, the panel code incorporates a nonlinear, deforming wake model, which allows for out-of-plane vorticity and nonuniform convection speeds for the large eddies. Results are also included in Figs. 6 and 7 for the panel code with all wake deformation eliminated; that is, the shed vorticity is forced to remain in the plane of the chord line much like Garrick's linear model. Most of the difference between the panel code and linear theory is due to the out-of-plane vorticity in the deforming wake model, with the finite airfoil thickness and finite plunge amplitude yielding a relatively minor influence on the performance. Note that the predicted thrust shown in Fig. 6 actually increases with the deforming wake model, but the efficiency shown in Fig. 7 decreases.

The thrust coefficient predicted by the Navier–Stokes solution (just the pressure contribution) is quite close to the panel-code results up to the point of dynamic stall, at roughly $hk = 0.35$. The Navier–Stokes simulations were run at $M_\infty = 0.3$ and $Re_c = 1 \times 10^6$ assuming a fully turbulent flow over a NACA 0012, with the Baldwin–Lomax turbulence model. Further details are available in the cited reference.

The panel-code results shown are for a NACA 0012 airfoil; however, simulations using NACA four-digit symmetrical airfoils with thicknesses between 1 and 15% are virtually identical, demonstrating that thrust production for a purely plunging airfoil is independent of airfoil thickness. While linear theory suggests no change in efficiency due to plunge amplitude, the panel code indicates a loss in efficiency for increasing h. Note, however, that although the loss in efficiency with increasing h is rather small, the increase in thrust with increasing h is large. As will be shown, for a given C_t, better performance is achieved with large h and small k.

Classically, the Strouhal number is thought to be the defining parameter for these wake-dominated flows; however, results from linear theory and the panel code have shown an additional dependence on k as indicated in Figs. 8 and 9. In Fig. 8 the wake topologies computed by the panel code are shown for three simulations of

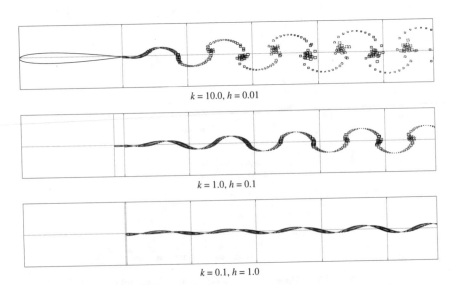

$k = 10.0, h = 0.01$

$k = 1.0, h = 0.1$

$k = 0.1, h = 1.0$

Fig. 8   Dependence of wake instability on k.

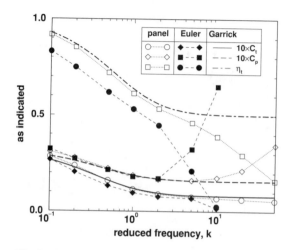

Fig. 9 Dependence of performance on k ($hk = 0.1$).

a NACA 0012 airfoil, flapping with the same Strouhal number, 0.032 ($hk = 0.1$), but with different reduced frequencies. The three plots are scaled by the wake wavelength ($\lambda = 2\pi/k$), with the vertical lines indicating the wake periods. The wake width (in terms of λ) and nonlinearity (roll up) increase with increasing k. The effect this has on the performance is plotted in Fig. 9, where the predictions of Garrick's linear theory are plotted against results from the panel code and the Euler code. The Euler code simulations are at $M_\infty = 0.3$, and the static drag has been subtracted.

As k approaches zero the motion approximates an airfoil plunging at a fixed speed, with an ideal efficiency of 1. Both codes agree with linear theory quite well at low k but diverge at high k, owing to the wake nonlinearity. Note a similar trend in thrust to that found in Figs. 6 and 7, with the panel code predicting higher thrust than linear theory at low k and high h, with the opposite being true at high k and low h, and with the the panel code always predicting a lower efficiency. Note also the rapid increase in the power coefficient at high k, which accounts for most of the decay in efficiency. This is most likely due to the lack of so-called added-mass terms in linear theory, which may become dominant at very high frequencies.

C. Experimental Investigations

Several experimental investigations of flapping wing dynamics have been carried out at the Naval Postgraduate School. Lai et al.[31] investigated the control of flow over a backward facing step using a flapping foil, Dohring et al.[32] investigated boundary-layer control using a flapping wing, Jones et al.[10] performed qualitative and quantitative investigations of the vortex streets behind flapping foils, and Lai and Platzer[33] investigated the jet characteristics of a plunging airfoil.

Figures 4 and 5 included photographs of wake structures visualized using two-color dye injection from Jones et al.[10] In that study several airfoils of varying chord length were driven in a pure plunge motion with variable amplitude and frequency, and both qualitative (flow-visualization) and quantitative (LDV) data were obtained. Chord Reynolds numbers ranged between about 5×10^2 for the small airfoils at low speeds to 5×10^4 for the big airfoils at high speeds. Reduced frequencies ranged from 0 to over 50.

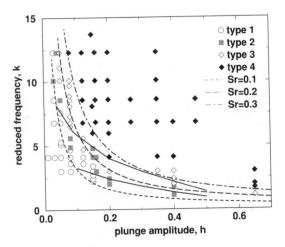

Fig. 10   Experimental wake classification.

In Fig. 10, based on the position and orientation of the vortical structures, and following the criteria of von Kármán and Burgers,[6] the data are classified as types 1–4, based on the photographed positions of the wake structures, where type 1 is indicative of drag (Fig. 4), type 3 is indicative of thrust (Fig. 5), type 2 is in between, at a point where all vortical structures are aligned along the symmetry plane, yielding neither drag nor thrust. The type 4 classification is a rather unusual formation where the wake is deflected from the symmetry plane, resulting in a net thrust and lift, as shown in Fig. 11. In Fig. 10 the data are plotted as a function of h and k along with lines of constant Strouhal number that approximate the boundaries separating the four types. The two thick, black lines in the figure are similar curve fits to the experimental data of Ohashi and Ishikawa[34] and Kadlec and Davis,[35] as published in Triantafyllou et al.[36] As indicated in the previous section, thrust and wake nonlinearity increase with increasing hk, slightly favoring k. This is reflected somewhat in our results and definitively in the two curves obtained from Triantafyllou et al.[36]

Several experimental and computational comparative studies have been performed by the present authors and associates at the Naval Postgraduate School (e.g., Refs. 10, 22, and 37). One of the most interesting findings of Ref. 10 is the *deflected wake* or *dual-mode* case, where a symmetrical problem results in a nonsymmetric solution. The example shown in Fig. 11 compares a numerical solution directly to an experimental photograph. The mode (deflected up/down) of the numerical simulations is determined by the initial conditions and remains fixed for all time. However, in the water tunnel experiments the wake occasionally changes mode at random, apparently due to minute fluctuations in the freestream flow and/or flapping motion. The case shown is for $hk = 1.5$, resulting in a maximum induced angle of attack of about 56 deg, clearly well beyond the dynamic stall limit for the airfoil.

Note that, even in these extreme conditions, the comparison between the inviscid computation and the experiment is quite good, indicating that the evolution of the shed vorticity is essentially an inviscid phenomenon. The secondary wake (deflected down in the figure) is much stronger in the experiment than in the numerical simulation and is thought to be due in large part to the dynamic stall

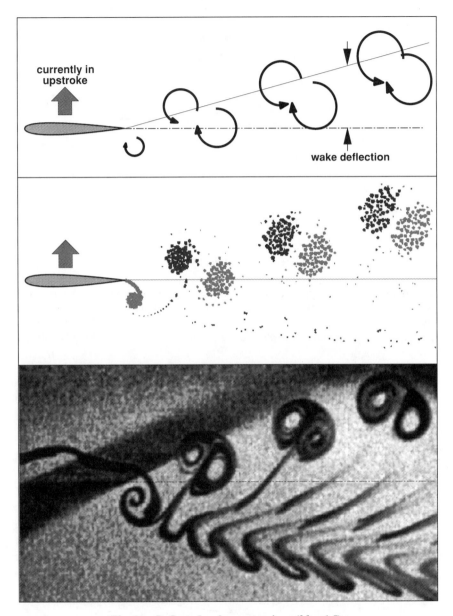

Fig. 11 Deflected wake comparison ($hk = 1.5$).

vortices, which act to further deflect the wake. This is indicated in Figs. 12 and 13, where the numerical and experimental streamwise velocity profiles downstream of flapping airfoils are compared.

For Fig. 12 a 10-cm-chord NACA 0012 was flapped at $k = 15.0$ with $h = 0.04$ ($hk = 0.60$, $Sr = 0.19$), and for Fig. 13 a 2-cm-chord NACA 0015 was flapped at $k = 26.1$ with $h = 0.088$ ($hk = 2.30$, $Sr = 0.73$). Numerical predictions for several step sizes are included to indicate step-size sensitivity of the code. The first

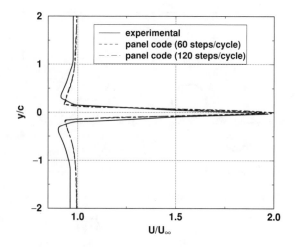

Fig. 12   Time-averaged velocity profiles ($hk = 0.6$).

case is well within the type 3 region, resulting in a symmetric velocity profile. The second case is far into the type 4 region, with a nonsymmetric wake both predicted by the panel code and measured with LDV. The wake deflection is larger in the experiment than predicted by the panel code, but the main features are well captured.

IV.   Pitching Airfoils

Not long after Knoller[1] and Betz[2] first explained flapping wing propulsion of a plunging wing, Katzmayr[3] experimentally verified the phenomenon. He actually performed two experiments. The first, holding an airfoil stationary in an oscillatory flow, was successful. The second, pitching an airfoil in a uniform flow, was not. That is, Katzmayr was unable to measure thrust for the pitching airfoil.

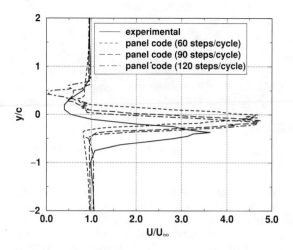

Fig. 13   Time-averaged velocity profiles ($hk = 2.3$).

Garrick[7] later showed that propulsion was possible for purely pitching airfoils, but only at high frequencies. His derivation using Theodorsen's method is given by

$$C_t = \pi k_G^2 \Delta\alpha^2 \left\{ (F^2 + G^2)\left[\frac{1}{k_G^2} + \left(\frac{1}{2} - a\right)^2\right] + \left(\frac{1}{2} - F\right)\left(\frac{1}{2} - a\right) \right.$$
$$\left. - \frac{F}{k_G^2} - \left(\frac{1}{2} + a\right)\frac{G}{k_G} \right\} \tag{5}$$

where a denotes the position of the pivot point, measured from the midchord, positive forward, in terms of half chords, and F and G are the real and imaginary parts of Theodorsen's lift-deficiency function, respectively.

In 1989, Koochesfahani[38] performed flapping wing experiments of a NACA 0012 pitching about its quarter chord. He took LDV measurements of the time-averaged velocity profile in the wake and used this to estimate the thrust using the equation

$$T = \rho_\infty \int_{-\infty}^{+\infty} U(y)[U(y) - U_\infty]dy \tag{6}$$

His results point out a common error in experimental thrust estimation. Equation (6) is a simplified version of the momentum integral, under the assumptions that at the downstream section where the velocity is measured, the pressure is freestream, the flow is parallel, and the time-fluctuating quantities are small. These assumptions are normally valid if the velocity profile is measured far downstream of the wing, where the eddies are essentially diffused. Apparently, Koochesfahani obtained his LDV data just one chord length downstream of the trailing edge, where the vortical structures are very much intact, and Eq. (6) is not expected to be valid.

In Fig. 14, the measurements of Koochesfahani are compared to the predictions of linear theory, Eq. (5), and the panel code, using both the conventional surface-pressure integration and the simplified momentum integral, Eq. (6), at the same downstream location as the LDV measurements. Because linear theory and the

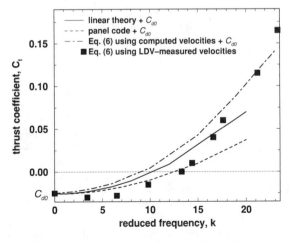

Fig. 14 Thrust production for pitching airfoils.

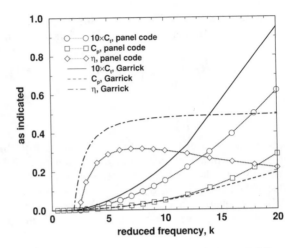

Fig. 15   Propulsion predictions for pitching airfoils.

panel code do not predict profile drag, for comparative purposes, the predictions of linear theory and the panel code have been shifted by the $k = 0$ drag measured by Koochesfahani, with this value called C_{d0} in the legend of Fig. 14. The panel code has no diffusion, and so Eq. (6) is never valid; however, for academic purposes, it is interesting to compare the two methods. Using surface-pressure integration to predict thrust, the panel code underpredicts linear theory over the full frequency range, but using the simplified momentum integral (6), the panel code overpredicts linear theory and seems to be in much better agreement with the experiment. It would seem that the application of Eq. (6) one chord length downstream of the trailing edge is inappropriate in this case.

In Fig. 15 the panel code and linear theory are compared for pitching airfoils. At low frequencies the thrust is negative, resulting in negative efficiencies. Note that linear theory predicts zero or negative efficiency at low k, asymptotically approaching 0.5 for high frequencies, whereas, for plunging airfoils, the efficiency approached 1 at low k and 0.5 at high k. The panel code showed some sensitivity to airfoil thickness, unlike the pure plunge case, where the performance was independent of thickness. The panel code results tend toward linear theory as the thickness is reduced.

V.   Pitching and Plunging Airfoils

Virtually all examples of flapping wing propulsion in nature combine both pitching and plunging, and thus it is often mistakenly thought that pitch and plunge are requirements for propulsion. Clearly, from the results presented in the previous sections, this is not the case.

The parameter space for combined pitching and plunging motions is considerably larger. In addition to the amplitudes, there is also the phase angle between pitch and plunge, ϕ. Some of this extensive parameter space has been explored and is briefly summarized in this section.

The key parameter for determining whether an airfoil creates thrust or extracts power from a flow is the effective angle of attack, as depicted in Fig. 16. In a)

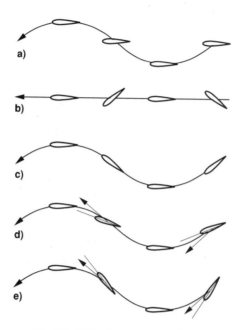

Fig. 16   Effective vs geometric α.

the airfoil is just plunged, creating an induced angle of attack $\alpha_i = \arctan(hk)$. If the airfoil is just pitched, then it may have an induced angle of attack at the leading edge, depending on where the pitch axis is located. If a flapping airfoil is pitched with a low amplitude, then thrust is generated, as depicted in Fig. 16d. If the airfoil is pitched with a large amplitude, greater than the induced angle of attack from the plunge motion, then power is extracted from the flow, as depicted in Fig. 16e. At some point in between, the induced angle of attack is just offset by the geometric angle of attack, and the wing is said to *feather* through the flow, essentially flapping without disturbing the flow, as depicted in Fig. 16c.

A.   Historical Perspective

Garrick's linear approach[7] was the first systematic approach for analyzing the flapping-wing propulsion of combined pitching and plunging airfoils. Garrick's full equation for two-degree-of-freedom motion is

$$C_t = \pi k_G^2 \left(C_{t_\alpha} + C_{t_h} + C_{t_c} \right) \tag{7}$$

where C_{t_α}, C_{t_h}, and C_{t_c} are the pitch, plunge, and cross-contributions to the total thrust, given by

$$C_{t_\alpha} = \Delta\alpha^2 \left[(F^2 + G^2)\left(\frac{1}{k_G^2} + \left(\frac{1}{2} - a \right)^2 \right) + \left(\frac{1}{2} - F \right)\left(\frac{1}{2} - a \right) - \frac{F}{k_G^2} - \left(\frac{1}{2} + a \right)\frac{G}{k_G} \right]$$

$$C_{t_h} = 4h^2(F^2 + G^2)$$

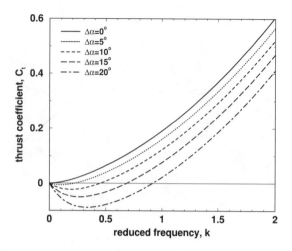

Fig. 17   Thrust coefficient for pitching/plunging motions.

and

$$C_{t_c} = 4\Delta\alpha h \left\{ \left(\frac{F}{2k_G} + \frac{G}{2} - \frac{F^2 + G^2}{k_G} \right) \sin\phi \right.$$

$$\left. + \left[(F^2 + G^2)\left(\frac{1}{2} - a\right) + \frac{1}{4} + \frac{G}{2k_G} - \frac{F}{2} \right] \cos\phi \right\}$$

The effect of pitch amplitude on the thrust coefficient is shown in Fig. 17, where the airfoil is flapped with $h = 0.4$ and $\phi = 90$ deg. In Fig. 18, the same data are dimensionalized and plotted as a function of U_∞, for a characteristic frequency and wing area that will show up in experimental data in a later section. Contrary to intuition, adding the pitch degree of freedom actually reduces thrust and sets

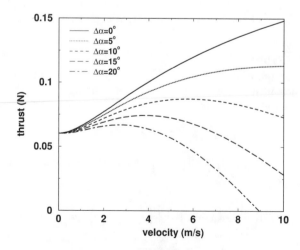

Fig. 18   Thrust for pitching/plunging motions.

a maximum speed for which thrust is produced. It would seem that if nature has found a combined pitch/plunge motion to be advantageous, then linear theory is a poor model of the real physics in those cases.

Many others have contributed to the knowledge base in this area, including De-Laurier and Harris[39] who performed experimental investigations of flapping wing propulsion, Liu[40,41] using vortex lattice and panel methods, Send[42,43] using linearized theory, Hall and Hall[44] and Hall et al.[45] using vortex lattice methods, Isogai et al.[46] performing Navier–Stokes simulations, and McKinney and DeLaurier[47] performing flutter-generator experiments.

B. Numerical Simulations

Several numerical investigations have been performed at the Naval Postgraduate School. These include an exploratory investigation of the parameter space by Jones and Platzer[20] and Navier–Stokes simulations by Tuncer et al.,[29] with results compared to the propulsive analysis of Isogai et al.[46] and power-extraction compared to the experimental results of McKinney and DeLaurier.[47]

The pitch/plunge parameter space was explored extensively using the panel code in Ref. 20. It was shown that by choice of the phase and pitch amplitude, thrust could be generated or power could be extracted from the flow. This is illustrated in Figs. 19 and 20, where a NACA 0012 airfoil is pitched about its quarter chord, while plunging with $k = 0.5$ and $h = 0.2$. In Fig. 19 the pitch amplitude is $\Delta\alpha = 4$ deg, and in Fig. 20 the pitch amplitude is $\Delta\alpha = 8$ deg, and in both cases the phase angle is swept through a full 360 deg. The results of Garrick's Eq. (7) are included.

C. Experimental Investigations

In an investigation by Jones et al.[48] the power-extraction capabilities were evaluated numerically and experimentally with comparisons to the experimental results of McKinney and DeLaurier,[47] leading to the development of a working

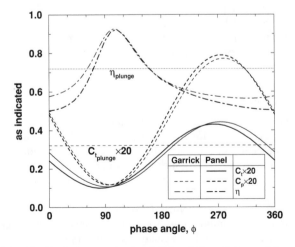

Fig. 19 Performance for $k = 0.5$, $h = 0.2$, and $\Delta\alpha = 4$ deg.

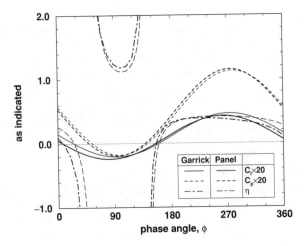

Fig. 20   Performance for $k = 0.5$, $h = 0.2$, and $\Delta\alpha = 8$ deg.

hydrodynamic flutter-engine. Numerical results compared reasonably well with those of McKinney and DeLaurier, and an exploration of the parameter space projected that much higher efficiencies were possible for a properly tuned system. Flapping wing propulsion experiments have been performed in the Naval Postgraduate School low-speed wind tunnel for multiwing configurations (discussed in a later section) and presently flapping wing propulsion of single airfoils undergoing combined pitch/plunge motions are being investigated in the wind tunnel.

VI.   Airfoil Combinations

One of the most interesting concepts, and most likely the area that will offer the greatest reward, is that of multiple wing configurations, which use interference effects to enhance performance. The most familiar examples of this are probably dragonflies, which use a pair of wings on either side to provide power and control verging on perfection. A less familiar, but quite advantageous use of multiple wings is the flight of single-winged vehicles near a ground plane, such as birds flying low over water. In this section, past and present research in multiple airfoil systems is reviewed.

A.   Historical Perspective

In 1965, Schmidt[49] described a tandem-wing configuration he coined the *wave-propeller* that placed a stationary wing in the oscillatory wake of a flapping wing to recapture some of the vortical energy shed into the flow by the flapping wing (see Fig. 3b). Schmidt built a shallow-water catamaran boat employing the wave-propeller for propulsion, and he claimed efficiencies close to conventional propellers.

In 1977, Bosch[50] developed a linear theory for predicting propulsion from flapping airfoils and airfoil combinations, for the first time including wake interference effects. However, as mentioned previously, linear theory does not include the effect of out-of-plane vorticity, which plays an important role here.

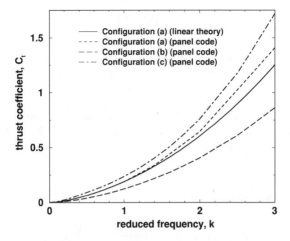

Fig. 21 Thrust coefficient vs reduced frequency.

B. Numerical Simulations

Several numerical studies have been performed for various multiple airfoil configurations, primarily configurations b and c in Fig. 3 (Refs. 20, 22, 26, 28, and 37). Some of the most interesting results are shown in Figs. 21 and 22, where the thrust coefficient and propulsive efficiency are plotted as functions of k for configurations a, b, and c of Fig. 3.

As previously mentioned, c is roughly equivalent to a single flapping wing near a ground plane, where the ground plane becomes a plane of symmetry between the airfoil and its reflection or image airfoil. Birds and insects often fly very close to the water surface, and from Figs. 21 and 22, it is easy to see why. The thrust is increased by 15–30% over the out-of-ground-effect value, and the efficiency is increased by about 10% over the full frequency spectrum. Note that these are

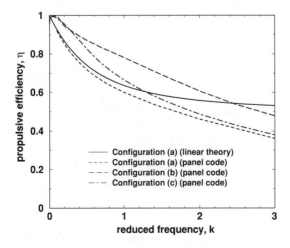

Fig. 22 Efficiency vs reduced frequency.

not optimized results. That is, the configuration is defined by an experimental apparatus to be discussed in a following section, and higher performance may be achieved by varying h and y_0. The configuration includes NACA 0014 wing section(s) flapping with a pure plunge amplitude of $h = 0.4$. For b, x_0 is 2.2 and for c, y_0 is 1.4. For both b and c the thrust coefficient is the average for both airfoils. For c both foils generate the same thrust, but for b most of the thrust is generated by the flapping airfoil, with only a small contribution by the trailing wing. Although the efficiency of b is higher than c over most of the frequency range, it is important to note that this is inviscid performance. Because b operates at roughly half the average C_t of c, profile drag will have a much larger effect on its real efficiency.

C. Experimental Investigations

To evaluate the application of the panel code as a flapping wing design tool, an experimental model capable of performing pitch/plunge oscillations of two wings with variable geometry was designed, as shown in Fig. 23. The model and the measured performance were detailed in Jones and Platzer[22] and Lund.[51] The model flapped two airfoils resembling NACA 0014s, with a chord length of 64 mm and an effective span of 1200 mm, with variable pitch and plunge amplitudes. The model was suspended by four cables from the tunnel ceiling, such that it could swing freely in the streamwise direction, as shown in Fig. 24.

Thrust was determined by measuring the streamwise deflection of the model when the wings were flapped. The deflection was measured by bouncing a laser sensor off a small notch on the back of the rear nacelle, as shown in Fig. 24. The laser analog sensor, an NAIS model ANL1651, is nominally 130 mm downstream of the model and measures distances accurately within the range 80 to 180 mm. The accuracy of the sensor is prescribed by the manufacturer as $\pm 100\ \mu$m $\pm 0.002 \times \Delta x$ for the range 130 ± 35 mm.

Results from Ref. 22 with velocities corrected by Lund[51] and improved numerical results are shown in Fig. 25 for (c) with $h = 0.4$ and $y_0 = 1.4$. In Fig. 26 results from Ref. 22 with velocities corrected by Lund[51] are shown for a combined pitch/plunge motion, with $\Delta\alpha = 3.6$ deg, $h = 0.316$, and $\phi = 0$ deg. Reynolds numbers for the experiment varied between 0 and about 45,000.

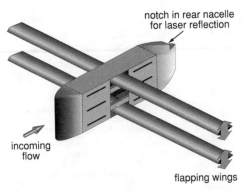

notch in rear nacelle
for laser reflection

incoming
flow

flapping wings

Fig. 23   Isometric view of the large model.

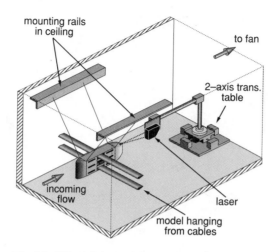

Fig. 24   View of the large model in the test section.

The agreement between the panel code and the experimental measurements is quite good, roughly 80% at higher velocities and frequencies. There is a notable dip in the measured thrust below about 5 m/s at 8 Hz, which may be an indication of the onset of dynamic stall. As the flow speed is reduced, the reduced frequency and the effective angle of attack are increased, and at 5 m/s and 8 Hz, $k = 0.64$, $hk = 0.26$, and the effective angle of attack exceeds 14 deg. It is likely that the wings are stalled for part or all of the cycle at lower speeds. Hopefully future flow visualization and LDV experiments will concur.

With the combined pitch/plunge motion, the decrease in performance at higher speeds is noted, but the measured thrust at higher speeds actually exceeds the panel-code predictions. The wings on the model were very high aspect ratio, and at the higher frequencies the wing flex was significant, such that the plunge amplitude at

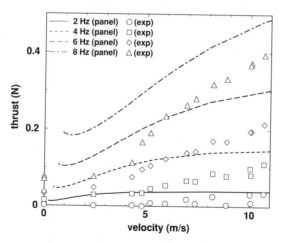

Fig. 25   Thrust for pure plunge motion of (c).

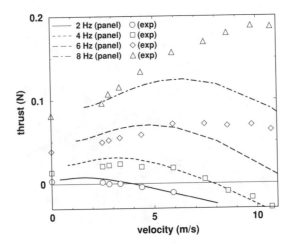

Fig. 26   Thrust for pitch/plunge motion of (c).

the tip greatly exceeded the specified value, and it is thought that this is the cause for the high thrust measurements. This may be modeled in the future by measuring the deflection of the wings as a function of spanwise position and phase, and using a strip-theory approach with the panel code.

Additional results in Ref. 22 showed little sensitivity to average angle-of-attack changes up to about 10 deg, where the wings appeared to stall at higher frequencies. Additionally, results of the configuration shown in Fig. 3d showed only marginal improvement in thrust with the addition of the stationary trailing airfoils (not enough additional thrust to offset the increase in profile drag) but a reduction in the drag-bucket region. With the addition of tip plates to minimize the flow three-dimensionality, the performance showed additional marginal improvement, virtually eliminating the drag-bucket region.

With the reasonable success at predicting propulsive performance on the large model, more recent experiments have been performed on much smaller, MAV-sized models, nominally with a 15-cm span and length. Details of the model design, construction, and testing can be found in Ref. 37; a brief summary is included here.

For mechanical simplicity, a slight variation of the flapping motion was adopted for the MAV model, depicted in Fig. 3e. Instead of a pure plunge motion, a pure pitch motion is used, with the pitch axis several chord lengths in front of the wing. A cross-sectional drawing of the MAV is shown in Fig. 27, where the flow goes from left to right. An exploded view with one wing omitted and isometric and side views of the complete model are shown in Figs. 28, 29, and 30, respectively.

The models are designed around exceptionally small, geared, stepping motors from RMB Smoovy®. The motors are 5 mm in diameter and, with a 25:1 planetary gear system, they are about 25-mm long. They weigh a meager 2.4 g and produce a torque of about 2.5 mN·m at speeds up to about 800 rpm, which yields a maximum power of roughly 0.21 Wa. The brushless motors are controlled by an oscillatory driver circuit, which allows for very precise, steady and reproducible rotational speeds. Presently, the control circuit and power supply are large and bulky and must

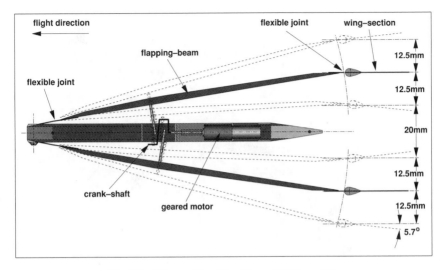

Fig. 27   Schematic of the 15-cm MAV model.

be external to the model. Although the motor has a reported maximum speed of around 100,000 rpm, the gearbox is only rated for 20,000 rpm input, which limits the flapping frequency to around 13 Hz; however, with some models, sustained frequencies of over 18 Hz have been reached.

A very small crankshaft is coupled to the motor using thin-wall silicon tubing. The crankshaft moves the flapping-beams via Scotch yokes constructed of very thin piano wire. The mechanism takes some effort to construct but has proven to be quite robust and may easily be disassembled for maintenance or part substitutions.

The three-pole power is fed into the model through four 0.076-mm-diam copper wires that support the model. The support wires are attached to the model via small gold-plated pins, at the nose and near the rear of the model. The mass of the wires is

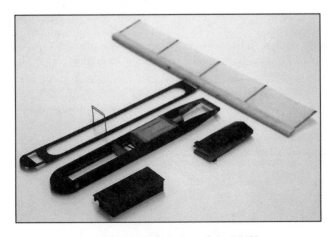

Fig. 28   Exploded view of the MAV.

Fig. 29   Isometric view of the MAV.

negligible compared to the mass of the model and the wires are flexible enough not to impede the model's motion. The model is suspended from the tunnel ceiling by the support wires such that the model may swing freely in the streamwise direction but is relatively steady in all other directions.

As the model flaps and creates thrust, the model is displaced in the streamwise direction, and the displacement is measured by bouncing the laser sensor off a small reflective surface on the back of the rear nacelle. The sensor is generally about 3 to 4 chord lengths downstream of the model and is not thought to create a significant flow interference effect. For some wing configurations, at higher frequencies the wings would completely come together in the middle, blocking the path of the laser to the reflective panel on the rear nacelle. In these cases an alternative rear nacelle was used that supported a small rectangular panel slightly downstream of the wing trailing edge.

The thrust is computed by measuring the precise weight of the model, the length of the pendulum, and the horizontal displacement resulting from thrust, and using

Fig. 30   Side view of the MAV.

the equation

$$T = \frac{W \Delta x}{\sqrt{L^2 - \Delta x^2}} \tag{8}$$

where W is the MAV weight, L is the pendulum length, and Δx is the horizontal deflection.

To reduce the deflection of the pendulum (in order to keep the model in the more accurate range of the laser sensor) ballast is added to the model. The dry mass of the MAV is typically about 6 g, depending on the set of wings used, and the ballast (the black box in the bottom left corner of Fig. 28) adds about 11 g.

The model has a composite construction, built primarily out of balsa wood and very thin graphite–epoxy laminates. The wings are constructed using tear-drop shaped balsa leading edges with thin carbon-fiber ribs, and the surfaces are made from very lightweight Japanese tissue. Typical wing masses are about 0.3 g. The wings are attached to the flapping-beams using thin carbon-fiber strips, with the length and width of the strip varied to control the elasticity of the joint. Super-glue is used to attach the carbon-fiber strips to the flapping-beams in such a way that the wings may easily be removed. The static, mean angle of attack of the wings is adjusted by heating the carbon-fiber strips with a soldering iron, which softens the epoxy and allows the strips to bend. Upon cooling they retain their new shape. The elasticity of the wing mount allows for a passive feathering mechanism. The wing deflects in pitch proportionally to the moment about the leading edge. The addition of this feature boosts static performance significantly and generally allows the motor to reach higher frequencies, since it reduces the peak torque requirements.

The initial wing geometry had a span of 150 mm and a chord of 25 mm, but a variety of additional wings have been built to investigate chord-length and aspect-ratio effects.

The flapping frequency is measured using a strobe light. The strobe light is set to a specific frequency, and the motor speed is adjusted until the wing motion appears frozen in the light of the strobe. The stepping-motor/controller circuit provides very precise, incremental speed control, and the motor speed is completely constant during a simulation, a feature that is typically not possible with conventional brushed motors.

Experiments are performed in the Naval Postgraduate School 1.5 × 1.5 m low-speed wind tunnel. The tunnel is a continuous, flow-through facility with an approximate flow speed range between 0 and 10 m/s. The speed is set by varying the pitch of a newly rebuilt fan that is driven by a constant-speed motor. The tunnel has a square, 4.5 × 4.5 m, bell-shaped inlet with a 9-to-1 contraction ratio to the 1.5 × 1.5 m test section. The turbulence level has been determined by Lund using LDV.[51] For speeds above 1.5 m/s the turbulence intensity was measured below 1.75% at the test section.

For the results presented here, flow speed in the tunnel was measured using a pitot-static tube at the upstream end of the test section, attached to an MKS Baratron type 223B differential pressure transducer. Presently LDV equipment is available for much more accurate measurements, especially in the low-speed range. Lund[51] validated the pitot/transducer system using LDV and found roughly a 2% underprediction of velocity.

For oscillatory motions the reduced frequency and/or Strouhal number are generally the significant nondimensional parameters. Reduced frequencies between

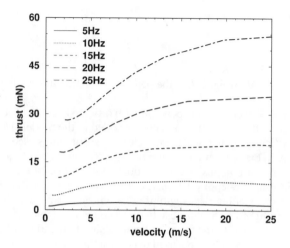

Fig. 31   Panel-code predictions for MAV.

about 0.1 and 10 are tested, as well as the limiting case of static thrust that
yields a theoretical reduced frequency of infinity (based on freestream speed). The
Reynolds number varied roughly between 0 and 17,500, based on chord length.

The panel-code predictions for thrust of the MAV are shown in Fig. 31 over the
desired speed range of the MAVs and for several flapping frequencies. The behavior is similar to that found for the purely plunging airfoil, with thrust increasing
roughly as the frequency squared. In Fig. 32 the predicted effect of the combined
pitch/plunge motion on the MAV configuration is plotted. Here the plunge amplitude of the leading edge was fixed at $h = 0.5$, the reduced frequency was 1, and the
pitch amplitude was varied between 0 and 10 deg. The panel code predicts a slight
increase in thrust with increased pitch amplitude, but there is a slight decrease in
efficiency.

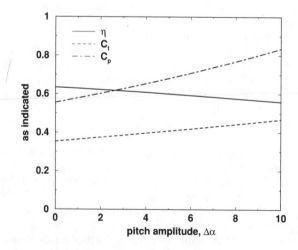

Fig. 32   Effect of pitch amplitude.

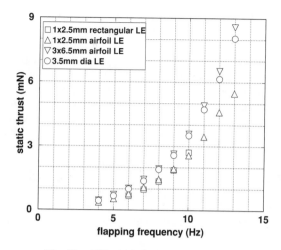

Fig. 33   Effect of leading-edge shape.

The airfoil sections used on the MAV are rather unconventional, not by choice but because of weight restrictions and the difficulty in producing more common airfoils. Of course, the Reynolds numbers are very low, and the dynamic angles of attack are very high, and so the choice of airfoils is not obvious. Therefore, a substantial effort was made to find a suitable airfoil shape and construction. In Fig. 33 the measured static thrust ($U_\infty = 0$) is plotted for a few of the airfoils tested.

The first had essentially a rectangular leading edge built from a thin layer of balsa wood sandwiched between laminations of carbon fiber. The leading edge was extremely stiff, but not all that light, and was not an ideal shape for high angles of attack. The second was the lightest wing, with an airfoiled balsa leading edge about the same size as the rectangular leading edge. Because of its light weight, the wings could be flapped faster, but it produced virtually identical thrust to the first case. The third case, which is now used on all wings, used a much larger airfoiled balsa leading edge. Balsa varies in density by several hundred percent from piece to piece, and by being very selective, very light wings with this airfoiled leading edge have been built. Since the balsa leading edge comprises around two-thirds of the wing weight, it was thought that a cylindrical leading edge might perform just as well and be a little lighter. However, the performance was not quite as good, and the weight difference was too small to help much.

It is thought that the improved performance for the larger airfoiled leading edge is due in part to a thicker, rounded leading edge, although the surface roughness is still quite large. Additionally, the present wing design has an aeroelastically changeable camber, owing to the flexibility of the graphite ribs. Unfortunately, the camber deflects in the less desirable direction, at least based on conventional airfoil theory. Some work has gone into the development of wings with passive camber changes, with the camber moving in the correct direction, but data are not yet available for those wings.

Some effort has been made to evaluate the flow quality over these sections, but on this scale, this is not a simple task. Use of standard fog machines has provided some insight; however, the fog particles alter the flow density, and if the fog hits the left or right or top or bottom of the MAV unevenly, the model is fairly violently thrown around as the result of the differential lift and thrust. The newest fuselage incorporates a rigid mount capability (as shown in Figs. 29 and 30), and this may provide a better means for future flow visualization studies.

Additionally, tufts have been used on some of the wings with varying success. With the very short lengths and the very energetic motions involved, the material selection for the tufts becomes difficult. Using a single strand of standard cotton thread (usually made of three or four thinner strands wound together), and varying lengths and placements, some data have been obtained. However, the cotton fibers tend to stick to the tissue surface, and they must be manually freed. Additionally, the proper length is an unknown. If the tufts are too short then the stiffness of the material may prevent the tufts from moving with the flow, and if they are too long they may indicate inertial effects rather than the local flow. One feature they have highlighted is a rather strong flow entrainment in the spanwise direction.

Wings of varying aspect ratio were tested, and at least in the static case (wind-off) the performance seemed to be independent of aspect ratio at values above about 4.5.

In Fig. 34 the thrust predicted by the panel code and measured experimentally are compared for the design configuration with $b = 150$ mm, $c = 25$ mm, and a fairly rigid wing mount for two frequencies, 8 and 12 Hz.

The panel code is unable to compute the static case for several reasons. First, the static case corresponds to an infinite reduced frequency, and second, the panel code assumes fully attached flow, which may be a poor assumption for the MAV altogether, and is certainly wrong at high reduced frequencies. Even at a reduced frequency of 2, the lowest velocity shown in the figure, the numerical results become questionable. Extrapolating the numerical values to the static case seems to agree pretty well with the measured thrust. However, when the tunnel is turned on, the performance of the MAV drops off quickly, but it begins to recover at higher speeds. The behavior is quite different from the panel-code predictions but may indicate something like the drag-bucket experienced in Ref. 22.

More complete data sets for the configuration described above and for a configuration with a span of 150 mm and a chord of 36 mm are shown in Figs. 35 and

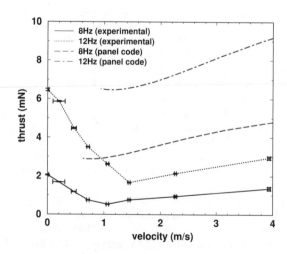

Fig. 34   Thrust for the 150 × 25 mm wing.

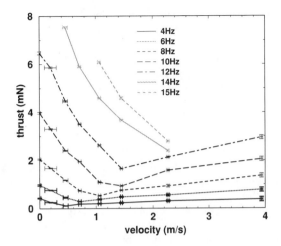

Fig. 35   Thrust for the 150 × 25 mm wing.

36, respectively. A similar shape is seen for all frequencies, with a rapid decline in thrust at low speeds, followed by a partial recovery after a minimum value.

The speed where the minimum thrust occurs increases with frequency but, interestingly, occurs at approximately the same reduced frequency for any configuration (about 1.2 for the first case and 1.1 for the second case), and in both configurations the Strouhal number at the minimum is about 0.23, approximately the Strouhal number of a naturally shed vortex street behind a cylinder ($Sr = 0.21$). Note that for the Strouhal number the reference length is the wake width, which must be approximated to include the elastic deflection of the wings. The induced angle of attack at the minimum is about 36 deg. It is fair to assume that the flow is fully detached, and most likely the convection rate of the dynamic-stall vortex places it in a highly disadvantageous position. In contrast, in the static case, the dynamic-stall vortex probably convects much more slowly, primarily because of the entrained

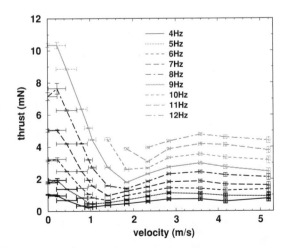

Fig. 36   Thrust for the 150 × 36 mm wing.

flow, and actually appears to aid in the propulsive performance. Clearly, a better understanding of the development and motion of the dynamic-stall vortices is required for the design of a successful flapping wing MAV.

The second configuration has a larger wing area and more flexible wing joints, and so it feathers much more than the first configuration. Consequently, in agreement with the panel code, its thrust is higher in the static case but diminishes at higher speeds.

VII. Summary and Prospective

In this chapter an attempt was made to review the current knowledge about the aerodynamics of flapping wings. To this end, pure plunge and pure pitch oscillations of single airfoils, combined pitch/plunge oscillations of single airfoils, and plunge or combined pitch/plunge oscillations of airfoil combinations were considered. The available experimental results and the two-dimensional panel-code and Euler/Navier–Stokes computations show that the major parameter for the classification of the flows induced by flapping airfoils is the Strouhal number. It is found that the prediction methods yield fairly good agreement as long as the flow over the flapping wing(s) remains attached, thus giving confidence in the use of these methods for design purposes. However, as soon as vortices are being shed from the airfoil leading edge, our incomplete understanding of these extremely complicated dynamic-stall phenomena and poor prediction capability become obvious, especially in the low Reynolds number flight regime. Micro air vehicles operate at very low Reynolds numbers. The design and development of micro air vehicles using flapping wings, therefore, requires additional systematic computational and experimental investigations of the dynamic-stall phenomena, which is bound to occur on such vehicles, in order to identify the most promising configurations.

Acknowledgments

This investigation was supported by the Naval Research Laboratory under project monitors Kevin Ailinger and Jill Dahlburg and by the Naval Postgraduate School direct research program.

References

[1]Knoller, R., "Die Gesetze des Luftwiderstandes," *Flug- und Motortechnik (Wien)*, Vol. 3, No. 21, 1909, pp. 1–7.

[2]Betz, A., "Ein Beitrag zur Erklärung des Segelfluges," *Zeitschrift für Flugtechnik und Motorluftschiffahrt*, Vol. 3, Jan. 1912, pp. 269–272.

[3]Katzmayr, R., "Effect of Periodic Changes of Angle of Attack on Behavior of Airfoils," NACA Rept. 147, Oct., 1922 (translated from *Zeitschrift für Flugtechnik und Motorluftschiffahrt*, 31 March 1922, pp. 80–82; and 13 April 1922, pp. 95–101).

[4]Birnbaum, W., "Das ebene Problem des schlagenden Flüels," *Zeitschrift für Angewandte Mathematik und Mechanik*, Vol. 4, No. 4, Aug. 1924, pp. 277–292.

[5]Birnbaum, W., "Der Schlagflügelpropeller und die kleinen Schwingungen elastisch befestigter Tragflügel," *Zeitschrift für Flugtechnik und Motorluftschiffahrt*, Vol. 15, 1924, pp. 128–134.

[6]von Kármán, T., and Burgers, J. M., "General Aerodynamic Theory—Perfect Fluids," *Aerodynamic Theory*, Division E, Vol. II, edited by W. F. Durand, Julius Springer, Berlin, 1934, p. 308.

[7]Garrick, I. E., "Propulsion of a Flapping and Oscillating Airfoil," NACA Rept. 567, 1936.

[8]Theodorsen, T., "General Theory of Aerodynamic Instability and the Mechanism of Flutter," NACA Rept. 496, 1935.

[9]Bratt, J. B., "Flow Patterns in the Wake of an Oscillating Airfoil," Aeronautical Research Council, R&M 2773, 1953.

[10]Jones, K. D., Dohring, C. M., and Platzer, M. F., "Experimental and Computational Investigation of the Knoller–Betz Effect," *AIAA Journal*, Vol. 36, No. 7, May 1998.

[11]Küchemann, D., and Weber, J., "Aerodynamic Propulsion in Nature," *Aerodynamics of Propulsion*, McGraw-Hill, New York, 1953, pp. 248–260.

[12]Teng, N. H., "The Development of a Computer Code for the Numerical Solution of Unsteady, Inviscid and Incompressible Flow over an Airfoil," M.S. Thesis, Naval Postgraduate School, Monterey, CA, June 1987.

[13]Pang, C. K., "A Computer Code for Unsteady Incompressible Flow past Two Airfoils," Aeronautical Engineer's Thesis, Dept. of Aeronautics and Astronautics, Naval Postgraduate School, Monterey, CA, Sept. 1988.

[14]Platzer, M. F., Neace, K. S., and Pang, C. K., "Aerodynamic Analysis of Flapping Wing Propulsion," AIAA Paper 93-0484, Jan. 1993.

[15]Neace, K. S., "A Computational and Experimental Investigation of the Propulsive and Lifting Characteristics of Oscillating Airfoils and Airfoil Combinations in Incompressible Flow," M.S. Thesis, Dept. of Aeronautics and Astronautics, Naval Postgraduate School, Monterey, CA, Sept. 1992.

[16]Riester, P. J., "A Computational and Experimental Investigation of Incompressible Oscillatory Airfoil Flow and Flutter Problems," M.S. Thesis, Naval Postgraduate School, Monterey, CA, June 1993.

[17]Turner, M., "A Computational Investigation of Wake-Induced Airfoil Flutter in Incompressible Flow and Active Flutter Control," M.S. Thesis, Naval Postgraduate School, Monterey, CA, March 1994.

[18]Jones, K. D., and Platzer, M. F., "Time-Domain Analysis of Low-Speed Airfoil Flutter," *AIAA Journal*, Vol. 34, No. 5, May 1996.

[19]Jones, K. D., and Center, K. B., "Numerical Wake Visualization for Airfoils Undergoing Forced and Aeroelastic Motions," AIAA Paper 96-0055, Jan. 1996.

[20]Jones, K. D., and Platzer, M. F., "Numerical Computation of Flapping Wing Propulsion and Power Extraction," AIAA Paper 97-0826, Jan. 1997.

[21]Jones, K. D., and Platzer, M. F., "Airfoil Geometry and Flow Compressibility Effects on Wing and Blade Flutter," AIAA Paper 98-0517, Jan. 1998.

[22]Jones, K. D., and Platzer, M. F., "An Experimental and Numerical Investigation of Flapping Wing Propulsion," AIAA Paper 99-0995, Jan. 1999.

[23]Ekaterinaris, J. A., Cricelli, A. S., and Platzer, M. F., "A Zonal Method for Unsteady, Viscous, Compressible Airfoil Flows," *Journal of Fluids and Structures*, Vol. 8, 1994, pp. 107–123.

[24]Tuncer, I. H., Platzer, M. F., and Ekaterinaris, J. A., "Computational Analysis of Flapping Airfoil Aerodynamics," American Society of Mechanical Engineers Fluids Engineering Division, Summer Meeting, June 1994.

[25]Ekaterinaris, J. A., and Platzer, M. F., "Numerical Investigation of Stall Flutter," *Journal of Turbomachinery*, Vol. 118, 1996, pp. 197–203.

[26]Tuncer, I. H., and Platzer, M. F., "Thrust Generation due to Airfoil Flapping," *AIAA Journal*, Vol. 34, No. 2, 1996, pp. 324–331.

[27]Ekaterinaris, J. A., and Platzer, M. F., "Computational Prediction of the Airfoil Dynamic Stall," *Progress in Aerospace Sciences*, Vol. 33, 1997, pp. 759–846.

[28]Tuncer, I. H., Lai, J., Ortiz, M. A., and Platzer, M. F., "Unsteady Aerodynamics of Stationary/Flapping Airfoil Combination in Tandem," AIAA Paper 97-0659, Jan. 1997.

[29]Tuncer, I. H., Walz, R., and Platzer, M. F., "A Computational Study of the Dynamic Stall of a Flapping Airfoil," AIAA Paper 98-2519, June 1998.

[30]Weber, S., Jones, K. D., Platzer, M. F., and Ekaterinaris, J. A., "Transonic Flutter Computations for a 2D Supercritical Wing," AIAA Paper 99-0798, Jan. 1999.

[31]Lai, J. C. S., Yue, J., and Platzer, M. F., "Control of Backward Facing Step Flow Using a Flapping Airfoil," ASME FEDSM97-3307, June 1997.

[32]Dohring, C. M., Fottner, L., and Platzer, M. F., "Experimental and Numerical Investigation of Flapping Wing Propulsion and Its Application for Boundary Layer Control," ASME Paper 98-GT-46, American Society of Mechanical Engineers, June 1998.

[33]Lai, J. C. S., and Platzer, M. F., "Jet Characteristics of a Plunging Airfoil," *AIAA Journal*, Vol. 37, No. 12, Dec. 1999, pp. 1529–1537.

[34]Ohashi, H., and Ishikawa, N., "Visualization Study of a Flow near the Trailing Edge of an Oscillating Airfoil," *Bulletin of the Japanese Society of Mechanical Engineers*, Vol. 15, 1972, pp. 840–845.

[35]Kadlec, R. A., and Davis, S. S., "Visualization of Quasiperiodic Flows," *AIAA Journal*, Vol. 17, 1979, pp. 1164–1169.

[36]Triantafyllou, G. S., Triantafyllou, M. S., and Grosenbaugh, M. A., "Optimal Thrust Development in Oscillating Foils with Application to Fish Propulsion," *Journal of Fluids and Structures*, Vol. 7, 1993, pp. 205–224.

[37]Jones, K. D., and Platzer, M. F., "Flapping Wing Propulsion for a Micro Air Vehicle," AIAA Paper 2000-0897, Jan. 2000.

[38]Koochesfahani, M. M., "Vortical Patterns in the Wake of an Oscillating Airfoil," *AIAA Journal*, Vol. 27, No. 9, Sept. 1989, pp. 1200–1205.

[39]DeLaurier, J. D., and Harris, J. M., "Experimental Study of Oscillating-Wing Propulsion," *Journal of Aircraft*, Vol. 19, No. 5, May, 1982, pp. 368–373.

[40]Liu, P., "Three-Dimensional Oscillating Foil Propulsion," M.S. Engineering Thesis, Univ. of Newfoundland, St. John's, NF, Canada, March 1991.

[41]Liu, P., "A Time-Domain Panel Method for Oscillating Propulsors with Both Chordwise and Spanwise Flexibility," Ph.D. Thesis, Univ. of Newfoundland, St. John's, NF, Canada, 1996.

[42]Send, W., "The Mean Power of Forces and Moments in Unsteady Aerodynamics," *Zeitschrift für Angewandte Mathematik und Mechanik*, Vol. 72, 1992, pp. 113–132.

[43]Send, W., "Otto Lilienthal und der Mechanismus des Schwingenflugs," DGLR-JT96-030, German Aerospace Congress, Dresden, Sept. 1996.

[44]Hall, K. C., and Hall, S. R., "Minimum Induced Power Requirements for Flapping Flight," *Journal of Fluid Mechanics*, Vol. 323, Sept. 1996, pp. 285–315.

[45]Hall, K. C., Pigott, S. A., and Hall, S. R., "Power Requirements for Large-Amplitude Flapping Flight," AIAA Paper 97-0827, Jan. 1997.

[46]Isogai, K., Shinmoto, Y., and Watanabe, Y., "Effects of Dynamic Stall Phenomena on Propulsive Efficiency and Thrust of a Flapping Airfoil," AIAA Paper 97-1926, June 1997.

[47]McKinney, W., and DeLaurier, J., "The Wingmill: An Oscillating Wing Windmill," *Journal of Energy*, Vol. 5, No. 2, March–April 1981, pp. 109–115.

[48] Jones, K. D., Davids, S., and Platzer, M. F., "Oscillating-Wing Power Generator," American Society of Mechanical Engineers, ASME Paper FEDSM99-7050, July 1999.

[49] Schmidt, W., "Der Wellpropeller, ein neuer Antrieb fuer Wasser-, Land-, und Luft-fahrzeuge," *Zeitschrift für Flugwissenschaften*, Vol. 13, 1965, pp. 472–479.

[50] Bosch, H., "Interfering Airfoils in Two-dimensional Unsteady Incompressible Flow," AGARD CP-227, Paper 7, Sept. 1977.

[51] Lund, T. C., "A Computational and Experimental Investigation of Flapping Wing Propulsion," M.S. Thesis, Dept. of Aeronautics and Astronautics, Naval Postgraduate School, Monterey, CA, March 2000.

Chapter 17

Aerodynamic Characteristics of Wings at Low Reynolds Number

Akira Azuma*
University of Tokyo, Tokyo, Japan

Masato Okamoto[†]
Wakayama Technical High School, Wakayama, Japan
and
Kunio Yasuda[‡]
Nihon University, Chiba, Japan

Nomenclature

a	= three-dimensional lift slope, nondimensional aerodynamic center	A	= acceleration of body
		A_{iq}	= added mass tensor
		AR	= aspect ratio
a_0	= two-dimensional lift slope	C	= attenuation coefficient, Theodorsen function
a_F	= mean slope of F with respect to k		
		C_d	= two-dimensional drag coefficient
b	= wingspan, half chord $= c/2$		
b_G	= mean slope of G with respect to k	C_D	= three-dimensional drag coefficient
c	= wing chord	C_F	= force coefficient
d	= two-dimensional drag	C_ℓ	= two-dimensional lift coefficient
f	= beating frequency		
g	= gravity acceleration	C_L	= three-dimensional lift coefficient
h	= heaving height		
h_e	= speed parameter $= 4\pi V_Z/bc$	C_m	= moment coefficient
k	= reduced frequency $= \omega c/2U$	C_n	= normal force coefficient
ℓ	= two-dimensional lift, reference length	C_P	= power coefficient
		C_t	= thrust force coefficient

*Professor Emeritus.
[†]Teacher.
[‡]Associate Professor, Department of Aerospace Engineering.

m = mass of body, two-dimensional moment

m_b = mass of blade

$m_{c/4}$ = two-dimensional moment about $c/4$ chord

m_e = frequency parameter = $\omega/\Omega b$

m_s = fluid mass passing through stroke plane

n = two-dimensional normal force, rate of spin

p = power, propulsive force

q = perpendicular force

r = spanwise distance

r = position vecter of a blade element

s = suction force, nondimensional distance = $Ut/(c/2)$

t = time, two-dimensional thrust, tangential force

v = induced flow speed = $|v|$

v = induced velocity

(x, y, z) = coordinate system

(x_i, y_i, z_i) = body frame of ith wing

α = angle of attack

β = side slip angle, flapping angle

ζ = lead–lag angle

η = Froude efficiency

θ = feathering or pitch angle

θ_t = wing twist

κ = yaw angle of stroke plane

λ = nondimensional inflow = $(V_{z_s} + v)/R\Omega$

μ = viscosity

ν = kinematic viscosity = μ/ρ, flapping angle with respect to stroke plane

ρ = density of fluid

ρ_b = density of wing or blade

σ = solidity

τ = time variable

ϕ = Wagner's function

ψ = Küsners function, phase difference, sweep or beating angle

ω = angular rate = $|\omega|$

ω = angular velocity of wing

D = three-dimensional drag

F = force = $|F|$, Theodorsen function

F = force

F_c = chordwise shear force

F_f = flatwise shear force

G = unit based on gravity acceleration

G = Theodorsen function

G = gravity acceleration

H_a = moment matrix of added mass

H_m = moment matrix of real mass

I_i = moment of inertia

J_{ij} = product of inertia

K_a = force matrix of added mass

K_m = force matrix of real mass

L = three-dimensional lift

M = three-dimensional moment

M_c = chordwise bending moment

M_f = flatwise bending moment

M_t = torsional moment about feathering axis

Q = torque = $|Q|$

Q = torque

R = position vector

R = wing or rotor radius

Re = Reynolds number = Ul/v

S = wing area

S_ℓ, S_r = transformation matrices

$T_b, T_i, T_s, T_t, T_{i\theta}$ = transformation matrices

U = relative inflow

V = forward speed = $|V|$

V = velocity

V_b = body volume

W = generalized Loewy's function

(X, Y, Z) = body frame

(X_I, Y_I, Z_I) = inertial frame

Γ = circulation

Θ = pitch angle of body

Θ_s = pitch angle of stroke plane

Θ_t = tilt angle of tergal plate

Φ = rolling angle of body

Ψ = yawing angle of body

Ω = angular rate = $|\Omega|$

Ω = angular velocity of body

Superscripts and Subscripts

$\overline{(\ )}$	= mean or average value	$(\ )_o$	= origin of frame
$(\dot{\ })$	= rate	$(\ )_2$	= intermediate frame
$\lvert(\ )\rvert$	= absolute value	$(\ )_{ac}$	= aerodynamic center
$(\ )^{-1}$	= inverse form	$(\ )_b$	= blade or body related
$(\ )^A$	= aerodynamic component	$(\ )_c$	= chordwise component
$(\ )^B$	= body component	$(\ )_e$	= elastic component
$(\ )^G$	= gravity component	$(\ )_f$	= flatwise component
$(\ )^i$	= ith wing	$(\ )_g$	= gravity component
$(\ )^I$	= inertial component	$(\ )_h$	= heaving component
$(\ )^j$	= jth virtual joint	$(\ )_l$	= liner component
$(\ )^w$	= wing component	$(\ )_{max}$	= maximum value
$(\ )^*$	= left-hand system	$(\ )_n$	= nth harmonics
		$(\ )_r$	= rotation component
		$(\ )_s$	= spanwise component, stroke plane
		$(\ )_t$	= tergal component
		$(\ )_\theta$	= feathering component

I. Introduction

T O EXPLAIN how an insect's flight relies on a high lift coefficient, Weis-Fogh proposed a new circulating lift mechanism consisting of clap-and-fling and flip. This mechanism is used by small insects characterized with low-aspect-ratio wings that beat at high reduced frequency. Without any concrete demonstration, many investigators have used the above mechanism to explain the effect of flow separation on the generation of extremely high lift to overcome the steady-state lift obtained by assuming a constant and homogeneous lift coefficient over the entire stroke plane.

However, recent experimental studies and numerical analyses based on the Navier–Stokes equations revealed that a periodic change of flow separation specifically observed at or near the leading edge of an oscillating wing, called *dynamic stall*, increased lift coefficient mainly during downstroke within a moderate maximum value. By analyzing filmed data of the free flight of dragonflies and damselflies, Azuma and his collaborators concluded that these insects can fly without generating an abnormally large lift coefficient.

The traditional analyses have mainly focused on the qualitative observation of vortical flow around a beating wing and, thus, have not taken up the quantitative calculation on the aerodynamic and inertial forces and moments acting on the beating wing. To perform the analysis using blade element theory, one must know exactly the relative flow velocity and the acceleration of a blade element affected by both the body motion in the inertial space and the wing motion with respect to the body.

II. Unsteady Wing Theory

A. Two-Dimensional Potential Flow

From potential theory, a thin airfoil or plate having either a sudden (step) or a sinusoidal change of angle of attack α generates an alleviating lift. As given in

Table 1, the indicial response is given by the lift alleviating functions, Wagner's function[1] $\phi(s)$ for a wing in stepping motion and Küssner's function[2] $\psi(s)$ for a sharp-edged gust where $s = Ut/(c/2)$ (Bisplinghoff et al.[3]). Similarly, as given in Table 1, the sinusoidal response is given by Theodorsen's function $C(\omega)$ for a sinusoidal oscillation of a wing (Theodorsen,[4] Theodorsen and Garrick[5]) and Sears's function for a sinusoidal gust (von Kármán and Sears,[6] Sears[7]). The lift alleviation results from a series of shed vortices left in the wake, which is assumed to be a flat surface parallel to the general flow V.

A thin plate (chord c) oscillating sinusoidally with a small coupled mode of feathering or pitching θ and heaving h motion (angular speed of ω) in an inviscid fluid generates the lift ℓ, thrust t, moment $m_{c/4}$, and power p as given by the coefficients in Tables 2 and 3 (Garrick,[8] Lighthill,[9] Wu,[10] Sato,[11] Azuma[12]). The interesting results for a beating wing assumed to be a flat plate and operating at low values of reduced frequency are given as follows:

1) The lift ℓ, which as shown in Fig. 1, consists of both the normal component n (with respect to the velocity U ($|U| = \sqrt{V^2 + \dot{h}^2}$) of the pressure integration over the plate surface) and the suction force component s (acting on the leading edge of the plate). The latter component is tangential to the plate and is given by using the Blasius formulae (Batchelor[13]) about the singular point at the leading edge.

2) The mean lift $\bar{\ell}$ is proportional to the angle of attack α; that is, $\bar{C}_\ell = 2\pi\alpha$, where α is given by $\alpha = \theta + \tan^{-1}(\dot{h}/V)$.

3) The thrust t, consisting of the parallel component of n and s with respect to the velocity V, is approximately the product of the lift and the nondimensional heaving velocity, $C_t \cong C_\ell(\dot{h}/V)$, except for very high values of k.

4) The mean moment about the three-quarter chord or the aerodynamic center is zero; that is, $\bar{C}_{m,c/4} = 0$.

5) The mean power is, by neglecting the feathering power, nearly equal to the product of the lift and the heaving velocity.

The above statements, specifically 3), also suggest that if the suction force is discounted, the aerodynamic force is tilted back far enough to have positive pitch ($\theta > 0$), the lift is reduced, the thrust is discarded, and the drag is, rather, generated in spite of the inviscid fluid. However, by taking negative pitch (or head-down attitude; $\theta < 0$) in the downward heaving, the thrust is assisted by the forward tilt of the normal force n. An adequate selection of the pitch angle θ is required to determine an optimal angle of attack α under a stall limit of the airfoil. The suction force may, approximately, be obtained by replacing the lift ℓ with the normal force n or $C_\ell \cong C_n$ and replacing the thrust t with the approximate expression of $s \cong t \cong n(\dot{h}/V)$ or $C_t \cong C_n(\dot{h}/V)$. The maximum thrust can be obtained when the phase difference $\phi_h - \phi_\theta$ is equal to $\psi_{h\theta}$, which, as shown in Table 3, increases from $\phi_h - \phi_\theta = 90$ deg to 0 deg as the reduced frequency increases. The Froude efficiency, the mean thrust times forward speed over the mean power, $\eta = tV/p = \bar{C}_t/\bar{C}_p$, is possible to have the maximum value and the optimal phase difference, both of which are, as shown in Table 2, given as a function of the reduced frequency under given parameters of $\bar{C}_t/(kh_1/c)^2$. Roughly speaking, as the reduced frequency increases, the thrust increases monotonically whereas the efficiency decreases from the maximum value of $\eta_{max} = 1.0$ with the phase difference of $\phi_h - \phi_\theta = 90$ deg to $\eta_{max} = 0.5$ with the phase difference of $\phi_h - \phi_\theta = 180$ deg.

This fundamental knowledge will be very important for understanding the physical characteristics of a wing in unsteady motion as well as steady flight. The

Table 1 Lift alleviating functions for an airfoil[a]

Item	Lift expression	Function	Parameter
Indicial response			
Wagner's function	$\ell = \frac{1}{2}\rho U^2 ca \int_0^1 \Phi(s-\sigma)\left(\frac{d\alpha}{d\sigma}\right) d\sigma$	$\Phi(s) = 1 - 0.165\exp(-0.0455s) - 0.335\exp(-0.3s)$ $\cong (s+2)/(s+4)$	$s = \dfrac{Ut}{(c/2)}$
Küssner's function	$\ell = \frac{1}{2}\rho U^2 ca \int_0^1 \Psi(s-\sigma)\left(\frac{d\alpha}{d\sigma}\right) d\sigma$	$\Psi(s) = 1 - 0.500\exp(-0.130s) - 0.500\exp(-s)$ $\cong (s^2+s)/(s^2 + 2.82s + 0.80)$	
Frequency response[b]			$k = \omega c/2U$
Theodorsen's function	$\ell = \frac{1}{2}\rho U^2 ca[\alpha_1 \exp(i\omega t)C(\omega)]$	$C(\omega) = i\omega \int_{-\infty}^{\infty} \Phi(t-\tau)\exp[-i\omega(t-\tau)]d\tau$ $= H_1^{(2)}(\omega)/\left[H_1^{(2)}(\omega) + iH_0^{(2)}(\omega)\right]$ $F \cong 1 - a_F k, \quad G \cong -b_G k^a$	$H_n^{(2)}$: Hankel function of the 2nd kind of order n $H_n^{(2)} = J_n - iY_n$ For small k
Sear's function	$\ell = \frac{1}{2}\rho U^2 ca\left[\dfrac{W_1}{U}\exp(i\omega t)S(\omega)\right]$	$S(\omega) = i\omega \int_{-\infty}^{\infty} \Psi(t-\tau)\exp[-i\omega(t-\tau)]d\tau$ $= C(\omega)[J_0(\omega) - iJ_1(\omega)] + J_1(\omega)$ $\cong \{1/\sqrt{1+2\pi\omega}\}\exp[-i(1/4)\pi(2\pi\omega)/(1+2\pi\omega)]$	J_n: Bessel function of the 1st kind Y_n: Bessel function of the 2nd kind
Loewy's function	$\ell = \frac{1}{2}\rho U^2 ca[\alpha_1 \exp(i\omega t)C(k, m_e, h_e)]$	$C(k, \bar{m}_e, \bar{h}_e) = \dfrac{H_1^{(2)}(k) + 2J_1(k)W(k, \bar{m}_e, \bar{h}_e)}{H_1^{(2)}(k) + iH_0^{(2)}(k) + [J_1(k) + J_0(k)]W(k, \bar{m}_e, \bar{h}_e)}$ $W(k, \bar{m}_e, \bar{h}_e) = \begin{cases} 1/[\exp(k\bar{h}_e + i2\pi\bar{m}_e) - 1]; & k > 0 \\ 0 & ; \quad k = 0 \end{cases}$	I_n: Modified Bessel function of the 1st kind K_n: Modified Bessel function of the 2nd kind $\bar{m}_e = \omega/\Omega b = k\bar{\tau}/bk\bar{\tau}_e$

(Cont.)

Table 1 (Continued)

Item	Lift expression	Function	Parameter[c]
Generalized Loewy's function	$\ell = \frac{1}{2}\rho U^2 ca[\alpha_1 \exp(i\omega t)C(k, m_e, h_e)]$	$C(\bar{s}, \bar{s}\bar{r}_e, \bar{s}\bar{h}_e) = \dfrac{K_1(\bar{s}) + \pi i I_1(\bar{s})W(\bar{s}\bar{r}_e, \bar{s}\bar{h}_e)}{K_1(\bar{s}) + K_0(\bar{s}) + \pi i[I_1(\bar{s}) - I_0(\bar{s})]W(\bar{s}\bar{r}_e, \bar{s}\bar{h}_e)}$ $W(\bar{s}\bar{r}_e, \bar{s}\bar{h}_e) = \begin{cases} 1/[\exp(-i\bar{s}\bar{h}_e + 2\pi\bar{s}\bar{r}_e) - 1]; & |\bar{s}| > 0 \\ 0 & ; \quad |\bar{s}| = 0 \end{cases}$	$\bar{h}_e = 4\pi V_Z/bc$ $\bar{r} = r/(c/2)$ $\bar{r}_e = \bar{r}/b$ $\bar{s} = s(c/2)/V_X^c$ $\sigma = bc/\pi R$ $\lambda = (V_Z + v)/R\Omega$
Miller's function	$\ell = \frac{1}{2}\rho U^2 ca[\alpha_1 \exp(i\omega t)C(k, m_e, h_e)]$	$C(k) = 1/[1 + (k\pi/2)]$ or $= 1/[1 + (\sigma\pi/4\lambda)]$	

[a]The values of F and G are assumed to be a linear function of k, and, thus, if the above linear relations are matching with the theoretical values at $k = 0.11$, the values of a_F and b_G should be $a_F = b_G = 1.7$.

[b]α_1 and W_1 are amplitude of sinusoidal inputs for angle of attack and gust respectively.

[c]V_X and V_Z are horizontal and vertical components of general velocity; v is the induced velocity; r, R, and c are radial distance, radius of rotor, and blade chord respectively; and b is number of blades.

Table 2   Aerodynamic force and moment coefficients for a thin airfoil on heaving and feathering oscillation (Sato,[11] Azuma[12])

Item	Expression	For small k
Coefficients		
C_ℓ	$C_{\ell,\theta_0}\theta_0 + [C_{\ell,\theta,c}\cos(\omega t + \phi_\theta) + C_{\ell,\theta,s}\sin(\omega t + \phi_\theta)]\theta_1 + [C_{\ell,h,c}\cos(\omega t + \phi_h)$	—
	$+ C_{\ell,h,s}\sin(\omega t + \phi_h)](kh_1/c)$	
$C_{m,c/4}$	$[C_{m,\theta,c}\cos(\omega t + \phi_\theta) + C_{m,\theta,s}\sin(\omega t + \phi_\theta)](k\theta_1) + [C_{m,h}\cos(\omega t + \phi_h)](kh_1/c)$	0
C_t	$[C_{t,\theta\theta,0} + C_{t,\theta\theta,c}\cos(2\omega t + 2\phi_\theta) + C_{t,\theta\theta,s}\sin(2\omega t + 2\phi_\theta)]\theta_1^2$	$C_\ell(\dot{h}/V)$
	$+ [C_{t,\theta h,0}\cos(\phi_h - \phi_\theta - \phi_\theta) + C_{t,\theta h,c}\cos(2\omega t + \phi_\theta + \phi_h) + C_{t,\theta h,c}\sin(2\omega t + \phi_\theta + \phi_h)]$	
	$\times \theta_0(kh/c)^2 + [C_{t,hh,0} + C_{t,hh,c}\cos(2\omega t + \phi_\theta) + C_{t,hh,c}\sin(\omega t + \phi_\theta)]\theta_0\theta_1$	
	$+ [C_{t,\theta\theta h,s}\cos(\omega t + \phi_h) + C_{t,\theta\theta h,c}\sin(\omega t + \phi_h)]\theta_0(kh_1/c)$	
Derivatives for C_t		
$C_{t,\theta\theta,0}$	$\pi[F^2 + G^2 - F + k^2(F^2 + G^2 - F + 1/2)]$	$-a_F\pi k$
$C_{t,\theta\theta,c}$	$\pi[F^2 - G^2 - F + 2kG(1 - 2F) + k^2(G^2 - F^2 + F)]$	$-a_F\pi k$
$C_{t,\theta\theta,s}$	$\pi[G(1 - 2F) + k(2G^2 - 2F^2 + 2F + 1/2) + k^2G(2F - 1)]$	$(1/2 + b_G)\pi k$
$C_{t,\theta h,0}$	$2\pi\sqrt{[G + k(2F^2 + 2G^2 - F + 1/2)]^2 + (2F^2 + 2G^2 - F - kG)^2}$	$2\pi(1 - 3a_F k)$
$C_{t,\theta h,c}$	$2\pi[G(1 - 4F) + k(2G^2 - 2F^2 + F + 1/2)]$	$-\pi k(1 - 3b_G k)$
$C_{t,\theta h,s}$	$2\pi[2G^2 - 2F^2 + F + kG(4F - 1)]$	$-2\pi(1 - 3a_F k)$
$C_{t,hh,0}$	$4\pi(F^2 - G^2)$	$4\pi(1 - 2a_F k)$
$C_{t,hh,c}$	$4\pi(G^2 - F^2)$	$-4\pi(1 - 2a_F k)$

(*Cont.*)

Table 2   *(Continued)*

Item	Expression	For small k
$C_{\ell,hh,s}$	$8\pi FG$	$-8\pi b_G k$
$C_{\ell,\theta\theta,c}$	$2\pi[F - kG + (k/2)^2]$	$2\pi(1 - a_F k)$
$C_{\ell,\theta\theta,s}$	$2\pi[-kF - G + (3k/2)]$	$\pi k(1 + 2b_G)$
$C_{\ell,\theta0h,c}$	$2\pi(k - 2G)$	$2\pi k(1 + 2b_G)$
$C_{\ell,\theta0h,s}$	$-4\pi F$	$-4\pi(1 - a_F k)$
Phase Lag $\phi_{h\theta}$	$\tan^{-1}\left[\dfrac{F - 2F^2 - 2G^2 + kG}{G + k(2F^2 + 2G^2 - F + 1/2)}\right]$	$\tan^{-1}\left[\dfrac{-(1 - 3a_F k)}{3/2 - a_F k}\right]$
Derivatives for C_ℓ		
$C_{\ell,\theta0}$	2π	2π
$C_{\ell,\theta,c}$	$2\pi[F - kG - (k/2)^2]$	$2\pi(1 - a_F k)$
$C_{\ell,\theta,s}$	$-2\pi(G + kF + k/2)$	$-3\pi k(1 - 2b_G/3)$
$C_{\ell,h,c}$	$-2\pi(k + 2G)$	$-2\pi k(1 - 2b_G)$
$C_{\ell,h,s}$	$-2\pi(2F)$	$-4\pi(1 - a_F k)$
Derivatives for C_m		
$C_{m,\theta,c}$	$(3/16)\pi k$	$(3/16)\pi k$
$C_{m,\theta,s}$	$\pi/2$	$\pi/2$
$C_{m,h}$	$k\pi/2$	$\pi k/2$

^{a}Input: $z = -h - \theta(x - ab)$, $\theta = \theta_0 + \theta_1\cos(\omega t + \phi_\theta)$, $h = h_0 + h_1\cos(\omega t + \phi_h)$, and $\alpha = \alpha_0 + \alpha_1\cos(\omega t + \phi_\alpha)$ with

$$\alpha_1 = \sqrt{\theta_1^2 + (h_1\omega/U)^2 + 2\theta_1(h_1\omega/U)\sin(\phi_\theta - \phi_h)}$$

and

$$\phi_\alpha = \tan^{-1}\{[\theta_1\sin\phi_\theta + (h_1\omega/U)\cos\phi_h]/[\theta_1\cos\phi_\theta - (h_1\omega/U)\cos\phi_h]\}.$$

Table 3   Aerodynamic power coefficients and efficiency for a thin airfoil on heaving and feathering oscillation (Sato,[11] Azuma[12])

Item	Expression	For small k
Power coefficients		
	$p/\frac{1}{2}\rho U^3 c = (\ell\dot{h} - m_{c/4})/\frac{1}{2}\rho U^3 c = C_\ell(\dot{h}/U) - C_{m_{c/4}}(\dot{\theta}/U)$	
C_p	$= \theta_1^2[C_{p\theta\theta,0} + C_{p\theta\theta,c}\cos(2\omega t + 2\phi_\theta)$	
	$\quad + C_{p\theta\theta,s}\sin(2\omega t + 2\phi_\theta)] + \theta_1(kh_1/c)[C_{p\theta h,0}\cos(\phi_h - \phi_\theta - \psi_{\theta h})$	—
	$\quad + C_{p\theta h,c}\cos(2\omega t + \phi_h + \phi_\theta) + C_{p\theta h,s}\sin(2\omega t + \phi_h + \phi_\theta)]$	
	$\quad + (kh_1/c)^2[C_{phh,0} + C_{phh,c}\cos(2\omega t + \phi_h)$	
	$\quad + C_{phh,s}\sin(2\omega t + \phi_h)] + \theta_0(kh_1/c)C_{p\theta h,s}\sin(\omega t + \phi_h + \phi_\theta)$	
$\bar{C}_p$	$\bar{C}_p \cong C_\ell\cos[\tan^{-1}(\dot{h}/V)](\dot{h}/V) \cong a\bar{\alpha}(\dot{h}/V)$	—
Derivatives for C_p		
$C_{p,\theta\theta,0}$	$(1/2)\pi k^2$	0
$C_{p,\theta\theta,c}$	$-(1/2)\pi k^2$	0
$C_{p,\theta\theta,s}$	$(3/16)\pi k^3$	0
$C_{p,\theta h,0}$	$2\pi\sqrt{(-F + kG)^2 + [G + k(F + 1/2)]^2}$	$2\pi(1 - a_F k)$
$C_{p,\theta h,c}$	$-2\pi[G + k(F + 1/2)]$	$-2\pi k(3/2 - b_G - a_F k)$
$C_{p,\theta h,s}$	$2\pi[-F + kG + (k^2/2)]$	$-2\pi(1 - a_F k)$
$C_{p,hh,0}$	$4\pi F$	$4\pi(1 - a_F k)$
$C_{p,hh,c}$	$-4\pi F$	$-4\pi(1 - a_F k)$
$C_{p,hh,s}$	$2\pi(k + 2G)$	$2\pi k(1 - 2b_G)$
$C_{p,\theta 0h,s}$	-4π	-4π

(Cont.)

Table 3 (*Continued*)

Item	Expression	For small k
Phase lag $\psi_{\theta h}$	$\tan^{-1}\left[\dfrac{-F+kG}{G+k(F+1/2)}\right]$	$\tan^{-1}\left[\dfrac{-(1-a_F k)}{k(3/2 - b_G)}\right]$
For the maximum thrust	$\phi_h - \phi_\theta = \phi_{h\theta}$; given in Table 2	
Efficiency η	$\eta = \bar{C}_t/\bar{C}_p$	

Graph (upper, right column): Maximum efficiency, $\eta_{\max}$ (vertical axis, values 0.5 and 1) versus Reduced frequency, k (horizontal axis, 0.01, 0.1, 1, 10). Curves labeled 0.1, 1, 2, 3, 4, 5. Labels: "For coupled single motion", "heaving motion only", "feathering motion only".

Graph (lower, right column): Phase difference, $\phi_h - \phi_\theta$ (deg) (vertical axis, -180, -90, 0, 90, 180) versus Reduced frequency, k (horizontal axis, 0.01, 0.1, 1, 10). Curves labeled 0.1, 1, 2, 3, 4, $\bar{C}_t/(kh_1/c)^2 = 5$. Labels: "For the minimum power", "For the maximum thrust".

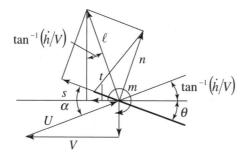

Fig. 1   Components of the aerodynamics forces and moment.

above potential theory was further extended to take into account large amplitude oscillations by Faigrieve and DeLaurier,[14] and airfoil thickness effects by Platzer et al.[15] and Jones and Platzer.[16]

In the case of a rotary (or beating) wing operating in hovering state, wake vortices are trailed in a continuous helical (or zigzagged helical) surface. Then, the numerical calculation based on the Biot-Savart law is required to get the effect of the wake vortices (Heyson[17,18]). However, by assuming that the wake vortices are consisting of either a series of vortex rings piled up in a semi-infinite vortex cylinder (Castles and DeLeeuw,[19] Heyson and Katzoff[20]) or a two-dimensional wake model represented by successive rows of distributed shed vortices (Loewy[21]), analytical treatments are possible and the results of the latter model are also given in Table 1, in which further simplified expressions for low value of reduced frequency are included (Miller[22,23]).

B.   Reduced Frequency

One of the notable indices for the unsteadiness of a beating wing is the reduced frequency. In hovering flight, the maximum tip speed of a sinusoidally beating wing is given by $V_{t,\max} = R\dot{\psi}_{\max} \cong \pi b f \psi_1$, where b, f, and ψ_1 are wingspan, beating frequency, and beating amplitude on the stroke plane, respectively. Then the reduced frequency at the wingtip is given by $k \cong \omega c/2V_{t,\max} = \pi c f /\pi b f \psi_1 = 1/\psi_1 AR \cong 1/AR$, where the approximation is obtained by assuming $\psi_1 = 1.0$ (Weis-Fogh[24]). This is a very important relation in the aerodynamics of hovering flight. The above relation states that any hovering object with a large aspect ratio wing operates the wings with low reduced frequency at which both the unsteady potential effect and dynamic stall effect are small and relies on a quasi-steady aerodynamic force to support its weight and that a hovering object with a small aspect ratio wing utilizes an unsteady aerodynamic force for this purpose (Azuma[12]).

C.   Two-Dimensional Viscous Flow

In the area of biofluid dynamics, nonlinear unsteady aerodynamics clearly plays a dominant role in the locomotion of aerial and aquatic animals (Wu and Hu-Chen[25]). In a viscous fluid, the aerodynamic characteristics of a two-dimensional airfoil in unsteady motion can be given by solving numerically the Navier–Stokes equation. In any computational code for solving the Navier–Stokes equation related to the same airfoil flying in a viscous fluid, if the viscosity of the fluid tends to decrease to zero, the numerically obtained aerodynamic forces and the moment must not contradict the theoretically obtained results for the inviscid flow. The accuracy of the calculation is strongly dependent on the degree of fineness of the computation mesh or grid, and because the suction force can mainly be obtained in the neighborhood of the leading edge, the length of the respective grid spacing must be as small as possible, specifically at the leading edge where the radius of curvature is very small.

Isogai and Shinmoto[26] made numerical calculations of the unsteady viscous flows around a single and around a tandem thin plate (with zero thickness) simulating a dragonfly's wings and oscillating in still air at a Reynolds number Re of 1.84×10^3. Their results of the single wing were not different from those obtained from potential theory. However, some errors resulting from the lack of suction force at the leading edge are unavoidable. For the tandem wings, they showed that enough lift could be obtained through the interference effect between fore- and hindwings.

Liu[27,28] also developed a computer code on airfoil aerodynamic characteristics to solve the Navier–Stokes equation. According to his private letter, responding to our inquiry on the possibility of applying the code to the unsteady aerodynamics of thin airfoils, he checked his computer code on the calculation of thin symmetric airfoils (NACA00 $t/c, t/c < 0.06$) operating in a steady state at a fixed angle of attack $\alpha = 4 \deg$ and $Re > 10^4$. Liu's results are as follows: 1) unsteady flow fluctuation is observed for thickness less than 1%, $t/c < 0.01$, caused by the flow separation at the leading edge; 2) the lift slopes of mean lift coefficient are small, for example, $a = 5.3$ and 3.6 for zero and 6% thickness respectively; 3) drag coefficients are a little larger than those estimated from the skin friction drag at the same Reynolds number; and 4) the tangential force generated by the pressure distribution is too small in comparison with the component $C_\ell \alpha$. These facts seem to show that, in the calculation of the pressure distribution on the airfoil surface, specifically, in the neighborhood of leading edge, a sufficiently dense grid is required for matching the radius of curvature.

Other examples of numerical computations based on the Navier–Stokes equations for solving the unsteady flowfield over an oscillating airfoil are found in many papers, for example, those by Johnson,[29] Visbal,[30] and Ekaterinaris.[31] One important finding is that for any thin wing having either a sharp or angular leading edge it is proper or natural to have a separation bubble at the leading edge even in a steady state except for the case of zero lift.

D. Added Mass

Sometimes, quasi-steady potential theory is applied to obtain the aerodynamic force acting on a blade element. However, if the added masses of fluid, related to the aerodynamic forces caused by the wing accelerations, are not negligible in comparison with the aerodynamic forces caused by the relative velocity of the wing, then inertial forces caused by the added masses of fluid must be introduced. When the wing is considered to be slender and thin, the added mass of the blade element is predominant only in the direction normal to the airfoil chord, and it is thus given by the mass of fluid enclosed in a circular cylinder, $(1/4)\rho \pi c^2 dr$, where the diameter and the streamwise length are respectively given by c and dr.

E. Unsteady Three-Dimensional Wing

Again two ways of numerical calculations are considered: 1) one based on the potential theory and 2) another based on the Navier–Stokes equation. In the former method, it is important to decide which end of a plate is the leading edge and which is the trailing edge. In aerodynamic sense, the leading edge is considered to be a terminal end of the chord at which either a velocity discontinuity exists in inviscid flow or a separate vortex called a *vortex bubble* is attached and moved with the leading edge in viscous flow, whereas the trailing edge is another end at which no velocity discontinuity exists (Kutta condition) and a series of shed vortices is generated and left in the wake of flow. This series of shed vortices makes an important contribution to the large lift generation (Edwards and Cheng[32]). It must, however, be mentioned that the leading and trailing edges are not always stuck onto the airfoil; it is possible for them to alternate from one end of the airfoil to the opposite end, as in the case in which von Kármán vortices are trailed periodically; the lift is then generated alternately at either side of the plate (except at very low Reynolds numbers). In such a case, however, the leading and trailing edges are determined automatically through numerical calculation.

F. Weis-Fogh Mechanism

Weis-Fogh[24,33] proposed a new circulating lift mechanism, called *clap-and-fling*, to explain the high lift coefficients required for flight of very small (1-mm) chalcid wasps. This insect has small aspect ratio wings, which operate at the low Reynolds number of 10^2 and at a fairly high reduced frequency of $k = 0.5$. The processes of their wing motion, called the *Weis-Fogh mechanism*, are as follows: Before beginning a downstroke the wings are *clapped* dorsally with the longitudinal axis. The wings then *fling* open about their trailing edges, and the flow of air into the opening gap creates equal but opposite circulation around the wings. At the final stage of the downstroke, the wing must be stopped and supinated for the subsequent upstroke phase. This motion is called *flip*.

Lighthill,[34] Maxworthy,[35] Edwards and Cheng,[32] and Ro and Tsutahara[36] analyzed the mechanical aspect of fling-and-flip for a pair of wings operating in an inviscid fluid. They stressed 1) the interference effects on the lift generation between two wings arranged symmetrically and, again, 2) the effects of feathering motions (pronation and spination) performed near the terminal ends of stroke motion.

Tsutahara and Kimura[37] analyzed numerically a lift-generating mechanism of two flat plates, which were assumed to be two dimensional and operating in either opening phase or closing phase. They found that a large lift coefficient C_ℓ (and thus the circulation Γ) was generated at the fling stage during the opening phase and at the clapping stage during the closing phase. However, it should be mentioned that the actual lift, $\ell = \frac{1}{2}\rho U^2 c C_\ell = \rho U \Gamma$, is not always large at these stages because the translational velocity, $U = \frac{1}{2} R \dot{\psi}$ (at spanwise station $R/2$), is very small. Tsutahara and his collaborators extended the discrete vortex method to a three-dimensional model of a rectangular plate (Tsutahara and Murakami,[38] Ro and Tsutahara[36]). The lift coefficient, which is based on the speed $U = c \dot{\theta}$ at fling stage and $U = \frac{1}{2} R \dot{\psi}$ at translating phase, is very large ($C_L = 3$–6), specifically in the translating (flapping-up and -down) stages, and is almost independent of the Reynolds number in the translating stage as well as in the fling stage. By applying a vortex lattice method, Sunada et al.[39,40] analyzed theoretically a mechanism of vortex and force generation for a pair of triangular wings and a pair of butterfly model wings operating near fling state. Although the analysis lacks the calculation of the suction force acting at leading edges, they could get the unsteady normal force corresponding to experimental results conducted by simulation models. Azuma[41] further obtained almost the same force variation of a butterfly during takeoff by using a much simpler method of calculation.

G. Other Computational Fluid Dynamic Studies

Smith et al.[42] presented an unsteady panel method based on potential theory and calculated the aerodynamic forces of flapping wings of a tethered sphingid moth. The results showed good agreement with the experimental data obtained by Wilkin and Williams[43] in the vertical force but not in the horizontal force. It is not clear whether the discrepancy in the horizontal force was due to the lack of suction force or not.

By using a computational fluid dynamic (CFD) modeling approach, Liu et al.[44] studied the unsteady aerodynamics of the beating wing of a hovering hawkmoth and showed well-simulated unsteady flow around the beating wing in comparison with the observed flow of a pair of robotic wings (Van Den Berg and Ellington[45]). They analyzed the mechanism of generation of the leading-edge vortex during one complete beating cycle and made clear the effect of axial flow caused by the velocity gradient along the span (Maxworthy[46]) and flow of the vortex system. The calculated vertical force was produced mainly during the downstroke and the latter half of the upstroke, with little force generated during pronation and supination. Its mean value was 1.41 times the weight before subtracting the body drag.

The above CFD-based result is very important for investigating the statement that the unsteady effect of the wing motion generates an extremely large lift, specifically during supination and pronation phases. This statement probably results from an expectation placed by supposing that a high value of lift coefficient must be generated during these phases, without performing direct measurements of pressure or force or numerical analysis of the force variations, but relying on vortical flow observations and/or experimental data obtained by assuming a constant and homogeneous lift coefficient over the entire stroke plane.

III. Experimental Aerodynamics

A. Steady Airfoil

Experimental measurements of the aerodynamic forces and moments acting on either a steady or an unsteady wing operating at low Reynolds numbers are very difficult because of the low air speed and thus low dynamic pressure. To get reliable data, it is necessary to have highly accurate test equipment and a well-established technique for correct data acquisition.

Measured data at low Reynolds numbers, which were obtained from the wind tunnel tests conducted by Okamoto et al.,[47] showed very interesting results. As shown in Table 4, a thin rectangular plate [Table 4, Ia)] shows that 1) the lift slope is nearly equal to the theoretical value of a two-dimensional plate flying in a inviscid fluid, $a_0 = 2\pi$, at the Reynolds number of 1.0×10^4; 2) the slope

Table 4 Steady aerodynamics characteristics of an airfoil at low Reynolds number

Item		Lift slope $a(a_0)^a$	Max. lift coeff., $C_{L,max}$	Min. drag coeff., C_{D_0}	Max. L/D, $(L/D)_{max}$	Reynolds number Re	Reference
I. Flat plate ($AR = 6$)							
a) Effect of Re							
$t/c = 0.01$		4.9(6.6)	0.80	0.03	7.10	1.0×10^4	Okamoto[105]
		3.8(4.8)	0.73	0.04	6.5	4.5×10^3	Okamoto et al.[47]
		3.7(4.6)	0.77	0.06	5.5	2.9×10^3	
b) Effect of thickness ratio							
$t/c = 1\%$		4.9(6.6)	0.80	0.03	7.0	1.1×10^4	Okamoto et al.[47]
$t/c = 3\%$		5.0(6.8)	0.75	0.04	5.9		
$t/c = 5\%$		5.6(8.0)	0.72	0.06	5.0		
c) Effect of turbulence							
Laminar		4.8(6.4)	0.84	0.03	7.2	1.1×10^4	
Turbulence		4.2(5.4)	0.87	0.03	6.3		
d) Effect of camber							
$h/c = 3\%$			0.93	0.03	11.0	1.0×10^4	
$h/c = 6\%$			1.18	0.06	10.2		
$h/c = 9\%$			1.41	0.07	8.1		
II. Airfoil section (Clark Y, $AR = 4$)							
a) Effect of thickness							
$t/c = 3\%$		4.0(5.9)	0.74	0.03	8.5	1.1×10^4	
$t/c = 6\%$		4.0(5.9)	0.77	0.04	7.3	1.4×10^4	
$t/c = 9\%$		4.0(5.9)	0.67	0.05	4.2		
b) Effect of turbulence, $t/c = 12\%$							
Laminar		4.1(6.1)	0.68	0.05	4.2	1.4×10^4	
Turbulence		3.4(4.7)	1.04	0.05	8.0		

(Cont.)

Table 4 *(Continued)*

Item	Lift slope $a(a_0)^a$	Max. lift coeff., $C_{L,max}$	Min. drag coeff., C_{D_0}	Max. L/D, $(L/D)_{max}$	Reynolds number Re	Reference
III. Model dragonfly forewing($AR = 6$)					1.0×10^4	Okamoto et al.[47]
0.2R	6.2(9.2)	1.40	0.10	7.4		
0.7R	5.6(8.0)	0.90	0.04	9.9		
IV. Real dragonfly[b]						
Forewing ($AR = 6$)	3.7(4.6)	0.95	0.08	5.6	1.2×10^3	
Hindwing ($AR = 4$)	3.9(5.7)	1.05	0.07	5.0	1.0×10^3	
V. Real dragonfly[c]						
Forewing ($AR = 14$)	3.7(4.0)	0.95	0.08	5.6	1.2×10^3	
Hindwing ($AR = 13$)	3.9(4.3)	0.95	0.07	5.0	1.0×10^3	
VI. Drosophila					4.3×10^2	Vogel[53]
Flat		0.60	0.240	1.18		
Cambered		0.80	0.265	1.87		
VII. Airfoil section						Gugliemo and Selig[48]
E374	6.0	1.0	0.017		6.1×10^3	
			0.009		3.0×10^3	
SD 6060	6.0	1.0	0.013		6.0×10^3	
			0.008		3.0×10^3	
VIII. Sharp-edged airfoil[d]	b/Re		$12/Re$	0.18	<2.0	Thom and Swart[55]

[a] Only lift slope a is corrected to two-dimensional slope a_0 by $a_0 = a/\{1 - (a/\pi\, AR)\}$.

[b] In this item, since the test is performed for a right wing, the aspect ratio is defined by that of single wing or $AR \cong (b/2)^2/(S/2) = (b^2/S)/2$, and the data of the real wing are deteriorated by the elastic deformation.

[c] These data are measured for a pair of right and left wings.

[d] These data are measured from a gliding test of a pair of hindwings.

decreases as the Reynolds number decreases; 3) the minimum drag coefficient increases as the Reynolds number decreases, because the minimum drag is mostly due to the skin friction drag; 4) the maximum lift coefficient is slightly changed; 5) as the thickness increases [Table 4, Ib)], the lift slope and the minimum drag coefficient increase whereas the maximum lift coefficient and the maximum lift-to-drag ratio decrease; and 6) the effect of turbulence [Table 4, Ic)] is to decrease the lift slope and the maximum lift-to-drag ratio.

For wings with circular arc section [Table 4, Id)], it can be seen that, as the camber increases, 1) the lift slope cannot be determined because of strong nonlinear behavior and 2) the maximum lift coefficient and the minimum drag coefficient slightly increase, but the maximum lift-to-drag ratio decreases.

In the airfoil section at $Re = 1.4 \times 10^4$ [Table 4, IIa)], as the thickness increases, the lift slope and the maximum lift coefficient do not undergo any striking change, but the minimum drag coefficient increases whereas the maximum lift-to-drag ratio decreases appreciably. The turbulence [Table 4, IIb)] decreases the lift slope but increases the maximum lift coefficient and the maximum lift-to-drag ratio. Other examples of airfoils tested by Guglielmo and Selig[48] at $Re = 6 \times 10^4$–3×10^5 (Table 4, VII) showed that as the Reynolds number increased, the lift slope and the maximum lift coefficient were not significantly affected, but the minimum drag coefficient decreased. They also found that the laminar-separation bubble and the developing wake play an important role.

For model dragonfly wings, which are rigidly made of paper to simulate real airfoil configuration at $0.2R$ and $0.7R$ of forewing (Table 4, III), although the determination of exact values of the lift slope is difficult, because of nonlinear characteristics, remarkably high values of the lift slope and of the maximum lift coefficient were obtained. The superior effect of corrugation over the flat plates is clearly observed (Newman et al.[49]). It is interesting to find that the aerodynamic characteristics of sharp-edged and corrugated plates are more favorable for flight at low Reynolds number than a smoothed airfoil section.

However, the lift slope, the maximum lift coefficient, and the lift-to-drag ratio of real dragonfly and damselfly wings (Table 4, IV and V) were too small, and the maximum lift coefficient could not be obtained within the measured range of angle of attack. Similar deteriorated data were reported for a gliding test of paired wings (Azuma and Watanabe[50]) and for the wind tunnel test of tethered locusts (Cloupeau et al.[51]) in comparison with that of detached wings (Jensen[52]). These facts are probably the results of the elastic deformation of the wing at the wind speed tested and the inadequate means for providing tension force caused by the centrifugal force in the spanwise direction as stated later.

In other models of insect wings such as drosophila (Table 4, VI; Vogel[53]), cranefly (Nachtigall[54]) and locust (Jensen[52]), the maximum lift coefficient is scattered over a wide range (0.6–1.3). At very low Reynolds numbers ($Re < 2$, Table 4, VIII), Thom and Swart[55] obtained the result that the steady aerodynamic forces acting on a wing (span b and chord c) moving with a very low speed V are given by $L = (3/2)\mu b \sin(2\alpha)$ and $D = 6\mu bV$. Hence, the aerodynamic forces are proportional to the fluid viscosity μ, wingspan b, and speed V but are independent of the chord c. The coefficients are, as shown in Table 4, given by $C_L = (3/Re)\sin(2\alpha)$ and $C_D = 12/Re$ and are inversely proportional to the Reynolds number based on the chord ($Re = Vc/\nu$).

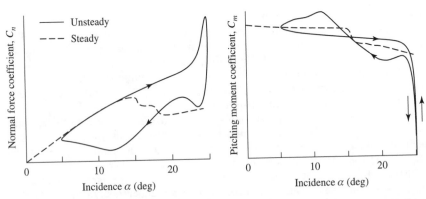

Fig. 2   Dynamic stall events on NACA 0012 airfoil: a) Normal force coefficient b) Pitching moment coefficient (Carr et al.[62]).

B.   Unsteady Airfoil

When a step increase or a quick oscillation is introduced in either pitching or heaving motion, and thus when a step increase or a sinusoidal oscillation of the angle of attack is imposed on an airfoil, the instantaneous values as well as the mean values of lift, drag, and moment coefficients increase to the values well beyond their equivalent steady values, in a condition called *dynamic stall*. In early stages of the study of dynamic stall, the stall flutter of rotary wings as well as fixed wings, coupled with lift and moment variations, were the main focus of attention (Ham,[56] Liiva and Davenport,[57] Gangwani,[58] Favier et al.,[59] Niven and McD. Galbraith,[60] Tang and Dowell[61]). However, the large increase in the lift coefficient caused by dynamic stall could not be overlooked. A typical example at $Re = 2.5 \times 10^6$ and $k = 0.15$ is shown in Fig. 2. This was presented by Carr et al.[62] and based on experimental tests with the NACA 0012 airfoil for a wide range of frequencies, Reynolds numbers, and amplitudes of oscillation. It can be seen from Fig. 2 that 1) for a sinusoidally oscillating wing at or near stall, the aerodynamic reactions vary in a periodic but definitely nonsinusoidal manner (Halfman et al.[63]), and 2) the normal force increases beyond its stationary value past the stalling angle of attack. The area inside the hysteresis indicates work per cycle. The favorable effect of unsteadiness, specifically on the mean lift coefficient, is obtained as a result of a cyclic phenomenon of boundary-layer separation and reattachment on the upper surface near the leading edge (Rainey,[64] Carta and Ham,[65] Karim and Acharya[66]).

Freymuth[67,68] showed the thrust generation experimentally for a wing undergoing a combined heaving and feathering oscillation. Maresca et al.[69] and Favier et al.[70] investigated experimentally the effects of simultaneous chordwise velocity and incidence fluctuations on the two-dimensional aerodynamic behavior of an airfoil. This kind of unsteady feature can be observed in beating wings as well as in the rotary wings of a helicopter in forward motion. Their results showed that 1) the unsteady effect was more predominant during the in-phase oscillation than during the out-of-phase oscillation, 2) through the stall, the combined oscillation significantly modified the vortex development and induced a lift hysteresis, and 3) increases of the amplitudes of the oscillation and the reduced frequency led to increases in the mean and the first-harmonic lift coefficients. Freymuth[71] also showed the effect of the acceleration on the formation and splitting of the vortex.

The utilization of the dynamic stall can be seen in some phases of a bird's flight. In very low speed flight, such as loitering flight or landing approach, many large birds, specifically sea birds having wings with pointed tips and of large aspect ratio such as the gannet and the albatross, perform mainly sinusoidal wing-pitching motions in addition to either slight flapping or lead–lag oscillations of small amplitude.

In the rotary wings of a hovering helicopter, 1) the aspect ratio $AR = R/c$ of the blade is large ($AR > 10$), 2) the reduced frequency $k = c\dot{\theta}/2U$ is small ($k < 0.1$), and 3) the Reynolds number is high ($Re > 10^6$). In contrast, in beating wings of hovering insects, these values are roughly estimated by $AR < 5$, $k > 0.2$, and $Re < 3 \times 10^4$. Furthermore, some other differences in the configuration of an airfoil section and a surface condition are clearly recognized. In operating condition, there are also some significant differences between a beating wing and a rotary wing: 1) the sweeping or plunging speed in the stroke plane of a beating wing is rather sinusoidally reciprocal than nearly constant, 2) the feathering motion is performed to get the lifting force generated mainly from the suction force acting on the top side of the wing in the power stroke and on the bottom side in the recovery stroke rather than on one side of the airfoil in the rotary wing, and 3) there are large angular rototations about the feathering hinge at the opposite ends of the respective half-stroke, pronation, and supination.

Thus, quantitative results obtained in rotary-wing aerodynamics (summarized, for example, by Johnson[72]) may not be applied directly. Dickinson[73] and Dickinson et al.[74] investigated the effects of wing rotation (flip) on unsteady aerodynamic performance at a low Reynolds number ($Re = 2.4 \times 10^3$) and confirmed that a high lift coefficient ($C_L = 2.7$) during flip phase was generated under the influence of an inter-vortex stream at a very high reduced frequency ($k \simeq 2.5$). As stated before, a thin wing always has a separation bubble at the leading edge. Complete stall emerges when a partly separated bubble detaches from the airfoil. The vortex further triggers the shedding of a counterrotating vortex near the trailing edge (Currier and Fung,[75] Shih et al.[76]).

Many papers, most of which are presented in biological journals, mainly pay attention to the flow observation of wake vortices (Spedding and Maxworthy,[77] Grodnitsky and Morozov[78]), and direct calculations of the aerodynamic force are performed based on this observation. In these calculations, a common conclusion is that insects cannot fly, according to the conventional laws of aerodynamics; during flapping or beating flight, their wings produce more lift than during steady motion at the same velocities and angles of attack (Weis-Fogh,[24] Norberg,[79] Savage,[80] Maxworthy,[46] Ellington,[81] Somps and Luttges,[82] Ennos,[83,84] Dudley and Ellington,[85] Dudley,[86,87] Willmott and Ellington,[88,89,90] Van Den Berg and Ellington,[45] Dickinson et al.[74]). It is strongly stressed that the generation of high lifting force by wing beating is the result of the effect of the flow separation or separation bubble or bubbles at or near the leading edge, specifically caused by the pitching rotation in the supination phase as a novel mechanism of insect flight. This mechanism is also observed in two-dimensional model experiments conducted by Newman et al.,[49] Saharon and Luttges,[91] Zanker and Götz,[92] and Dickinson and

Götz[93] at appropriate Reynolds numbers. However, in insect flight, because the wings are not streamlined but corrugated and angular at the leading edge, any insect is always flying by relying on partly separated wings. Specifically, at hovering flight, which requires high lift coefficient because of the low wing speed, insects sometimes utilize dynamic stall as they need to have a large mean lift coefficient.

Savage et al.[94] have found that, unlike pronation, supination can generate a large lift if it is performed with the high sweeping speed near the terminal end of the power or downstroke of the beating. In many papers, as already depicted, it is stressed that a large circulation Γ is generated around the blade element, mainly caused by the fast flow around the leading edge. However, if the translational speed U at the aerodynamic center or nearly $c/4$ is small, such as at pronation, then, as stated earlier, the airloading ℓ should be small because it is given by $\ell = \rho U \Gamma$.

C. Dragonfly and Damselfly

By making force measurement of tethered dragonflies, Somps and Luttges[82] reported that the simple large lift peaks of 15–20 times body weight or 15–20G occurred once in each stroke period and suggested that the lift generation was dominated by integrated interactions between wings rather than by the unsteady effects elicited independently by each of the four wings. The result of the above lift generation has sometimes been quoted in papers by Wakeling and Ellington[95–97] and Ellington et al.,[98] who stress the favorable effect of the unsteady separated flow. However, this extraordinary large lift is considered to be unrealistic for the flight of living creatures, except when rapidly maneuvering (Azuma and Watanabe[50]). Reavis and Luttges[99] presented later that the maximum lift obtained from the tethered dragonfly was 5.5G in escape mode and 3.7G in hovering mode.

Based on the data obtained by Norberg,[79] Savage et al.[94] calculated the aerodynamic forces acting on the hovering wings of a dragonfly using two-dimensional unsteady potential-flow theory. They obtained a very large mean lift, over four times the weight or 4G, in spite of Norberg's result of 0.4G based on the steady-state aerodynamics for the same data. In the calculation of Savage et al.,[94] the wing was subjected only to its own shed vortex and, thus, the effect of the induced velocity generated by trailing and shed vortices of the other wings was neglected.

Wakeling and Ellington[95–97] filmed and analyzed the free flight of dragonflies and damselflies. However, because the greater part of their data were obtained from unsteady flight accompanied by large accelerations, their data were scattered so much that, except for the wind tunnel data, the in-flight data could not be used to establish any reliable polar curve from the measured lift coefficient and the drag coefficient calculated from the lift-to-drag ratio (as given in Table 2 in their first paper). These problems probably resulted from several factors: 1) the acceleration (and thus the aerodynamic forces) and the orientation of the stroke plane were specified only for the inertial coordinate system, 2) unlike the description given later, the angles of body attitude or Euler angles of the body coordinate system and their rates were not defined correctly, and, thus, 3) not only the angle of attack, the side slip angle of the dragonfly's body, and geometrical information on flight path but also the blade motion and the attitude of the stroke plane with respect to the body were not determined. In conclusion, their results seem to lack reliability and their criticism of the local circulation method (LCM) described in their paper is probably due to the misinterpretation of the LCM.

Azuma et al.,[100] Azuma and Watanabe,[50] and Sato and Azuma[101] also filmed or taped and analyzed free flights of dragonflies and damselflies by applying the blade element theory and the LCM considering the feathering motion. Their detailed description is presented later. By introducing attenuation coefficients into the estimation of the induced velocity, they could obtain spanwise and time or azimuthwise variations of the airloading and the performance of the insects to the degree of accuracy of vortex theory by including effects of interference caused by vortices generated by the four wings. In these analyses, it was not necessary to introduce any extremely large lift coefficient ($C_{\ell,\max} = 1.2$) for the flight of dragonflies and damselflies, even at the very low flight speed near hovering flight.

The flight of the damselfly has interesting differences from that of the dragonfly: 1) the stroke plane is shallow, near horizontal in cruising flight, 2) the forewings lead the hindwings, 3) the beating amplitude is large, and 4) the contribution of the drag forces to the aerodynamic forces is a little higher. The above items 1) through 3) do not contradict the results obtained by Rüppell.[102]

D. Other Insects

In insects having low-aspect-ratio wings, as stated before, the effect of unsteady wing motion on the aerodynamic forces and moments becomes predominant at low flying speed. The maximum lift-to-drag ratio occurs at a large lift coefficient, and normally insect flight relies on drag as well as lift at high angle of attack. It is supposed that, unlike the flight of Odonata, the strong shed vortex coupled with the separation bubble at the leading edge generates an extremely large aerodynamic force. Sunada et al.[39,40] showed clearly, theoretically and experimentally, the generation of a large aerodynamic force, mainly drag, at takeoff flight of a butterfly using a triangular plate moving perpendicularly to the plate surface ($\alpha = 90$ deg). However, it must be mentioned that a low-aspect-ratio wing can generate a much larger aerodynamic force consisting of the lift and drag when the wing is translating with an adequate angle of attack, in the range $30 < \alpha < 60$ deg (Azuma,[103] Ito and Azuma[104]). The force augmentation caused by the shed vortex is also treated theoretically by Smith et al.[42]

E. Surface Roughness

It is a well-known fact that as the Reynolds number increases, the aerodynamic characteristics of a wing flying at low Reynolds number (below the critical value) is improved (Okamoto[105]) and the surface roughness also improves the aerodynamic characteristics as if the Reynolds number increases equivalently. Many examples exist: the seeds of plants or samara (Azuma and Okuno[106]) and insect wings (Rees,[107,108] Okamoto et al.[47]). Yasuda and Azuma[109] demonstrated that 1) samaras are autorotating with a high lift coefficient, which is close to the maximum coefficient of $C_{\ell,\max} = 1.2–1.5$, 2) by smoothing the surface irregularities on the samara wing of a maple, the rate of spin was reduced and the falling rate increased, and 3) by attaching thin columns near the leading edge of a model samara made of balsa wood, the falling rate was remarkably reduced. Some changes in performance are shown in Table 5.

Shown in Fig. 3 is an example of wind-tunnel-test results for the degeneration of the aerodynamic characteristics of a corrugated airfoil with sharp leading edge by modifying the upper surface with a cellophane tape. It can be seen that by

Table 5 Experimental flight data of real samaras and magnified models (Yasuda and Azuma[109])

Item	Rate of fall V (m/s)	Rate of spin n (rpm)	Coning angle β_0 (deg)	Feathering angle θ_0 (deg)
Real samara				
Natural	0.90	1000	10	0
Modified to smooth surface	1.10	850	18	6
Magnified balsa model				
Simulated model	1.24	680	12	2
Flat plate	3.51	U[a]	U[a]	U[a]

[a] U = unmeasurable.

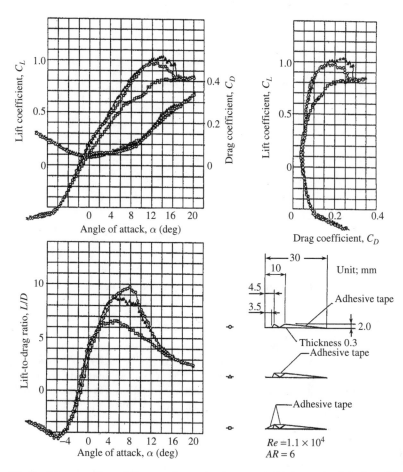

Fig. 3   Degeneration of the aerodynamic characteristics for a rippled plate.[105]

reducing the degree of roughness, the lift slope, the maximum lift coefficient, and the lift-to-drag ratio clearly deteriorate while drag changes are insignificant.

IV.   Geometrical Consideration of Blade Element Theory

The fundamental mechanism of natural beating flight of insects flying in a symmetric plane was first examined by Osborne.[110] Since then, few significant contributions to the field of flight (or swimming) dynamics of living creatures have been published. In this section the flight (or swimming) dynamics of beating wings will be presented in detail.

A.   Inertial and Body Coordinate Systems

To describe the motion of a flying object, two Cartesian coordinate systems or frames may be introduced: an *inertial frame* (X_I, Y_I, Z_I) and a *body frame* (X, Y, Z). As shown in Fig. 4, the plane consisting of X_I and Y_I axes is a horizontal plane, and thus the Z_I axis is directed parallel to the gravity force. The

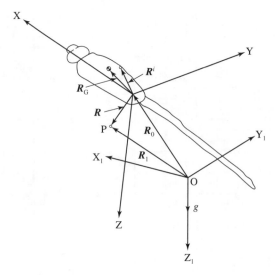

Fig. 4   Inertial and body coordinate systems.

specific value of the gravity force in the Z_I direction for a unit mass is given by g. The body frame is fixed to the body of the flying object and the origin of the frame is located at a point (not necessarily always the center of gravity CG) of the flying object. The (X, Z) plane consisting of X and Z axes is embedded in a symmetric plane of the flying object. The X axis is directed forward and is parallel to the longitudinal axis of the object, and the Y axis is directed to form the right-hand Cartesian coordinate system. The body frame can, as shown in Fig. 5,

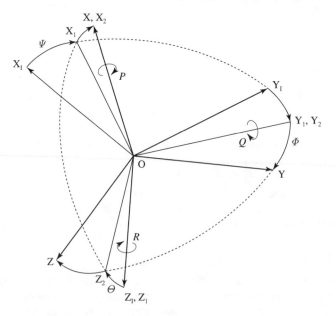

Fig. 5   Transformation between inertial and body coordinate systems.

be obtained by a successive rotation of the inertial frame through Euler's angles: 1) rotation about the Z_I axis (*yawing* motion, Ψ), $(X_I \rightarrow X_1, Y_I \rightarrow Y_1)$; 2) rotation about the Y_I axis (*pitching* motion, Θ), $(X_1 \rightarrow X_2, Z_I \rightarrow Z_2)$; and 3) rotation about the X axis (*rolling* motion, Φ), $(Y_2 \rightarrow Y, Z_2 \rightarrow Z)$. Then the transformation matrix T_b is given by Eq. (1) in Table 6. If the origin of the body frame is located at $R_0 = (X_0, Y_0, Z_0)^T$ in the inertial frame, then the position, the velocity V_0, and the acceleration A_0 of the origin, given by $\dot{R}_0$ and $\ddot{R}_0$, can be transformed into those of the body frame as given respectively by Eqs. (6–8) in Table 7.

The position of any fixed point P in the body frame, such as the position of the center of gravity $R^G = (X^G, Y^G, Z^G)^T$ and the position of fulcrum or joint of the ith wing $R^i = (X^i, Y^i, Z^i)^T$ with respect to the origin of the frame, is affected by the angular velocity of the body $\Omega = (P, Q, R)^T$, which is determined from the rate of change in Euler's angle by Eq. (9) in Table 7. Then, the linear velocity V, acceleration A, and the angular acceleration $\dot{\Omega}$ of the point P are respectively given by Eqs. (10–12) in Table 7.

B. Wing or Blade Motion

Assume that an individual beating wing of right-hand side is rigid or inflexible and, as shown in Fig. 6, attached to the body as an articulated blade at the wingroot or the joint specified by a position vector of R^i for the ith blade. A coordinate system (x_0^i, y_0^i, z_0^i), which is oriented to be parallel to the body frame, the origin of which is located at the joint of ith wing, can be introduced. Then the blade motion in the body frame is defined by three successive angular motions: 1) *lead–lag* or forward-and-backward motion ζ^i about the $-z_0^i$ axis or lead–lag hinge, which is parallel to the Z axis and is normal to the feathering axis, which is assumed to be a straight line passing through the joint and the aerodynamic center (ac) of the airfoil at any spanwise station; 2) *flapping* or up-and-down motion β^i about the $-x_1^i$ axis or the flapping hinge; and 3) *feathering* or blade pitching motion to change the blade pitch θ^i about the y_2^i or y^i axis or the feathering hinge (the axis of pronation and supination). A reference airfoil is defined by a chord line at a spanwise location r^i, where the chord line is perpendicular to the feathering axis, connected with the leading and trailing edges of the airfoil, before starting the feathering motion.

The right-hand blade coordinate system or the blade frame (x^i, y^i, z^i) can further be defined such that the origin of the frame is at the universal joint where the x^i axis coincides with the chord line at the wingroot, the y^i axis coincides with the feathering axis, and the z^i axis is perpendicular to the above x^i and y^i axes to form a Cartesian coordinate system. The transformation matrix T_i, which transforms from the frame (X, Y, Z) or (x_0^i, y_0^i, z_0^i) to the frame (x^i, y^i, z^i), is given by Eq. (2) in Table 6. Since the (x^i, y^i) plane of the blade frame is parallel to the blade chord at the wingroot, every quantity expressed in this frame is, as stated later, conveniently used to obtain the shear force and the bending moment and to analyze the aeroelastic characteristics of the blade. Because the matrices T_b and T_i are unitary, their inverse matrices T_b^{-1} and T_i^{-1} are transposed matrices obtained simply by replacing their columns with rows.

The angular velocity of the blade frame $\omega^i = (p^i, q^i, r^i)^T$ is, then, given by Eq. (13) in Table 8. The position r^i of the aerodynamic center of a blade element at spanwise station r^i on the feathering axis of the ith blade, its velocity

Table 6 Transformation matrices

Item	Expression	Eq. No.
Inertial frame to body frame	$T_b = \begin{pmatrix} \cos\Psi\cos\Theta & \sin\Psi\cos\Theta & -\sin\Theta \\ \cos\Psi\sin\Theta\sin\Phi - \sin\Psi\cos\Phi & \sin\Psi\sin\Theta\sin\Phi + \cos\Psi\cos\Phi & \cos\Theta\sin\Phi \\ \cos\Psi\sin\Theta\cos\Phi + \sin\Psi\sin\Phi & \sin\Psi\sin\Theta\cos\Phi - \cos\Psi\sin\Phi & \cos\Theta\cos\Phi \end{pmatrix}$	(1)
Body frame to blade frame	$T_i = \begin{pmatrix} \cos\zeta^i\cos\theta^i - \sin\zeta^i\sin\beta^i\sin\theta^i & -\sin\zeta^i\cos\theta^i - \cos\zeta^i\sin\beta^i\sin\theta^i & -\cos\beta^i\sin\theta^i \\ \sin\zeta^i\cos\beta^i & \cos\zeta^i\cos\beta^i & -\sin\beta^i \\ \cos\zeta^i\sin\theta^i + \sin\zeta^i\sin\beta^i\cos\theta^i & -\sin\zeta^i\sin\theta^i + \cos\zeta^i\sin\beta^i\cos\theta^i & \cos\beta^i\cos\theta^i \end{pmatrix}$	(2)
Body frame to stroke plane	$T_s = \begin{pmatrix} \cos\kappa\cos\Theta_s & -\sin\kappa\cos\Theta_s & \sin\Theta_s \\ \sin\kappa & \cos\kappa & 0 \\ -\cos\kappa\sin\Theta_s & \sin\kappa\sin\Theta_s & \cos\Theta_s \end{pmatrix}$	(3)
Left-handed frame Linear quantity	$S_l = \begin{pmatrix} 1 & 0 & 0 \\ 0 & -1 & 0 \\ 0 & 0 & 1 \end{pmatrix}$	(4a)
Rotation quantity	$S_r = \begin{pmatrix} -1 & 0 & 0 \\ 0 & 1 & 0 \\ 0 & 0 & -1 \end{pmatrix}$	(4b)
Intermediate transformation matrix	$T_{i,\theta} = T_i \cdot T_i^{-1}(\theta = 0) = \begin{pmatrix} \cos\theta^i & 0 & -\sin\theta^i \\ 0 & 1 & 0 \\ \sin\theta^i & 0 & \cos\theta^i \end{pmatrix}$	(5)

Table 7 Position, velocity, and acceleration in the body frame

Item	Expression	Eq. No.
Position at the origin of the body frame	$T_b \cdot R_0$	(6)
Velocity at the origin of the body frame	$V_0 = T_b \cdot \dot{R}_0 = (V_X,\ V_Y,\ V_Z)^T$	(7)
Acceleration at the origin of the body frame	$A_0 = T_b \cdot \ddot{R}_0 = (A_X,\ A_Y,\ A_Z)^T$	(8)
Angular velocity of the body frame	$\Omega = \begin{pmatrix} P \\ Q \\ R \end{pmatrix} = \begin{pmatrix} \dot{\Phi} - \dot{\Psi}\sin\Theta \\ \dot{\Theta}\cos\Phi + \dot{\Psi}\cos\Theta\sin\Phi \\ \dot{\Psi}\cos\Theta\cos\Phi - \dot{\Theta}\sin\Phi \end{pmatrix}$	(9)
Velocity of a fixed point at R	$V = V_0 + \Omega \times R$	(10)
Acceleration of a fixed point at R	$A = A_0 + \dot{\Omega} \times R + \Omega \times (\Omega \times R)$	(11)
Angular acceleration of the body frame	$\dot{\Omega} = \begin{pmatrix} \dot{P} \\ \dot{Q} \\ \dot{R} \end{pmatrix} = \begin{pmatrix} \ddot{\Phi} - \ddot{\Psi}\sin\Theta - \dot{\Psi}\dot{\Theta}\cos\Theta \\ \ddot{\Theta}\cos\Phi + \ddot{\Psi}\cos\Theta\sin\Phi - \dot{\Theta}\dot{\Phi}\sin\Phi - \dot{\Psi}\dot{\Theta}\sin\Theta\sin\Phi + \dot{\Psi}\dot{\Phi}\cos\Theta\cos\Phi \\ \ddot{\Psi}\cos\Theta\cos\Phi - \ddot{\Theta}\sin\Phi - \dot{\Psi}\dot{\Theta}\sin\Theta\cos\Phi - \dot{\Psi}\dot{\Phi}\cos\Theta\sin\Phi - \dot{\Theta}\dot{\Phi}\cos\Phi \end{pmatrix}$	(12)

Table 8 Position, velocity, and acceleration of a blade element

Item	Expression	Eq. No.
Angular velocity of ith blade frame	$\boldsymbol{\omega}^i = \begin{pmatrix} p^i \\ q^i \\ r^i \end{pmatrix} = \begin{pmatrix} -\dot{\beta}^i \cos\theta^i + \dot{\zeta}^i \cos\beta^i \sin\theta^i \\ \dot{\theta}^i - \dot{\zeta}^i \sin\beta^i \\ -\dot{\beta}^i \sin\theta^i - \dot{\zeta}^i \cos\beta^i \cos\theta^i \end{pmatrix}$	(13)
Position of the blade element	$\boldsymbol{r}^i = (0,\ r^i,\ 0)^T$	(14)
Velocity of the blade element	$\boldsymbol{V}^i = (V_{x^i},\ V_{y^i},\ V_{z^i})^T = \boldsymbol{T}_i \cdot (\boldsymbol{V}_0 + \boldsymbol{\Omega} \times \boldsymbol{R}^i) + (\boldsymbol{T}_i \cdot \boldsymbol{\Omega} + \boldsymbol{\omega}^i) \times \boldsymbol{r}^i$	(15)
Acceleration of the blade element	$\boldsymbol{A}^i = (A_{x^i},\ A_{y^i},\ A_{z^i})^T = \dot{\boldsymbol{V}}^i + (\boldsymbol{T}_i \cdot \boldsymbol{\Omega} + \boldsymbol{\omega}^i) \times \boldsymbol{V}^i$	(16)
Angular acceleration of the blade frame	$\dot{\boldsymbol{\omega}}^i = \begin{pmatrix} \dot{p}^i \\ \dot{q}^i \\ \dot{r}^i \end{pmatrix} = \begin{pmatrix} -\ddot{\beta}^i \cos\theta^i + \dot{\beta}^i\dot{\theta}^i \cos\theta^i + \ddot{\zeta}^i \cos\beta^i \sin\theta^i - \dot{\zeta}^i\dot{\beta}^i \sin\beta^i \sin\theta^i + \dot{\zeta}^i\dot{\theta}^i \cos\beta^i \sin\theta^i \\ \ddot{\theta}^i - \ddot{\zeta}^i \sin\beta^i - \dot{\zeta}^i\dot{\beta}^i \cos\beta^i \\ -\ddot{\beta}^i \sin\theta^i - \dot{\beta}^i\dot{\theta}^i \cos\theta^i - \ddot{\zeta}^i \cos\beta^i \cos\theta^i + \dot{\zeta}^i\dot{\beta}^i \sin\beta^i \cos\theta^i + \dot{\zeta}^i\dot{\theta}^i \cos\beta^i \sin\theta^i \end{pmatrix}$	(17)
Blade element in stroke plane frame	$\boldsymbol{r}_s = r\,(\cos\gamma \sin\psi,\ \cos\gamma \cos\psi,\ -\sin\nu)^T$	(18)
Blade element in body frame	$\boldsymbol{r} = \boldsymbol{T}_i^{-1}(\theta = 0) \cdot \boldsymbol{r}^i = r(\sin\zeta^i \cos\beta^i,\ \cos\zeta^i \cos\beta^i,\ -\sin\beta^i)^T$	(19)
Equation of identity for position vector	$\boldsymbol{r} = \boldsymbol{T}_s^{-1} \cdot \boldsymbol{r}_s$	(20)

(*Cont.*)

Table 8 (*Continued*)

Item	Expression	Eq. No.
Equation of identity in blade frame and stroke plane frame	$\sin\zeta^i\cos\beta^i = \cos\kappa\cos\Theta_s\cos\gamma\sin\psi + \sin\kappa\cos\gamma\cos\psi + \cos\kappa\sin\Theta_s\sin\nu$	(21a)
	$\cos\zeta^i\cos\beta^i = -\sin\kappa\cos\Theta_s\cos\gamma\sin\psi + \cos\kappa\cos\gamma\cos\psi - \sin\kappa\sin\Theta_s\sin\nu$	(21b)
	$-\sin\beta^i = \sin\Theta_s\cos\gamma\sin\psi - \sin\Theta_s\sin\nu$	(21c)
	$\theta = e_{y^i}\cdot T_i\,(\theta=0)\cdot T_s^{-1}\cdot\begin{pmatrix}\theta_s\cos\nu\sin\psi \\ \theta_s\cos\nu\cos\psi - \Theta_s \\ \theta_s\sin\nu\end{pmatrix}$	
	$= -\Theta_s\cos\beta^i\cos(\zeta^i-\kappa) + \theta_s\{\cos\beta^i\sin(\zeta^i-\kappa)(\cos\Theta_s\cos\gamma\sin\psi + \sin\Theta_s\sin\gamma)$ $+ \sin\beta^i(-\sin\Theta_s\cos\gamma\sin\psi + \cos\Theta_s\sin\gamma) + \cos\beta^i\cos\gamma\cos(\zeta^i-\kappa)\}$	(21d)
Gravity acceleration	$G = (0,\ 0,\ -g)^T$	(22a)
	$G^i = (G_{x^i},\ G_{y^i},\ G_{z^i})^T = T_i\cdot T_b\cdot G$	(22b)

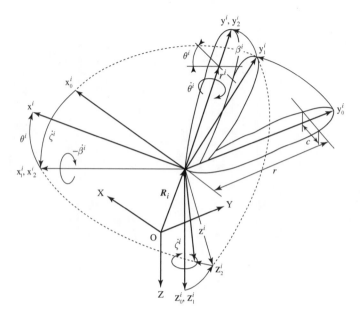

Fig. 6   Blade motion.

$V^i = (V_{x^i}, V_{y^i}, V_{z^i})^T$, its acceleration $A^i = (A_{x^i}, A_{y^i}, A_{z^i})^T$, and the angular acceleration $\dot{\omega}^i$ in the blade frame are given through Eqs. (14–17) in Table 8.

C.   Stroke Plane

A *stroke plane* is considered to be a virtual plane in which a wing is assumed to beat periodically. This is confined only when the beating is, as stated later, performed by the first harmonic motion (Azuma[12]). The attitude of the plane with respect to the (x_0^i, y_0^i, z_0^i) frame, the origin of which is at the wingroot or the joint and the axes of which are parallel to the body frame (X, Y, Z), can be given as follows: By referring to Fig. 7, 1) rotation κ about the $-z_0^i$ or $-Z$ axis results in an intermediate coordinate system (X_1', Y_1', Z_1') and 2) rotation of the Θ_s about the $-Y_1'$ axis results in a new coordinate system (X_s, Y_s, Z_s) of the stroke plane, called *stroke-plane frame*. The origin of the (x^i, y^i, z^i) frame is considered to be located at the joint of the respective wing R^i. In the subsequent description, however, it will be assumed that a pair of joints of left- and right-hand wings are located at a jth virtual joint in the symmetric plane $\bar{R}^j = (R^i + R^{i*})/2 = (\bar{X}^j, 0, \bar{Z}^j)^T$, where $R^{i*} = (X^i, -Y^i, Z^i)^T$ is, as will be stated later again, the joint of the left-hand wings. Thus, the ith wing is considered to be $i = 1$ for diptera of insect and birds and $i = 1$ and 2 for other insects.

The transformation matrix T_s, from the body frame to the stroke-plane frame, is then given by Eq. (3) in Table 6. As shown in Fig. 7, the position vector r_s of a blade element at spanwise location r and sweep angle ψ in the stroke plane $(X_s, Y_s, 0)$ and elevation angle ν out of the stroke plane is also represented by r^i at the same radius station r in the (x^i, y^i, z^i) frame or by $r = T_i^{-1}(\theta = 0) \cdot r^i$ in

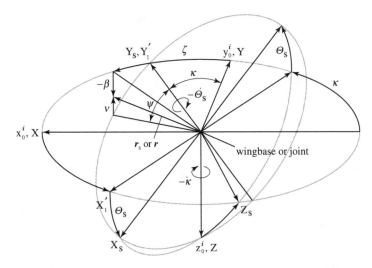

Fig. 7   Wing motion in the body frame (X, Y, Z) and in the stroke plane (X_s, Y_s).

the (x_0^i, y_0^i, z_0^i) frame. Then Eqs. (18) and (19) are obtained, and the equation of identity given by (20) in Table 8 is established. These equations yield the relations given by Eqs. (21) in Table 8, in which e_y is a unit vector oriented along the y_2^i or y^i axis. If the beating is performed symmetrically with respect to the (X, Z) plane, then the angle κ should be zero.

D.   Harmonic Analysis

Generally, wing motion can be expressed by a Fourier series including higher harmonics as given by Eqs. (23a,b,c) and (24a,b,c) in Table 9. For the analysis

Table 9   Harmonic analysis

Item	Expression	Eq. No.
Fourier series of the blade motion in the (x_0^i, y_0^i, z_0^i) frame	$\zeta = \zeta_0 + \sum_{n=1}^{\infty} \zeta_n \sin(n\omega t + \phi_{\zeta n})$	(23a)
	$\beta = \beta_0 + \sum_{n=1}^{\infty} \beta_n \sin(n\omega t + \phi_{\beta n})$	(23b)
	$\theta = \theta_0 + \sum_{n=1}^{\infty} \theta_n \sin(n\omega t + \phi_{\theta n})$	(23c)
Fourier series of the blade motion in the (X_s, Y_s, Z_s) frame	$\psi = \psi_0 + \sum_{n=1}^{\infty} \psi_n \sin(n\omega t + \phi_{\psi n})$	(24a)
	$\theta_s = \theta_{s,0} + \sum_{n=1}^{\infty} \theta_{s,n} \sin(n\omega t + \phi_{s,\theta n})$	(24b)
	$v = v_0 + \sum_{n=1}^{\infty} v_n \sin(n\omega t + \phi_{vn})$	(24c)

of aerodynamic forces and moments, usually, only first harmonics of the lead–lag angle, flapping angle, sweep angle, and elevation angle are required but more than third harmonics of the feathering angles are necessary. In the symmetric beating ($\kappa = 0$), whenever the phase angle difference between lead–lag and flapping angles is given by $\phi_{\zeta 1} - \phi_{\beta 1} = \pi$, the blade tip path plane, which is parallel ($v_0 > 0$) or equivalent ($v_0 = 0$) to the stroke plane, tilts by the angle of $\Theta_s = \tan^{-1}(\beta_1/\zeta_1)$ (Azuma[12]).

E. Left Wing

Similar treatments can be performed for the left wing by introducing the left-hand blade coordinate system or frame expressed with an asterisk($*$), where all linear and angular velocities are conducted in the left-handed way so that positive is counterclockwise rotation viewed from above. Then, the linear and angular transformation matrices between the right-hand frame and the left-hand frame, S_l and S_r, can be given respectively by Eqs. (4a) and (4b) in Table 6. The position R^*, V^*, Euler's angles Φ^*, Θ^*, and Ψ^*, and angular velocity Ω^* in the body frame are given by Eqs. (25a,b) and (26a,b) in Table 10. The transformation matrix T_i^* has the same form as T_i.

The velocity V^{i*} and the acceleration A^{i*} of a blade element of the left wing in the blade frame are given respectively by Eqs. (27a) and (27b) in Table 10, where the position vector R^{i*} at a specified point, or ith wing joint, can be given by Eqs. (28) in Table 10. However, the angles of blade orientation (ζ^*, β^*, θ^*) in the left-handed blade frame and its rate ω^* are treated the same as in the case of the right-handed frame and, thus, given by Eqs. (29a,b) in Table 10.

F. Local Angle of Attack

Another intermediate transformation matrix $T_{i,\theta}$ is introduced. This transforms the (x_2^i, y_2^i, z_2^i) frame to the (x^i, y^i, z^i) frame such that $T_i = T_{i,\theta} \cdot T_i(\theta = 0)$, as given by Eq. (5) in Table 6. If, as stated later, the induced velocity generated by the wing (bound vortices) and wake (free vortices) motions is specified either in the (x_2^i, y_2^i, z_2^i) frame or in the blade frame (x^i, y^i, z^i), v_2 or v then, referring to Fig. 8, the total inflow velocity with respect to a blade element U_2 can be given by $U_2 = V_2^i - v_2 = T_{i,\theta}^{-1} \cdot U$, where $U = T_i \cdot (V^i - v^i)$ and $|U_2| = |U| = U$ as shown by Eqs. (30a,b) in Table 11. The angle of attack is, thus, expressed by $\alpha = \theta + \tan^{-1}(U_{z_2^i}/U_{x_2^i})$ as given by Eqs. (31) and (32a,b,c) in Table 11.

G. Gravity and Centripetal Accelerations

Since the gravity acceleration G is, as shown in Fig. 4, given along the $-Z_I$ axis, as $G = (0, 0, -g)^T$, its expression in the blade frame can be given by Eqs. (22a,b) in Table 8.

In the acceleration A^i given by Eq. (16) in Table 8, the most dominant part is the centripetal acceleration along the wingspan, A_{y^i}, because this acceleration is given approximately by $A_{y^i} = r\dot{\psi}^2$, the maximum values of which are, in the case of the dragonfly (*Anax parthenope julius*) and chalcid wasp (*Encarsia formosa*), about 170G and 350G respectively at the wingtip. This is roughly a fourth or a fifth that

Table 10   Variables related to the left wing in the body frame and the blade frame

Item	Expression	Eq. No.
Position of lift wing	$R^* = S_\ell \cdot R = (X, -Y, Z)^T$	(25a)
Velocity of lift wing	$V^* = S_\ell \cdot V = (V_X, -V_Y, V_Z)^T$	(25b)
Euler's angle of left wing	$(\Phi^*, \Theta^*, \Psi^*)^T = S_\ell \cdot (\Phi, \Theta, \Psi)^T = (-\Phi, \Theta, -\Psi)^T$	(26a)
Angular velocity of left wing	$\Omega^* = (P^*, Q^*, R^*)^T = S_\ell \cdot (P, Q, R)^T = (-P, Q, -R)^T$	(26b)
Velocity of blade element	$V^{i*} = T_i^* \cdot (V^* + \Omega^* \times R^*) + (T_i^* \cdot \Omega^* + \omega^{i*}) \times r^{i*} = (V_{X^i}, -V_{Y^i}, V_{Z^i})^T$	(27a)
Acceleration of blade element	$A^{i*} = \dot{V}_i^* + (T_i^* \cdot \Omega^* + \omega^{i*}) \times V_i^* = (A_{X^i}^*, A_{Y^i}^*, A_{Z^i}^*)^T$	(27b)
Position of joint	$R^{i*} = (X^i, -Y^i, Z^i)^T$	(28)
Blade orientation and its angular velocity	$(\zeta^*, \beta^*, \theta^*)^T = (\zeta, \beta, \theta)^T$	(29a)
	$\omega^{i*} = (p^{i*}, q^{i*}, r^{i*})^T = (p^i, q^i, r^i)^T$	(29b)

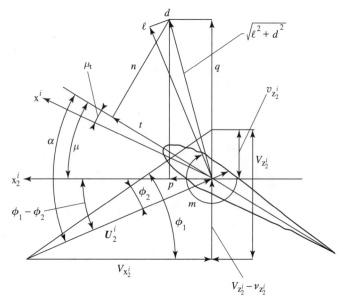

Fig. 8   Inflow velocity and aerodynamic forces and moment acting on a blade element.

of a helicopter rotor but is marvelously high. The centrifugal force in the spanwise direction does not have any influence on the aerodynamic behavior around the beating wing except for the air in the boundary layer, because it rotates or swings with the blade element. However, this spanwise tension force contributes significantly to the large increase in wing stiffness as stated later.

V.   Forces and Moments Acting on Beating Wings

The instantaneous aerodynamic forces and moments at a wing root can be obtained by integrating the elemental (or two-dimensional) forces and moment along a blade span from wing root to wingtip. To perform this integration, the relative velocity of the blade element with respect to the air must be known as a function of time or azimuthwise position. The absolute velocity of the blade element with respect to the inertial frame was obtained in the previous section, and so we need here, to obtain the induced velocity generated by the wing motion.

A.   Induced Velocity

Although the induced velocity is determined exactly from a free-wake vortex system of wings through numerical computation based on the Biot–Savart law, many simple ways of estimation have been presented as described in the following.

1) *The simple momentum theory.* The theory requires a balance between the mean thrust $\bar{T}$ acting on a stroke plane, which is assumed to be an actuator disk of area $S = \frac{1}{4}\pi b^2$, and the change of momentum in a slip stream, $\bar{T} = -2m_s v$, where m_s and v are respectively the mass and the induced velocity of the fluid passing through the disk as shown in Fig. 9 and given by Eqs. (33–35) in Table 12. Because the induced velocity v and the mean thrust $\bar{T}$ are considered to be

Table 11 Inflow angle and local angle of attack

Item	Expression	Eq. No.
Total inflow velocity	$U_2 = (U_{x_2^i}, U_{y_2^i}, U_{z_2^i}) = V_2^i - v_2^i$	(30a)
	$\quad = (V_{x_2^i} - v_{x_2^i},\ V_{y_2^i} - v_{y_2^i},\ V_{z_2^i} - v_{z_2^i})$	
	$\quad = T_i(\theta = 0) \cdot (V_0 + \Omega \times R^i) + [T_i(\theta = 0) + \omega^i(\dot{\theta} = 0)] \times (T_{i,\theta}^{-1} \cdot r^i)$	
	$U = (U_{x^i},\ U_{y^i},\ U_{z^i})^T = V^i - v^i = (V_{x^i} - v_{x^i},\ V_{y^i} - v_{y^i},\ V_{z^i} - v_{z^i})^T$	(30b)
Angle of attack	$\alpha = \theta + \phi = \theta_0 + \theta_t + \phi_1 - \phi_2$	(31)
Inflow angle	$\phi = \tan^{-1}(U_{z_2^i}/U_{x_2^i}) \cong \phi_1 - \phi_2$	(32a)
	$\phi_1 = \tan^{-1}(V_{z_2^i}/U_{x_2^i})$	(32b)
	$\phi_2 = \tan^{-1}(v_{z_2^i}/U_{x_2^i})$	(32c)

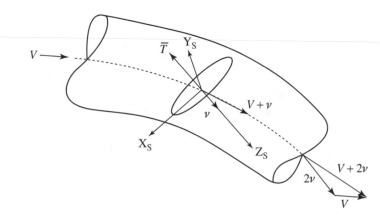

Fig. 9   Inflow velocity through an actuator disk.

perpendicular to the actuator disk embedded in the stroke plane or parallel to the Z_s axis, they are expressed in the body frame by Eqs. (35d) and (36) in Table 12.

By referring to Eq. (10) in Table 7, the general inflow velocity at the virtual joint $\bar{R}^j$ is given by $V_j = V_0 + \Omega \times \bar{R}^j$, and the total inflow velocity is given by $U_j = V_j - v$ in the body frame. Then, by assuming that the rotational component $\Omega \times \bar{R}^j$ is small, the absolute value of the total inflow in the body frame can be given by $|U_j| = |V_0 + \Omega \times \bar{R}^j - v_j|$. The induced velocity expressed in the (x_2^i, y_2^i, z_2^i) frame v_2 and (x^i, y^i, z^i) frame v^i are given respectively by Eqs. (35e) and (35f) in Table 12.

2) *Dynamic inflow.*   When the mean thrust $\bar{T}$ of the actuator disk increases suddenly by $\Delta\bar{T}$, the induced velocity increases as if the disk shifted like a circular disk in its perpendicular direction (Pitt and Peters,[111] Gaonker and Peters[112]). The added mass of the circular disk of radius R and diameter or wingspan b, moving in its normal direction, is equivalent to the mass of a sphere surrounding the disk times $(2/\pi)$ or $A_{33} = \rho \frac{8}{3} R^3 = \frac{4}{3\pi}\rho Sb$, which is independent of the motion in the parallel direction (Landweber[113]). Therefore, the equation of motion simulating the unsteady induced velocity, called *dynamic inflow*, and its solution can then be given respectively by Eqs. (37) and (38) in Table 12, where V_t and $\dot{\psi}_{max}$ are expressed respectively by Eqs. (39a) and (39b) and where the inflow mass m_s is defined for the trimmed state. It can then be said that the time constant of this first-order system $\tau = \frac{4}{3\pi}(\rho Sb/m_s)$ is proportional to the wingspan and inversely proportional to the inflow mass.

In the beating flight of flying animals, a sudden increase of thrust can result from either an impulsive increase of the wing pitch angle or a penetrating vertical gust. Two examples are considered. For a dragonfly having wingspan of $b = 0.1$ m and beating its wings at the frequency of $f = 30$ Hz, the time constant is $\tau = 0.03$ s for a flight speed of $V = 4$ m/s. Then the induced velocity reaches the terminal value within the time of one stroke, $1/f = 0.03$ s. For an osprey having wingspan of $b = 1.6$ m and beating its wings at frequency of $f = 3$ Hz, the time constant is $\tau = 0.2$ s for a flight speed of 10 m/s. This is also on the order of one stroke.

Table 12 Aerodynamic forces and moments as an actuator disk

Item	Expression	Eq. No.
Fluid mass	$m_s = \rho S \sqrt{V_{X_s}^2 + V_{Y_s}^2 + (v - V_{Z_s})^2}$	(33)
General velocity at jth virtual joint		
In body frame	$V_j = V_\theta + \Omega \times \bar{R}^j$	(34a)
In jth stroke plane frame	$V_s = (V_{X_s},\ V_{Y_s},\ V_{Z_s})^T = T_s \cdot V_j$	(34b)
Induced velocity based on momentum theory		
In stroke plane frame	$\sqrt{|\bar{T}|/2\rho S}\,(-\bar{T}/|\bar{T}|) = (0,\ 0,\ v_h)^T$ at hovering	(35a)
	$v_{n+1} = |\bar{T}|/2\rho S \sqrt{V_{X_s}^2 + V_{Y_s}^2 + (v_n - V_{Z_s})^2}$ in forward flight	(35b)
In body frame	$v_j = -\bar{T}/2m_s = T_s^{-1} \cdot v$	(35c)
	$= (-v\sin\Theta_s \cos\kappa,\ v\sin\Theta_s \sin\kappa,\ v\cos\Theta_s)^T$	(35d)
In blade frame	$v_2 = T_i(\theta = 0) \cdot v_j$	(35e)
	$v_i^i = T_i \cdot v_j$	(35f)
Mean thrust		
In body frame	$\bar{T} = (\bar{T}\sin\Theta_s\cos\kappa,\ -\bar{T}\sin\Theta_s\sin\kappa,\ -\bar{T}\cos\Theta_s)^T$	(36)
Equation of dynamic inflow		
In stroke plane frame	$|\Delta\bar{T}| = \left(\dfrac{4}{3\pi}\rho Sb\right)\dot{v} + 2m_s v$	(37)
Induced velocity of dynamic inflow		
In stroke plane frame	$v = \left(\dfrac{|\Delta\bar{T}|}{2m_s}\right)\left\{1 - \exp\left[\dfrac{3}{4}\pi\left(\dfrac{m_s}{\rho Sb}\right)t\right]\right\}$	(38)
Parameter used in equation (38)		
In stroke plane frame	$V_t = \frac{1}{2}b\dot\psi_{max}$ $\dot\psi_{max} = \dot\psi\left(\omega t + \kappa = \frac{\pi}{2}\right)$	(39a) (39b)

3) Vortex theory.   In vortex theory, a vortex system of beating wings as well as rotary wings is represented by a set of bound vortices that are attached to the wing or the blade and the wake vortices that are embedded in the wake. When the wake vortices are distributed in a fixed wake prescribed by the general velocity including the induced velocity determined by simple momentum theory, the vortex system is called a *fixed-wake vortex system*. When the wake vortices are distributed in a free wake distorted by mutual interference among the wake vortices, the vortex system is called a *free-wake vortex system*.

If the wing is slender (high aspect ratio), it is effectively represented by a single vortex line along the $c/4$ axis or feathering axis of the wing. The wake vortices are comprised of *shed vortices*, which, at the moment of leaving the wing, are parallel to the trailing edge, and *trailing vortices*, springing along the wingspan in the original direction either exactly or approximately perpendicular to the blade feathering axis. The following are examples of fixed-wake systems: In the case of a rotary wing operating in a hovering state, a vortex sheet filled with wake vortices is trailed in a spiral plane confined within a semi-infinite circular cylinder (Lock,[114] Moriya[115]). If the helical pitch, which is determined by $v/r\Omega$, is small, the vortex sheet is replaced by a set of piled-up vortex rings of radius r $(0 < r < R)$ in the cylinder, and usually, the vortex rings are assumed to be composed of tip vortices of the blade only (Castles and De Leeuw,[19] Heyson and Katzoff[20]).

The normal component of the induced velocity generated by a vortex ring was analytically expressed by Lamb.[116] Then, Castles and De Leeuw[19] presented the numerical chart of the normal induced velocity distribution generated by the piled-up vortex rings. When the rotor has some forward velocity, the wake cylinder changes from circular to elliptic (Heyson and Katzoff[20]). Heyson and others conducted detailed analyses on such fixed wakes for a single rotor and multiple rotors (Heyson[117–119]).

A similar treatment can be applied for the vortex wakes of the beating wings (Azuma et al.[100]). Because of the cyclic changes of the sweeping velocity and of the feathering motion, the effect of shed vortices, which exist on the radial lines, cannot be discarded.

4) Free wake.   The rotor wake cannot be stable in far wake for all modes and all wavelengths of harmonic airloading, except in a few cases, because of mutual interference among helical wake vortices (Gupta and Loewy,[120] Gupta and Lessen,[121] Pouradier and Horowitz,[122] Rand and Rosen[123]). It is, thus, necessary to determine the rotor wake positions or geometry for the prediction of higher-harmonic airloads acting on the blades (Greenberg and Powers,[124] Landgrebe[125]). With mutual-inductance instability of helical vortices (Widnall[126]), the vortex sheet emanated from the tip region of the blade is rapidly rolled up, and within two chord lengths behind the blade (Tangler et al.[127]) it makes a strong tip vortex that extends downstream somewhat inboard of the blade tip under the strong interaction caused by adjacent vortices, generated by neighboring blades, and sometimes brings vibratory airloading on the succeeding blade (Cone,[128] Levinsky and Strand[129]).

Determination of the rotor's wake geometry and the strength of the vorticity for such free wakes and the prediction of the resulting blade airload distribution have recently begun by using high-speed and large-capacity digital computers.

5) Local momentum theory.   The local momentum theory (LMT) and its extension, the LCM, are based on three fundamental considerations: 1) A wing is composed of n elliptic wings, the induced velocity (or the circulation) of which is constant (elliptic) along the wingspan; 2) the induced velocity (or the circulation) generated by a wing or a blade element at any spanwise station is determined by counterbalance with the lift acting on that element at any time t and place r; and 3) the induced velocity $v(t, r)$, which is left in the wake, is determined simply by multiplying the *attenuation coefficient* C by the induced velocity generated at the time τ and position s, $v(\tau, s)$, just when the wing element has passed, $v(t, r) = C \cdot v(\tau, s)$. Because the coefficient is assumed to be diagonal ($C_{ij} = 0$ for $i \neq j$) the attenuation coefficient is given by a function of the time difference $(t - \tau)$ and the relative position of $r - s$ and is, thus, determined by the induced velocity ratio between two positions, $C_{ij}(t - \tau, r - s) = v_i(t, r)/v_j(\tau, s)$ for $i, j = X, Y, Z$ (Azuma and Kawachi,[130] Azuma et al.[131]). For the calculation of $v_i(t, r)$ and $v_i(\tau, s)$, any aforementioned method can be utilized in correspondence with the accuracy of the computation. However, the accuracy of LMT is nearly equal to that of the vortex theory utilized to determine the attenuation coefficient because the coefficient can be adjusted by iterating the calculation in the respective time step, and the effect of the induced velocity $v_i(\tau, s)$ is primarily reserved. The computation is also very manageable in comparison with those of complicated vortex theories, specifically the free-wake analysis, which requires laborious computation accompanied by lengthy time calculation. Furthermore, the method can be applied to various problems including unsteady rotary and beating wing motions. Thus, in the aerodynamic study of rotary wings, many applications of the LMT and LCM have been devised (Azuma and Saito,[132] Saito and Azuma,[133] Nasu and Azuma,[134] Azuma et al.,[135,136] Azuma and Yasuda[137]). The methods have also been extended to analyze the beating wings in which aerodynamic modifications caused by very low Reynolds numbers, flow separation, unsteady motion with very high reduced frequency, and other factors can be introduced without any accompanying laborious calculations (Azuma et al.,[100] Azuma and Watanabe,[50] Sato and Azuma[101]).

B. Blade Element Theory

By referring to Fig. 9, aerodynamic forces, lift ℓ, drag d, and moment m, acting on a blade element, are given by Eqs. (40 a,b,c) in Table 13. The propulsive force p and perpendicular force q with respect to the (x_2^i, y_2^i, z_2^i) frame and tangential force t and normal force n with respect to the blade chord frame are respectively given by Eqs. (41a,b) and (42a,b) in Table 13.

The aerodynamic forces and feathering moment acting on a wing can be given by Eqs. (43a,b,c) in Table 13, where the spanwise aerodynamic force F_s^A is assumed to be zero as given by Eq. (44b). The chordwise and flatwise shear forces F_c^A, F_f^A and chordwide and flatwise bending moments M_c^A, M_f^A at spanwise station s are respectively given by Eqs. (44a,b) and (45a,b) in Table 13. By assuming that the elastic axis of the wing is located forward in the chord line at a distance x_e^i from the feathering axis, the torsional moment along the elastic axis M_t^A is given by Eq. (45c) in Table 13.

Similarly, the inertial forces (F_c^I, F_s^I, F_f^I)^T and moments (M_c^I, M_f^I, M_t^I)^T including the gravity acceleration acting at the center of gravity (x_g^i, r, z_g^i) of the blade element at span r of the wing are given by Eqs. (46a,b,c) and (47a,b,c) in Table 14, in which the mass of blade element is defined later by Eq. (51b)

Table 13 Aerodynamic forces and moments

Item	Expression	Eq. No.
Blade element		
Lift	$\ell = (1/2)\rho U^2 c C_\ell(\alpha, \dot\alpha)$	(40a)
Drag	$d = (1/2)\rho U^2 c C_d(\alpha, \dot\alpha)$	(40b)
Moment	$m = (1/2)\rho U^2 c^2 C_m(\alpha, \dot\alpha)$	(40c)
Propulsive force	$p = \ell \sin(\phi_1 - \phi_2) - d\cos(\phi_1 - \phi_2)$	(41a)
Perpendicular force	$q = \ell\cos(\phi_1 - \phi_2) + d\sin(\phi_1 - \phi_2)$	(41b)
Tangential force	$t = \ell\sin\alpha - d\cos\alpha$	(42a)
Normal force	$n = \ell\cos\alpha + d\sin\alpha$	(42b)
Single wing		
Propulsive force	$F_{x_2^i}^A = \int_0^R p\,dr$	(43a)
Perpendicular force	$F_{z_2^i}^A = -\int_0^R q\,dr$	(43b)
Feathering moment	$M_{y_2^i}^A = \int_0^R m\,dr$	(43c)
Chordwise shear force	$F_c^A(s) = \int_s^R \{t(r)\cos[\theta_t(r) - \theta_t(s)] - n(r)\sin[\theta_t(r) - \theta_t(s)]\}\,dr$	(44a)
Spanwise tension	$F_s^A(s) = 0$	(44b)
Flatwise shear force	$F_f^A(s) = -\int_s^R \{n(r)\cos[\theta_t(r) - \theta_t(s)] + t(r)\sin[\theta_t(r) - \theta_t(s)]\}\,dr$	(44c)
Chordwise bending moment	$M_c^A(s) = -\int_s^R \{t(r)\cos[\theta_t(r) - \theta_t(s)] - n(r)\sin[\theta_t(r) - \theta_t(s)]\}(r-s)\,dr$	(45a)
Flatwise bending moment	$M_f^A(s) = -\int_s^R \{n(r)\cos[\theta_t(r) - \theta_t(s)] + t(r)\sin[\theta_t(r) - \theta_t(s)]\}(r-s)\,dr$	(45b)
Torsional moment about elastic axis	$M_t^A(s) = -\int_s^R [m(r) - x_e^i n(r)]\,dr$	(45c)

Table 14 Inertial forces and moments

Item	Expression	Eq. No.
Chordwise shear force	$F_c^I = -\int_s^R [(A_{x^i} + G_{x^i})\cos\theta_t(s) - (A_{z^i} + G_{z^i})\sin\theta_t(s)]\left(\dfrac{dm_b}{dr}\right)dr$	(46a)
Spanwise tension force	$F_s^I = -\int_s^R (A_{y^i} + G_{y^i})\left(\dfrac{dm_b}{dr}\right)dr$	(46b)
Flatwise shear force	$F_f^I = -\int_s^R [(A_{x^i} + G_{x^i})\sin\theta_t(s) + (A_{z^i} + G_{z^i})\cos\theta_t(s)]\left(\dfrac{dm_b}{dr}\right)dr$	(46c)
Chordwise bending moment	$M_c^I = -\int_s^R \left\{(x_g^i - x_e^i)(A_{y^i} + G_{y^i}) - (r - s)[(A_{x^i} + G_{x^i})\cos\theta_t(s) - (A_{z^i} + G_{z^i})\sin\theta_t(s)]\right\}\left(\dfrac{dm_b}{dr}\right)dr$	(47a)
Flatwise bending moment	$M_f^I = -\int_s^R \left\{(r - s)[(A_{x^i} + G_{x^i})\sin\theta_t(s) + (A_{z^i} + G_{z^i})\cos\theta_t(s)] - z_g^i(A_{y^i} + G_{y^i})\right\}\left(\dfrac{dm_b}{dr}\right)dr$	(47b)
Torsional moment about elastic axis	$M_t^I = -\int_s^R \left\{z_g^i[(A_{x^i} + G_{x^i})\cos\theta_t(s) - (A_{z^i} + G_{z^i})\sin\theta_t(s)] - (x_g^i - x_e^i)[(A_{x^i} + G_{x^i})\sin\theta_t(s) + (A_{z^i} + G_{z^i})\cos\theta_t(s)]\right\}\left(\dfrac{dm_b}{dr}\right)dr$	(47c)

in Table 16. It should be noted again that the spanwise tension force F_s^I, or the spanwise acceleration A_{y^i}, is very large such that the maximum value is (1–2) × 10^2 G. It is further necessary to mention that if the buoyant force is considered, the density of wing ρ_b should be replaced with $\rho_b - \rho$, such that $dm_b/dr = \int (\rho_b - \rho) t(x, r) dx$, only for the terms related to the gravity acceleration $\boldsymbol{G}$.

If the added mass is necessary to calculate the inertial force because of very thin and light weight wing, then, another inertial force, $\frac{1}{4}\rho\pi c^2 A_{z^i}$, should be added only along the z^i axis, and the blade mass dm_b should be replaced with $\{(dm/dr) + \frac{1}{4}\rho\pi c^2\}$ along the z^i direction only.

C. Centrifugal Force

As stated before, the maximum centripetal acceleration reaches a very large value during the beating action at high frequency. In such case, under the assumption of small feathering angle, the acceleration $\boldsymbol{A}^i$ at a point $\boldsymbol{r}^i = (x, r, z)^T$ in the wing given by Eq. (16) in Table 8 can be approximated by Eqs. (48a,b) in Table 15, where $\dot{\psi} = (\dot{\zeta}^2 + \dot{\beta}^2)1/2$ is an angular rate of the wing beating in the ith stroke plane. The inertial forces and moments acting on the wing, which is assumed to have homogeneous density ρ_b and to be nontwist for simplicity, can respectively be given by Eqs. (49a,b) and (50a,b) in Table 15. The symbols used in Eqs. (49) and (50) are defined by Eqs. (51–54) in Table 16.

The centrifugal force, which is proportional to the product of the blade mass m_b and the angular rate of beating $\dot{\psi}$ squared, acts in every direction, but the largest one is given by $m_b \bar{r} \dot{\psi}^2$ along the span. Like the blade of a helicopter rotor, the inertial moment in every direction, which is also proportional to $\dot{\psi}^2$, acts to strengthen the wing stiffness. Although the elastic deformation of the wing has not been taken into consideration here, as supposed from Eq. (50b), the effect of wing flexibility is simulated equivalently by assuming some shifts of the center of gravity, $\bar{x}$, $\bar{z} \neq 0$. The pleating in insect wings is known to be effective for reinforcement of their stiffness (Newman and Wootton[138]), but this inertial effect in beating wings, as far as we know, has not been indicated in any reference on biokinetics. The torsional moment is proportional to the product of inertia J_{xz} plus $(I_x - I_z)\theta^i$; the latter gives a restoring moment for keeping the wing parallel in the stroke plane or for reducing the feathering angle. Whenever two moments of inertia are not equal or $I_x \neq I_z$, because the wing is thin, $I_x - I_z$ may be replaced with $-I_y$. The above inertial contribution to the restoring moment, derived from the centrifugal force acting on a thin plate (like a tennis racket or a helicopter rotor blade) is called the *tennis racket effect*.

D. Power

Torques or moments acting on the ith joint $\boldsymbol{Q}^i = \boldsymbol{M}^i(s = 0) = (Q_{x^i}, Q_{y^i}, Q_{z^i})^T$ and the power P^i required to operate the beating motion against the moments can be given respectively by Eqs. (56a,b,c) and (57) in Table 17. If the beating motion is conducted through an elastic device and if the frequency is tuning with the resonant frequency of this dynamic system, then the power may be given by the aerodynamic components only.

E. Forces and Moments Generated by Wings and Body

The shear force $\boldsymbol{F}^i = (F_{x^i}, F_{y^i}, F_{z^i})^T$ and the moment or torque $\boldsymbol{Q}^i = (Q_{x^i}, Q_{y^i}, Q_{z^i})^T$ acting at the ith joint are, as given by Eqs. (58a,b), expressed in the

Table 15 Quantities related to the tennis racket effect

Item	Expression	Eq. No.
Inertial acceleration	$A^i = (A_{x^i}, A_{y^i}, A_{z^i})^T$	(48a)
	$\cong \omega^i \times (\omega^i \times r^i)$	
	$\cong \dot{\psi}^2 (\theta^i, 0, -1)^T \times [(\theta^i, 0, -1) \times (x, r, z)^T]$	(48b)
	$= -\dot{\psi}^2 [(x + z\theta^i), r, x\theta^i]^T$	
Centrifugal force	$F^i = -\iiint \rho_b A^i \, dx \, dr \, dz$	(49a)
	$= -\dot{\psi}^2 \iint \dfrac{d}{dx}\left(\dfrac{dm_b}{dr}\right)[(x + \bar{z}\theta^i), r, x\theta^i]^T \, dx \, dr$	
	$= \dot{\psi}^2 \int (dm_b/dr)[(\bar{x} + \bar{z}\theta^i), r, \bar{x}\theta^i]^T \, dr$	(49b)
	$= m_b \dot{\psi}^2 [(\bar{\bar{x}} + \bar{\bar{z}}\theta^i), \bar{r}, \bar{\bar{x}}\theta^i]^T$	
Moment generated by centrifugal force	$M^i = -\iiint \rho_b r^i \times A^i \, dx \, dr \, dz$	
	$= \dot{\psi}^2 \iint \dfrac{d}{dx}\left(\dfrac{dm_b}{dr}\right) \rho_b t\,(x, r) \begin{bmatrix} rx\theta^i - r\bar{z} \\ x\bar{z} + (\bar{z^2} - x^2)\,\theta^i \\ -r\bar{z}\theta^i \end{bmatrix} dx \, dr$	(50a)
	$= \dot{\psi}^2 \int \left(\dfrac{dm_b}{dr}\right) \begin{bmatrix} r\bar{x}\theta^i - r\bar{z} \\ \bar{x}\bar{z} + (\bar{z^2} - \bar{x^2})\theta^i \\ -r\bar{z}\theta^i \end{bmatrix} dr$	(50b)
	$= m_b \dot{\psi}^2 \begin{bmatrix} \overline{r\bar{x}}\theta^i - \overline{r\bar{z}} \\ \overline{\bar{x}\bar{z}} + (\overline{\bar{z^2}} - \overline{\bar{x^2}})\theta^i \\ -\overline{r\bar{z}}\theta^i \end{bmatrix}$	(50c)
	$= \dot{\psi}^2 \begin{bmatrix} J_{yx}\theta^i - J_{yz} \\ J_{xz} + (I_x - I_z)\,\theta^i \\ -J_{yz}\theta^i \end{bmatrix}$	

body frame by $T_i^{-1} \cdot F^i$ and $T_i^{-1} \cdot Q^i$, respectively. Then, the total force F^W and the moment M^W about the origin of the body frame, generated by both pairs of wings, can respectively be given by Eqs. (59a) and (59b) in Table 18, in which the summation $\sum_{i=1}^{2}$ is performed for two pairs of wings and is discarded for one pair of wings.

Both the body's angles of attack and side slip, α and β, can be given by Eqs. (60a,b) in Table 18, in which the induced velocity v may be determined simply from the simple momentum theory given by Eqs. (35a) or (35b) and may be kept constant at the body position, which is close to the stroke plane. Then the aerodynamic forces and moments can be given by a function of these angles and their rates as given by Eqs. (61a,b,c) and (62a,b,c) in Table 18, respectively, and the coefficients are also a function of Reynolds number. The gravity and buoyant forces, and the gravity and buoyant moments are respectively

Table 16 Definition of symbols used in Table 15

Item	Expression	Eq. No.
Wing thickness and mass	$t(x,\ r) = \int dz$	(51a)
	$\rho_b = d[d(dm_b/dr)/dx]/dz$	(51b)
	$dm_b/dr = \iint \rho_b\, dz\, dx$	(51c)
	$m_b = \int_0^R (dm_b/dr)\, dr$	(51d)
Center of gravity and radius of gyration	$\bar{z} = \int \rho_b z\, dz/d(dm_b/dr)/dx$	(52a)
	$\overline{z^2} = \int \rho_b z^2 dz/d(dm_b/dr)/dx$	(52b)
	$\bar{\bar{z}} = \int [d(dm_b/dr)/dx]\bar{z}\, dx/(dm_b/dr)$	(53a)
	$\bar{x} = \int [d(dm_b/dr)/dx]x\, dx/(dm_b/dr)$	(53b)
	$\overline{\overline{z^2}} = \int [d(dm_b/dr)/dx]\overline{z^2}\, dx/(dm_b/dr)$	(53c)
	$\overline{x^2} = \int [d(dm_b/dr)/dx]x^2\, dx/(dm_b/dr)$	(53d)
	$\overline{x\bar{z}} = \int [d(dm_b/dr)/dx]x\bar{z}\, dx/(dm_b/dr)$	(53e)
	$\bar{\bar{x}} = \int (dm_b/dr)\bar{x}\, dr/m_b$	(54a)
	$\bar{\bar{\bar{z}}} = \int (dm_b/dr)\bar{\bar{z}}\, dr/m_b$	(54b)
	$\bar{r} = \int (dm_b/dr)r\, dr/m_b$	(54c)
Moments and products of inertia	$\overline{r\bar{x}} = \int (dm_b/dr)r\bar{x}\, dr/m_b = J_{yx}/m_b$	(55a)
	$\overline{r\bar{\bar{z}}} = \int (dm_b/dr)r\bar{\bar{z}}\, dr/m_b = J_{yz}/m_b$	(55b)
	$\overline{\overline{x\bar{z}}} = \int (dm_b/dr)\overline{x\bar{z}}\, dr/m_b = J_{xz}/m_b$	(55c)
	$\overline{\overline{\overline{z^2}}} = \int (dm_b/dr)\overline{\overline{z^2}}\, dr/m_b = \left[I_x - \int (dm_b/dr)r^2\, dr\right]/m_b$	(55d)
	$\overline{\overline{x^2}} = \int (dm_b/dr)\overline{x^2}\, dr/m_b = \left[I_z - \int (dm_b/dr)r^2\, dr\right]/m_b$	(55e)

Table 17   Torques and power

Item	Expression	Eq. No.
Torque	$Q_{x^i} = M_f^A(s=0) + M_f^I(s=0) + M_f^G(s=0)$	(56a)
	$Q_{y^i} = M_t^A(s=0) + M_t^I(s=0) + M_t^G(s=0)$	(56b)
	$Q_{z^i} = M_c^A(s=0) + M_c^I(s=0) + M_c^G(s=0)$	(56c)
Power	$P^i = \boldsymbol{Q} \cdot \boldsymbol{\omega}^i$	(57)
	$= Q_{x^i} \cdot \dot{\theta}^i + (Q_{x^i} \cos \beta^i \sin \theta^i + Q_{y^i} \sin \beta^i$	
	$- Q_{z^i} \cos \beta^i \cos \theta^i) \dot{\zeta}^i + (Q_{x^i} \cos \theta^i - Q_{z^i} \sin \theta^i) \dot{\beta}^i$	

given by Eqs. (63a,b,c) and (64a,b,c) in Table 18, where V_b is the volume of body and $\boldsymbol{R}^B = (X^B, Y^B, Z^B)^T$ is the position vector of the buoyant center from the origin of the body frame. Finally, the total external forces and moments acting on the body can be given by $\boldsymbol{F}^W + \boldsymbol{F}^B + \boldsymbol{F}^G$ and $\boldsymbol{M}^W + \boldsymbol{M}^B + \boldsymbol{M}^G$, respectively.

Because the muscle forces are required to generate the above shear forces and moments acting on the wing in the body frame, the wing forces and moments should be expressed in a new coordinate system fixed to, for an example, the tergal plate. As shown in Fig. 10, the transformation matrix $\boldsymbol{T}_t$ between the body frame (X, Y, Z) and the tergal plate frame (X_t, Y_t, Z_t), which is inclined at angle Θ_t, is given by Eq. (65) and the transformed forces and moments are expressed by Eqs. (66) and (67), respectively, in Table 19.

Then, the shear forces other than $F_{Z_t}^W$, which acts as tension force of the muscles in relation to the beating motion, are mechanically supported by other parts of the body, and the moments act against driving moments generated by mutually cooperated tension forces related to the beating muscles.

F.   Equations of Motion

Equations of motion of the body can be obtained by balancing the external forces and moments with the inertial forces and moments of the body and the added mass of fluid, as given respectively by Eqs. (68) and (69) in Table 20. The detailed expressions of the inertial matrix for forces and moments of the body, K_m, K_a and H_m, H_a, are given by Eqs. (70a,b,c) and (71a,b,c) in Table 21 for the body itself and Eqs. (72a,b,c) and (73a,b,c) in Table 22 for added mass respectively, some of which can be discarded by the symmetric configuration of the body. Like the buoyant force and moment, the inertial force generated by the added mass of fluid resulting from the body acceleration is usually neglected for flying objects in the air, but it cannot be neglected for objects swimming in water, except for nondiagonal elements.

The nonlinear equations in Table 20 can be solved by numerical computation, applying, for example, the Runge–Kutta method (Kreyszig[139]). It is also possible to express the results as motion in the inertial frame.

VI.   Conclusion

For low Reynolds number and small-aspect-ratio wings, the drag coefficient of insects and samaras are, in comparison with other flying objects, so large that the maximum lift-to-drag ratio takes place at large lift coefficient. In the beating flight

Table 18　Forces and moments generated by wings and fuselage

Item	Expression	Eq. No.
Force acting at ith joint	$\boldsymbol{F}^i = [F_c(s=0),\ F_s(s=0),\ F_f(s=0)]^T$	(58a)
Torque generated by all wings	$\boldsymbol{Q}^i = (Q_{x^i},\ Q_{y^i},\ Q_{z^i})^T$	(58b)
Force generated by all wings	$\displaystyle \boldsymbol{F}^W = \sum_{i=1}^{2} (\boldsymbol{T}_i^{-1} \cdot \boldsymbol{F}^i + \boldsymbol{S}_\ell \cdot \boldsymbol{T}_i^{-1*} \cdot \boldsymbol{F}^{i*})$	(59a)
Moment generated by all wings about the origin of body frame	$\displaystyle \boldsymbol{M}^W = \sum_{i=1}^{2} \left[(\boldsymbol{T}_i^{-1} \cdot \boldsymbol{Q}^i + \boldsymbol{S}_\ell \cdot \boldsymbol{T}_i^{-1*} \cdot \boldsymbol{Q}^{i*}) \right.$ $\left. + \boldsymbol{R}^i \times (\boldsymbol{T}_i^{-1} \cdot \boldsymbol{F}^i) + \boldsymbol{R}^{i*} \times (\boldsymbol{S}_\ell \cdot \boldsymbol{T}_i^{-1*} \cdot \boldsymbol{F}^{i*}) \right]$	(59b)
Angle of attack of body	$\alpha = \tan^{-1}\left[(V_Z - \boldsymbol{T}_s^{-1} \cdot \boldsymbol{v})/V_X\right]$	(60a)
Side slip angle of body	$\beta = \tan^{-1}(V_Y/V_X)$	(60b)
Aerodynamic forces acting on body	$\boldsymbol{F}^B = (F_X^B,\ F_Y^B,\ F_Z^B)^T$ $= \dfrac{1}{2}\rho U^2 A \begin{bmatrix} C_{F,X}(\alpha,\ \beta) \\ C_{F,Y}(\alpha,\ \dot{\alpha}) \\ C_{F,Z}(\beta,\ \dot{\beta}) \end{bmatrix}$	(61a) (61b) (61c)
Aerodynamic moments acting on body	$\boldsymbol{M}^B = (M_X^B,\ M_Y^B,\ M_Z^B)^T$ $= \dfrac{1}{2}\rho U^2 A\ell \begin{bmatrix} C_{M,X}(\alpha,\ \beta) \\ C_{M,Y}(\alpha,\ \dot{\alpha}) \\ C_{M,Z}(\beta,\ \dot{\beta}) \end{bmatrix}$	(62a) (62b) (62c)

(Cont.)

Table 18 *(Continued)*

Item	Expression	Eq. No.
Gravity and buoyant forces acting on body	$F^G = (F_X^G,\ F_Y^G,\ F_Z^G)^T = -(m - \rho V_b) T_b \cdot G$ $$= -(m - \rho V_b) g \begin{pmatrix} \sin\Theta \\ -\cos\Theta\sin\Phi \\ -\cos\Theta\cos\Phi \end{pmatrix}$$	(63a) (63b) (63c)
Gravity and buoyant moments acting on body	$M^G = (M_X^G,\ M_Y^G,\ M_Z^G)^T = -m R^G \times (T_b \cdot G) + \rho V_b R^B \times (T_b \cdot G)$ $$= -mg \begin{pmatrix} -Y^G\cos\Theta\cos\Phi + Z^G\cos\Theta\sin\Phi \\ Z^G\sin\Theta + X^G\cos\Theta\cos\Phi \\ -X^G\cos\Theta\sin\Phi - Y^G\sin\Theta \end{pmatrix}$$ $$+ \rho V_b g \begin{pmatrix} -Y^B\cos\Theta\cos\Phi + Z^B\cos\Theta\sin\Phi \\ Z^B\sin\Theta + X^B\cos\Theta\cos\Phi \\ -X^B\cos\Theta\sin\Phi - Y^B\sin\Theta \end{pmatrix}$$	(64a) (64b) (64c)

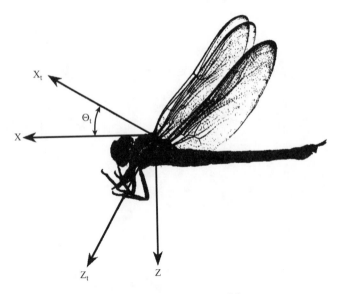

Fig. 10   Body frame and tergal frame.

Table 19   Transformation matrix, force, and moments in tergal plate frame

Item	Expression	Eq. No.
Body frame to tergal plate frame	$T_t = \begin{pmatrix} \cos\Theta_t & 0 & -\sin\Theta_t \\ 0 & 1 & 0 \\ \sin\Theta_t & 0 & \cos\Theta_t \end{pmatrix}$	(65)
Force $T_t \cdot F^W$		
$\quad F_{X_t}^W$	$F_X^W \cos\Theta_t - F_Z^W \sin\Theta_t$	(66a)
$\quad F_{Y_t}^W$	F_Y^W	(66b)
$\quad F_{Z_t}^W$	$F_X^W \sin\Theta_t + F_Z^W \cos\Theta_t$	(66c)
Moment $T_t \cdot M^W$		
$\quad M_{X_t}^W$	$M_X^W \cos\Theta_t - M_Z^W \sin\Theta_t$	(67a)
$\quad M_{Y_t}^W$	M_Y^W	(67b)
$\quad M_{Z_t}^W$	$M_X^W \sin\Theta_t + M_Z^W \cos\Theta_t$	(67c)

Table 20   Inertial force and moment balance

Item	Expression	Eq. No.
Force balance	$F^W + F^B + F^G = K \cdot [\dot{V} + \Omega \times V + \dot{\Omega} \times R^G + \Omega \times (\Omega \times R^G)]$	(68)
Moment balance	$M^W + M^B + M^G = H \cdot [\dot{V} + \Omega \times V + \dot{\Omega} \times R^G + \Omega \times (\Omega \times R^G)]$	(69)

Table 21   Inertial tensors of body

Item	Linear acceleration			Angular acceleration			Linear velocity times angular velocity[a]									Angular velocity times angular velocity						Eq. No.
	$\dot{U}_X$	$\dot{U}_Y$	$\dot{U}_Z$	ω_X	ω_Y	ω_Z	$U_X\omega_X$	$U_X\omega_Y$	$U_X\omega_Z$	$U_Y\omega_X$	$U_Y\omega_Y$	$U_Y\omega_Z$	$U_Z\omega_X$	$U_Z\omega_Y$	$U_Z\omega_Z$	ω_X^2	ω_Y^2	ω_Z^2	$\omega_X\omega_Y$	$\omega_Y\omega_Z$	$\omega_Z\omega_X$	
Force K_m	M	0	0	0	MZ_G	$-MY_G$	0	0	0	0	0	$-M$	0	M	0	0	$-MX_G$	$-MX_G$	MY_G	0	MZ_G	(70a)
	0	M	0	$-MZ_G$	0	MX_G	0	0	M	0	0	0	$-M$	0	0	$-MY_G$	0	$-MY_G$	MX_G	MZ_G	0	(70b)
	0	0	M	MY_G	$-MX_G$	0	0	$-M$	0	M	0	0	0	0	0	$-MZ_G$	$-MZ_G$	0	0	MY_G	MX_G	(70c)
Moment H_m	0	$-MZ_G$	MY_G	I_X	$-J_{XY}$	$-J_{XZ}$	0	$-MY_G$	$-MZ_G$	MY_G	0	0	MZ_G	0	0	0	$-J_{YZ}$	J_{YZ}	$-J_{XZ}$	I_Z-I_Y	J_{XY}	(71a)
	MZ_G	0	$-MX_G$	$-J_{XY}$	I_Y	$-J_{YZ}$	0	MX_G	0	$-MX_G$	0	$-MZ_G$	0	MZ_G	0	J_{XZ}	0	$-J_{XZ}$	J_{YZ}	$-J_{XY}$	I_X-I_Z	(71b)
	$-MY_G$	MX_G	0	$-J_{XZ}$	$-J_{YZ}$	I_Z	0	0	MX_G	0	0	MY_G	$-MX_G$	$-MY_G$	0	$-J_{XY}$	J_{XY}	0	I_Y-I_X	J_{XZ}	$-J_{YZ}$	(71c)

[a] The linear velocity times linear velocity values $(U_X^2, U_Y^2, U_Z^2, U_X U_Y, U_X U_Z,$ and $U_Z U_X)$ are all zero.

Table 22 Inertial tensors of added mass number

Item	Linear acceleration			Angular acceleration			Linear velocity times angular velocity									Eq. No.
	$\dot{V}_X$	$\dot{V}_Y$	$\dot{V}_Z$	$\dot{\Omega}_X$	$\dot{\Omega}_Y$	$\dot{\Omega}_Z$	$V_X\Omega_X$	$V_X\Omega_Y$	$V_X\Omega_Z$	$V_Y\Omega_X$	$V_Y\Omega_Y$	$V_Y\Omega_Z$	$V_Z\Omega_X$	$V_Z\Omega_Y$	$V_Z\Omega_Z$	
Force K_a	A_{11}	A_{12}	A_{13}	A_{14}	A_{15}	A_{16}	0	A_{13}	$-A_{12}$	0	A_{23}	$-A_{22}$	0	A_{33}	$-A_{23}$	(72a)
	A_{12}	A_{22}	A_{23}	A_{24}	A_{25}	A_{26}	$-A_{13}$	0	A_{11}	$-A_{23}$	0	A_{12}	$-A_{33}$	0	A_{13}	(72b)
	A_{13}	A_{23}	A_{33}	A_{34}	A_{35}	A_{36}	A_{12}	$-A_{11}$	0	A_{22}	$-A_{12}$	0	A_{23}	$-A_{13}$	0	(72c)
Moment H_a	A_{14}	A_{24}	A_{34}	A_{44}	A_{45}	A_{46}	$-A_{16}-A_{34}$	A_{16}	$-A_{15}$	A_{34}	$A_{26}+A_{35}$	A_{24}	$-A_{24}$	$A_{36}-A_{25}$	$-A_{36}-A_{26}$	(73a)
	A_{15}	A_{25}	A_{35}	A_{45}	A_{55}	A_{56}	$A_{15}+A_{24}$	$-A_{35}$	$A_{14}-A_{36}$	$A_{25}-A_{14}$	0	$A_{36}-A_{25}$	$A_{14}-A_{36}$	A_{15}	$A_{34}+A_{16}$	(73b)
	A_{16}	A_{26}	A_{36}	A_{46}	A_{56}	A_{66}	A_{26}	$A_{25}-A_{14}$	A_{26}	$A_{25}-A_{14}$	$-A_{24}-A_{15}$	$-A_{16}$	A_{35}	$-A_{34}$	0	(73c)

Item	Angular velocity times angular velocity						Linear velocity times velocity						Eq. No.
	Ω_X^2	Ω_Y^2	Ω_Z^2	$\Omega_X\Omega_Y$	$\Omega_Y\Omega_Z$	$\Omega_Z\Omega_X$	U_X^2	U_Y^2	U_Z^2	U_XU_Y	U_YU_Z	U_ZU_X	
Force K_a	0	A_{35}	$-A_{26}$	A_{34}	$A_{36}-A_{25}$	$-A_{24}$	0	0	0	0	0	0	(72a)
	$-A_{34}$	0	A_{16}	$-A_{35}$	A_{15}	$A_{14}-A_{36}$	0	0	0	0	0	0	(72b)
	A_{24}	$-A_{15}$	0	$A_{25}-A_{14}$	$-A_{16}$	A_{26}	0	0	0	0	0	0	(72c)
Moment H_a	0	A_{56}	$-A_{56}$	A_{46}	$A_{66}-A_{55}$	$-A_{45}$	0	A_{23}	$-A_{23}$	A_{13}	$A_{33}-A_{22}$	$-A_{12}$	(73a)
	$-A_{46}$	0	A_{46}	$-A_{56}$	A_{45}	$A_{44}-A_{66}$	$-A_{13}$	0	A_{13}	$-A_{23}$	A_{12}	$A_{11}-A_{33}$	(73b)
	A_{45}	$-A_{45}$	0	$A_{55}-A_{44}$	$-A_{46}$	A_{56}	A_{12}	$-A_{12}$	0	$A_{22}-A_{11}$	$-A_{13}$	A_{23}	(73c)

of small insects a strong circulation is generated at the pronation and supination phases, and thus a large lift is generated at high speed during the downstroke with increased maximum lift coefficient caused by dynamic stall.

The centripetal acceleration caused by the beating motion of a wing is very high, and thus, the centrifugal force is very effective in increasing the stiffness of the wing.

The mathematical formulations of aerodynamic and inertial forces and moments based on blade element theory are presented. The equations of body motion in inertial space, blade motion with respect to the body, and the structural dynamics of the wing coupled with the body motion in maneuvering flight are established in the sense of flight dynamics.

References

[1] Wagner, H., "Über die Entstehung des Dynamischen Auftriebes von Tragflubeln," *Zeitschrift für Angewandte Mathematik und Mechanik*, Vol. 5, 1925, pp. 17–35.

[2] Küssner, H. G., "General Airfoil Theory," NACA TM 979, 1941.

[3] Bisplinghoff, K. L., Ashley, H., and Halfman, R. L., *Aeroelasticity*, Addison-Wesley, Reading, MA, 1955.

[4] Theodrsen, T., "General Theory of Aerodynamic Instability and the Mechanism of Flutter," NACA Rept. 496, 1934.

[5] Theodrsen, T., and Garrick, I. E., "General Potential Theory of Arbitrary Wing Sections," NACA Rept. 452, 1933.

[6] von Kármán, T., and Sears, W. R., "Airfoil Theory for Nonuniform Motion," *Journal of Aeronautical Sciences*, Vol. 5, 1938, pp. 379–390.

[7] Sears, W. R., "Some Aspect of Non-Stationary Airfoil Theory and Its Practical Application," *Journal of Aeronautical Sciences*, Vol. 8, No. 3, 1941, pp. 104–108.

[8] Garrick, I. E., "Propulsion of Flapping and Oscillating Airfoil," NACA Rept. 567, 1936, pp. 419–472.

[9] Lighthill, M. I., "Large-Amplitude Elongated-Body Theory of Fish Locomotion," *Proceedings of the Royal Society of London, Series B*, No. 179, 1971, pp. 125–138.

[10] Wu, T. Y.-T., "Hydromechanics of Swimming Propulsive. Part 2: Some Optimum Shape Problems," *Journal of Fluid Mechanics*, Vol. 46, No. 3, 1971, pp. 521–544.

[11] Sato, M., "Aerodynamic Analysis of Wing Motion in Living Creatures," Ph.D. Thesis, Engineering Univ. of Tokyo, 1980 (in Japanese).

[12] Azuma, A., *The Biokinetics of Flying and Swimming*, Springer-Verlag, Tokyo, 1992.

[13] Batchelor, G. K., *An Introduction to Fluid Dynamics*, Cambridge Univ. Press, Cambridge, U.K., 1991.

[14] Faigrieve, J. D., and DeLaurier, J. D., "Propulsive Performance of Two-Dimensional Thin Airfoil Undergoing Large-Amplitude Pitch and Plunge Oscillations," Univ. of Toronto, Inst. of Aerospace Studies, UTIAS TN 226, Downsview, ON, Canada, 1982.

[15] Platzer, M. F., Neace, K. S., and Pang, C. K., "Aerodynamic Analysis of Flapping Wing Propulsion," AIAA Paper 93-0484, Jan. 1993.

[16] Jones, K. D., and Platzer, M. F., "Numerical Computations of Flapping Propulsion and Power Extraction," AIAA Paper 97-0026, Jan. 1997.

[17] Heyson, H. H., "A Note on the Mean Value of Induced Velocity for a Helicopter Rotor," NASA TN D-240, 1960.

[18] Heyson, H. H., "Equation for the Induced Velocities near a Lifting Rotor with Nonuniform Azimuthwise Vorticity Distribution," NASA TN D-394, 1960.

[19]Castles, W., Jr., and De Leeuw, J. H., "The Normal Component of the Induced Velocities in the Vicinity of a Lifting Rotor and Some Examples of Its Application," NACA Rept. 1184, 1954 (supersedes NACA TN 2912).

[20]Heyson, H. H., and Katzoff, S., "Induced Velocities near a Lifting Rotor with Nonuniform Disc Loading," NACA Rept. 1319, 1957 (supersedes NACA TN 3190 and 3691).

[21]Loewy, R. G., "Two-Dimensional Approximation to the Unsteady Aerodynamics of Rotary Wings," *Journal of Aeronautical Sciences*, Vol. 24, No. 82, 1959, pp. 81–92 and p. 144.

[22]Miller, R. H., "On the Computation of Airloads Acting on Rotor Blades in Forward Flight," *Journal of the American Helicopter Society*, Vol. 7, No. 2, 1962, pp. 56–66.

[23]Miller, R. H., "Rotor Blade Harmonic Air Loading," *AIAA Journal*, Vol. 2, No. 7, 1964, pp. 1254–1269.

[24]Weis-Fogh, T., "Quick Estimates of Flight Fitness in Hovering Animals Including Novel Mechanisms for Lift Production," *Journal of Experimental Biology*, Vol. 59, 1973, pp. 169–230.

[25]Wu, J. C., and Hu-Chen, H., "Unsteady Aerodynamics of Articulate Lifting Bodies," AIAA Paper 84-2184, Aug. 1984.

[26]Isogai, K., and Shinmoto, Y., "On Generation of Hovering Force by Mutual Interactions of Oscillating Tandem Airfoils," *Forum of 15th Symposium on the Comp. Fluid Dynamics of Aircraft*, NAL No. 37, 1999, pp. 191–203 (in Japanese).

[27]Liu, H., "Unsteady Solutions to the Incompressible Navier–Stokes Equations with the Pseudo-Compressibility Method," *Proceedings 1995 ASME/ISMI Fluids Engineering Annual Fluid Engineering Division*, edited by M. N. Dhaubhale, FED 215, 1995, pp. 105–121.

[28]Liu, H., "Computation of Unsteady Flow around a Rigid/Flexible Oscillating Body with the MUSCL Method," *Proceedings Int. Conf. on Comp. Eng. Sci. 1*, edited by S. N. Alturi, Springer-Verlag, 1995, pp. 817–832.

[29]Johnson, W., "The Effects of Dynamic Stall on the Response and Airloading of Helicopter Rotor," *Journal of the American Helicopter Society*, April 1969, pp. 68–79.

[30]Visbal, M. R., "Dynamic Stall of a Constant Rate Pitching Airfoil," *Journal of Aircraft*, Vol. 27, No. 5, May 1990, pp. 400–406.

[31]Ekaterinaris, J. A., "Numerical Investigation of Dynamic Stall of an Oscillating Wing," *AIAA Journal*, Vol. 33, No. 10, October 1995, pp. 1803–1808.

[32]Edwards, R. H., and Cheng, H. K., "The Separation Vortex in the Weis-Fogh Circulation-Generation Mechanism," *Journal of Fluid Mechanics*, Vol. 120, 1982, pp. 463–473.

[33]Weis-Fogh, T., "Flapping Flight and Power in Birds and Insects, Conventional and Novel Mechanism," *Swimming and Flying in Nature*, Vol. 2, edited by T. Y.-T Wu, C. J. Brokaw, and C. Brennen, New York, 1975, pp. 729–762.

[34]Lighthill, M. J., "On the Weis-Fogh Mechanism of Lift Generation," *Journal of Fluid Mechanics*, Vol. 60, 1973, pp. 1–17.

[35]Maxworthy, T., "The Fluid Dynamics of Insect Flight," *Annual Review of Fluid Mechanics*, Vol. 13, 1981, pp. 329–350.

[36]Ro, K., and Tsutahara, M., "Numerical Analysis of Unsteady Flow in the Weis-Fogh Mechanism by Discrete Vortex Method with GRAPE3A," Transactions of the ASMF, Vol. 119, March 1997, pp. 96–102.

[37]Tsutahara, M., and Kimura, T., "Aerodynamic Characteristics of the Weis-Fogh Mechanism (II) Numerical Computations by the Discrete Vortex Method," *Journal of the Japan Society of Aeronautical and Astronautical Sciences*, Vol. 35, No. 407, 1987, pp. 596–604 (in Japanese).

[38]Tsutahara, M., and Murakami, H., "Analysis of Flowfield in the Weis-Fogh Mechanism by the Three-Dimensional Discrete Vortex Method," *Journal of the Japan Society of Aeronautical and Astronautical Sciences*, Vol. 45, No. 523, 1997, pp. 19–27 (in Japanese).

[39]Sunada, S., Kawachi, K., Watanabe, I., and Azuma, A., "Fundamental Analysis of Three-Dimensional 'Near Fling'," *Journal of Experimental Biology*, Vol. 183, 1993, pp. 217–248.

[40]Sunada, S., Kawachi, K., Watanabe, I., and Azuma, A., "Performance of a Butterfly in Take-Off Flight," *Journal of Experimental Biology*, Vol. 183, 1993, pp. 249–277.

[41]Azuma, A., *"Encyclopedia of Locomotion in Living Creatures,"* Asakura, Tokyo, 1997 (in Japanese).

[42]Smith, M. J. C., Wilkin, P. J., and Williams, M. II., "The Advantage of an Unsteady Panel Method in Modelling the Aerodynamic Forces on Rigid Flapping Wings," *Journal of Experimental Biology*, Vol. 199, 1996, pp. 1073–1083.

[43]Wilkin, P. J., and Williams, M. H., "Comparison of the Aerodynamic Forces on a Flying Sphingied Moth with Those Predicted by Quasi-Steady Theory," *Journal of Physiology and Zoology*, Vol. 66, 1993, pp. 1015–1044.

[44]Liu, H., Ellington, C. P., Kawachi, K., Berg, C. V. D., and Willmott, A. P., "A Computational Fluid Dynamic Study of Hawkmoth Hovering," *Journal of Experimental Biology*, Vol. 201, pp. 461–477.

[45]Van Den Berg, C., and Ellington, C., "The Vortex Wake of a 'Hovering' Model Hawkmoth," *Philosophical Transactions of the Royal Society of London*, Series B, Vol. 352, 1997, pp. 317–328.

[46]Maxworthy, T., "Experiments on the Weis-Fogh Mechnism of Lift Generation by Insects in Hovering Flight. Part 1. Dynamics of the 'Fling'," *Journal of Fluid Mechanics*, Vol. 93, 1979, pp. 47–63.

[47]Okamoto, M., Yasuda, K., and Azuma, A., "Aerodynamic Characteristics of the Wings and Body of a Dragonfly," *Journal of Experimental Biology*, Vol. 199, 1996, pp. 281–294.

[48]Guglielmo, J. J., and Selig, M. S., "Spanwise Variations in Profile Drag for Airfoils at Low Reynolds Number," *Journal of Aircraft*, Vol. 33, No. 4, July–Aug. 1996, pp. 699–707.

[49]Newman, B. G., Savage, S. B., and Schouella, D., "Model Tests on a Wing Section of an Aeschna Draognfly," *Scale Effect in Animal Locomotion*, edited by T. J., Pedley, Academic Press, London, 1997, pp. 445–477.

[50]Azuma, A., and Watanabe, T., "Flight Performance of a Dragonfly," *Journal of Experimental Biology*, Vol. 137, 1988, pp. 221–252.

[51]Cloupeau, M., Devilliers, J. F., and Devezeaux, D., "Direct Measurements of Instantaneous Lift in Desert Locust Comparison with Jensen's Experiments and Detached Wings," *Journal of Experimental Biology*, Vol. 80, 1979, pp. 1–15.

[52]Jensen, M., "Biology and Physics of Locust Flight. III. The Aerodynamics of Locust Flight," *Philosophical Transactions of the Royal Society of London*, Series B, Vol. 239, 1956, pp. 511–552.

[53]Vogel, S., "Flight in Drosophila. III. Aerodynamic Characteristics of Fly Wings and Wing Models," *Journal of Experimental Biology*, Vol. 46, 1967, pp. 431–443.

[54]Nachtigall, W., "Die Aerodynamische Polare des Tipula-Flugels und cine Einrichtung zur Malbautomatischen Polarenaufnahme," *The Physiology of Movement Biomechanics*, edited by W. Nachtigall, Fischer, Stuttgart, Germany, 1977, pp. 347–352.

[55]Thom, A., and Swart, P., "The Forces on an Aerofoil at Very Low Speeds," *Royal Aeronautical Society*, Vol. 44, 1940, pp. 761–770.

[56]Ham, N. D., "An Experimental Investigation of Stall Flutter on a Model Helicopter Rotor in Forward Flight," Aeroelastic and Structural Research Lab., MIT TN 86-1, March 1960.

[57]Liiva, J., and Davenport, F. J., "Dynamic Stall of Airfoil Sections for High Speed Rotors," *Journal of the American Helicopter Society*, Vol. 14, No. 2, April 1969.

[58]Gangwani, S. T., "Synthesized Airfoil Data Method for Prediction of Dynamic Stall and Unsteady Airloads," *Vertica*, Vol. 8, No. 2, 1984, pp. 93–118.

[59]Favier, D., Maresca, C., and Rebont, J., "Dynamic Stall due to Fluctuations of Velocity and Incidence," *AIAA Journal*, Vol. 20, No. 7, 1982, pp. 865–871.

[60]Niven, A. J., and McD. Galbraith, R. A., "The Effect of Pitch Rate on the Dynamic Stall of a Modified NACA 23012 Airfoil and Comparison with the Unmodified Case," *Vertica*, Vol. 11, No. 4, 1987, pp. 751–759.

[61]Tang, D. M., and Dowell, E. H., "Experimental Investigation of Three-Dimensional Dynamic Stall Model Oscillating Mechanisms," *Journal of Aircraft*, Vol. 32, No. 5, Sept.–Oct. 1995, pp. 1062–1071.

[62]Carr, L. W., McAlister, K. W., and McCroskey, W. J., "Analysis of the Development of Dynamic Stall Based on Oscillating Airfoil Experiments," NASA TN D 8382, 1977.

[63]Halfman, R. M., Johnson, H. C., and Haley, S. M., "Evaluation of High-Angle-of-Attack Aerodynamic-Derivation Data and Stall-Flutter Prediction Techniques," NACA TN 2533, 1951.

[64]Rainey, G. A., "Measurement of Aerodynamic Forces for Various Mean Angles of Attack on an Airfoil Oscillating in Pitch and on Two Finite-Span Wings Oscillating in Bending with Emphasis on Damping in the Stall," Langley Aero. Lab., NACA TN 3642, 1956.

[65]Carta, F. O., and Ham, N. D., "An Analysis of the Stall Flutter Instability of Helicopter Rotor Blades," *American Helicopter Society 23rd Annual National Forum Proceedings*, Vol. 130, May 1967.

[66]Karim, M. A., and Acharya, M., "Suppression of Dynamic-Stall Vortices over Pitching Airfoils by Leading-Edge Suction," *AIAA Journal*, Vol. 32, No. 8, Aug. 1994, pp. 1647–1655.

[67]Freymuth, P., "Propulsive Vertical Signature of Plunging and Pitching Airfoil," *AIAA Journal*, Vol. 26, 1988, pp. 881–883.

[68]Freymuth, P., "Thrust Generation by an Airfoil in Hover Modes," *Experiments in Fluids*, Vol. 9, 1990, pp. 17–24.

[69]Maresca, C., Favier, D., and Rebont J., "Experiments on an Aerofoil at High Angle of Incidence in Longitudinal Oscillations," *Journal of Fluid Mechanics*, Vol. 92, No. 4, June 1979, pp. 671–690.

[70]Favier, D., Agnes, A., Barbi, C., and Maresca, C., "Combined Translation/Pitch Motion: A New Aircraft Dynamic Stall Simulation," *Journal of Aircraft*, Vol. 25, No. 9, Sept. 1988, pp. 805–814.

[71]Freymuth, P., "The Vortex Patterns of Dynamic Separation: A Parametric and Comparative Study," *Progress in Aerospace Sciences*, Vol. 22, No. 3, 1985, pp. 161–208.

[72]Johnson, W., "The Effect of Dynamic Stall on the Response and Airloading of Helicopter Rotor Blades," *Journal of the American Helicopter Society*, Vol. 14, No. 2, April 1969, pp. 68–79.

[73]Dickinson, M. H., "The Effects of Wing Rotation on Unsteady Aerodynamic Performance at Low Reynolds Number," *Journal of Experimental Biology*, Vol. 192, 1994, pp. 179–206.

[74]Dickinson, M. H., Lehmann, F. O., and Sane, S. P., "Wing Rotation and the Aerodynamic Basis of Insect Flight," *Science*, Vol. 284, 18 June 1999, pp. 1954–1960.

[75]Currier, J. M., and Fung, K. Y, "Analysis of the Onset of Dynamics Stall," *AIAA Journal*, Vol. 30, No. 10, Oct. 1992, pp. 2469–2477.

[76]Shih, C., Lourenco, L., Van Dommelen, L., and Krothapalli, A., "Unsteady Flow Past an Airfoil Pitching at a Constant Rate, "*AIAA Journal*, Vol. 30, No. 5, May 1992, pp. 1153–1161.

[77]Spedding, G. R., and Maxworthy, T., "The Generation of Circulation and Lift in a Rigid Two-Dimensional Fling," *Journal of Fluid Mechanics*, Vol. 165, 1986, pp. 247–272.

[78]Grodnitsky, D. L., and Morozov, P. P., "Vortex Formation during Tethered Flight of Functionally and Morphologically Two-Winged Insects, Including Evolutionary Considerations on Insect Flight,"*Journal of Experimental Biology*, Vol. 182, 1993, pp. 11–40.

[79]Norberg, R. A., "Hovering Flight of the Dragonfly *Aschna junces* L., Kinematics and Aerodynamics," *Swimming and Flying in Nature*, edited by T. Y.-T. Wu, C. Brokaw, and C. Brennen, Plenum, New York, 1975, pp. 763–781.

[80]Savage, S. B., "The Role of Vortices and Unsteady Effects during the Hovering Flight of Dragonflies," *Journal of Experimental Biology*, Vol. 83, 1979, pp. 59–77.

[81]Ellington, C. P., "The Aerodynamics of Hovering Insect Flight. (a) I. The Quasi-Steady Analysis; (b) II. Morphological Parameters; (c) III. Kinematics; (d) IV. Aerodynamic Mechanisms; (e) V. A Vortex Theory; (f) VI. Lift and Power Requirement," *Philosophical Transactions of the Royal Society of London*, Series B, Vol. 35, No. 1122, 1984.

[82]Somps, C., and Luttges, M., "Dragonfly Flight: Novel Uses of Unsteady Separated Flows," *Science*, Vol. 228, 14 June 1985, pp. 1326–1329.

[83]Ennos, A. R., "The Kinematics and Aerodynamics of the Free Flight of Some Diptera," *Journal of Experimental Biology*, Vol. 142, 1989, pp. 49–85.

[84]Ennos, A. R., "Inertial and Aerodynamic Torques on the Wings of Diptera in Flight," *Journal of Experimental Biology*, Vol. 142, 1989, pp. 87–95.

[85]Dudley, R., and Ellington, C. P., "Mechanics of Forward Flight in Bumblebees. II. Quasi-Steady Lift and Power Requirements," *Journal of Experimental Biology*, Vol. 148, 1990, pp. 87–95.

[86]Dudley, R., "Extraordinary Flight Performance of Orchid Bees (*Apidae; Euglossini*) Hovering in Heliox ($80\%He/20\%O_2$)," *Journal of Experimental Biology*, Vol. 198, 1995, pp. 71–91.

[87]Dudley, R., "Unsteady Aerodynamics," *Science*, Vol. 284, No. 5422, 18 June 1999, pp. 1937–1939.

[88]Willmott, A. P., and Ellington, C. P., "Measuring the Angle of Attack of Beating Insect Wings: Robust Three-Dimensional Reconstruction from Two-Dimensional Images," *Journal of Experimental Biology*, Vol. 200, 1997, pp. 2693–2704.

[89]Willmott, A. P., and Ellington, C. P., "The Mechanics of Flight in the Hawkmoth *Manduca sexta*. I. Kinematics of Hovering and Forward Flight," *Journal of Experimental Biology*, Vol. 200, 1997, pp. 2693–2704.

[90]Willmott, A. P., and Ellington, C. P., "The Mechanics of Flight in the Hawkmoth *Manduca sexta*. II. Aerodynamic Cosequences of Kinematic and Morphological Variation," *Journal of Experimental Biology*, Vol. 200, 1997, pp. 2723–2745.

[91]Saharon, D., and Luttges, M. W., "Visualization of Unsteady Separated Flow Produced by Mechanically Driven Dragonfly Wing Kinematics Model," AIAA Paper 88-0569, 1988.

[92]Zanker, J. M., and Götz, K., "The Wing Beat of *Drosophila melanogaster*. II. Dynamics," *Philosophical Transactions of the Royal Society of London*, Series B, Vol. 327, 1990, pp. 19–44.

[93]Dickinson, M. H., and Götz, K. C. "Unsteady Aerodynamic Performance of Model Wings at Low Reynolds Numbers," *Journal of Experimental Biology*, Vol. 174, 1993, pp. 45–64.

[94]Savage, S. B., Newman. B. C., and Wong, D. T. M., "The Role of Vortices and Unsteady Effects during the Hovering Flight of Dragonflies," *Journal of Experimental Biology*, Vol. 83, 1979, pp. 59–77.

[95]Wakeling, J. M., and Ellington, C. P., "Dragonfly Flight. I. Gliding Flight and Steady State Aerodynamic Forces," *Journal of Experimental Biology*, Vol. 200, 1997, pp. 543–556.

[96]Wakeling, J. M., and Ellington, C. P., "Dragonfly Flight. II Velocities Accelerations and Kinematics of Flapping Flight," *Journal of Experimental Biology*, Vol. 200, 1997, pp. 557–582.

[97]Wakeling, J. M., and Ellington, C. P., "Dragonfly Flight. III. Lift and Power Requirements," *Journal of Experimental Biology*, Vol. 200, 1997, pp. 583–600.

[98]Ellington, C. P., Berg, C. V. D., Willmott, A. P., and Thomas, A. L. R., "Leading Edge Vortices in Insect Flight," *Nature*, Vol. 384, No. 1, Dec. 1996, pp. 626–630.

[99]Reavis, M., and Luttges, M. W., "Aerodynamic Forces Produced by a Dragonfly," AIAA Paper 88-0030, Jan. 1988.

[100]Azuma, A., Azuma, S., Watanabe, I., and Furuta, T., "Flight Mechanics of a Dragonfly," *Journal of Experimental Biology*, Vol. 116, 1985, pp. 79–107.

[101]Sato, M., and Azuma, A., "The Flight Performance of a Damselfly; *Cerigraion melanurum* Selys," *Journal of Experimental Biology*, Vol. 200, 1997, pp. 1765–1779.

[102]Rüppell, C., "Kinematic Analysis of Symmetrical Flight Manoeuvrcs of Odonata," *Journal of Experimental Biology*, Vol. 144, 1989, pp. 13–42.

[103]Azuma, A., "A View on the Ways of Locomotion and Their Biomechnical Characteristics," *1st International Symposium on Aqua Bio-Mechanisms*, edited by M. Nagai, Tokai Univ. Pacific Central, Honolulu, HI, 28–30 Aug. 2000.

[104]Ito, S., and Azuma, A., "Mechanical Analysis of Paddling Concerning Turtle's Locomotion," *1st International Symposium on Aqua Bio-Mechanisms*, edited by M. Nagai, Tokai Univ. Pacific Central, Honolulu, HI, 28–30 Aug. 2000.

[105]Okamoto, M., "The Aerodynamic Characteristics of Airfoil for Small Model Airplane," *The 1st Sky Sport Symposium*, 2B6, Japan Society of Aeronautical and Space Sciences, 2–3 Dec. 1995, Tokyo (in Japanese).

[106]Azuma, A., and Okuno, Y., "Flight of Samara, *Alsomitra macrocarpa*," *Journal of Theoritical Biology*, Vol. 129, 1987, pp. 263–274.

[107]Rees, C. J. C., "Form and Function in Corrugated Insect Wings," *Nature*, Vol. 256, 1975, pp. 200–203.

[108]Rees, C. J. C., "Aerodynamic Properties of an Insect Wing Section and a Smooth Aerofoil Compared," *Nature*, Vol. 258, 1975, pp. 141–142.

[109]Yasuda, K., and Azuma, A., "The Autorotation Boundary in the Flight of Samaras," *Journal of Theoretical Biology*, Vol. 185, 1997, pp. 313–320.

[110]Osborne, M. F. M., "Aerodynamics of Flapping Flight with Application to Insects," *Journal of Experimental Biology*, Vol. 28, 1951, pp. 221–245.

[111]Pitt, D. M., and Peters, D. A., "Rotor Dynamic Inflow Derivatives and Time Constants from Various Inflow Models," *9th European Rotorcraft Forum*, Stresa, Italy, 13–15 Sept. 1983.

[112]Gaonkar, G. H., and Peters, D. A., "Effectiveness of Current Dynamic Inflow Models in Hover and Forward Flight," *Journal of the American Helicopter Society*, Vol. 31, No. 2, April 1986, pp. 47–57.

[113]Landweber, L., "On a Generalization of Taylor's Virtual Mass Relation for Rankine Bodies," *Quarterly Journal of Applied Mathematics*, Vol. 14, No. 1, 1956, pp. 51–56.

[114]Lock, C. N. H., "The Application of Goldstein's Theory to the Practical Design of Airscrews," British A.R.C. R&M 1377, 1931.

[115]Moriya, T., "On the Integration of Biot–Savart's Law in Propeller Theory," *Journal of the Japan Society of Aeronautical and Space Sciences*, Vol. 9, No. 89, 1942, pp. 1051–1020.

[116]Lamb, H., *Hydrodynamics*, 5th ed., Cambridge Univ. Press, Cambridge, UK, 1930, p. 218.

[117]Heyson, H. H., "An Evaluation of Linearized Vortex Theory as Applied to Single and Multiple Rotors Hovering in and out of Ground Effect," NASA TN D-43, 1959.

[118]Heyson, H. H., "Measurements of the Time-Averaged and Instantaneous Induced Velocities in the Wake of a Helicopter Hovering at High Tip Speeds," NASA TN D-393, 1960.

[119]Heyson, H. H., "Tables and Charts of the Normal Component of Induced Velocity in the Lateral Plane of Rotor with Harmonic Azimuthwise Vorticity Distribution," NASA TN D-809, 1961.

[120]Gupta, B. P., and Loewy, R. G., "Theoretical Analysis of the Aerodynamic Stability of Mutiple Interdigitated Helical Vortices," *AIAA Journal*, Vol. 12, No. 10, 1974, pp. 1381–1387.

[121]Gupta, B. P., and Lessen, M., "Hydrodynamic Stability of the Far Wake of a Hovering Rotor," *AIAA Journal*, Vol. 13, No. 6, 1975, pp. 766–769.

[122]Pouradier, J. M., and Horowitz, E., "Aerodynamic Study of a Hovering Rotor," *Vertica*, Vol. 5, 1981, pp. 301–315.

[123]Rand, O., and Rosen, A., "Efficient Method for Calculating the Axial Velocities Induced along Rotation Blades by Trailing Helical Vortices," *Journal of Aircraft*, Vol. 21, No. 6, 1984, pp. 433–435.

[124]Greenberg, M. D., and Powers, S. R., "Nonlinear Actuator Disk Theory and Flow Field Calculation Including Nonuniform Loading," NASA CR 1672, Sept. 1970.

[125]Landgrebe, A. J., "An Analytical and Experimental Investigation of Helicopter Rotor Hover Performance and Wake Geometry Characteristics," USAAMRDL Tech. Rept. 71-24, 1971.

[126]Widnall, S. E., "The Stability of a Helical Vortex Filament," *Journal of Fluid Mechanics*, Vol. 54, No. 4, 1972, pp. 641–663.

[127]Tangler, J. L., Wohlfeld, R. M., and Miley, S. J. Y., "An Experimental Investigation of Vortex Stability Tip Shapes, Compressibility and Noise for Hovering Model Rotors," NASA CR 2305, 1973.

[128]Cone, C. D., Jr., "A Theoretical Investigation of Vortex-Sheet Deformation behind a Highly Loaded Wing and Its Effect on Lift," NASA TN D 657, 1961.

[129]Levinsky, E. S., and Strand, T., "A Method for Calculating Helicopter Vortex Paths and Wake Velocities," AFFDL-TR-69-113, AD 710694, 1970.

[130]Azuma, A., and Kawachi, K., "Local Momentum Theory and Its Application to the Rotary Wing," *Journal of Aircraft*, Vol. 16, No. 1, 1979, pp. 6–14.

[131]Azuma, A., Nasu, K, and Hayashi, T., "An Extension of the Local Momentum Theory to the Rotors Operating in Twisted Flow Field," *Vertica*, Vol. 7, No. 1, 1983, pp. 45–59.

[132]Azuma, A., and Saito, S., "Study of Rotor Gust Response by Mean of the Local Momentum Theory," *Journal of the American Helicopter Society*, Jan. 1982, pp. 58–72.

[133]Saito, S., and Azuma, A., "A Numerical Approach to Co-Axial Rotor Aerodynamics," *Vertica*, Vol. 6, No. 4, 1982, pp. 253–266.

[134]Nasu, K., and Azuma, A., "An Experimental Verification of the Local Circulation Method for a Horizontal Axis Wind Turbine," *Research on Natural Energy*, Vol. 8, No. 8, The Ministry of Education, Science and Culture, Japan, March 1984, pp. 245–252.

[135] Azuma, A., Saito, S., and Kawachi, K., "Response of a Helicopter Penetrating the Tip Vortices of a Large Airplane," *Vertica*, Vol. 11, No. 1/2, 1987, pp. 65–76.

[136] Azuma, A., Furuta, T., Iuchi, M., and Watanabe, I., "Hydrodynamic Analysis of the Sweeping of "Ro"—An Oriental Scull," *Journal of Ship Research*, Vol. 33, No. 1, March 1989, pp. 47–62.

[137] Azuma, A., and Yasuda, K., "Flight Performance of Rotary Seeds," *Journal of Theoretical Biology*, Vol. 138, 1989, pp. 23–54.

[138] Newman, D. J. S., and Wootton, R. J., "An Approach to the Mechanics of Bleating in Dragonfly Wings," *Journal of Experimental Biology*, Vol. 125, 1986, pp. 361–372.

[139] Kreyszig, E., *Advanced Engineering Mathematics*, 4th ed., Wiley, New York, 1979, p. 95.

Chapter 18

A Nonlinear Aeroelastic Model for the Study of Flapping Wing Flight

Rambod F. Larijani* and James D. DeLaurier[†]

University of Toronto, Downsview, Ontario, Canada

Nomenclature

A	= element cross-sectional area
AR	= aspect ratio
a	= mass-proportional damping constant
b	= stiffness-proportional damping constant
C_d	= drag coefficient
C_{df}	= skin-friction drag coefficient
C_{mac}	= airfoil moment coefficient about its aerodynamic center
C_n	= normal force coefficient
c	= wing segment chord length
D_c	= drag due to camber
D_f	= friction drag
E	= modulus of elasticity
F_y	= total chordwise force
$F'(k), G'(k)$	= terms for modified Theodorsen function
G	= shear modulus of elasticity
g_r	= acceleration due to gravity
h	= total plunging displacement
$\tilde{h}$	= elastic component of plunging displacement
h_0	= imposed displacement
I	= moment of inertia
J	= polar moment of inertia
k	= reduced frequency based on $\frac{1}{2}c$
L	= total lift
M	= total twisting moment acting on a wing segment

*Ph.D. Student, Institute for Aerospace Studies.

[†]Professor, Institute for Aerospace Studies. Associate Fellow AIAA.

m	= distributed moment per unit length
$\overline{m}$	= mass per unit length
N	= total normal force acting on a wing segment
n	= distributed normal force per unit length
P_1, P_2	= element boundary conditions for torsion
p	= contact pressure
$Q_1, \ldots, Q_4$	= element boundary conditions for bending
R	= total thrust
T_s	= leading edge suction force
U	= freestream velocity
V	= relative velocity at $\frac{1}{4}$-chord location
V_y	= relative velocity tangential to a wing segment
v_r	= relative velocity of upper vs lower surface of the wing
x	= distance from flapping axis to middle of segment
y	= distance from the leading edge

Greek

α	= relative angle of attack at $\frac{3}{4}$-chord point due to wing segment motion
α'	= the flow's relative angle of attack at $\frac{3}{4}$-chord point
α_0	= wing segment's angle of zero-lift line
α_{stall}	= segment stall angle
Γ_0	= magnitude of flapping dihedral angle
Δx	= length of an element
$\Delta \overline{h}$	= nonflapping plunging displacement
$\Delta \overline{\theta}$	= nonflapping elastic twist
η_s	= leading-edge suction efficiency
θ	= total segment twist angle with respect to U
$\tilde{\theta}$	= elastic twist angle
$\overline{\theta}_a$	= angle of flapping axis with respect to U
$\overline{\theta}_{\text{wash}}$	= built-in pretwist
ρ	= atmospheric density
ζ	= decay constant
ψ	= transverse twist angle (along local y axis)
ω	= flapping frequency

Subscripts

a	= apparent mass
ac	= aerodynamic center
aero	= aerodynamic
ave	= average
c	= circulatory
cf	= crossflow
cs	= center section
con	= contact
damp	= damping
dsr	= shear rate dependent damping

ea	= elastic axis
ef	= effective
f	= friction
fr	= fabric and rib
inertia	= inertial
rs	= rigid section
s	= spar
sep	= separated flow
sr	= super rib
st	= static
te	= trailing edge

Superscripts

-	= mean value
$n, n + 1$	= time level
·	= time derivative

I. Introduction

I N SEPTEMBER 1991 an engine-powered remotely piloted ornithopter flew successfully for 2 min 46 s. This quarter-scale proof-of-concept model (Fig. 1) is described in Ref. 1. A second ornithopter model was built for the Canada Pavilion at Expo 1992 in Seville, Spain. This aircraft, appropriately named "Expothopter," was similar to the quarter-scale model in that it was completely flightworthy, but owing to the success of the quarter-scale model it was not necessary to fly the Expothopter.

Using the knowledge gained during the construction and testing of the quarter-scale model and Expothopter, a full-scale engine-powered human-carrying ornithopter (Fig. 2) was designed and built at the University of Toronto's Institute for Aerospace Studies. Details of the design, construction, and initial testing are described in Ref. 2. Results from the initial static-flapping (no forward speed) tests as well as some of the low-speed taxi tests are presented by Mehler.[3] The 1997 and 1998 taxi tests and the main bulk of the experimental data are discussed in detail by Fenton.[4]

An S1020 airfoil used in the quarter-scale model's wing was designed by Professor Michael Selig of the University of Illinois. This thick airfoil provides very high leading-edge suction efficiency as well as considerable structural depth. The full-scale ornithopter's wing also incorporates the S1020 airfoil. The inner portion of the wing uses this airfoil up to the "knuckle" (Fig. 4), and the outer tapered portion linearly transforms the S1020 to a Selig and Donovan SD8020 symmetrical airfoil at the tip. Figure 3 shows the airfoil shapes along the span of the full-scale ornithopter's wing. A rigid section, which is also known as the "SuperBox" (Fig. 4), is a closed structure made up of thin composite panels, internal ribs, a D-nose spar, and a rear shear web. Its function is to support the wing and transfer the loads to the outrigger struts. The outer rib of the SuperBox is referred to as the "SuperRib." The rest of the wing consists of 20 full-length ribs, made with a foam core and capped with basswood strips. The wing is covered with lightweight polyester fabric.

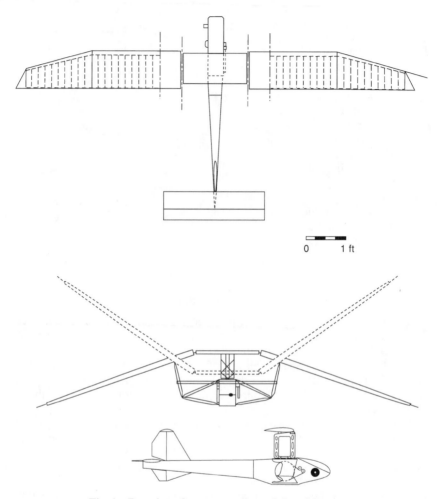

Fig. 1   Drawing of quarter-scale model ornithopter.

The center section is made up of composite panels, fore and aft shear webs, and nine internal ribs. The aerodynamic shape is created with a hollow blue-foam D-nose and nine half-length ribs. The cross section of the full-scale ornithopter wing spar (Fig. 5) is similar to that for the quarter-scale model. It is comprised of a carbon-fiber reinforced shear web for bending stiffness and strength, and a Kevlar® D-nose shell over a foam core for a structure of specified torsional compliance.[5]

A special feature of this wing's structure is that it is able to provide the required torsional compliance while at the same time incorporating the efficient and thick S1020 airfoil. Figure 6 shows how this is accomplished, where the closed torsion box normally formed by the thick airfoil is opened by splitting the trailing edge. This feature, patented as the "Shearflex Principle," allows a double-surface wing to have the high torsional compliance of two single-surface wings joined at the

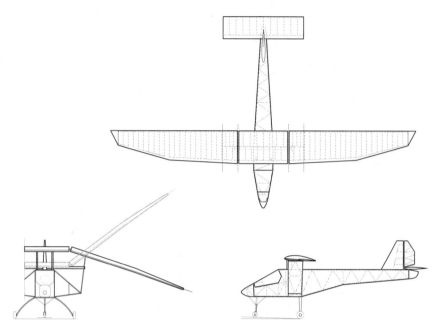

Fig. 2   Drawing of full-scale ornithopter.

leading edge. Therefore the ornithopter wing is able to twist freely even though its lightweight covering does not stretch. In fact, shearflexing would work well even if the skin were thick and relatively inflexible.

Analytical work done for the initial stages of the ornithopter project was based on a program called "Fullwing," which was developed to predict the performance of a flapping wing in steady flight.[6] Fullwing has gone through several revisions and it was most recently modified by the first author to improve its accuracy.

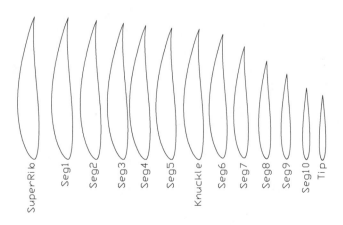

Fig. 3   Airfoils along the span.

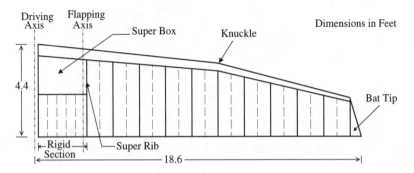

Fig. 4   Top view of ornithopter wing.

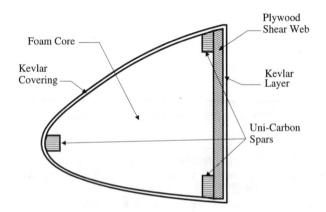

Fig. 5   Wing spar cross section.

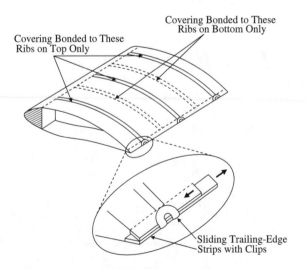

Fig. 6   Shearflexing principle.

The most significant assumption made in Fullwing is that of fully attached flow to linearize the equations. This gives inaccurate results as stalling dominates in the low forward-speed, high flapping-frequency regime, which became apparent during the static tests in 1996. It was noted that the phase angle between pitching and plunging was nowhere near the design value of approximately -90 deg. A "full-stall" formulation should give a more accurate representation of the problem. A second assumption is that the response has a simple harmonic motion. Although this is a reasonable description of the actual response, the presence of nonharmonic motion cannot be predicted. Also, its use of structural influence coefficients instead of a stiffness formulation has the disadvantage that for every new geometry the coefficients have to be rederived. An algorithm that uses transformation matrices can handle various geometric configurations more easily.

II.   Structural Analysis

This section presents the linear ordinary differential equations governing the flapping motion of a wing and describes the matrices and vectors associated with them. The two principal modes of motion for a flapping wing are *bending* and *torsion*. A finite element discretization breaks the wing into spar elements with bending and torsional degrees of freedom and fabric and rib elements that have torsional degrees of freedom only (Fig. 7). For bending, the computational domain is divided into a set of elements, with each having two nodes (Fig. 8). A single element is then isolated and the Galerkin method[7] is applied, using a set of Hermite cubic interpolation functions, to derive the finite element formulation of the problem.

For torsion, the domain is also divided into a set of elements, with each element having two nodes as shown in Fig. 9. The Galerkin method is applied using a set of linear interpolation functions.

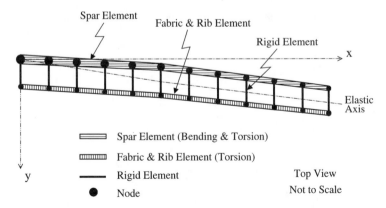

Fig. 7   Finite element discretization of ornithopter wing.

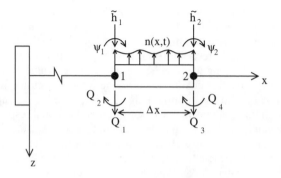

Fig. 8 Bending element.

The system dynamic equilibrium equations, including damping and neglecting external reactions, which are accounted for by boundary conditions, are

$$[K]\{y\} + [D]\{\dot{y}\} + [M]\{\ddot{y}\} = \{F\} \tag{1}$$

The stiffness matrix for an element having both bending and torsional degrees of freedom is given by

$$[K] = \begin{bmatrix} \frac{12EI}{\Delta x^3} & 0 & \frac{-6EI}{\Delta x^2} & \frac{-12EI}{\Delta x^3} & 0 & \frac{-6EI}{\Delta x^2} \\ 0 & \frac{GJ}{\Delta x} & 0 & 0 & \frac{-GJ}{\Delta x} & 0 \\ \frac{-6EI}{\Delta x^2} & 0 & \frac{4EI}{\Delta x} & \frac{6EI}{\Delta x^2} & 0 & \frac{2EI}{\Delta x} \\ \frac{-12EI}{\Delta x^3} & 0 & \frac{6EI}{\Delta x^2} & \frac{12EI}{\Delta x^3} & 0 & \frac{6EI}{\Delta x^2} \\ 0 & \frac{-GJ}{\Delta x} & 0 & 0 & \frac{GJ}{\Delta x} & 0 \\ \frac{-6EI}{\Delta x^2} & 0 & \frac{2EI}{\Delta x} & \frac{6EI}{\Delta x^2} & 0 & \frac{4EI}{\Delta x} \end{bmatrix} \tag{2}$$

where EI is the bending stiffness parameter, GJ is the corresponding torsional stiffness parameter for an element, and Δx is the length of a particular element. The overall consistent mass matrix for an element with total mass per unit length $\overline{m}$ is

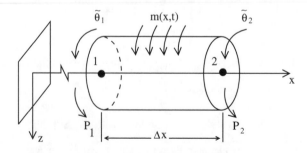

Fig. 9 Torsional element.

given by

$$[\mathbf{M}] = \overline{m}\,\Delta x \begin{bmatrix} \frac{156}{420} & 0 & \frac{-22\Delta x}{420} & \frac{54}{420} & 0 & \frac{13\Delta x}{420} \\ 0 & \frac{J}{3A} & 0 & 0 & \frac{J}{6A} & 0 \\ \frac{-22\Delta x}{420} & 0 & \frac{4\Delta x^2}{420} & \frac{-13\Delta x}{420} & 0 & \frac{-3\Delta x^2}{420} \\ \frac{54}{420} & 0 & \frac{-13\Delta x}{420} & \frac{156}{420} & 0 & \frac{22\Delta x}{420} \\ 0 & \frac{J}{6A} & 0 & 0 & \frac{J}{3A} & 0 \\ \frac{13\Delta x}{420} & 0 & \frac{-3\Delta x^2}{420} & \frac{22\Delta x}{420} & 0 & \frac{4\Delta x^2}{420} \end{bmatrix} \qquad (3)$$

Assuming that the normal force and moment are constant over an element [i.e., $n(x, t) = n(t)$ and $m(x, t) = m(t)$] the force vector and the vector of independent variables are given by

$$\{\mathbf{F}\} = \begin{Bmatrix} -\frac{N(t)}{2} \\ \frac{M(t)}{2} \\ \frac{N(t)\Delta x}{12} \\ -\frac{N(t)}{2} \\ \frac{M(t)}{2} \\ -\frac{N(t)\Delta x}{12} \end{Bmatrix}, \qquad \{\mathbf{y}\} = \begin{Bmatrix} \tilde{h}_1 \\ \tilde{\theta}_1 \\ \psi_1 \\ \tilde{h}_2 \\ \tilde{\theta}_2 \\ \psi_2 \end{Bmatrix} \qquad (4)$$

where $N(t) = n(t) \times \Delta x$ and $M(t) = m(t) \times \Delta x$. The aerodynamic forces and moments associated with a flapping wing in steady flight are nonlinear functions of the twisting and plunging deflections of the wing and their first and second derivatives. Equation (1) is a second-order nonlinear equation that is solved by using a Taylor series expansion to approximate the nonlinear components of the force vector. Several time-marching methods were considered and the nonlinear Newmark method was chosen because of its stability and ease of use. The nonlinear Newmark method is described in detail in Ref. 8. In the present study, since a temporal approximation has been used to obtain a set of linear second-order equations, an iteration must be performed at each time step to ensure that the equilibrium equations are satisfied. This iteration procedure is outlined by Owen.[9] The performance of the Newmark algorithm has been studied extensively and it is known to be unconditionally stable.[10]

III. Aerodynamic and Inertial Forces and Moments

The unsteady aerodynamic model for the study of a flapping wing is based on a modified strip theory approach. Vortex wake effects are accounted for as well as partial leading-edge suction and poststall behavior along with sectional mean angle of attack, camber, and friction drag. This model is then used for the calculation of average lift and thrust, power required, and propulsive efficiency of a flapping wing in equilibrium flight. A detailed treatment of this is given in Refs. 11 and 12.

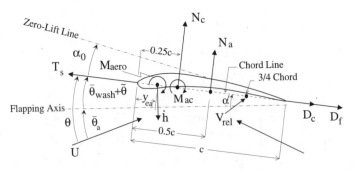

Fig. 10 Wing section aerodynamic forces and motion variables.

The starting point for the aerodynamic analysis is to determine the normal force N and moment M in Eq. (4) acting on a wing segment. For any element, M and N consist of aerodynamic and inertial terms:

$$N = N_{\text{aero}} + N_{\text{inertia}} \tag{5}$$

$$M = M_{\text{aero}} + M_{\text{inertia}} + M_{\text{damp}} \tag{6}$$

A. Aerodynamic Forces and Moments

The aerodynamic forces and moments are introduced in this section and the damping moment is treated separately in the next section along with the spar's structural damping. Figure 10 shows the aerodynamic forces on a representative segment (element) of the wing. Aerodynamic loads depend on whether the flow over a segment is attached or stalled.

1. Attached Flow

In the attached-flow regime the aerodynamic normal force is given by

$$N_{\text{aero}} = N_c + N_a \tag{7}$$

where N_c is the circulatory normal force and N_a is the apparent-mass normal force. The circulatory normal force on the wing segment is given by

$$N_c = \tfrac{1}{2}\rho U V C_n c \Delta x \tag{8}$$

where U is freestream air speed, c is the chord length of a particular segment, and Δx is the segment's length. V is the flow velocity at the $\frac{1}{4}$-chord location. Using small-angle assumptions, the normal force coefficient is shown to be[11]

$$C_n = 2\pi(\alpha' + \alpha_0 + \bar\theta_a + \bar\theta_{\text{wash}}) \tag{9}$$

where α_0 is the zero-lift angle, $\bar\theta_a$ is the flapping axis angle of attack, and $\bar\theta_{\text{wash}}$ is the built-in pretwist of the wing. α' is the flow's relative angle of attack at the $\frac{3}{4}$-chord point and is given by[11]

$$\alpha' = \frac{AR}{2+AR}\left[F'(k)\,\alpha + \frac{c}{2U}\frac{G'(k)}{k}\,\dot\alpha \right] - \frac{2[\alpha_0 + \bar\theta_a + \bar\theta_{\text{wash}}]}{2+AR} \tag{10}$$

Equation (10) is predicated on simple harmonic motion for α. In this case the motion can be periodic but not necessarily simple harmonic. Therefore, the use of this equation is considered to be an approximation to the unsteady shed wake effects. Each chordwise strip on the wing is assumed to act as if it were part of an elliptical planform wing executing simple harmonic whole-wing motion identical to that of the strip's. AR is the wing's aspect ratio and k is the reduced frequency, which is given by

$$k = \frac{c\omega}{2U} \tag{11}$$

Using a simplified formulation of the modified Theodorsen function, which was originally presented by Jones,[13] $F'(k)$ and $G'(k)$ are given by[11]

$$F'(k) = 1 - \frac{C_1 k^2}{k^2 + C_2^2}, \qquad G'(k) = -\frac{C_1 C_2 k}{k^2 + C_2^2}$$

$$C_1 = \frac{0.5\,AR}{2.32 + AR}, \qquad C_2 = 0.181 + \frac{0.772}{AR} \tag{12}$$

The equations for α and $\dot{\alpha}$ are (from Ref. 11)

$$\alpha = [\dot{h} \cos(\tilde{\theta} + \bar{\theta}_{\text{wash}}) + (0.75c - y_{ea})\dot{\tilde{\theta}}]/U + \tilde{\theta} \tag{13}$$

$$\dot{\alpha} = [(\ddot{h}_0 + \ddot{h}) \cos(\tilde{\theta} + \bar{\theta}_{\text{wash}}) - \dot{h}\dot{\tilde{\theta}} \sin(\tilde{\theta} + \bar{\theta}_{\text{wash}}) + (0.75c - y_{ea})\ddot{\tilde{\theta}}]/U + \dot{\tilde{\theta}} \tag{14}$$

The Fullwing code uses a linear version of the above expression for $\dot{\alpha}$ by simply ignoring the second term in the numerator. The nonlinear Newmark code includes the complete expression.

It is appropriate at this point to introduce the total twist angle θ of a segment about its elastic axis, which is a combination of elastic and constant parts:

$$\theta = \tilde{\theta} + \bar{\theta}_a + \bar{\theta}_{\text{wash}} \tag{15}$$

Similarly, the total plunging displacement is a combination of an imposed motion h_0 and an elastic component $\tilde{h}$:

$$h = h_0 + \tilde{h} \tag{16}$$

The imposed motion for a given wing segment is defined as

$$h_0 = \Gamma_0 x \, \cos(\omega t) \tag{17}$$

where Γ_0 is the maximum flapping amplitude (which for the full-scale ornithopter is about 31 deg) and x is the distance from the center of a wing segment to the flapping axis. Returning to Eq. (8), we note that the flow velocity V must include the downwash as well as the wing's motion relative to the freestream velocity U. This is done by including α' along with the kinematic parameters:

$$V = \sqrt{[U\cos\theta - \dot{h} \, \sin(\tilde{\theta} + \bar{\theta}_{\text{wash}})]^2 + [U(\alpha' + \bar{\theta}_a + \bar{\theta}_{\text{wash}}) - (0.5c - y_{ea})\,\dot{\tilde{\theta}}]^2} \tag{18}$$

An additional normal-force contribution comes from the apparent-mass effect, which acts at the midchord location (Fig. 10) and is given by

$$N_a = \tfrac{1}{4} \rho \, \pi c^2 \left(U\dot{\alpha} - \tfrac{1}{4} c \, \ddot{\bar{\theta}} \right) \Delta x \tag{19}$$

A section's circulation distribution generates forces in the chordwise direction as shown in Fig. (10). From DeLaurier,[14] the chordwise force due to camber is given by

$$D_c = -2\pi \alpha_0 (\alpha' + \bar{\theta}_a + \bar{\theta}_{\text{wash}}) \frac{\rho UV}{2} c \Delta x \tag{20}$$

The leading-edge suction force is obtained from Garrick[15] as

$$T_s = \eta_s 2\pi \left(\alpha' + \bar{\theta}_a + \bar{\theta}_{\text{wash}} - \frac{c\dot{\bar{\theta}}}{4U} \right)^2 \frac{\rho UV}{2} c \Delta x \tag{21}$$

The only change to Garrick's formulation is the addition of the η_s term, which is referred to as the leading-edge suction efficiency factor and is determined experimentally. This efficiency factor is required since Garrick's formulation is based on ideal potential flow.

Viscous drag on the airfoil due to skin friction is found by using the skin-friction drag coefficient C_{df} for which an expression may be found in Hoerner.[16] Reference 11 presents this drag as

$$D_f = C_{df} \frac{\rho V_y^2}{2} c \Delta x \tag{22}$$

where V_y is the relative flow speed tangent to the section, which can be approximated by

$$V_y = U \cos\theta - \dot{h} \sin(\tilde{\theta} + \bar{\theta}_{\text{wash}}) \tag{23}$$

Therefore the total chordwise force is

$$F_y = T_s - D_c - D_f \tag{24}$$

2. Stalled Flow

When the attached-flow range is exceeded, totally separated flow is assumed to abruptly occur, for which the contribution of chordwise forces is negligible:

$$T_s = D_c = D_f = 0 \tag{25}$$

and the normal force is given by

$$N = (N_c)_{\text{sep}} + (N_a)_{\text{sep}} \tag{26}$$

$(N_c)_{\text{sep}}$ acts at the midchord point and is due to crossflow drag and is assumed to be

$$(N_c)_{\text{sep}} = (C_d)_{cf} \frac{\rho \hat{V} V_n}{2} c \Delta x \tag{27}$$

where

$$\hat{V} = \sqrt{V_y^2 + V_n^2} \tag{28}$$

V_n is the midchord normal velocity component due to the wing's motion given by

$$V_n = \dot{h}\cos(\tilde{\theta} + \bar{\theta}_{\text{wash}}) + \tfrac{1}{2}c\dot{\theta} + U\sin\theta \tag{29}$$

and V_y is given by Eq. (23). It is evident from Eq. (28) that $\hat{V}$ is a nonlinear function of the independent variables $\tilde{\theta}$ and $\tilde{h}$. This shows that both the stalled and attached-flow aerodynamic formulations are indeed nonlinear.

Experiments conducted at the Institute for Aerospace Studies showed the value of the separated-flow apparent-mass normal force to be about half of that for attached flow. $(N_a)_{\text{sep}}$ is therefore assumed to be half of the value given by Eq. (19):

$$(N_a)_{\text{sep}} = \frac{N_a}{2} = \frac{1}{8}\rho\pi c^2\left(U\dot{\alpha} - \frac{1}{4}c\ddot{\theta}\right)\Delta x \tag{30}$$

3. Aerodynamic Moments

The attached-flow aerodynamic moment about the elastic axis is a function of the circulatory and apparent-mass normal forces and is given by

$$M_{\text{aero}} = M_{ac} - N_c(0.25c - y_{ea}) - N_a(0.5c - y_{ea}) - \frac{1}{16}\rho\pi c^3 U\Delta x\dot{\theta} - \frac{1}{128}\rho\pi c^4\Delta x\ddot{\theta} \tag{31}$$

N_c and N_a are given by Eqs. (8) and (19) respectively. The fourth and fifth terms in Eq. (31) account for the apparent camber and apparent inertia moments respectively.[11] The moment about the aerodynamic center is given by[17]

$$M_{ac} = C_{\text{mac}}\frac{\rho U V}{2}c^2\Delta x \tag{32}$$

The chordwise forces do not contribute to these moments as they essentially pass through the elastic axis of a wing segment.

The stalled aerodynamic moment is given by

$$(M_{\text{aero}})_{\text{sep}} = -[(N_c)_{\text{sep}} + (N_a)_{\text{sep}}](0.5c - y_{ea}) \tag{33}$$

where $(N_c)_{\text{sep}}$ and $(N_a)_{\text{sep}}$ are given by Eqs. (27) and (30) respectively. The moments about the aerodynamic center due to apparent camber and apparent inertia effects are negligible because these quantities are defined for attached flow only.

4. Stall Criterion

Prouty[18] has shown that a pitching airfoil can retain attached flow at angles greatly exceeding the airfoil's static stall angle. An advantage of a strip-theory model is that it allows for an approximation to the localized stall behavior. Prouty uses a dynamic stall-delay effect, represented by an angle $\Delta\alpha$, to account for the difference between the static and effective stall angles:

$$(\alpha_{\text{stall}})_{ef} - (\alpha_{\text{stall}})_{st} = \Delta\alpha = \xi\sqrt{\frac{c\dot{\alpha}}{2U}} \tag{34}$$

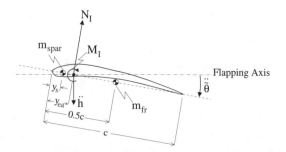

Fig. 11   Inertial loads and moments on a wing segment.

where ξ is found experimentally and depends on the local Mach number. In this case it was determined that $\Delta\alpha$ is given by[11]

$$\Delta\alpha = 0.51\left(\frac{\dot{\alpha}}{\dot{\alpha}_{\text{mag}}}\right)\sqrt{\frac{c\,\dot{\alpha}_{\text{mag}}}{2U}} \qquad (35)$$

where $\dot{\alpha}_{\text{mag}} = \text{abs}(\dot{\alpha})$. The magnitude of $\dot{\alpha}$ is used to ensure that the term under the square root is positive and the term in the brackets ensures that the correct sign is used. Therefore, the criterion for attached flow over a wing segment is

$$(\alpha_{\text{stall}})_{\text{min}} \le \left[\alpha' + \bar{\theta}_a + \bar{\theta}_{\text{wash}} - \frac{3}{4}\left(\frac{c\dot{\bar{\theta}}}{U}\right)\right] \le (\alpha_{\text{stall}})_{\text{max}} \qquad (36)$$

B.   Inertial Forces and Moments

Figure 11 shows the inertial forces and moments acting on a wing segment. The Fullwing code treats the masses as inertial loads but the Newmark code breaks down the inertial loads into reactions that involve the elastic components and external forces that are a result of the imposed motion and gravity. The inertial reactions have already been considered by the consistent mass matrix formulation, and the external inertial normal force is given by[19]

$$N_{\text{inertia}} = (m_{\text{spar}} + m_{fr})(g_r - \ddot{h}_0) \qquad (37)$$

Further, the external inertial moment is given by[19]

$$M_{\text{inertia}} = [(y_s - y_{ea})m_{\text{spar}} + (0.5c - y_{ea})m_{fr}](g_r - \ddot{h}_0) \qquad (38)$$

where g_r is acceleration due to gravity.

C.   Temporal Approximation for Nonlinear Terms

The most straightforward solution procedure is to use a nonlinear time-marching algorithm such as the fourth-order Runge–Kutta method. This would avoid the requirement of finding a temporal approximation to the nonlinear terms. An attempt was made to use this method to solve the dynamic equilibrium equations. However, because of the presence of two distinct flow regimes and inherent instability, the solution would diverge after a few time steps.

It was noted in the previous section that the circulatory and apparent-mass normal forces are nonlinear functions of $\tilde{\theta}$, $\dot{\tilde{\theta}}$, $\ddot{\tilde{\theta}}$, h, and $\ddot{h}$:

$$N_c = f(\tilde{\theta}, \dot{\tilde{\theta}}, \ddot{\tilde{\theta}}, h, \ddot{h}), \qquad N_a = f(\tilde{\theta}, \dot{\tilde{\theta}}, \ddot{\tilde{\theta}}, h, \ddot{h})$$

$$(N_c)_{\text{sep}} = f(\tilde{\theta}, \dot{\tilde{\theta}}, h), \qquad (N_a)_{\text{sep}} = f(\tilde{\theta}, \dot{\tilde{\theta}}, \ddot{\tilde{\theta}}, h, \ddot{h}) \tag{39}$$

The value of the circulatory normal force for attached flow at time $n + 1$ is found by using a Taylor series approximation:

$$N_c^{n+1} = N_c^n + \frac{\partial N_c^n}{\partial \tilde{\theta}}[\tilde{\theta}^{n+1} - \tilde{\theta}^n] + \frac{\partial N_c^n}{\partial \dot{\tilde{\theta}}}[\dot{\tilde{\theta}}^{n+1} - \dot{\tilde{\theta}}^n] + \frac{\partial N_c^n}{\partial \ddot{\tilde{\theta}}}[\ddot{\tilde{\theta}}^{n+1} - \ddot{\tilde{\theta}}^n]$$

$$+ \frac{\partial N_c^n}{\partial h}[\dot{h}_0^{n+1} - h_0^n] + \frac{\partial N_c^n}{\partial \dot{h}}[\dot{h}^{n+1} - \dot{h}^n] + \frac{\partial N_c^n}{\partial \dot{h}}[\dot{h}_0^{n+1} - \dot{h}_0^n]$$

$$+ \frac{\partial N_c^n}{\partial \ddot{h}}[\ddot{h}^{n+1} - \ddot{h}^n] \tag{40}$$

Other normal-force components [Eqs. (19), (27), and (30)] and aerodynamic moments [Eqs. (31) and (33)] are also expanded in a similar way.

Terms involving the nth time step are moved to the right-hand side of the equilibrium equations [Eq. (1)] and terms involving the $(n + 1)$'st time step are moved to the left-hand side of Eq. (1). Essentially, stiffness, mass, and damping matrices are augmented by terms that account for the nonlinearity of the forcing functions. Some authors (Refs. 10 and 20) refer to the augmenting matrices as tangent stiffness and tangent mass matrices.

D. Rigid and Center Sections

Aerodynamic forces and moments acting on the rigid and center sections of the wing are identical to those acting on the outer portion of the wing, and the same equations and the same criterion for stall can be used. The only difference is that the elastic variables are taken to be zero ($\tilde{\theta} = \tilde{h} = 0$). Furthermore, the forcing function for the rigid section is given by

$$(h_0)_{rs} = -\frac{(\Delta x)_{rs}}{2} \Gamma_0 \cos(\omega t) \tag{41}$$

where $(\Delta x)_{rs}$ is the width of the rigid section. Likewise, the center section forcing function is given by

$$(h_0)_{cs} = -(\Delta x)_{rs} \Gamma_0 \cos(\omega t) \tag{42}$$

E. Average Lift and Thrust

The total lift and thrust of a flapping wing at any given time is the sum of the contributions from the rigid, center, and flexible sections. The loads outlined in the previous section are normal N and tangential F_y to a given segment. Lift L and thrust R for a particular segment of the flexible wing are calculated as follows:

$$R = F_y \cos(\theta) - N \sin(\theta), \qquad L = N \cos(\theta) + F_y \sin(\theta) \qquad (43)$$

Lift and thrust produced by the rigid and center sections are calculated in a similar way. The average thrust and lift generated by the whole wing over N_l time intervals are

$$R_{\text{ave}} = \frac{1}{N_l} \sum_{i=1}^{N_l} \left(2 \sum_{j=1}^{N_e} R_j + 2R_{rs} + R_{cs} \right)_i,$$

$$L_{\text{ave}} = \frac{1}{N_l} \sum_{i=1}^{N_l} \left(2 \sum_{j=1}^{N_e} L_j + 2L_{rs} + L_{cs} \right)_i \qquad (44)$$

where N_e is the total number of elements per wing.

F.   Bending and Twisting Moments

The bending moments encountered at each wing segment are calculated by transforming the forces in the wing's frame of reference to the flapping-axis frame of reference. The flapping-axis angle of attack with respect to the freestream velocity can thus be accounted for. Hence, a new normal force F_{normal} must be defined:[5]

$$F_{\text{normal}} = N \cos(\theta - \bar{\theta}_a) + (N_{\text{spar}} + N_{fr}) \cos(\bar{\theta}_a)$$

$$+ (T_s + D_f + D_c) \sin(\theta - \bar{\theta}_a) \qquad (45)$$

where normal forces acting on the spar and fabric and rib components are[19]

$$N_{\text{spar}} = m_{\text{spar}}[\ddot{h} + (y_s - y_{ea})\ddot{\bar{\theta}} - g_r]$$

$$N_{fr} = m_{fr}[\ddot{h} + (0.5c - y_{ea})\ddot{\bar{\theta}} - g_r] \qquad (46)$$

The bending moment at the flapping axis is thus calculated by adding up the contributions of all the segments on the flexible portion:

$$(M_{\text{bend}})_{sr} = \sum_{i=1}^{N_e} (F_{\text{normal}})_i \, x_i \qquad (47)$$

where x_i is the distance of a given segment to the flapping axis. Similarly, the twisting moment is given by

$$(M_{\text{twist}})_{sr} = \sum_{i=1}^{N_e} (M_{ac})_i - (F_{\text{normal}})_i \, y_i \qquad (48)$$

where y_i is the distance of the elastic axis for a given segment from a line perpendicular to the flapping axis and passing through the elastic axis at the flapping-axis location.

IV. Damping

Early in the development of the Newmark code it became apparent that a reliable damping model is required. Without a damping model and when using the Expothopter data file, the solution would diverge numerically. The reason was found to be that the undamped differential equations had positive eigenvalues. This showed that the initial theoretical model did not represent the actual system. However, this was in contradiction to what the Fullwing code predicted, as it did not have a damping model either. Furthermore, the Expothopter did perform reasonably well in wind tunnel tests conducted in Ottawa in 1995. At first it was believed that the cause was a set of stiff differential equations. However, examination of the system eigenvalues showed that the equations are not stiff. Finally, it was concluded that a damping model was required as that is the only component not accounted for by Fullwing. Furthermore, the shearflexing action of the wing creates damping forces along the trailing edge that must be accounted for.

The reason for Fullwing's predictions lies in the assumptions that were made when the code was developed, the most critical of which is that the output is assumed to be harmonic (sinusoidal) with a frequency equal to that of the forcing function. This basically guarantees that the output will not diverge. Therefore, when using a time-marching method, one must account for damping to obtain a stable solution.

Besides aerodynamic damping, there are two other forms of damping associated with a flapping wing:

1) *Fabric and rib damping.* This is a result of the frictional losses from the rubbing of the wing's fabric covering against the ribs caused by the shear flexing action. This action also contributes to damping as the trailing-edge strips rub against one another.

2) *Structural damping.* This is due to the viscoelastic losses in the spar. In homogeneous materials such as metals, this type of damping is relatively well understood and damping coefficients are known. In the case of the ornithopter wing, which is a nonhomogeneous material, experiments were conducted to determine the free-decay constant ζ for the overall wing.

A. Fabric and Rib Friction Damping

Experiments carried out in 1997 showed that the total damping moment is a combination of a constant value (M_{d0}) and a moment that is a function of the shear rate (M_{dsr}):

$$M_{\text{damp}} = M_{d0} + M_{dsr} \tag{49}$$

In terms of the friction force:

$$F_f = C_{f0} p A_{\text{con}} + C_f p A_{\text{con}} v_r \tag{50}$$

The first term on the right-hand side represents the shear-rate-independent friction force and the second term represents the shear-rate-dependent friction force. C_{f0} is the friction coefficient for the shear-rate-independent motion, and C_f is the friction, coefficient for the shear-rate-dependent motion; p is the contact pressure, and A_{con}, is the contact area between the surfaces. Also, v_r is the spanwise velocity of the upper surface relative to the lower surface of the wing.

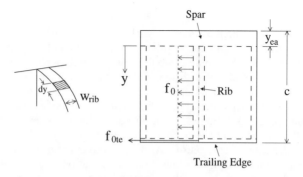

Fig. 12   Shear-rate-independent friction terms.

Considering the shear-rate-independent term and referring to Fig. 12, we can split the total friction force into a uniformly distributed fabric and rib force f_0 and a trailing-edge force f_{0te}:

$$f_0 = C_{f0} p w_{\text{rib}} \, dy \tag{51}$$

$$f_{0te} = C_{f0} p A_{te} \tag{52}$$

For a single wing segment the total shear-rate-independent damping moment is found by summing the moments about the elastic axis, which in Fig. 12 is shown to be the back of the spar. A_{te} is the contact area of the trailing-edge surface and w_{rib} is the width of a single rib. One thus has

$$M_{d0} = C_{f0} p A_{te}(c - y_{ea}) + C_{f0} p w_{\text{rib}} \int_0^{c-y_{ea}} dy \tag{53}$$

where c is the chord length of a particular wing section and y_{ea} is the distance from the leading edge to the elastic axis. M_{d0} is therefore given by

$$M_{d0} = C_{f0} p (c - y_{ea}) \left[A_{te} + \frac{w_{\text{rib}}}{2} (c - y_{ea}) \right] \text{sgn}(\dot{\theta}) \tag{54}$$

For the shear-rate-dependent damping moment, it is assumed that the friction force is related to velocity as shown by the second term on the right-hand side of Eq. (50). For each segment of the wing there is a twist/shear rate expressed as $\partial \theta / \partial x$, which is shown in Fig. 13. At this point we assume that the relative velocity of the upper surface to the lower surface v_r has a linear distribution, going from zero at the elastic axis to $v_r = v_{te}$ at the trailing edge. This gives rise to a friction force distribution as shown in Fig. 14. The two force components $f(y)$ and f_{te} are given by

$$f(y) = C_f p v_r(y) w_{\text{rib}} \, dy \tag{55}$$

$$f_{te} = C_f p v_{te} A_{te} \tag{56}$$

The shear-rate-dependent damping moment is thus given by

$$M_{dsr} = F_{te} (c - y_{ea}) + \int_0^{c-y_{ea}} f(y) y \, dA$$

$$= C_f p v_{te} \left[A_{te} (c - y_{ea}) + \frac{w_{\text{rib}}}{(c - y_{ea})} \int_0^{c-y_{ea}} y^2 \, dy \right] \tag{57}$$

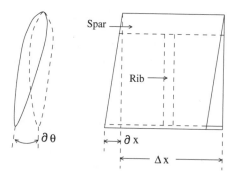

Fig. 13   Shear rate illustration.

Performing the integral on the right-hand side and simplifying gives

$$M_{\text{dsr}} = C_f \, p v_{te} \, (c - y_{ea}) \left[A_{te} + \frac{w_{\text{rib}}}{3} (c - y_{ea}) \right] \tag{58}$$

The velocity at the trailing edge can be written as

$$v_{te} = \frac{\partial x}{\partial t} = \frac{\partial x}{\partial \theta} \frac{\partial \theta}{\partial t} = \frac{\partial x}{\partial \theta} \dot{\theta} \tag{59}$$

so that M_{dsr} is finally given by

$$M_{\text{dsr}} = C_f \, p \, (c - y_{ea}) \left[A_{te} + \frac{w_{\text{rib}}}{3} (c - y_{ea}) \right] \left(\frac{\partial x}{\partial \theta} \right) \dot{\theta} \tag{60}$$

B.   Structural Damping

Structural damping is primarily due to mechanisms such as hysteresis in the material and slip in connections. These mechanisms are not well understood and they are awkward to incorporate into the equilibrium equations. Therefore, the actual mechanism is usually approximated by viscous damping. Comparisons of theory and experiment show that this approach is sufficiently accurate in most cases.[20] With such approximate methods, experimental observations of the vibratory response of structures are used to assign a fraction of critical damping as a

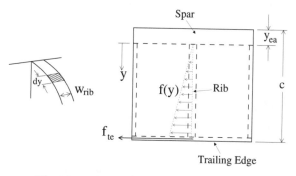

Fig. 14   Shear-rate-dependent friction forces.

function of frequency or, more commonly, a single decay constant ζ for the frequency range.

A popular damping scheme, called Rayleigh or proportional damping, is to form a damping matrix [**D**] as a linear combination of the stiffness and mass matrices, that is,

$$[\mathbf{D}] = a[\mathbf{M}] + b[\mathbf{K}] \tag{61}$$

where a and b are called the mass- and stiffness-proportional damping constants respectively. The damping matrix [Eq. (61)] is orthogonal because it permits modes to be uncoupled by eigenvectors associated with the undamped eigenvalues. The relationship among a, b, and decay constants ζ at a frequency ω is given by[20]

$$\zeta = \frac{a + \omega^2 b}{2\omega} \tag{62}$$

Damping constants a and b are determined by choosing two distinct decay constants (ζ_1 and ζ_2) at two different frequencies (ω_1 and ω_2) and solving simultaneous equations for a and b. Thus,

$$a = 2\omega_1\omega_2 \frac{\zeta_1\omega_2 - \zeta_2\omega_1}{\omega_2^2 - \omega_1^2}, \qquad b = 2\frac{\zeta_2\omega_2 - \zeta_1\omega_1}{\omega_2^2 - \omega_1^2} \tag{63}$$

As part of the work leading to the design of the full-scale ornithopter, several sample spars were constructed. One of these, known as Spar 6, was used to perform fatigue testing (Fig. 15). Spar 6's bending and torsional stiffness characteristics are identical to the actual wing spar's characteristics at the flapping-axis location.

In 1998, Spar 6 was used to determine its bending and torsional decay constants. Figure 16 shows Spar 6 in its bending configuration and Fig. 17 shows Spar 6 in its torsional configuration.

The testing procedure involved adding weights to the spar and striking it with a hammer while its vibratory response was measured using an accelerometer. The

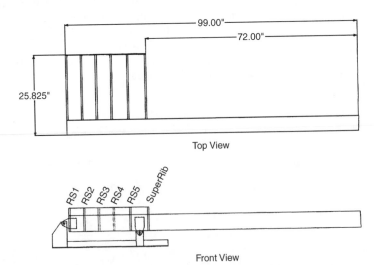

Fig. 15   Spar 6 schematic.

Spar 6 Top View

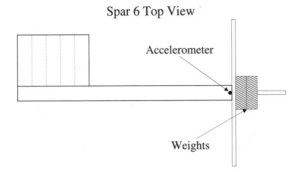

Fig. 16   Bending configuration.

weights were varied to produce a range of natural frequencies. Figure 18 shows the experimental results as well as a theoretical decay constant obtained using Rayleigh constants $a = 0.0136$ and $b = 0.0008$. It is evident from the graph that decay constants for bending and torsion are fairly constant over the range of frequencies tested. Basically, the Rayleigh model assumes a damping ratio that is within the limits prescribed by the two main damping modes. The Rayleigh model is also reasonable considering that the decay constants are on the order of 6×10^{-3}, or 0.6% of critical damping.

V.   Results and Discussion

A large quantity of experimental data have been obtained during the design and testing of the two models as well as the full-scale ornithopter. The results of tests on the Expothopter are presented in a thesis by Fowler[5] and the bulk of experimental data for the full-scale ornithopter is contained in two M.S. theses by Mehler[3] and Fenton.[4] The objective of this section is to present a comparison between the nonlinear Newmark code and experimental results in all flight regimes, and to compare the Newmark and Fullwing codes at near-flight conditions.

Spar 6 Top View

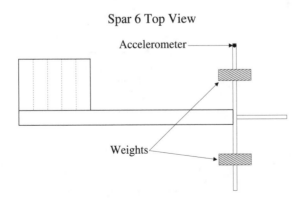

Fig. 17   Torsional configuration.

Spar 6

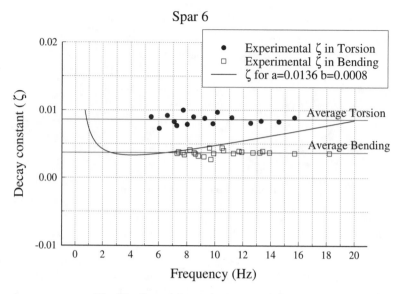

Fig. 18   Spar 6 decay constant variation.

A.   Quarter-Scale Model Wing Tests

One of the wings of the quarter-scale model was attached to a flapping mechanism in the subsonic wind tunnel at the University of Toronto's Institute for Aerospace Studies.[12] The wing was also attached to scales that measured the generated average lift and thrust values at different flapping frequencies. Figures 19 and 20 show a comparison between theoretical results from Newmark and Fullwing codes as well as experimental data obtained during the tests. The discrepancy between theoretical and experimental lift is mainly due to the difference between

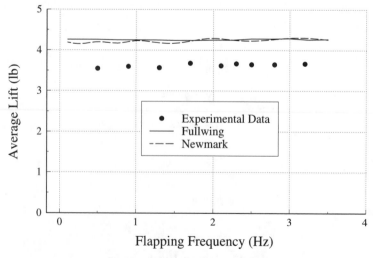

Fig. 19   Quarter-scale lift performance. $U = 45$ ft/s; $\bar{\theta}_a = 6$ deg.

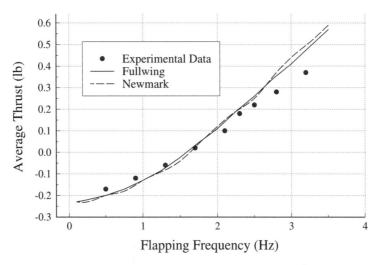

Fig. 20 Quarter-scale thrust performance. $U = 45$ ft/s; $\bar{\theta}_a = 6$ deg.

the actual bending and torsional stiffness properties of the wing and the theoretical values used in the analysis.

B. 1996 Full-Scale Static-Flapping Tests

In 1996, a series of static-flapping tests were conducted. Strain gauges were attached to the wing at four locations as shown in Fig. 21 and twisting and bending moments were measured at different flapping frequencies. Figures 22 and 23 show a summary of maximum and minimum twisting and bending moment values for flapping frequencies from 0.4 to 1.1 Hz. Newmark's prediction is also shown. It should be recalled that the aerodynamic model is based on a strip theory. Such an approximation is limited, particularly in a fully-stalled flow regime.

Video footage of the wingtip motion was used to assess the twisting behavior of the wing and compare it to Newmark's predictions. Figures 24 and 25 show the wingtip twist at 0.91 and 0.97 Hz, respectively. It is evident from these figures that the nonlinear aerodynamic formulation is capable of predicting the wing's twisting behavior at conditions of zero forward speed.

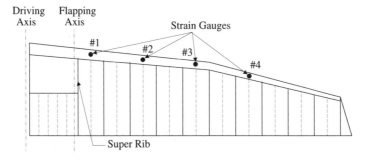

Fig. 21 Full-scale ornithopter strain gauge locations.

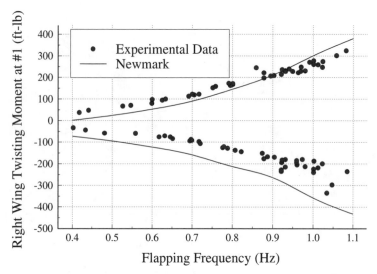

Fig. 22   Static twisting moment vs flapping frequency. $U = 0.01$ ft/s; $\bar{\theta}_a = 0$ deg.

C.   1997 and 1998 Taxi Tests

In the summer of 1997 and 1998, extensive taxi trials were conducted at Downsview airport in Toronto. Figure 26 shows the ornithopter during a liftoff test in 1999. Unfortunately, all throughout 1997 and 1998 some strain gauges failed progressively and only a limited amount of data was collected. This is sufficient, however, to provide an opportunity for comparing the Newmark predictions with experimental data. Figures 27 and 28 show a summary of the maximum and

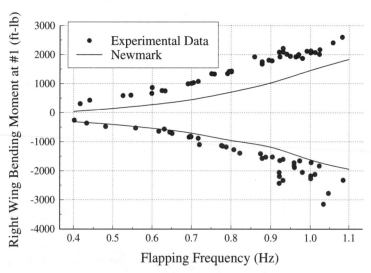

Fig. 23   Static bending moment vs flapping frequency. $U = 0.01$ ft/s; $\bar{\theta}_a = 0$ deg.

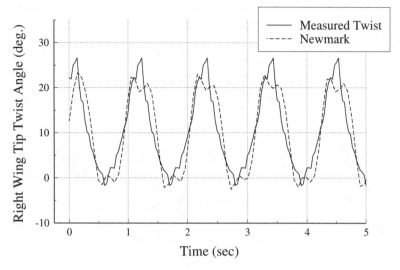

Fig. 24   Wing tip twist angle variation (ω = 0.91 Hz).

minimum twisting and bending moments during the 1997 and 1998 seasons. It should be noted that the reason for the limited amount of twisting-moment data is that all twisting-moment strain gauges had failed by the end of the 1997 season. Furthermore, the results presented here represent only portions of taxi tests where steady-state conditions were present. A steady-state condition in a taxi test would occur only when the throttle was maintained at a constant level and air speed was not rapidly changing. The reason for not connecting the data points on Figs. 27

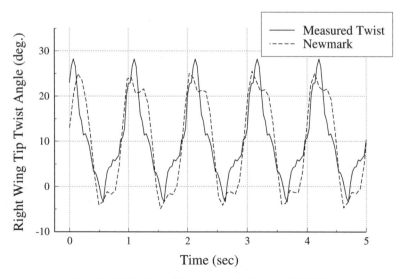

Fig. 25   Wing tip twist angle variation (ω = 0.97 Hz).

Fig. 26   Liftoff test in 1999.

and 28 is that each data point represents a run with different air speed U and angle of attack $\overline{\theta}_a$ values.

The strain gauges provided a significant quantity of instantaneous twisting- and bending-moment data as well. Figures 29 and 30 show the instantaneous twisting and bending moments during one of the runs in 1997. There is a maximum error of 20% associated with the twisting-moment data and about 15% for the bending-moment data.

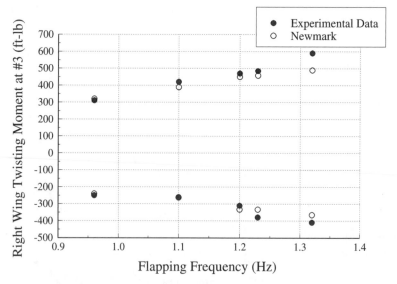

Fig. 27   Taxi test twisting moment vs flapping frequency.

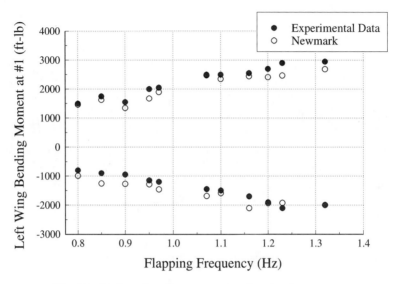

Fig. 28   Taxi test bending moment vs flapping frequency.

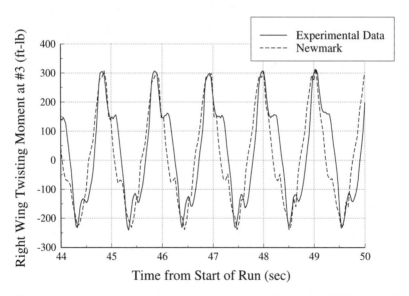

Fig. 29   Instantaneous twisting moment for run 1 on 15 September 1997. $U = 44$ ft/s; $\bar{\theta}_a = 3$ deg; $\omega = 0.96$ Hz.

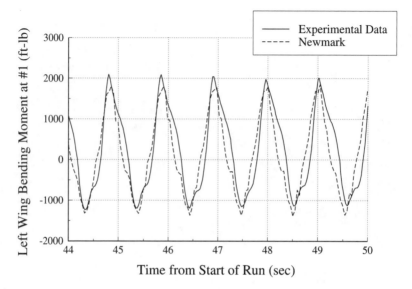

Fig. 30 Instantaneous bending moment for run 1 on 15 September 1997. U = 44 ft/s; $\bar{\theta}_a$ = 3 deg; ω = 0.96 Hz.

The Newmark code is very valuable for predicting lifting and thrusting performance of the wings, as shown in Figs. 31 and 32 for a flapping frequency of 1.2 Hz and an angle of attack $\bar{\theta}_a$ of zero. Although there is good agreement between Fullwing and Newmark average lift results, the static thrusting values are considerably different. As part of the 1996 static tests the value of average thrust at 1 Hz was measured to be about 25 lb. This, and the fact that in the 1997 and 1998 taxi trials the ornithopter was able to start its ground roll under its own power, shows that the Newmark thrust predictions are more accurate.

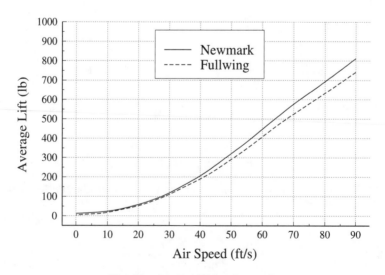

Fig. 31 Average lift vs air speed.

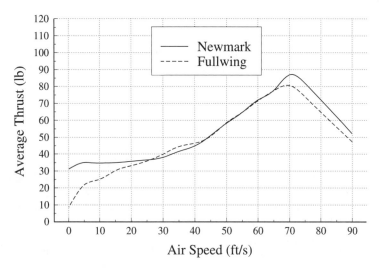

Fig. 32 Average thrust vs air speed.

VI. Conclusions

This chapter has presented an updated numerical method for predicting the performance of a flapping wing in steady flight. This is a design-oriented analysis using a nonlinear strip theory model for attached flow and poststall behavior. The finite element structural model that is used is both practical and versatile enough to handle different geometrical configurations with little or no modifications. A translational matrix formulation eliminates the need for cross members that would have otherwise been required to connect the spar and fabric and rib elements to one another. The Rayleigh damping model is a convenient way for modeling the damping characteristics of the spar, and experimental data were used to determine approximate values for bending and torsional decay constants associated with a sample spar. The Rayleigh model is also suitable because the spar's decay constant is about 0.6% of critical damping. The fabric and rib friction-damping model account for the energy loss caused by the shearflexing action of the wing. This damping model is necessary since shearflexing is a unique feature of the ornithopter's wing.

The Newmark predictions for maximum and minimum twisting moment during static flapping are quite accurate and much better than expected considering the relatively simple aerodynamic model used. There is a maximum error of about 25% associated with the maximum and minimum bending moments at the flapping-axis location for static flapping.

One of the issues raised during the 1996 static-flapping tests was the unexpected twisting behavior of the wings. This was particularly worrisome at the time because the phase angle between flapping and twisting was not close to the optimum value of approximately −90 deg. By properly modeling the fully stalled characteristics of the wing it became possible to predict the correct wing twisting behavior at near-static conditions. This basically shows that severe stalling at near-static conditions produces a twisting behavior that is far from optimum. This is possibly the most important contribution from the analysis.

There is better agreement between theory and experimental data for near-flight conditions. The maximum error for maximum and minimum bending and torsion is about 15%. This is quite reasonable considering that by 1998 the strain gauges still functioning were more than two-years old and were beginning to suffer from adhesive deterioration.

This analysis has not only helped the research team at the Institute for Aerospace Studies better understand the aerodynamic characteristics of the ornithopter's wings, but it will also provide a valuable tool for future design of flapping wing aircraft.

References

[1]DeLaurier, J. D., and Harris, J. M., "A Study of Mechanical Flapping Wing Flight," *Aeronautical Journal*, Vol. 97, No. 968, Oct. 1993, pp. 277–286.

[2]DeLaurier, J. D., "The Development and Testing of a Full-Scale Piloted Ornithopter," *Canadian Aeronautics and Space Journal*, Vol. 45, No. 2, June 1999, pp. 72–82.

[3]Mehler, F. E., "The Structural Testing and Modification of a Full-Scale Ornithopter's Wing Spars," M.A.Sc. Thesis, Faculty of Applied Science and Engineering, University of Toronto, Downsview, ON, Canada, 1997.

[4]Fenton, B. A., "The Testing and Modification of a Full-Scale Ornithopter," M.A.Sc. Thesis, Faculty of Applied Science and Engineering, University of Toronto, Downsview, ON, Canada, 1999.

[5]Fowler, S. J., "The Design and Development of a Wing for a Full-Scale Piloted Engine-Powered Flapping Wing Aircraft (Ornithopter)," M.A.Sc. Thesis, Faculty of Applied Science and Engineering, University of Toronto, Downsview, ON, Canada, 1995.

[6]DeLaurier, J. D., "An Ornithopter Wing Design," *Canadian Aeronautics and Space Journal*, Vol. 40, No. 1, March 1994, pp. 10–18.

[7]Reddy, J. N., *An Introduction to the Finite Element Method*, McGraw-Hill, New York, 1984.

[8]Bathe, K. J., *Finite Element Procedures*, Prentice-Hall, Englewood Cliffs, NJ, 1996.

[9]Owen, D. R. J., "Implicit Finite Element Methods for the Dynamic Transient Analysis of Solids with Particular Reference to Non-Linear Situations," *Advanced Structural Dynamics*, Applied Science Publishers, London, 1978, pp. 123–152.

[10]Humar, J. L., *Dynamics of Structures*, Prentice-Hall, Englewood Cliffs, NJ, 1990.

[11]DeLaurier, J. D., "An Aerodynamic Model for Flapping Wing Flight," *Aeronautical Journal*, Vol. 97, April 1993, pp. 125–130.

[12]DeLaurier, J. D., "The Development of an Efficient Ornithopter Wing," *Aeronautical Journal*, Vol. 97, May 1993, pp. 153–162.

[13]Jones, R. T., "The Unsteady Lift of a Wing of Finite Aspect Ratio," NACA Rept. 681, 1940.

[14]DeLaurier, J. D., "Drag of Wings with Cambered Airfoils and Partial Leading Edge Suction," *Journal of Aircraft*, Vol. 20, Oct. 1983, pp. 882–886.

[15]Garrick, I. E., "Propulsion of a Flapping and Oscillating Aerofoil," NACA Rept. 567, 1936.

[16]Hoerner, S. F., *Fluid Dynamic Drag*, published by the author, Brick Town, NJ, 1965, pp. 2-1 to 2-16.

[17]DeLaurier, J. D., "Time Marching Solution for Elastic Flapping Wings," 1997 (unpublished personal notes).

[18]Prouty, R. W., *Airfoils for Rotor Blades, Helicopter Performance, Stability and Control*, PWS Engineering, Boston, 1986.

[19]Larijani, R. F., "Inertial Forces and Moments for a Second Order Non-Linear Formulation," 2000 (unpublished personal notes).

[20]Cook, R. D., Malkus, D. S., and Plesha, M. E., *Concepts and Applications of Finite Element Analysis*, Wiley, New York, 1996.

Chapter 19

Euler Solutions for a Finite-Span Flapping Wing

M. F. Neef[*]and D. Hummel[†]

Technical University of Braunschweig, Braunschweig, Germany

Nomenclature

A	=	aspect ratio, $A = 2\,s/c$
b_{ref}	=	width of two-dimensional airfoil section
c	=	chord length
$\bar{c}_L$	=	mean lift coefficient per motion cycle, according to Eq. (12)
c_p	=	pressure coefficient, $c_p = (p - p_\infty)/(0.5\,\rho U_\infty^2)$
$\bar{c}_T$	=	mean thrust coefficient per motion cycle, according to Eq. (13)
c_x, c_y, c_z	=	horizontal, lateral, and vertical force coefficients in the inertial coordinate system, nondimensionalized with $(1/2)\,S_{\text{ref}}\,U_\infty^2$
$\bar{c}_{\pi,\text{in}}, \bar{c}_{\pi,\text{out}}$	=	mean coefficient for power input and output per motion cycle, nondimensionalized with $(1/2)\rho\,S_{\text{ref}}\,U_\infty^3$
E	=	specific total energy
F^c, F^v	=	tensors of convective and viscous fluxes, Eq. (1)
F_x, F_z	=	forces in horizontal and vertical direction
f	=	frequency
g	=	source term, according to Eqs. (1) and (4)
I	=	unit matrix in Eq. (3)
i, j, k	=	curvilinear grid coordinates; see Fig. 11
k	=	reduced frequency, $k = \pi f c / U_\infty$
M_∞	=	freestream Mach number
n	=	normal unit vector
p	=	pressure
q	=	velocity vector
S	=	cell surface area, Eq. (1)
S_{ref}	=	reference area for aerodynamic coefficients, two-dimensional: $S_{\text{ref}} = b_{\text{ref}}c$, three-dimensional: $S_{\text{ref}} = 2sc$

*Graduate Research Student, Institute of Fluid Mechanics.
†Professor, Institute of Fluid Mechanics. Senior Member AIAA.

s	= half-span of the wing
T	= time for one motion cycle, $T = 1/f$
t	= time
U_∞	= freestream velocity
u, v, w	= cell face velocities in Cartesian coordinates x, y, z
V	= cell volume
w	= vector of dependent variables, according to Eq. (1)
x, y, z	= Cartesian coordinates of the inertial system; see Figs. 1 and 8
y^*	= coordinate along the span of the three-dimensional wing; see Fig. 8
α	= pitching angle (two-dimensional), twisting angle (three-dimensional)
α_{tip}	= twisting angle at wingtip
α_0	= mean angle of attack (=mean pitching angle)
γ	= angle of attack through plunging only
λ	= amplitude ratio, $\lambda = \alpha_1 c/(2 z_1 k)$
ρ	= density
Φ	= phase shift
Ψ	= flapping angle
ω, ω_x	= vorticity vector, vorticity component in x direction
$\dot{\varphi}$	= vector of angular velocities of a rotating coordinate system

Subscripts and Superscripts:

1	= amplitude
b	= quantity of cell boundary
n	= quantity of body surface cell boundary
$(\bar{\ })$	= time-averaged mean value

I. Introduction

AT THE Institute of Fluid Mechanics of the Technical University of Braunschweig previous work has focused on slotted wingtips,[1] formation flight,[2] and the aerodynamics of the tail in birds.[3] An approach to flapping wing flight has also been made by estimating the forces acting on a bird's wing during its downstroke.[4] The present calculation of the flowfield around flapping wings has only become possible with the development of efficient and time-accurate flow solvers. With the help of these tools the aim is now to calculate the flow around flapping wings and to evaluate the thrust output as well as the propulsive efficiency for various wing motions.

The flow physics involved is relatively complicated for the following reasons. During the flapping cycle, the moving wing continually sheds starting and stopping vortices from its trailing edge. The effect of these vortices on the wing is to cause a phase shift between wing motion and aerodynamic forces. In three-dimensional flow, the finite span of the wing causes additional trailing vortices to occur, which interfere with the system of starting and stopping vortices. All of these effects are based on the pressure distribution and can therefore be computed by solving the Euler equations for a flapping wing. The consideration of visous effects, flow separation, and the corresponding special problems for low Reynolds number flight are beyond the scope of the present study. Regarding the flapping flight of birds, the current investigation is therefore restricted to the cruising of large species.

This implies wings of large aspect ratio ($A \approx 8$) and slow flapping frequencies at moderate amplitudes. For these conditions, the flow around the wing remains in the regime of attached flow.

Some previous work on flapping flight is worth noting. At the end of the 19th century, the rising interest in aviation led to Lilienthal's first book on bird flight,[5] where he already describes basic motion strategies of birds. However, it took another three decades to develop the first valuable analytical approach to the problem of flapping wing flight by Birnbaum.[6] For moderate reduced frequencies he suggests an approximate solution for calculating thrust and efficiencies of flat plates in plunging and pitching motion. The exact solution for the general theory of oscillating airfoils in two-dimensional flow became well established by the works of Theodorsen[7] and Garrick,[8] while the same results were achieved independently by Küssner.[9] Although other analytical approaches were presented later by Bosch[10] and Send,[11] the main interest in the study of oscillating airfoils had already shifted from thrust generation to flutter analysis. All analytical approaches capable of predicting thrust output give highest propulsive efficiencies when the pitching motion precedes the plunging motion by 90 deg in phase. This is in agreement with observations of birds in nature, which was already noted by Lilienthal.[5]

Experimental investigations in flapping wing flight are relatively rare with respect to quantitative results on thrust generation and efficiencies, although thrust was already measured for an oscillating flow past a rigid wing by Katzmayr[12] in 1922. More recently, experiments have been carried out by DeLaurier and Harris[13] and other research groups including Favier et al.[14] and Chandrasekhara et al.,[15] the latter mainly focusing on dynamic stall phenomena. Recent results on the flowfield behind oscillating airfoils by Jones and Platzer[16] are carried out in water tunnels, where propulsion characteristics are derived from vortex analysis. Apart from studying rigid wings in aerodynamic research laboratories, a number of biologists have collected valuable material by observing birds in nature or even in wind tunnel flight. Slow-motion photography as well as the analysis of particles in the flowfield behind the bird can be found in publications of Nachtigall[17] and Rayner and Gordon.[18]

In recent years, computational fluid dynamics (CFD) has also been used to investigate the unsteady flow phenomena in oscillating wings. Again, Jones and Platzer[19] have made their contributions by calculating the inviscid flow for two-dimensional airfoils in plunging and pitching motion in comparison with analytical results. Calculations of the three-dimensional flow around flapping wings are also available with respect to time-dependent forces (Smith,[20] Vest and Katz[21]), but results for mean thrust output and efficiency are rarely found. Navier–Stokes solutions for the viscous flow around two-dimensional wings have been presented by Isogai et al.[22] and a three-dimensional approach has been aimed at the hovering flight of insects (Liu and Kawachi[23]).

Flapping wing propulsion has recently gained attraction by the rising interest in alternative propulsion systems for micro air vehicles (MAVs). A number of small aircraft with flapping wings, known as ornithopters, have already made their first flight. As Shyy et al.[24] have pointed out, the specification for MAVs restricts the flight regime of such vehicles to low Reynolds numbers by enforcing similarity to birds with respect to size and cruising speed. Flapping wings are therefore assumed to be an appropriate means to efficiently propel such vehicles.

II. Numerical Method

All numerical calculations were carried out with the CFD code FLOWer,[25–27] which has been developed by the DLR Braunschweig (Deutsches Zentrum für Luft- und Raumfahrt), Germany. It is capable of solving the three-dimensional Euler equations as well as the Reynolds-averaged Navier–Stokes equations in integral form on structured meshes. The spatial discretization is characterized by a second-order cell-vertex method for finite volumes. A five-stage Runge–Kutta scheme is employed to achieve convergence of the solution by integration with respect to time. For unsteady flows an implicit dual time-stepping scheme is used, which is based on the method of Jameson.[28] Implicit residual smoothing and a multigrid scheme can be employed to accelerate convergence.

To allow arbitrary motions of the computational grid, the governing equations are written for a moving grid in a Cartesian coordinate system[29] as

$$\frac{\partial}{\partial t} \int_V w \, dV + \int_{\partial V} (F^c - F^v) n \, dS + \int_V g \, dV = 0 \qquad (1)$$

where w is the vector of conservative variables,

$$w = \begin{bmatrix} \rho \\ \rho q \\ \rho E \end{bmatrix} \qquad (2)$$

consisting of the density ρ, the specific total energy E, and the velocity vector q within the moving coordinate system. Equation (1) accounts for arbitrary motions of the grid by correcting the local velocity vector q with the velocity vector of the moving cell boundary q_b, which becomes apparent in the definition of the convective part of the flux density

$$F^c = \begin{bmatrix} \rho(q - q_b) \\ \rho q \cdot (q - q_b) + pI \\ \rho E(q - q_b) + pq \end{bmatrix} \qquad (3)$$

The components of the viscous flux-density tensor F^v are set to zero here and hence the Euler equations are obtained for inviscid flow. The source term g of Eq. (1) accounts for the time derivatives of the unity direction vectors in the case of a rotating coordinate system. It, g, is proportional to the cross product of the angular velocity vector $\dot{\varphi}$ of the rotation and the velocity vector q,

$$g = \begin{bmatrix} 0 \\ \rho \dot{\varphi} \times q \\ 0 \end{bmatrix} \qquad (4)$$

In Eq. (1) the control volume V with cell boundary ∂V and the outer normal n do not need to be constant with time. If flexible meshes are used, the time-dependent change of local cell volumes is corrected within the CFD code FLOWer by the implementation of a geometric conservation law, following the suggestion of Thomas and Lombard.[30]

Because the compressible flow equations are solved, acceptable convergence rates are obtained for freestream Mach numbers of $M_\infty \geq 0.2$. In all calculations for this chapter, a Mach number of $M_\infty = 0.3$ was used. This not only ensures good convergence of the solutions but it is also low enough to assume incompressible flow behavior, since the compressibility effects are still negligible.

Throughout the chapter, all results are given for an advanced number of motion cycles after the transients from starting the calculation have died out. In most cases, calculation of two or three motion cycles was found to be sufficient to neglect further changes in the results from one period to another. Since an implicit dual time-stepping scheme is employed, the choice of the global time step in the calculation is not restricted by numerical stability. For third-order accuracy of the scheme, 40 time steps per cycle are sufficient to resolve one motion cycle and this resolution was used in all calculations.

III.   Investigations for Two-Dimensional Flow

To simplify the problem of flapping wing propulsion, some first steps are taken with the analysis of the flow around a two-dimensional airfoil in a combined plunging and pitching motion. Although the final aim is the calculation of three-dimensional flow around a flapping and twisting wing of finite span, the two-dimensional calculations were necessary to validate the present method by comparison with analytical results. Futhermore, the use of flexible meshes, which is necessary for the analysis of a twisting wing, was tested and validated with help of the two-dimensional cases.

A.   Description of Airfoil Motion and Two-Dimensional Grid

For the two-dimensional calculatons, a NACA 0012 airfoil section was considered. This is shown in Fig. 1 together with a description of the motion. The plunging $z(t)$ is defined as the leading motion within the inertial coordinate system x, z, which can be regarded as fixed to the birds body cruising with freestream

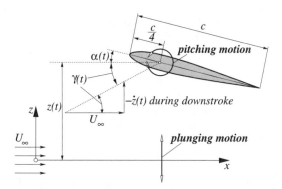

Fig. 1   Schematic of a two-dimensional airfoil in plunging/pitching motion around the $\frac{c}{4}$ axis. Angle of attack γ through plunging is indicated for the downstroke.

velocity U_∞,

$$z(t) = z_1 \cos(2\pi f t) \tag{5}$$

The pitching motion is defined by

$$\alpha(t) = \alpha_0 + \alpha_1 \cos(2\pi f t + \Phi) \tag{6}$$

where the momentary angle of pitch $\alpha(t)$ is counted from the horizontal parallel to the x axis. It consists of a time-independent part α_0, which is equivalent to the mean angle of attack, and a time-dependent part with amplitude α_1. The pitching can vary in phase Φ relative to the plunging motion. The axis of pitch is located 0.25 chord lengths behind the nose. The oscillation frequency f is often expressed in nondimensional form as the reduced frequency

$$k = \frac{\pi f c}{U_\infty} \tag{7}$$

Considering large birds in cruising flight restricts the charcteristic value of k to an order of magnitude of $k \approx 0.1$. For example, a gull (*Larus marinus*[31]) with a root chord length of about 0.16 m flaps its wings with a frequency of $f \approx 3\,\mathrm{s}^{-1}$. The cruising speed ranges from $U_\infty = 12$ to 21 m/s, which leads to corresponding reduced frequencies of $k = 0.126$ and $k = 0.072$.

Together with the dimensionless plunging amplitude z_1/c, the reduced frequency influences the angle of attack caused by pure plunging (see Fig. 1), which is given by

$$\gamma(t) = \arctan \frac{-\dot{z}(t)}{U_\infty} \tag{8}$$

Therefore, the maximum angle of attack through plunging only is approximately

$$\gamma_1 \approx 2 k z_1/c \tag{9}$$

which holds for small values of k and z_1/c, where $\gamma \approx \arctan(\gamma)$. A superimposed pitching motion can enlarge or diminish the momentary effective angle of attack $(\gamma + \alpha)$, depending on the phase shift Φ. To ensure attached flow throughout the whole flapping cycle, the angle $(\gamma + \alpha)$ is always kept below 12 to 15 deg.

Figure 2 shows a special case of the airfoil motion where the pitching leads the plunging by $\Phi = 90$ deg. In this case, the amplitude of the angle of attack through plunging γ_1 is reduced by the pitching, thus resulting in smaller forces in down- and upstroke compared to pure plunging. If the pitching amplitude α_1 is large enough it can almost outweigh the effect of plunging on the effective angle of attack $(\gamma + \alpha)$, thus leading to a motion that is nearly free of aerodynamic forces and close to steady flow conditions. Some unsteady effects remain, however, owing to the effect of dynamic cambering through pitching, phase shifts between motion and aerodynamic coefficients, and the trigonometric approximation made for γ [see Eqs. (8) and (9)].

For reduction of redundant motion parameters, it is useful to define an amplitude ratio λ as

$$\lambda = \frac{\alpha_1 c}{2 k z_1} \approx \frac{\alpha_1}{\gamma_1} \tag{10}$$

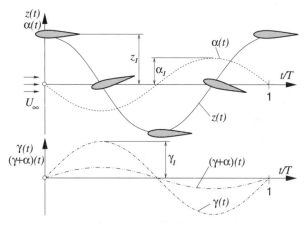

Fig. 2   Time history of plunging/pitching motion for phase shift $\Phi = 90$ deg and mean angle of attack $\alpha_0 = 0$ deg.

This relationship was also used by Birnbaum[6] and it has a geometric meaning, since it relates the pitching amplitude to the maximum angle of attack through plunging. If $\lambda = 0$, there is plunging only. An amplitude ratio of $\lambda = 1$ means that the amplitudes of the angles through plunging and pitching are about equal. It is important to note that λ is itself a function of k [see Eq. (10)]. This has convinced others to define the amplitude ratio as a relation between z_1 and α_1 only, but then the geometric meaning of $\lambda = 1$ gets lost.

For the two-dimensional calculations, two structured grids with O-topology were used, one with a resolution of $160 \times 32 = 5120$ grid points and the other one with $320 \times 64 = 20{,}480$ grid points. They will be referred to as coarse and fine grid respectively. The far-field boundary was separated by 50 chord lengths from the airfoil.

B.   Results and Validation

Figure 3 shows the result of a two-dimensional Euler calculation for a combined plunging and pitching motion with a phase shift $\Phi = 90$ deg. The reduced frequency is $k = 0.1$, and plunging and pitching amplitudes are $z_1/c = 1.0$ and $\alpha_1 = 4$ deg respectively. Horizontal and vertical force coefficients c_x and c_z according to

$$c_x = \frac{F_x}{\frac{1}{2}\rho S_{\mathrm{ref}} U_\infty^2}, \qquad c_z = \frac{F_z}{\frac{1}{2}\rho S_{\mathrm{ref}} U_\infty^2} \tag{11}$$

are plotted against time. In the two-dimensional case, the width b_{ref} of the considered airfoil section is chosen; thus $S_{\mathrm{ref}} = b_{\mathrm{ref}} c$. The force coefficient in the vertical direction c_z is a sinusoidal function of the same period as the motion. A phase shift between the aerodynamic forces and the wing motion is apparent by the fact that $c_{z,\max}$ occurs some time after $t/T = 0.25$ (see Fig. 3). For the horizontal force coefficient c_x there exist two periods in one cycle (see case $\alpha_0 = 0$ deg, solid line in Fig. 3). The coefficient c_x is a measure of the momentary thrust output of

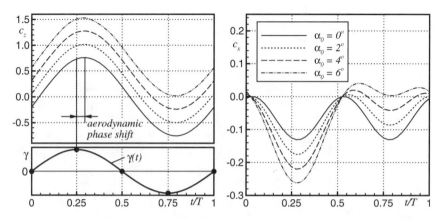

Fig. 3   Time-dependent force coefficients for different mean angles of attack α_0. Plunging/pitching motion with $k = 0.1$, $z_1/c = 1.0$, $\alpha_1 = 4$ deg, $\Phi = 90$ deg, NACA 0012 airfoil, and fine grid.

the motion, where $c_x < 0$ means that thrust is produced. If the mean angle of attack α_0 is zero, the thrust output during the downstroke ($t/T = 0-0.5$) and the upstroke ($t/T = 0.5-1.0$) are symmetrical. This changes with increasing mean angle of attack α_0, where more thrust is produced during the downstroke and drag instead of thrust, can be found during the upstroke. The consequence of the asymmetric thrust distribution plays an important role in bird flight, where the wings are used to produce thrust and lift simultanuously. Depending on the motion parameters, birds may produce nearly all of their thrust during the downstroke. This coincides with the different size of the muscles employed in the down- and upstroke (*musculus pectoralis major* and *minor* respectively).

The effect of a mean angle of attack $\alpha_0 > 0$ deg on the progression of the vertical force coefficient c_z is a constant shift, as shown in Fig. 3. It causes a mean lift $\bar{c}_L$, which is proportional to α_0,

$$\bar{c}_L = \frac{1}{T} \int_0^T c_z \, \mathrm{d}t \tag{12}$$

By integrating the horizontal force coefficient c_x with respect to time, the mean thrust output of the motion is obtained:

$$\bar{c}_T = -\frac{1}{T} \int_0^T c_x \, \mathrm{d}t \tag{13}$$

In Table 1 lift and thrust output are given for all cases of Fig. 3. It can be seen that the mean thrust is almost independent of the mean angle of attack α_0 and, therefore, independent of the generation of a mean lift. This agrees with the assumptions made in all analytical approaches to the calculation of oscillating airfoils, where lift and thrust generation are regarded as mutually independent; that is, additional lift has no influence on thrust output and efficiency of the motion. The slight decrease of thrust in Table 1 can be accredited to the increasing influence of numerical dissipation, which is experienced when the airfoil encounters higher effective angles of attack throughout the motion.

Table 1   Mean lift, mean thrust, and efficiency for a combined plunging/pitching motion at different mean angles of attack[a]

α_0	$\bar{c}_L$	$\bar{c}_T$	η
0 (deg)	0.000	0.065297	0.89556
2 (deg)	0.258	0.065270	0.89540
4 (deg)	0.515	0.065158	0.89443
6 (deg)	0.769	0.064441	0.88945

[a]NACA 0012 airfoil, Euler calculation on fine grid, $k = 0.1$, $z_1/c = 1.0$, $\alpha_1 = 4$ deg, $\Phi = 90$ deg.

The calculation of the propulsive efficiency affords the determination of the power input needed to sustain the motion of the airfoil against the surrounding pressure distribution. In terms of the numerical computation each surface cell face contributes to the input power with the product of its velocity and the actual force on the cell surface. The sum over the complete number of N surface cells and subsequent integration in time finally yields the input power per cycle. In dimensionless form, this can be written as

$$\bar{c}_{\pi,\text{in}} = \frac{1}{TU_\infty} \int_0^T \sum_{n=1}^{N} (c_{x,n} u_n + c_{y,n} v_n + c_{z,n} w_n)\, dt \qquad (14)$$

where u_n, v_n, and w_n are the velocities of the surface cell faces in the Cartesian coordinates of the inertial system. All force coefficients of the surface cells $c_{x,n}$, $c_{y,n}$, and $c_{z,n}$ are referred to the same quantities as the global force coefficients, that is, $(1/2)\rho S_{\text{ref}} U_\infty^2$.

In the same way, the mean power output can be expressed in dimensionless form, which is identical to the definition of the mean thrust coefficient:

$$\bar{c}_{\pi,\text{out}} = \frac{1}{TU_\infty} \int_0^T c_x\, U_\infty\, dt \equiv -\bar{c}_T \qquad (15)$$

The propulsive efficiency is defined as the ratio of power output to power input, and thus

$$\eta = \frac{-\bar{c}_{\pi,\text{out}}}{\bar{c}_{\pi,\text{in}}} \qquad (16)$$

The importance of phase shifts near 90 deg becomes obvious from Fig. 4, where efficiency and mean thrust are plotted against varying phase angle Φ. At $\Phi = 90$ deg, the angle of attack caused by the plunging is in phase with the pitching motion, leading to the highest efficiencies as already explained above. Simultaneously, the mean thrust output is at a minimum. It should be noted that the extreme values for thrust and efficiency occur somewhat offset from $\Phi = 90$ and 180 deg, which is again assigned to the dynamic cambering effect by means of pitching and is caused by the related aerodynamic phase shift.

If similar line plots as in Fig. 4 for various motion amplitudes are combined to produce a three-dimensional surface plot, the efficiency can be shown as a function of amplitude ratio *and* phase shift (see Fig. 5). If λ exceeds 1 near $\Phi = 90$ deg, no

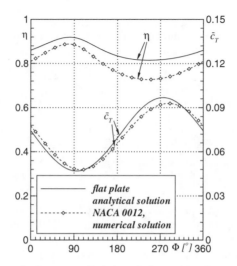

Fig. 4 Efficiency and thrust coefficient varying with phase shift. Comparison between Küssner's theory[9] (flat plate) and Euler solutions on the coarse grid for a NACA 0012 airfoil. Plunging/pitching motions with $k = 0.1$, $\lambda = 0.39$, and $\alpha_0 = 0$ deg.

thrust is produced and, as others have pointed out, flutter occurs.[6,11] No efficiency values are given for this regime. It can also be seen from Fig. 5 that pure plunging ($\lambda = 0$) is also capable of producing thrust. Additional pitching with phase shifts near 90 deg can help to increase the efficiency. Conversely, pure pitching does not produce a mean thrust output, as noted by Birnbaum.[6]

Another way of plotting efficiency results, which is often found in literature, is shown in Fig. 6 for a pure plunging motion. Faster plunging yields more thrust

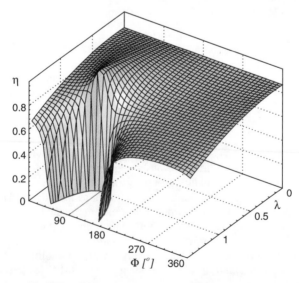

Fig. 5 Efficiency varying with amplitude ratio and phase shift for combined plunging/pitching motions of a flat-plate solution according to Küssner's theory.[9] $k = 0.1$, $\Phi = 90$ deg, and $\alpha_0 = 0$ deg.

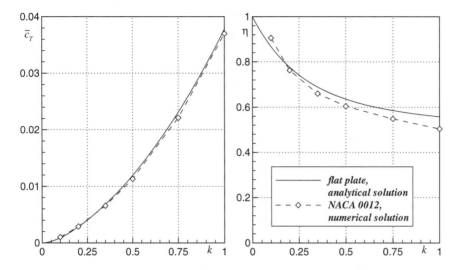

Fig. 6   Thrust coefficient and efficiency varying with reduced frequency. Comparison between Küssner's theory[9] (flat plate) and Euler solutions on the fine grid for a NACA 0012 airfoil. Pure plunging motions with $z_1/c = 0.1$, $\alpha_0 = 0$ deg, and $\alpha_1 = 0$ deg.

while efficiency decreases. Since the flow becomes more and more unsteady with increasing reduced frequency, the propulsive efficiency drops owing to a larger amount of vorticity shed from the trailing edge. The small relative plunging amplitude of $z_1/c = 0.1$ in Fig. 6 was chosen to ensure attached flow at the higher frequencies.

In Figs. 4 and 6 the results are given for a flat plate and the NACA 0012 airfoil. The flat-plate results are obtained from Küssner's theory,[9] which solves the two-dimensional problem analytically, whereas the numerical calculations for an airfoil of finite thickness were carried out with the Euler code described above. The comparison between both solutions was used to show qualitative congruity. Finer computational grids could slightly improve the agreement, but then the effect of a finite airfoil thickness compared to the flat plate must also be considered. Because most of the thrust is produced in the nose region, the grid resolution needs to be extremely fine to properly capture the thrust peak at the leading edge of a thin airfoil.

C.   Implementation of Wing Motion

The numerical results for a two-dimensional plunging and pitching motion can be obtained by defining a rigid body motion for the complete computational grid. However, this only works as long as the moving body is a rigid structure. If three-dimensional motions are considered, a dynamic twisting of the wing along its span affords the use of flexible meshes. The application of such techniques was validated for a two-dimensional test case, which can be compared to the established solution for the rigid body motion of the grid. The available strategies are sketched in Fig. 7 for the simple example of a pure plunging motion of a two-dimensional airfoil.

Case a) represents the technique used for the two-dimensional results, where only one grid needs to be generated. The mesh is indicated by its far-field and

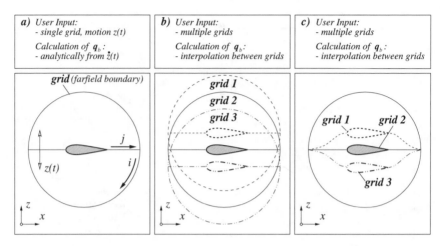

Fig. 7   Different grid motion strategies for a two-dimensional plunging motion.

surface boundary and two connecting lines. To solve Eq. (1) for the moving coordinate system, the velocity vector q_b of each cell boundary needs to be known.
Because the grid is regarded as rigid, and if only plunging is applied, q_b is equal
for every cell surface and can be obtained by deriving the law of motion according
to Eq. (5). In any case of a rigid body motion of the whole grid, the law of motion
is given analytically. Note that if pitching is added, q_b is no longer equal for each
cell surface, but it can still be calculated with the help of Eqs. (5) and (6) and the
distance of the cell face from the center of rotation. As the result of a rotation of
the coordinate system, the source term g of Eq. (1) comes into effect.

The use of flexible meshes implies the preparation of a new computational grid
for each time step of the calculation. Over one flapping cycle the pregenerated
meshes must represent the desired motion of the wing. This is shown in case
b) of Fig. 7, where only three meshes are indicated. Although the same motion
is modeled as in case *a*), it is now defined in terms of multiple meshes rather
than in terms of a single grid and the analytically prescribed motion $z(t)$. The
computational flow solver still needs the input for q_b for each cell. Since the
locations of the cell faces are given at distinct times, the velocities of the cell
faces are obtained by a backward-difference scheme, which is consistent with the
formulation of the time discretization of the conservative variables. Therefore,
cases *a*) and *b*) are only different in terms of the way q_b is calculated.

Three pregenerated meshes are also indicated for case *c*), but here the airfoil
is moving within fixed far-field boundaries. Therefore, the flexible meshes are
changing in volume from one time step to another. The law of motion for each cell
face is not given analytically but is again determined by backward differencing
over two or more time steps as in case *b*).

A comparison of results for all three motion strategies was carried out for a combined plunging and pitching motion of the airfoil with amplitudes and frequency
representative for later test cases. If aerodynamic coefficients are plotted against
time for the motion realized with methods *a*) to *c*), the difference between the
cases is negligible. If the force coefficients are integrated for one period in time,
the error is still less than 1%.

Because a resolution of 40 global time steps per cycle was chosen for the computation of a motion cycle, 40 meshes are needed to represent the complete flapping cycle in cases b) and c) of Fig. 7. However, the generation of large numbers of meshes prior to the solution of the flow equations is costly in terms of computation time and memory requirement and, therefore, another method was chosen for preprocessing. Only seven meshes need to be generated in advance and in accordance with the motion. Seven points for one cycle are sufficient to develop the time history of every grid coordinate into a third-order Fourier series. Instead of the pregenerated meshes, the resulting coefficients are stored for later use during the calculation. The motion of each grid node can therefore be expressed in terms of the Fourier series. Alternatively, the grid in all of the 40 time steps can be determined from the Fourier coefficients. This not only saves memory and computation time but also ensures that the developed time history for each grid point is smooth. The procedure works for all sorts of periodic motions and was used for the calculation of the flapping and twisting three-dimensional wing.

IV.   Investigations for Three-Dimensional Flow

The flapping wing flight of birds involves a complicated motion in three-dimensional space and is hence governed by a large number of parameters. For the calculations subsequently described, a simple test case with respect to wing geometry has been chosen. By limiting the range of the selected motion parameters the flow remains attached throughout the flapping cycle.

A.   Description of Wing Motion and Three-Dimensional Grid

A rectangular wing with aspect ratio of $A = 8$ and a NACA 0012 cross section was chosen (see Fig. 8). The combined flapping and twisting motion of the wing is defined in the same way as the plunging and pitching in the two-dimensional case, but the amplitudes of both are now functions of the lateral coordinate y. The leading edge of the wing is represented by a straight line, which is inclined against the horizontal $x–y$ plane at the flapping angle

$$\Psi(t) = \Psi_1 \cos(2\pi f t) \qquad (17)$$

For convenience, the twisting angle is defined using the spanwise coodinate y^*, which is undergoing the flapping motion,

$$\alpha(y^*, t) = \alpha_1(y^*) \cos(2\pi f t + \Phi) \qquad (18)$$

$$\alpha_1(y^*) = \alpha_{1,\text{tip}} \frac{y^*}{s} \qquad (19)$$

The twisting angle increases linearly along the span with the leading edge being the axis of rotation. The amplitude of the twisting motion at the wing tip, $\alpha_{1,\text{tip}}$, is where $y^* = s$. The two motions can vary in phase and, for the presented calculations, the phase angle was chosen to be constant along the span with $\Phi = 90$ deg. Apart from the amplitudes Ψ_1 and $\alpha_{1,\text{tip}}$ and the phase shift Φ, the motion is governed by the reduced frequency k, which is defined in the same way as in the two-dimensional case [see Eq. (7)]. In addition to the time-dependent motion parameters, a constant angle of attack α_0 can be applied. If $\alpha_0 = 0$ deg, no mean lift is produced with the

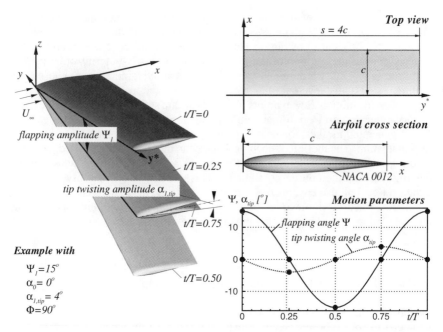

Fig. 8   Wing geometry and motion parameters for the three-dimensional rectangular wing.

symmetric airfoil. In nature, this can be found in the swimming of penguins, where flapping wings are used for propulsion only while the lift in water is provided by the floating body. If $\alpha_0 > 0$ deg, a mean lift will be achieved, corresponding to the conditions of bird flight in air.

From the preceding two-dimensional studies it was assumed that the most interesting range of motion parameters would be the regime at phase shifts of $\Phi = 90$ deg. The reduced frequency was again fixed at $k = 0.1$ in accordance with the in-flight conditions of large birds. The flapping amplitude was chosen to be $\Psi_1 = 15$ deg, which, in combination with $k = 0.1$, is low enough to assure attached flow conditions throughout the flapping cycle. Starting from pure flapping, the twist amplitude $\alpha_{1,\text{tip}}$ was varied as well as the mean angle of attack α_0. All test cases are listed in Table 2.

The numerical method used to calculate the flow around the three-dimensional wing is exactly the same as outlined above for the two-dimensional airfoil. Again, a resolution of 40 time steps per cycle was used and the analysis of the aerodynamic

Table 2   Parameters varied in three-dimensional test cases[a]

Case	a	b	c	d	e	f
α_0 (deg)	0.0	0.0	0.0	4.0	4.0	4.0
$\alpha_{1,\text{tip}}$ (deg)	0.0	4.0	8.0	0.0	4.0	8.0

[a]Common motion parameters: $k = 0.1$, $\Psi_1 = 15$ deg, and $\Phi = 90$ deg.

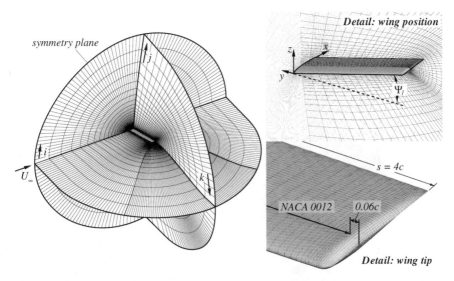

Fig. 9 O-O-grid topology around the flapping three-dimensional rectangular wing at $t/T = 0$.

coefficients was carried out for the second or third cycle to avoid transient effects. Because of the symmetry of the problem only one-half of the wing is modeled, and the surrounding grid in O-O topology can be seen in Fig. 9. It consists of $160 \times 32 \times 40 = 204,800$ grid points and the far-field boundary is separated from the wing by at least 10 chord lengths c. Figure 9 also shows a detail of the mesh at the surface of the wingtip. The NACA 0012 airfoil section ends 0.06 chord lengths before the wingtip is reached. From there, wing thickness decreases while the wing extends to a half-span of 4 chord lengths, with a tip-rounding in the $x-y^*$ plane of radius $0.06c$. The cross sections of the wingtip perpendicular to the $x-y^*$ plane are round shaped near the leading edge and a half-circle is reached at maximum profile thickness. Up to the trailing edge, the cross sections become more and more sharp edged. Although the planform of the wing is rounded at the tip, the reference area of the wing is given by $S_{ref} = 2sc$ (see also Fig. 8, top view). As mentioned before, the motion is defined in terms of meshes around seven different wing positions according to seven points in time during the cycle, during which the wing is moved within fixed far-field boundaries. Computation of two flapping cycles took about 17 hours on an HP®-9000/J5000 workstation. Therefore, only a few test cases were selected for calculation.

B. Results for Pressure Distribution and Aerodynamic Coefficients

The immediate result obtained from the Euler solution is the unsteady pressure distribution around the wing. It was calculated for a flapping and twisting wing with zero mean angle of attack according to Case b) from Table 2. For four different spanwise locations the c_p distribution is plotted in Fig. 10 for the wing in horizontal position, that is, halfway through the downstroke, where maximum flapping-induced velocities occur. Since the effective angle of attack increases with the span of the wing, the outer locations show higher pressure differences.

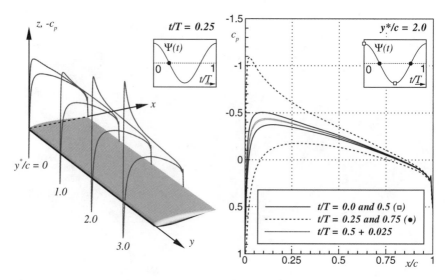

Fig. 10 Pressure distributions for different spanwise locations of the three-dimensional wing at constant time $t/T = 0.25$ (left) and for different times at constant spanwise location $y^*/c = 2.0$ (right) where $k = 0.1$, $\Psi_1 = 15$ deg, $\alpha_{1,\text{tip}} = 4$ deg, $\Phi = 90$ deg, and $\alpha_0 = 0$ deg.

For the midspan location of the wing ($y^*/c = 2.0$) some pressure distributions have been plotted for different times t/T. Because the presented case is symmetric, the pressure distributions for $t/T = 0.0$ and 0.5 as well as for $t/T = 0.25$ and 0.75 are identical with reversed c_p for upper and lower sides of the airfoil. At $t/T = 0.5$ the wing is in its lowest position during flapping and the speed of motion caused by flapping is zero. The momentary angle of attack resulting from twisting is zero at the same time, so that the pressure distribution would be symmetric for a quasi-steady case. However, because of unsteady effects mentioned earlier, this happens some time later, when the wing has already begun the upstroke ($t/T = 0.525$).

In Fig. 11 the time history of the force coefficients is shown for combined flapping and twisting with $\alpha_{1,\text{tip}} = 4$ deg. Two different cases from Table 2 with $\alpha_0 = 0$ deg (Case b) and $\alpha_0 = 4$ deg (Case e) are plotted. The vertical force coefficient c_z is important for the mean lift, which only occurs for the case with $\alpha_0 = 4$ deg, and is a measure of the momentary thrust output. The lateral force coefficient c_y is zero for the entire configuration. However, local wing surface elements are experiencing nonzero lateral forces, while the flapping wing is working against the local pressure distribution. Therefore, local lateral forces are contributing to the input power $c_{\pi,\text{in}}$ according to Eq. (14).

In Fig. 12 thrust output and efficiency are plotted for all considered test cases. In general, values for η are lower in the three-dimensional cases compared to two-dimensional results. Since the wing is experiencing effective angles of attack $(\alpha + \gamma) \neq 0$ during the motion, instantaneous tip vortices are formed behind the wingtip. The presence of such momentary vortices and the corresponding loss in energy decreases the overall efficiency, even for the cases with a mean angle of attack $\alpha_0 = 0$ deg. The additional generation of mean lift at angles of attack $\alpha_0 > 0$ deg leads to a large induced drag and a strong tip vortex, which causes a

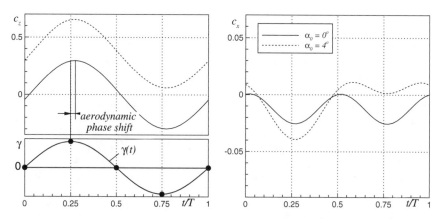

Fig. 11   Time-dependent force coefficients for a three-dimensional wing motion at mean angles of attack $\alpha_0 = 0$ deg and 4 deg where $k = 0.1$, $\Psi_1 = 15$ deg, $\alpha_{1,\text{tip}} = 4$ deg, and $\Phi = 90$ deg.

further decrease in thrust output and efficiency. This can be clearly seen in Fig. 12, if the solid line for $\alpha_0 = 0$ deg is compared to the dashed line for $\alpha_0 = 4$ deg.

Furthermore, the effect of inreasing twist amplitude $\alpha_{1,\text{tip}}$ is apparent in Fig. 12. Since the phase shift is $\Phi = 90$ deg, $\alpha_{1,\text{tip}}$ reduces the angle of attack through flapping in the same way as pitching reduces the angle of attack through plunging γ in the two-dimensional case. This results in a reduced energy loss in the unsteady wake and thus leads to higher efficiencies (see Fig. 5 for $\Phi = 90$ deg and increasing λ). In three-dimensional flow, this only holds for $\alpha_0 = 0$ deg, where the influence of the instantaneous tip vortices is temporary. However, for larger mean angles of

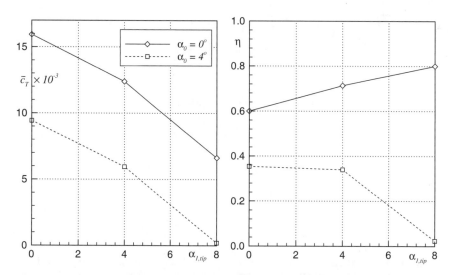

Fig. 12   Effect of twisting on thrust coefficient and efficiency for a three-dimensional wing motion at mean angles of attack $\alpha_0 = 0$ deg and $\alpha_0 = 4$ deg, $k = 0.1$, $\Psi_1 = 15$ deg, and $\Phi = 90$ deg.

attack ($\alpha_0 > 0$ deg) the influence of the trailing vortices becomes dominant and, while thrust output decreases with increasing twist, efficiency drops to zero for the test case with $\alpha_0 = 4$ deg and $\alpha_{1,\text{tip}} = 8$ deg. Further variation of parameters such as phase shift Φ and twist distribution is needed to fully understand the impact of the trailing vortices on thrust and efficiency.

Nonetheless, the proper choice of the motion parameters allows a generation of thrust output in the three-dimensional case. Subsequent investigation will also focus on achieving Navier–Stokes results for such motions, where the excess thrust is consumed to overcome the friction drag of the moving wing. The full account of all aerodynamic phenomena then allows one to find the motion parameters that propel the flapping wing configuration with the cruising speed U_∞. The final aim is to provide a tool that can evaluate the effect of arbitrary wing motions on the performance of flapping wing propulsion for attached-flow conditions.

C. Flowfield Results

Because the Euler equations are solved for the complete flowfield around the wing, the flow conditions are known everywhere and can be visualized (e.g., in planes $x/c = const$ behind the flapping wing). For a single position during the flapping cycle at $t/T = 0.25$ the wing and the planes used for visualization are shown in Fig. 13. The curved arrows indicate the motion of the trailing edge at the wingtip.

In Fig. 14 the velocity vectors are plotted for four different times within the flapping cycle in a plane directly behind the trailing edge. The gray-scale layer

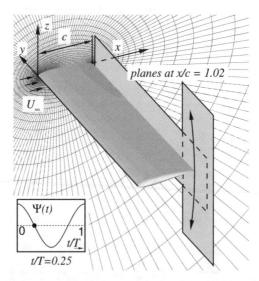

Fig. 13   Orientation of planes $x/c = const$ behind the flapping wing at $t/T = 0.25$. Velocity vectors and vorticity component ω_x are plotted for these planes in subsequent figures.

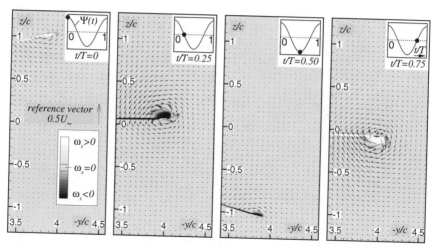

Fig. 14   Velocity vectors and vorticity ω_x in a cross section $x/c = 1.02$ for $\alpha_0 = 0$ deg and $\alpha_{1,\text{tip}} = 4$ deg (Case b of Table 2; see also Fig. 13) with $t/T = 0, 0.25, 0.50,$ and 0.75.

indicates the sign and strength of the vorticity component ω_x in the x direction

$$\omega_x = \frac{1}{2}\left(\frac{\partial w}{\partial y} - \frac{\partial v}{\partial z}\right) \tag{20}$$

The plots of Fig. 14 are shown for $\alpha_0 = 0$ deg and $\alpha_{1,\text{tip}} = 4$ deg (Case b of Table 2); that is, the mean lift is zero. Therefore, the flow conditions are virtually the same in the up- and downstroke. At $t/T = 0.25$, the flapping wing reaches its maximum vertical velocity. This results in a large effective angle of attack, which causes the formation of a momentary tip vortex. Although the wing is then decelerated to its lowest position at $t/T = 0.5$ some remains of the vortex are still present behind the wing because of a lag between the motion and the time history of the aerodynamic quantities. During the upstroke a tip vortex in reverse direction is formed (see $t/T = 0.75$) and is still present in the upper position of the wing ($t/T = 0$).

Near the core of the trailing vortex in mid-downstroke and mid-upstroke a small region of vorticity appears that is reversed to the general direction of rotation of the vortex. This effect can be explained if the vorticity behind the whole wing is considered. This is plotted in Fig. 15 for maximum flapping velocities ($t/T = 0.25$ and 0.75) in a plane $x/c = 1.02$ behind the tailing edge (see also Fig. 13). During the downstroke, the tip vortex is rotating counterclockwise ($\omega_x < 0$). Behind the trailing edge, in the region $0 \leq -y/c \leq 3$, clockwise rotating vorticity with $\omega_x > 0$ is present (white). In the steady flow about a rectangular wing the vorticity shed from the trailing edge has the same sense of rotation as the tip vortex [$\omega_x < 0$, (black)]. In the case of a flapping rectangular wing, the vorticity shed from the inner part of the trailing edge is rotating in reverse direction compared to the wingtip vortex. Both effects interfere with each other while the tip vortex dominates the vorticity near the wingtip. However, the small region of opposite vorticity in the trailing vortex is a remainder of the vorticity generated by the flapping wing.

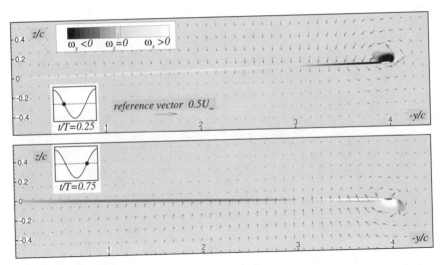

Fig. 15 Velocity vectors and vorticity ω_x in a cross section $x/c = 1.02$ for $\alpha_0 = 0$ deg and $\alpha_{1,tip} = 4$ deg (Case b of Table 2) with $t/T = 0.25$ and 0.75.

In Fig. 16 results for a test case with $\alpha_0 = 4$ deg are given (Case e of Table 2). The figure can be compared to Fig. 14 for $\alpha_0 = 0$ deg. The mean angle of attack causes a tip vortex with counterclockwise direction of rotation, even for a rigid wing without flapping. This vortex is now undergoing a periodic change in magnitude caused by the flapping and twisting motion. In the downstroke, the strength of the vortex is increased, whereas in the upstroke the vortex almost diminishes. The flow conditions are no longer symmetric in both phases, which is already known from the time history of c_x in Fig. 11.

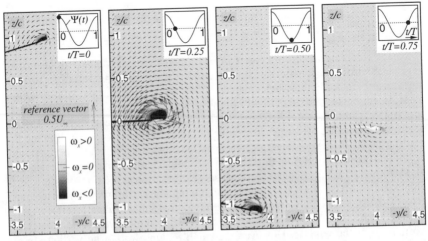

Fig. 16 Velocity vectors and vorticity ω_x in a cross section $x/c = 1.02$ for $\alpha_0 = 4$ deg and $\alpha_{1,tip} = 4$ deg (Case b of Table 2) with $t/T = 0$, 0.25, 0.50, and 0.75.

V. Conclusions

Euler solutions for a two-dimensional plunging and pitching NACA 0012 airfoil have been calculated in agreement with analytical results. The two-dimensional calculations were used to check the available CFD code and the application of flexible grids for unsteady motions of the airfoil. A resolution of 40 time steps per cycle turned out to be sufficient. To provide the meshes for each of the 40 time steps an interpolation procedure based on only seven pregenerated grids has been developed and tested.

Calculations for a three-dimensional flapping and twisting motion have been carried out for a rectangular wing of aspect ratio $A = 8$. Flapping (plunging) and twisting (pitching) varied linearly along the span at a constant phase shift of $\Phi = 90$ deg. Solutions have been provided for attached-flow conditions at a reduced frequency of $k = 0.1$. This led to the choice of a moderate flapping amplitude $\Psi_1 = 15$ deg, while the twisting amplitude was varied, $\alpha_{1,\text{tip}} = 0, 4$, and 8 deg. Mean angles of attack were $\alpha_0 = 0$ deg (no lift, thrust only) and $\alpha_0 = 4$ deg (lift and thrust).

In two-dimensional unsteady flow, the efficiency of thrust generation is governed by the shedding of vorticity from the trailing edge as the result of the time-dependent circulation of the wing. Additionally, unsteady trailing vortices behind the wingtip cause an induced drag in three-dimensional flow. Therefore, the efficiency of thrust generation is considerably reduced compared to the two-dimensional case. For low reduced frequencies and attached-flow conditions, both the interference between the starting and stopping vortices caused by the unsteadiness and the unsteady trailing-edge vortex system resulting from the finite span are weak.

Acknowledgments

This investigation is part of the Graduate College "Structure–Fluid Interaction" of the Technical University of Braunschweig and is supported by the DFG (Deutsche Forschungsgemeinschaft), Germany. The authors would like to thank the Institute of Design Aerodynamics of the DLR Braunschweig, Germany, for providing the source code of the Euler/Navier–Stokes solver FLOWer as well as the MegaCads grid generation tool.

References

[1]Hummel, D., "The Aerodynamic Characteristics of Slotted Wing-Tips in Soaring Birds," *Acta XVII Congressus Internationalis Ornithologici*, edited by R. Nöhring, Vol. 1, Verlag der deutschen Ornithologen-Gesellschaft, Berlin, 1980, pp. 391–396.

[2]Hummel, D., "The Use of Aircraft Wakes to Achieve Power Reductions in Formation Flight," *AGARD Symposium, The Characterisation and Modification of Wakes from Lifting Vehicles in Fluids*, CP-584, AGARD, 1996, pp. 36-1–36-13.

[3]Hummel, D., "Aerodynamic Investigations on Tail Effects in Birds," *Zeitschrift für Flugwissenschaften und Weltraumforschung*, Vol. 16, No. 3, 1992, pp. 159–168.

[4]Hummel, D., and Möllenstädt, W., "On the Calculation of the Aerodynamic Forces Acting on a House Sparrow (Passer domesticus L.) during Downstroke by Means of Aerodynamic Theory," *Fortschritte der Zoologie*, Vol. 24, No. 2/3, 1977, pp. 235–256.

[5]Lilienthal, O., *Der Vogelflug als Grundlage der Fliegekunst*, R. Gaertners Verlagsbuchhandlung, Berlin, 1889.

[6]Birnbaum, W., "Das ebene Problem des schlagenden Flügels," *Zeitschrift für Angewandte Mathematik und Mechanik*, Vol. 4, No. 4, 1924, pp. 277–292.

[7]Theodorsen, T., "General Theory of Aerodynamic Instability and the Mechanisms of Flutter," NACA Rept. 496, 1934.

[8]Garrick, I. E., "Propulsion of a Flapping and Oscillating Airfoil," NACA Rept. 567, 1936.

[9]Küssner, H. G., "Zusammenfassender Bericht über den instationären Auftrieb von Flügeln," *Luftfahrtforschung*, Vol. 13, No. 12, 1936, pp. 410–424.

[10]Bosch, R., "Interfering Airfoils in Two-Dimensional Unsteady Incompressible Flow," *AGARD Symposium, Unsteady Aerodynamics*, CP-227, AGARD, 1978, pp. 7-1–7-15.

[11]Send, W., "The Mean Power of Forces and Moments in Unsteady Aerodynamics," *Zeitschrift für Angewandte Mathematik und Mechanik*, Vol. 72, No. 2, 1992, pp. 113–132.

[12]Katzmayr, R., "Über das Verhalten von Flügelflächen bei periodischen Änderungen der Geschwindigkeitsrichtung," *Zeitschrift für Flugtechnik und Motorluftschiffahrt*, Vol. 13, No. 6, 1922, pp. 80–83.

[13]DeLaurier, J. D., and Harris, J. M., "Experimental Study of Oscillating Wing Propulsion," *Journal of Aircraft*, Vol. 19, No. 5, 1982, pp. 368–373.

[14]Favier, D., Agnes, A., Barbi, C., and Maresca, C., "Combined Translation/Pitch Motion: A New Airfoil Dynamic Stall Simulation," *Journal of Aircraft*, Vol. 25, No. 9, 1988, pp. 805–814.

[15]Chandrasekhara, M. S., Carr, L. W., and Wilder, M. C., "Interferometric Investigations of Compressible Dynamic Stall over a Transiently Pitching Airfoil," *AIAA Journal*, Vol. 32, No. 3, 1994, pp. 586–593.

[16]Jones, K. D., and Platzer, M. F., "An Experimental and Numerical Investigation of Flapping Wing Propulsion," AIAA Paper 99-0995, Jan. 1999.

[17]Nachtigall, W., "Bird flight: Kinematics of Wing Movement and Aspects of Aerodynamics," *Acta XVII Congressus Internationalis Ornithologici*, edited by R. Nöhring, Vol. 1, Verlag der deutschen Ornithologen-Gesellschaft, Berlin, 1980, pp. 377–383.

[18]Rayner, J. M. V., and Gordon, R., "Visualization and Modelling of the Wakes of Flying Birds: Vortices, Gaits and Flapping Flight," *Motion Systems*, edited by W. Nachtigall and A. Wisser, BIONA Report, Vol. 13, Gustav Fischer Verlag, Stuttgart, 1999, pp. 165–173.

[19]Jones, K. D., and Platzer, M. F., "Numerical Computation of Flapping Wing Propulsion and Power Extraction," AIAA Paper 97-0826, Jan. 1997.

[20]Smith, M. J. C., "Simulating Moth Wing Aerodynamics: Towards the Development of Flapping Wing Technology," *AIAA Journal*, Vol. 34, No.7, 1996, pp. 1348–1355.

[21]Vest, M. S., and Katz, J., "Unsteady Aerodynamic Model of Flapping Wings," *AIAA Journal*, Vol. 34, No. 7, 1996, pp. 1435–1440.

[22]Isogai K., Shinmoto, Y., and Watanabe, Y., "Effects of Dynamic Stall on Propulsive Efficiency and Thrust of Flapping Airfoil," *AIAA Journal*, Vol. 37, No. 10, 1999, pp. 1145–1151.

[23]Liu, H., and Kawachi, K., "A Numerical Study of Insect Flight," *Journal of Computational Physics*, Vol. 146, No. 1, 1998, pp. 124–156.

[24]Shyy, W., Berg, M., and Ljungqvist, D., "Flapping and Flexible Wings for Biological and Micro Air Vehicles," *Progress in Aerospace Sciences*, Vol. 35, 1999, pp. 455–505.

[25]Radespiel, R., Rossow, C., and Swanson, R. C., "Efficient Cell Vertex Multigrid Scheme for the Three-Dimensional Navier–Stokes Equations," *AIAA Journal*, Vol. 28, No. 8, 1990, pp. 1464–1472.

[26] Kroll, N., Radespiel, R., and Rossow, C.-C., "Accurate and Efficient Flow Solvers for 3D Applications on Structured Meshes," *Computational Fluid Dynamics*, VKI Lecture Series 1994-04, Brussels, Mar. 1994.

[27] Rossow, C., Kroll, N., Radespiel, R., and Scherr, S., "Investigation of the Accuracy of Finite Volume Methods for 2- and 3-Dimensional Flow," *AGARD Symposium, Validation of Computational Fluid Dynamics*, CP-437, Vol. 2, AGARD, 1988, pp. P14-1–P14-11.

[28] Jameson, A. J., "Time Dependent Calculations Using Multigrid with Appplications to Unsteady Flows Past Airfoils and Wings," AIAA Paper 91-1596, June 1991.

[29] Heinrich, R., and Bleecke, H., "Simulation of Unsteady, Three-Dimensional Viscous Flows Using a Dual-Time Stepping Method," *Notes on Numerical Fluid Mechanics*, Vol. 60, Vieweg Verlag, Braunschweig, 1997, pp. 173–180.

[30] Thomas, P. D., and Lombard, C. K., "Geometric Conservation Law and Its Application to Flow Computations on Moving Grids," *AIAA Journal*, Vol. 17, No. 10, 1979, pp. 1030–1037.

[31] Herzog, K., *Anatomie und Flugbiologie der Vögel*, Gustav Fischer Verlag, Stuttgart, 1968.

Chapter 20

From Soaring and Flapping Bird Flight to Innovative Wing and Propeller Constructions

Rudolf Bannasch*

Technische Universität Berlin, Berlin, Germany

Nomenclature

b = wingspan
C_l = section lift coefficient
C_d = section drag coefficient
c = wing chord
l = characteristic length
Re = Reynolds number ($\rho U_\infty l / v$)
S = wing area
U_∞ = freestream velocity
α = angle of attack in degrees
α_{stall} = stall angle of attack
Γ = circulation
Γ_0 = maximum circulation of the bound vortex
Λ = aspect ratio (b^2/S)
ρ = air density
v = kinematic viscosity

I. Introduction

B IONICS means the use of results of biological evolution in engineering. One century ago, the bionic approach enabled the breakthrough to modern aerodynamics. Studies on the flight apparatus of birds enabled Otto Lilienthal to discover and to utilize the principle of aerodynamic lift generation. Soon after the success of the first man-made gliders, Gustav Weisskopf and the Wright brothers managed the first motor flight, and aerodynamic engineering started its own powerful career. However, the bionics engineer knows that evolution optimizes ingeniously. In contrast to engineering, nature had a huge experimental ground. Over millions

*Senior Scientist.

of years, a wealth of designs have been created, tested, and optimized. In that process, nature had to solve a variety of problems relevant to modern aircraft design as well. The sometimes spectacular achievements of animal locomotion clearly indicate that, in view of the complex fluid-dynamic refinements, engineers can still learn from nature. Indeed, a comparison of the costs of transport of animal flight with that of aircrafts and helicopters in a dimensionless way shows that nature has found much more economic solutions.[1] But animals fly at quite low Reynolds numbers Re, ranging from just under 200 for small insects to less than 10^6 for the fastest large birds. Scaling rules predict that they may not deal with the same flow and drag problems as does engineering. For certain, with respect to the manned flight, it makes little sense and, in many cases, it is virtually impossible to copy the structural solutions developed by nature. Organic life cannot provide "blueprints," for example, for a huge transport aircraft. However, studies on the real achievements in biopropulsion and natural drag reduction show that, especially in fluid dynamics, many details and fundamental effects still remain undiscovered or poorly understood. It is very important to understand what happens at various scales. The aim of the bionic approach is to single out the mechanisms governing biological adaptation, to understand the general idea and direction of optimization, and to have a further look at how to transfer these principles or how to extend these developments into the dimensions and materials applicable in engineering. The latter is a creative process, which may eventually lead to completely new structural solutions as well.

Nature's method of optimization is a further domain of the bionics engineer. By means of the Evolution Strategy[2] optimal configurations for a new parameter constellation can be found and, sometimes, it can also help to prepare a breakthrough into completely innovative concepts.

II. Bionic Airfoil Construction

A. Feathers Help to Prevent Flow Separation

As in engineering, flight safety represents the strongest criterion for selection in flying animals as well. Considering that flow separation represents one of the major reasons for crashes in aviation, one may wonder how birds are able to manage crucial flight situations. Apart from the pilot's skill and favorable scale effects certainly involved, some explanation can also be found in the structure of the bird's flight apparatus. It is well known in ornithology that, for example, during the landing approach or in gusty winds, the feathers covering the upper surface of the bird wings may pop up. In analogy to the behavior of wool threads often used in aerodynamics for flow visualization, biologists interpreted the coverts in conjunction with the mechano-receptors in the skin as a sensor system indicating flow detachment from the bird.[3,4] However, in the 1930s, the aircraft designer W. Liebe became puzzled by the idea that there might be some functional analogy to the boundary layer fences he was working on. In 1938 he created an experiment in which a piece of leather was attached to the upper side of one wing of a fighter airplane, a Messerschmitt Me 109, to simulate bird feathers. Liebe reported (personal communication) that the leather flap caused a dramatic asymmetry to the aerodynamic behavior of the airplane by increasing the lift at the respective wing, especially at higher angles of attack. Since the pilot had difficulties in handling the aircraft while landing, the experiment was considered to be too risky to be repeated again. Forty years later, Liebe explained his ideas in a journal article.[5] Separation starts from the trailing edge where the flow may become unstable at a certain location along the span. Following the negative pressure gradient at the suction side, it spreads out toward the rear of the wing. As soon as the separation reaches the low pressure zone at the leading edge, the lift suddenly breaks down, and the wing stalls. In birds, however, the reverse flow causes the light feathers to pop up, where after they act like a fence or brake, preventing the separation of the flow from spreading out any further. Considering that flow separation is a three-dimensional effect and for the most part begins as a local event, it seems to be an advantage of feathers that they can react locally by forming so-called reverse flow bags just marking off the endangered zone, so that the flow over the other parts of the wing can remain undisturbed, preserving lift generation by the wing.

Liebe's ideas stimulated some rather tentative flight experiments conducted by a group of pilots in Aachen.[6] But that report only gives brief information about a movable narrow plastic strip attached to a glider wing close to the trailing edge.

In 1995, a special research project on this issue was begun by our Bionic Department together with three other research partners: The DLR Berlin, Hermann-Foettinger Institute of Fluid Mechanics at the TU Berlin (HFI) and the STEMME Aircraft Company in Strausberg near Berlin.[7–10] Prior to this project, by the opportunity offered by our Antarctic expedition, we were able to make detailed observations on the biological aspects involved. While some of us were working on penguins, another expedition member, Ingo Rechenberg, investigated the flight maneuvers of the skuas nesting near the colony. Using a telephoto lens with fast autofocus, he could take hundreds of photographs documenting the dynamic behavior of the wing feathers in any possible flight situation (see, e.g., Fig. 1). Thus,

Fig. 1   Antarctic skua in a landing approach (low speed, high angle of attack). The coverts of the right wing form a "reverse flow bag," restricting flow separation to a small, noncrucial area close to the trailing edge. (Photo by Ing. Rechenberg.)

we could gain a deep understanding on what happens on the bird wing *in situ*. These studies became extremely useful for the technical experiments carried out later on.

To get maximum benefit from these observations, our group decided to concentrate on wind tunnel experiments simulating flow regimes close to those observed in nature, whereas the group of Dietrich Bechert (DLR) conducted studies on a section of an original STEMME airfoil in the large wind tunnel of the HFI to learn about the effects of movable flaps at flow conditions close to technical applications and to prepare a construction to be eventually tested on a STEMME S10 motor glider in real flight experiments.

The model wing that we used was rectangular ($b = 0.70$ m, $c = 0.20$ m, $\Lambda = 3.5$) without endplates to allow the development of three-dimensional effects. A NACA 2412 profile was chosen because it shows a well-pronounced decrease in lift beyond the critical angle of attack α_{stall} (solid lines in Fig. 2). The wind speed

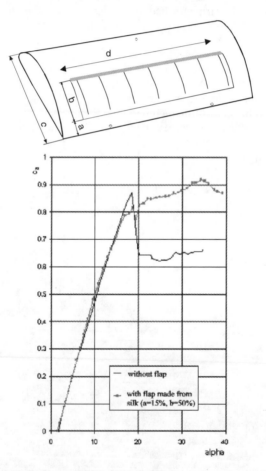

Fig. 2   Silk flap installed on a NACA 2412 airfoil (top); experimental data (bottom). The solid lines indicate data curves without a flap. The distance from the trailing edge, a, is 0.03 m; the length of the flap in the chordwise direction, b, is 0.1 m; the wing chord c is 0.2 m, and the silk flap span d is 0.6 m.

in front of our freestream wind tunnel was adjusted to 10 m/s. Two electronic balances were used to measure the lift and drag force separately.

Numerous materials and flap designs were tested to find the right properties, size, and optimal attachment position to achieve the desired effect. The flaps could be easily changed. In the given flow regime a simple strip of Scotch® tape could be used to fix their leading edge to the model airfoil so that the flap could pivot on this line. From this first series of experiments, we found that a thin plastic material did the best job. It was crucial that the trailing edge be made flexible enough to be sensitive to the reverse flow. This facilitates the whole flap to pop up in time to prevent further disturbances. The flap must be also flexible in the spanwise direction, for otherwise it may cause a dramatic increase in drag by disturbing the flow, for example, in the region close to the wingtip, which (in a three-dimensional case) is usually much more stable in view of separation.

An undesirable effect was, however, that the flaps became slightly raised, even under attached-flow conditions. A typical engineering approach to fight the resulting drag penalty would be to lock the flap onto the airfoil surface in an appropriate way and to release it when necessary. But such an attempt would contradict our "bionic philosophy." Why do not birds need these mechanisms? To answer this question, W. Müller and G. Patone, who did most of the experiments, designed a series of additional investigations to learn more about bird feathers. They came up with a surprising result: The coverts are not airproof. They have tiny pores allowing a small air flow to penetrate through. This small "leakage" can be neglected when the feathers are exposed to a large volume stream as occurs in a separated-flow regime. But it plays an important role in keeping the coverts in the profile line under attached-flow conditions. Because the static pressure gradient is in the chordwise direction on the upper surface of the airfoil, the leading edge of the feathers or flaps protrudes into the minimum pressure zone, whereas the trailing edge experiences a relatively higher static pressure. Since the trailing edge is open, the pressure beneath the flap equals the static pressure at the downstream part of the airfoil. As a result of the pressure difference between its lower and upper side, the flap becomes slightly lifted. In a porous flap, however, the static pressure at both sides becomes balanced. Implementing this "trick" substantially improved our constructions. Later, Bechert and his team found out that a similar effect can also be achieved by making the trailing edge of the flap jagged. This second method can be interpreted to mimic the serial arrangement of the coverts. It facilitates an exchange of pressure as well.

Eventually, in our small-scale experiments, the best results were achieved by a flap made from silk, using a few steel wires to keep it in shape in the chordwise direction but allowing a high degree of freedom in view of the flexibility in the spanwise direction (Fig. 2). By applying this construction, the stall angle of attack could shift from initially 18 deg to more than 40 deg. As desired, this flap was very sensitive, popped up and reattached automatically, and exhibited a small hysteresis. The lift measurements show that apart from stall prevention, those self-activated flaps may also serve as high-lift devices. Interestingly, at least at the relatively small Reynolds numbers used in our experiments, such an effect can be achieved by just a tiny membrane. However, it must be large enough so that the trailing edge stays in contact with the outer flow (to prevent being "washed over" by the reverse flow), the position at which this flap becomes self-stabilized.

Visualization experiments using a smoke sonde showed that a large steady vortex is formed in front of the flap, whereas behind or underneath the flap, a zone of

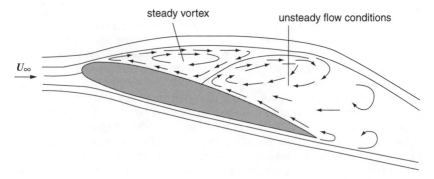

Fig. 3   Scheme of the flow pattern observed in smoke visualization experiments.[10]

highly unsteady flow was observed (Fig. 3). Measurements of the static pressure distribution indicated further that the steady vortex in front of the flap has to be considered as an integrated part of what is usually described as the bound vortex in wing theory. Although this vortex is located on top of the profile, it makes a major contribution to the lift generated at that part of the wing. As this compensates for the decreased lift in the downstream part of the wing, a momentum shift must be evident (unfortunately, those measurements were not considered in our experiments).

Considering the typical arrangement of feathers in a bird wing (Fig. 4), one becomes aware of the following aspects: 1) coverts are considerably longer than the flap we used, 2) they are attached to the very rear part of the wing, and 3) birds possess several consecutive rows of coverts that overlap each other so that a smooth contour of the wing profile is formed. If the caudal coverts pop up, the anterior coverts will be elevated as well. Consequently, the whole nose of the wing obtains another profile. It is possible that the formation of a free steady vortex on top of the profile (Fig. 3) was an effect of our simplified technical arrangement. However, experiments performed by the DLR team with more than one movable flap turned out to be tricky. We could gain little insight, and further research is required to obtain a deeper understanding of the highly sophisticated mechanisms of self-adapting wings developed in birds.

A major problem in transferring those mechanisms into engineering is that the aerodynamic forces increase drastically at larger dimensions. That is why the DLR team did not follow our "membrane theory." Another substantial limitation resulted from the fact that in the STEMME S10, laminar glider airfoils (Horst and Quast profile HQ 41) are used. To maintain laminar flow over the first 70% of the chord

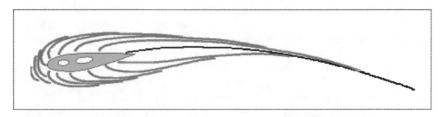

Fig. 4   Cross section of a bird wing (redrawn from Ref. 11).

length, Bechert and his team have designed comparatively small flaps having a length of about 12% of the airfoil chord length. These were attached to the airfoil so that their trailing edge was located slightly upstream of the trailing edge of the airfoil. In this arrangement, limiting strings had to be attached to the flaps to ensure that they do not tip over into the forward direction when exposed to the reverse flow. The beneficial effect of the flap was limited to its full opening. In the given arrangement, the lift increased by up to 10%. Beyond that point, a further increase of the angle of attack may not prevent the separation from jumping over, with the subsequence that the effect of the movable flaps may vanish. A subsequent implementation of all these aspects would have required a completely new wing design. But even by the largely provisional setup tested at the STEMME S10 aircraft, remarkable positive effects could be achieved in real flight tests.

More detailed information on the experiments carried out by the DLR/HFI/STEMME can be found in Refs. 7 and 8; for the studies conducted in our department see Refs. 9 and 10.

In the given context, it is also worth mentioning another promising approach. In his M.S. thesis, one of our students, B. Göksel, applied the electro-aerodynamic effect to control separation and to enhance lift production.[12,13] In this experiment, Plexiglas® end disks were attached to the rectangular airfoil to maintain two-dimensional flow. A thin corona discharge wire (diameter = 0.15 mm) was arranged in the spanwise direction in front of the conducting leading edge of a dielectric airfoil. Additional wire electrodes are located on the upper surface of the wing.

When applying high voltage (16–17 kV) with a maximum corona current of 0.5 mA (nonlethal power range), the electrostatic field causes a weakly ionized air flow. The tangential ionic wind accelerates the boundary layer. As it contributes to the circulation, the bound vortex becomes strengthened, and the lift produced by the airfoil increases (Fig. 5). It should be noted, however, that in the given experimental setup, the ionic wind effect was limited to rather small Reynolds numbers. Nevertheless, in a certain range of supercritical angles of attack the separated flow could be fully reattached by switching the electrostatic field on. Further improvements are in preparation. It turned out that the active electro-aerodynamic lift and separation control and the passive movable flaps may work well together.

B. Bionic Wingtips and Their Potential for Induced Drag Reduction

Aerodynamic lift generation by an airfoil with limited span unavoidably leads to the production of a vortex sheet in the wake (Fig. 6), which rolls up into so-called tip vortices. These vortices induce a cross-stream velocity field that causes a backward force on the airfoil. Apart from the work against friction and parasite drag, additional work is required to compensate for the energy loss resulting from this induced force. In each of the concentrated trailing tip vortices, the circulation equals the maximum circulation of the bound vortex. The total energy contained in the flowfield induced by such a vortex system may vary since it also depends on the distance between the vortices.

The easiest and most effective means to reduce the induced drag is to increase the wingspan. This strategy is widely used in aircraft design (e.g., gliders) as well as

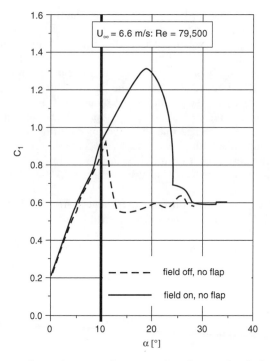

Fig. 5 Electro-aerodynamic separation control and supercirculation effect at low Reynolds number.[12]

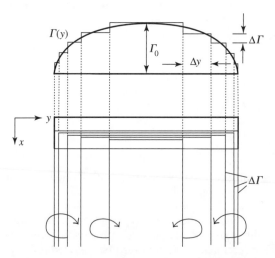

Fig. 6 Vortex model of a planar airfoil with finite span (redrawn from Ref. 14). Vorticity is shed from the trailing edge mainly in the wake of the tip region where the circulation distribution in the bound vortex shows the steepest gradient. At each side, the total circulation in the wake is equal to Γ_0, the maximum circulation of the bound vortex. (The roll-up mechanism is not shown in this graph.)

Fig. 7   Black vulture (Brazil) in soaring flight. (Photo by Ing. Rechenberg.)

biology (e.g., the wandering albatross). But often, the wingspan becomes limited by other considerations (ground handling, etc.). Similarly in nature, there are many environmental situations and functional restrictions qualifying shorter wings to be more applicable. So how did nature answer the induced drag problem in this case?

Most of the birds soaring over land (e.g., eagles, vultures, storks, and kites) show characteristically slotted wingtips (Fig. 7). In response to aerodynamic forces acting on them during flight, the primaries (i.e., the feathers of the hand or, in technical terms, winglets) bend up and become staggered in height. Those multiple-winglet configurations are thought to reduce the induced drag. Such effects are well known from other nonplanar lifting systems in engineering, for example, those described by Prandtl.[15,16] In Prandtl's pioneering studies, he showed that in biplanes and multiple planes, the kinetic energy in the trailing vortex sheet, and hence the induced drag, can be reduced by the spatial spreading of the vorticity in the wake. However, in multiplanes, this beneficial effect becomes somewhat counterbalanced by an increase of the friction drag. But as most of the vorticity is shed from the tip region, there is no need for a nonplanar arrangement in the inner part of the lift producing system.[17] In this respect, the multiple-winglet configuration developed by birds can be seen as a synthesis of a multiple plane reducing the induced drag by spreading the vorticity in the tip region[18,19] and a planar wing keeping the friction drag low at least in the central part of the lift-generating system.[20]

In practice, however, the construction of a multiple-winglet configuration is rather complicated. Owing to their complex aerodynamic interaction, the various parts must be carefully adjusted to obtain the desired effect in drag reduction.[21] To facilitate this, a wing was designed with slotted wing tips that could be varied in respect of their vertical angle and their angle of attack. This was achieved by making the joining base out of lead. In wind tunnel experiments, the geometry of this wing (with five winglets at each end) could then be optimized by means of the Evolution Strategy.[2] In the course of the experiment conducted by M. Stache

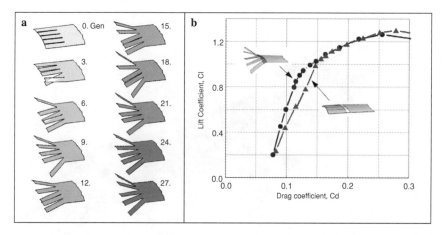

Fig. 8   a) Experimental optimization of a multiple winglet configuration. b) Data of the initial and final wingtip; U_∞ = 7 m/s, Re = 85,000, Λ = 3.9. (Experiments and graphs by M. Stache.)

in our department,[22,23] a winglet configuration with a geometry very similar to that of soaring land birds has evolved (Fig. 8a). In this configuration, the gliding ratio was improved by 11% compared to a planar wing of equal area and span (Fig. 8b). Numerical flow computations confirmed that the induced drag and also the total drag were lower than in a conventional mono-winglet wing. Theoretically, the effect can be enhanced by increasing the number of winglets, but the greater the number of winglets, the more pronounced the interference between them and also the friction drag become. Eventually, the flow through that array becomes totally blocked.

Because of these constrains as well as in an attempt to simplify the construction, a new wing configuration was developed. During numerical optimization of the vortex distribution in the wake of a three-dimensional lift-generating configuration, also by means of the Evolution Strategy, most of the vorticity concentrated along the vertical outlines of that field where a continuous vortex sheet was formed.[22,23] When discussing the results of both optimization experiments in conjunction, M. Stache and I came to the idea[24] to totally remove the inner part of the multiple-winglet configuration and to develop the enveloping curve of the whole configuration as a lifting line, instead (Figs. 9 and 10).

In practice, such a configuration can be achieved by making the now remaining upper and lower winglets broader so that the lift of the planar base part of the wing fully splits up between the two branches. The ends of these branches must then be extended and bent to be connected. In this way, a continuous closed split-wing loop will be obtained. By adjusting the twist, camber, and chord length along that configuration, an optimal vortex distribution (continuous sheet) in the wake can be achieved.[20,22,23] The arrows in Fig. 10a illustrate that at the connecting point of the two branches, the lift gently switches from one surface to the opposite surface, the circulation of the bound vortex changes its sign, the suction side converts into the pressure side, and vice versa, like in a Möbius strip. It does not matter whether the circulation along the airfoil changes from Γ to 0 or from 0.5 Γ to

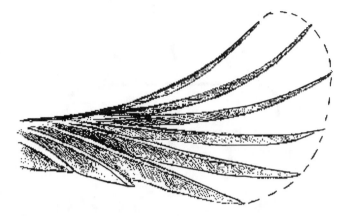

Fig. 9   Primaries of a stork wing: Theoretically, an optimal wake configuration with minimum induced drag can be obtained by a lifting line shaped like the enveloping curve and by an appropriate arrangement of the circulation distribution along that line.

-0.5 Γ; the vortex filaments induced in the wake all rotate in the same direction because their spin depends only on the local gradient in the bound vortex circulation, not on its absolute value.

In addition to reducing the induced drag, the drag penalty caused by friction can also be minimized. Compared to the initial multiple-winglet configuration, the total length of the trailing edge of the wing becomes reduced. Keeping the central part of the wing planar allows one to take advantage of the higher local Reynolds number (reduction of the local wall shear stress) in this area. Additionally, the chord length at the outer part of the loop can be reduced to the degree that the mechanical stability is still guaranteed, but this refers only to the area in which the circulation changes its sign (Fig. 10b). Experimental investigations to verify the theoretical assumptions are in preparation.

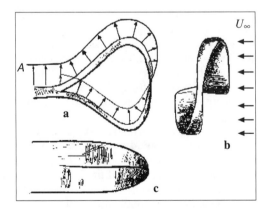

Fig. 10   Sketch of the bionic wingtip configuration developed from the approach illustrated in Fig. 9 for the 'split-wing loop': a) frontal view with lift distribution, b) side view, c) top view.[24]

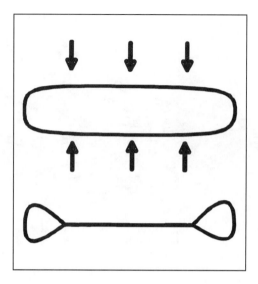

Fig. 11 A conceivable alternative method, which may lead to the same principal construction is to take a box wing or circular wing and squeeze the central part together to make it planar.

The split-wing loop can probably be handled better than a wing with many winglets at its tip. Furthermore, it may help to reduce aerodynamic noise and to facilitate some faster dissipation of the tip vortices, which represent a serious consideration, for example, for the takeoff and landing frequency, especially when huge aircraft become involved. A wide field of other applications (propellers, wind turbines, swords of sailing boats, etc.) is conceivable.

Figure 11 illustrates that a similar construction can also be obtained in other ways. If that, indeed, represents an optimum, say the top of the fitness hill, then one can imagine that it can be approached from various sides. And, there were other "hill climbers" as well. Soon after our bionic invention called "Schlaufenförmiger Quertriebskörper/Split-wing loop"[24] was submitted to the patent office, we found out that before us, Luis Grazer, the former Boeing chief of aerodynamics, came to a very similar solution from another side. He developed a construction he called "Spiroid Wing Tip."[25] Some minor disadvantages of the construction described in his patent could, obviously, be eliminated when a practical application is developed. Figure 12 shows that Spiroids eventually applied to a Gulfstream II (G-II) aircraft (www.aviationpartners.com) became more "organic." It converged to our bionic solution.

On its home page, the Aviation Partners, Inc. claims that initial flight tests of the Spiroid concept on a G-II carried out in 1998 yielded a reduction of the cruise fuel consumption by more than 10%. The Spiroid eliminated concentrated wingtip vortices, which represent nearly half the induced drag generated during cruise. Vorticity is gradually shed from the trailing edge. No doubt, engineers may not need to study biology to find optimal structural solutions. However, this particular example shows that the various lines of development converge when approaching the optimum. Meanwhile, we have heard (personal communication with J. Rayner) that another construction called "Moebius Wing" was developed

Fig. 12   Gulfstream II aircraft with Spiroid Wing Tips developed by Dr. Luis Gratzer. (Photographs taken from the home page of Aviation Partners, Inc.)

in Russia, possibly even earlier. Unfortunately, we were unable to obtain a copy of that patent, as yet. So, it remains a mystery to us what is behind.

III.   Bionic Propeller

So far, only the fixed wing situation has been considered. It is well known, however, that in a rotating airfoil, the problem of concentrated wingtip vortices becomes much more pronounced. In aircraft using propellers, the amount of power and therefore the overall performance of an airplane is often limited not by the engine but by the amount of power that can be converted to thrust within the limitation of propeller size, propeller efficiency, and noise produced. All three factors are strongly related to the strength of the vortices generated at the tips of the propeller blades. For obvious reasons, the propeller size is often limited in practice. The higher the power level relative to the propeller diameter the stronger the vortices become and efficiency is sacrificed and noise levels are raised. In this respect, the performance can be improved by enclosing the propeller in a shroud. Shrouds tend to disperse the tip vortices, but the vortices reform at virtually full strength some distance downstream from the propeller plane, limiting the benefits of the shrouds. Structural weight, manufacturing costs, and a frictional drag penalty are other considerations limiting a widespread commercial application of shrouds.

L. Gratzer puzzled about this problem as well. He came up with a "Ring-Shrouded Propeller" in which the shroud is attached to the propeller blades and rotates with them.[26] The shroud was shaped to produce countervortices close to the locations where the propeller blades shed their tip vortices. The idea of this invention was that in each pair, the two opposite rotating vortices will eliminate each other and that, as a consequence, only minute vortices are shed from the trailing edges of the propeller blades and the shroud. A major disadvantage of this construction seems to be that complicated adjustments are required to adapt the shroud to variable loadings of the propeller, and the concept of vortex killing by a secondary system does not really convince me. One can consider a propeller simply as a pump producing a momentum flux with vortices appearing as a side effect, or one can use the vortex theory to calculate the flowfield involved. Independently from what researchers would like to give priority, both approaches must conclusively lead to the same result.

In the flapping flight of birds, at least the downstroke can be considered to represent a certain analogy to a propeller. Smaller bird species especially use a considerably high wingbeat frequency. As one can see in slow-motion films, many of these birds own wings with deeply split primaries, obviously for the same reason discussed above for soaring birds. This represents a further indication that the multiple-winglet concept and consequently also the idea of the split-wing loop are general principles that may work equally well under various boundary conditions, also including different scales. Thus, we may have a chance to apply our bionic approach to propellers as well, possibly with some modifications.

The split-wing loop concept is intriguing. It demonstrates that it is possible to split and bend the respective lifting lines in any appropriate way and to adjust the circulation distribution along those lines to distribute the vorticity in the wake so that the least amount of energy is lost. In this respect, an optimal solution for a propeller would be to envelop the stream downstream from the propeller plane in a continuous vortex sheet. Such a wake configuration would be similar to that formed close to the trailing edge downstream of a conventional shrouded propeller but without the respective disadvantages listed above. But how can this be achieved?

Let us take a usual propeller with, say, four blades. In a somewhat simplified model, each of theses blades can be represented by a lifting line with a given circulation distribution. The solid line in Fig. 13 shows the optimal circulation distribution found by A. Betz for such a configuration.[27]

Let each line split into two branches of about the same strength, with the split point at a position, say, where the circulation achieves its maximum. So, each of the two branches carries just half of the circulation of the basic part. With respect to the wake, this results in the formation of two smaller tip vortices, each having just half the strength of the initial large one. Now, in a highly abstract approach ignoring the flow conditions at a real propeller for a while, we lengthen the branches of the lifting line with the consequence that the circulation distribution along these parts flattens out. This should lead to the effect that vorticity is shed more gradually from these parts. However, as long as the respective vortex sheets have a free margin, they tend to roll up. The only measure to restrain them from reforming to a concentrated tip vortex is to connect the vortex sheets so that the velocities the vortex filaments induce on each other become balanced across the whole area. In contrast to the split-wing loop (or Gratzer's Spiroid Wing Tip) developed for

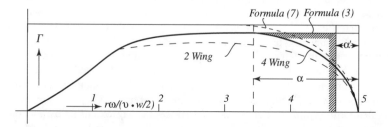

Fig. 13 Optimal circulation distribution at a propeller blade calculated by Betz.[27]

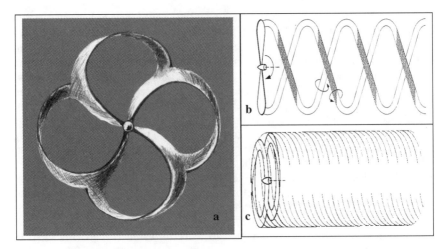

Fig. 14 a) Sketch of a bionic propeller with four blades. b) A usual propeller generates concentrated tip vortices. c) A continuous thin vortex sheet is shed from the trailing edge of the ringlike outer part of the bionic propeller.

fixed wing aircraft, we can take advantage of the serial arrangement of blades in the propeller. Here, the desired configuration can be obtained in another way.[28] In each blade, with respect to the direction of rotation, we bend one branch forward and the other one backward so that, the leading one connects to the trailing branch coming from the blade in front and the second one connects to the leading branch coming from the blade behind. Eventually, the outer part of the propeller will form a ringlike structure like that shown in Fig. 14. Consequently, the vortex sheet shed from the trailing edge of this outer part forms a tube with a considerably thin wall. In the cross section, it is not an ideal circle, but it is not far from circular. All minute vortices in this tube rotate in the same direction. Their strength and special distribution can be optimized by adjusting the chord length, twist, and camber along the close-looped outer part of the propeller. Hence, there is a real chance of eventually achieving a stable wake configuration.

The form of flow from the propeller sketched in Fig. 14c is known to involve minimum energy loss (i.e., maximum efficiency and also minimum noise generation for a given propeller loading[26]). Thereby, the small penalty in friction drag owing to the increased overall length of the structure becomes more than compensated. Nevertheless, it may be appropriate to gradually reduce the chord in the middle of the loop sections in a similar way as was discussed for the split-wing loop.

Although it looks rather simple, several attempts were required to find out how to build such a three-dimensional configuration. Our first model was made out of paper. Experiments with this bionic propeller showed that the close-looped, ringlike structure develops an enormous mechanical stability. Even the paper model was able to withstand considerably high loads. This can also be advantageous for keeping the structural weight low. Moreover, it turned out that the construction is self-stabilizing to a certain degree. The latter property was further investigated using a second model made from flexible plastic bands, which were flat (without profile) but twisted in an appropriate way. If one changes the angle of attack by turning the blades at their point of attachment to the hub, the whole configuration becomes twisted. The twist angle gradually declines from the root toward the periphery. Certainly, this behavior can be influenced or steered by choosing the right stiffness distribution along that structure. Apart from that active manipulation, some self-adjustment of the angles of attack, which means a certain passive adaptation to variable loads, also could be observed in the second construction. Even at this very preliminary stage of development, the bionic propeller (second construction) could well compete with a professionally designed propeller for model aircraft (Graupner) with similar design parameters. In our first wind tunnel tests, although the bionic propeller had no profile, it already produced twice the thrust. The total efficiency suffered somewhat from some flow separation at the inner part, but, nevertheless, it was also a bit higher. There is no doubt that this concept works well, and with some optimization, it may find a wide range of practical applications. Highly efficient aircraft propellers, low-noise ventilators, fans, exhausters, blowers, windmills, turbines, etc. may benefit from this invention.

The bionic propeller can be adapted to work in water as well (Fig. 15). In liquid environments, another interesting aspect comes up. Because of the reduced tip loading, the highly undesired cavitation effect can probably be delayed. This would help to avoid material erosion and noise emission. Ships can be made faster. Bionic propellers may also have a better "grip," and so their application can be useful to improve maneuverability. At the rear of a ship, propellers operate in a highly turbulent flow. Here, the close-looped construction will be useful to reduce vibration and thereby the load on the structure as well as noise emission. Taken together these properties should yield lower costs for maintenance and a longer life-time. We have just started to explore the interesting properties of this new construction. What we have learned so far is very promising.

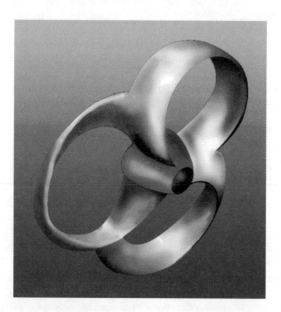

Fig. 15   Draft of a bionic ship propeller with three blades (not yet optimized).

The considerations made before apply to a single propeller. Although it is obvious to those skilled in propellers, it might be worth mentioning here as well that further improvements can be achieved when using, for example, a dual propeller arrangement with contrarotation about a single axis. In such an arrangement, the slipstream rotation can also be eliminated. The same effect can also be achieved by keeping the first propeller fixed (stator) and by arranging one or several bionic propellers in a line downstream. In such an assembly, propellers with different blade numbers can be used to avoid vibration caused by interference. And, in an optimal arrangement, the subsequent propellers may not have the same diameter. In a good design, their vortex systems should superimpose on each other. Elimination of the slipstream rotation also involves the elimination of the vortex components rotating in the plane perpendicular to the main flow. This refers to the vortices shed from the inner parts of the blades close to the hub as well as to the vortex sheet forming the envelope of the wake. Eventually, the vortex system in the wake of such an assembly would consist of a thin vortex tube represented by a continuous sheet of minute vortices that rotate around circular lines. This vortex tube would enclose a linear jet stream with constant velocity over the whole cross section. Such a flow configuration would involve the absolute minimum energy loss achievable. The aim of practical optimization is to approximate this ideal as closely as possible.

Finally, it should be noted as well that reduced fuel consumption, pollutants, noise emission, reduced total vorticity, and faster dissipation of the vortices in the wake are also important considerations in view of environmental protection. Consequently, there may be some positive feedback to nature as well.

IV. Conclusions

Although birds fly at relatively low Reynolds numbers, they show a number of adaptations that are interesting not only from the biological point of view. Some of the fluid-dynamic mechanisms involved may work at other flow regimes as well. A combined biological and engineering approach is useful for identifying the problem possibly solved by a given structural adaptation. Creative interdisciplinary work is required to design and to conduct experiments with appropriately scaled models to explore the phenomena. However, often it is not that easy to get the desired result. Nature does not provide "blueprints" for practical applications in engineering. The aim is to find the underlying principle and to learn how to "translate" it into technology. At an early stage of development, some of the questions can be referred back to nature. Later, the technological application may lose any similarity to the biological example. For those not familiar with the history of its development, it might be hard to believe that, for example, the bionic propeller represents an implementation of a fundamental principle derived from a flapping bird wing, albiet at a fairly abstract level.

Grazer's Spiroid Wing Tip shows that engineering may not need inspiration from biology. If there is an optimal solution, sooner or later engineers will find it. But sometimes the bionic approach can help reveal a shortcut in those developments. I am sure, when biologists and engineers learn to understand each other, there will be a large benefit for all of us as well as for the environment. In this sense, there may be a good chance to combine our effort to develop more "organic" aircraft and other useful constructions in the near future.

Acknowledgments

Our research was supported by grants given by the Deutsche Forschungsgemeinschaft, Volkswagen-Stiftung, Bundesministerium für Forschung und Technologie, and INTAS.

References

[1]Videler, J., "Comparing the Cost of Flight: Aircraft Designers Can Still Learn from Nature," *BIONA Report 8 (1. Bionik-Kongress, Wiesbaden)*, edited by W. Nachtigall, Akad. Wiss. u. Lit., Mainz; G. Fischer, Stuttgart/New York, 1992, pp. 53–72.

[2]Rechenberg, I., *Evolutionsstrategie '94*, Problemata, Frommann-Holzboog, Stuttgart, 1994.

[3]Rüpell, G., *Vogelflug*, Rowolth Taschenbuch Verlag, Reinbeck, 1980.

[4]Nachtigall, W., *Warum die Vögel fliegen*, Rasch und Röhring, Hamburg/Zürich, 1985.

[5]Liebe, W., "Der Auftrieb am Tragflügel: Entstehung und Zusammenbruch," *Aerokurier* Vol. 12, 1979, pp. 1520–1523.

[6]Malzbender, B., "Projekte der FV Aachen, Erfolge im Motor- und Segelflug," *Aerokurier* Vol. 1, 1984, p. 4.

[7]Bechert, D. W., Bruse, M., Hage, W., and Meyer, R., "Biological Surfaces and Their Application—Laboratory and Flight Experiments on Drag Reduction and Separation Control," AIAA Paper 97-1960, 1997.

[8]Meyer, R., Bechert, D. W., Hage, W., and Montag, P., "Aeroflexible Oberflächenklappen als 'Rückstrombremse' nach dem Vorbild der Deckfedern des Vogelflügels," Abschlußbericht DLR-IB 92517-97/B5, 1997.

[9]Rechenberg, I., Bannasch, R., Patone, G., and Müller, W., "Aeroflexible Oberflächenklappen als 'Rückstrombremse' nach dem Vorbild der Deckfedern des Vogelflügels," Abschlußbericht 1997 für das BMBF-Vorhaben 13N6536/7, FG Bionik und Evolutionstechnik, Technische Universität Berlin, 1997.

[10]Patone, G., "Deckfedern als Vorbild für technische Rückstromtaschen an Tragflügeln," Ph.D. Dissertation, FG Bionik und Evolutionstechnik, Technische Universität Berlin, Berlin, 2000.

[11]Burton, R., *Bird Flight*, Facts on File, New York, 1990.

[12]Göksel, B. "Experimente zur Strömungsbeeinflussung mittels elektrischer Felder am Tragflügeprofil E338," Diplomarbeit, FG Bionik und Evolutionstechnik und Institut für Luft- und Raumfahrt, Technische Universität Berlin, Berlin, 2000.

[13]Göksel, B., "Verbesserung der aerodynamischen Effizienz und Sicherheit von Mikro-Flugzeugen durch Ablösekontrolle in teilionisierter Luft," DGLR-JT2000-203, *Proceedings DGLR-Jahrestagung*, Leipzig, 2000, pp. 1317–1331.

[14]Schlichting, H., and Truckenbrodt, E., *Aerodynamik des Flugzeuges*, 2 Bände, 2. Auflage, Springer-Verlag, Berlin, 1967 and 1969.

[15]Prandtl, L., *Ergebnisse der Aerodynamischen Versuchsanstalt zu Göttingen. II. Lieferung*, R. Oldenburg, München/Berlin, 1923.

[16]Prandtl, L., "Tragflügeltheorie. II. Mitteilung," *Vier Abhandlungen zur Hydrodynamik und Aerodynamik*, Kaiser Wilhelm Instituts für Strömungsforschung, Göttingen, 1927.

[17]Kroo, I., McMasters, J., and Smith, S. C., "Highly Nonplanar Lifting Systems," *Transportation Beyond 2000: Technologies Needed for Engineering Design*, NASA Langley Research Center, 26–28 September 1995.

[18]Tucker, V. A., "Drag Reduction by Wing Tip Slots in a Gliding Harris' Hawk, *Parabuteo Unicintus*," *Journal of Experimental Biology*, Vol. 198, 1995, pp. 775–781.

[19]Tucker, V. A., "Gliding Birds: Reduction of Induced Drag by Wing Tip Slots between the Primary Feathers," *Journal of Experimental Biology*, Vol. 180, 1993, pp. 285–310.

[20]Stache, M., and Bannasch, R., "Bionische Tragflügelenden zur Minimierung des induzierten Widerstandes," *BIONA Report 12*, edited by W. Nachtigall and A. Wisser, Akad. Wiss. u. Lit., Mainz; G. Fischer, Stuttgart, 1998, pp. 211–224.

[21]Hummel, D., "Recent Aerodynamic Contributions to Problems of Bird Flight," *11th JCAS Congress*, edited by J. Singer and R. Staufenbiel, Vol. 1, Paper A 1-05, Lisbon, 1978, pp. 115–129.

[22]Stache, M., "Evolutionsstrategische Optimierung eines Tragflügels mit aufgespreizten Flügelenden," Diplomarbeit, Bionik und Evolutionstechnik, Technische Universität Berlin, Berlin, 1992.

[23]Stache, M., "Entwicklung von Flügelenden nach dem Vorbild der Vögel," Ph.D. Dissertation, FG Bionik und Evolutionstechnik, Technische Universität Berlin, Berlin, 2000.

[24]Bannasch, R., and Stache, M., "Schlaufenförmiger Quertriebskörper ('Split Wing Loop')," Deutsche Patentanmeldung No. 19752820.1, 1997.

[25]Gratzer, L. B., "Spiroid-Tipped Wing," U.S. Patent No. 5102068, 1992.

[26]Gratzer, L. B., "Ring-Shrouded Propeller," U.S. Patent No. 5096382, 1992.

[27]Betz, A., "Schraubenpropeller mit geringstem Energieverlust," *Vier Abhandlungen zur Hydrodynamik und Aerodynamik*, edited by L. Prandtl and A. Betz, Kaiser Wilhelm-Institut für Strömungsforschung, Göttingen, 1927.

[28]Bannasch, R., "Rotor mit gespaltenem Rotorblatt," Deutsche Patentanmeldung No. 19931035.1, 1999, PCT/EP00/06412, 2000.

Chapter 21

Passive Aeroelastic Tailoring for Optimal Flapping Wings

Kenneth D. Frampton,[*] Michael Goldfarb,[†] Dan Monopoli,[‡]
and Dragan Cveticanin[‡]
Vanderbilt University, Nashville, Tennessee

Nomenclature

C_T = coefficient of thrust = $T/\frac{1}{2}\rho v^2 s^2$
c = chord
Re = Reynolds number = $\rho v s/\mu$
s = span
T = thrust (in horizontal plane)
v = wingtip velocity = $\omega_e s$
γ = wing frequency ratio = ω_b/ω_t
η = excitation frequency ratio = ω_e/ω_t
ρ = density of air
ω_b = wing bending natural frequency
ω_t = wing torsion natural frequency
ω_e = excitation or flapping frequency

I. Introduction

I N AN effort to overcome many of the Reynolds number limitations associated with micro air vehicles (MAVs) much attention has turned to the investigation of flapping flight. Biologists have studied bird and insect flight empirically for quite some time. A good review of this material can be found in Refs. 1, 2, and 3 while some interesting recent work can be found in Refs. 4, 5, and 6. One thing that is clear is that all of these creatures use two specific mechanisms to overcome the small-scale aerodynamic limitations: flexible wings and flapping wings. Birds and insects exploit the coupling between flexible wings and aerodynamic

*Assistant Professor, Department of Mechanical Engineering.
†Associate Professor, Department of Mechanical Engineering.
‡Graduate Research Assistant, Department of Mechanical Engineering.

Fig. 1   Ornithoptic MAV prototype developed at Vanderbilt University.

forces (termed aeroelasticity) such that the aeroelastic wing deformations improve aerodynamic performance.[4–6] It is interesting that, historically, mankind has sought to minimize aircraft wing susceptibility to these very same aeroelastic responses.[7,8] In addition, by flapping their wings, birds and insects effectively increase the Reynolds number seen by the wings without increasing their forward flight speed.[1–3] Current programs aimed at developing ultra-efficient actuation systems that will enable the realization of small-scale, flying robotic insects (such as the example in Fig. 1) have identified aeroelastic wings as a critical hurdle in the development of these devices.

Our understanding of the fundamental physics behind low Reynolds number, small-scale aerodynamics is limited as is our understanding of aeroelastic effects at small scales. The analytical studies that have sought to understand flapping flight employed either rigid wings or wings with a prescribed motion.[9–12] The inclusion of aeroelastic coupling, which accounts for the aerodynamically induced wing deformation, has not been modeled for small-scale wings or for flapping wings. Yet, the aerodynamic performance of birds and insects cannot be predicted or understood without considering aeroelastic effects. If our understanding of the physics involved in low Reynolds number flight is limited, then our ability to construct aircraft of this scale will also be limited.

The object of this work is to quantify some of the fundamental parametric influences on the performance of flapping wings. In particular, we are interested in the effect that the phase angle between translational and rotational motion have on wing thrust production. Dickenson et al.[6] performed an experiment that highlighted the importance of these effects. Namely, the efficiency with which flapping wings

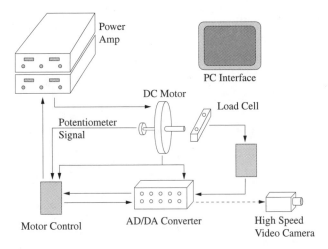

Fig. 2   Schematic of the experimental rig.

generate thrust forces is strongly dependent on the phase lag between translational and rotational wing articulation. Our primary objective is to investigate a method of wing construction that results in an optimal relationship between flapping wing bending and twisting such that optimal thrust forces are generated. The desirable aeroelastic coupling between wing bending and twisting is established through wing construction techniques and materials. Experiments aimed at measuring the wing thrust and power were carried out to quantify the wing performance.

II.   Experimental Setup

A test bed, shown in Fig. 2, was built to measure the instantaneous thrust forces and power usage of different wing designs. This rig consisted of a low-inertia dc motor to create flapping motions with preselected frequency and amplitude. Wings were connected to the motor shaft by a load cell and were articulated in the vertical plane. The rig permitted the variation of flapping frequency and amplitude. The system was managed with a PC that performed experimental control and data collection. The data that were collected included load cell readings (detailed later), motor shaft position, and motor shaft velocity. The PC was also used to trigger a high-speed video camera. Post-test data processing included low-pass filtering of all signals, the calculation of average thrust, and the synchronization of quantitative data with the high-speed video. The most critical data that were collected came from the load cells as detailed in the following section.

A.   Load Cell Configuration

Two types of load cells were used to collect data in the tests. One was a binocular type of design as shown in Fig. 3. The binocular load cell was used to measure the thrust created by the wing in the horizontal plane (wing motion occurred in the vertical plane). This type of load cell was desirable because it is only sensitive to force and not moment. One end of the load cell was attached to the motor shaft while

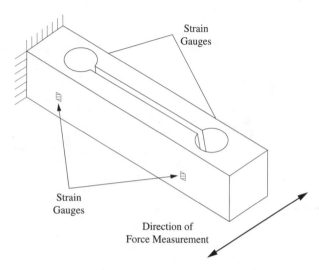

Fig. 3   Depiction of the binocular load cell.

the wing was attached to the other end. A full-bridge strain-gauge circuit was used, consisting of four semiconductor gauges bonded to the aluminum device as shown in Fig. 3. During thrust testing, the load cell was oriented to measure horizontal forces while the wing was articulated (flapped) in the vertical plane. Thus, the load cell was insensitive to any vertical forces (be they inertial or otherwise).

The second load cell employed was of a simple beam type. This device consisted of an aluminum bar with two strain gauges attached opposite to each other. Again, one end of the load cell was attached to the motor shaft and the other was attached to the wing. A half-bridge conditioning circuit was used in this case. This load cell measured the torque applied to the base of the wing (as opposed to the force measurement of the binocular load cell). Knowing the torque and motor shaft velocity permitted the calculation of the input mechanical power.

B.   Video Synchronization

Wing motion and thrust data were correlated by synchronizing high-speed video recordings with instantaneous thrust data. Synchronization was accomplished by triggering the high-speed camera to begin recording at a known point in the thrust data collection process. Video of the flapping wing was recorded at 500 frames per second and matched with data for three full stroke cycles. Using Adobe® Premiere video editing software, the high-speed video of the wing was synchronized with an animation of the thrust data plot, including the timeline to denote the location of the data corresponding to the wing video. The two videos were superimposed upon one another to create a single video file.

C.   Wing Construction

All wings were constructed with a similar planform to preserve the geometric similarity. These wings had a roughly rectangular shape with a span of 6 in. and an aspect ratio of 2.3. The wings were made with Mylar® sheet, which is stiffened by carbon-fiber ribs. The Mylar sheet was essentially rigid to in-plane motion while the carbon-fiber ribs provided out-of-plane or bending stiffness. One rib comprised the leading edge of the wings while four other ribs ran in the chord direction. The primary change in wing construction was the width and thickness of the leading-edge spar. By varying the spar width and thickness the bending natural frequency could be tuned over a wide range.

The primary objective in wing construction was to develop wings that possess specific relationships between their translational and rotational motion. This desire was based on findings that indicated that the phase lag between wing translation and rotation is critical to the efficient production of thrust.[6] However, the articulation of these wings is constrained to a single degree of freedom (i.e., wing root rotation). Therefore, the specified articulation would have to be achieved through passive elastic design. This was done by tailoring the natural frequencies of the fundamental wing bending and torsion modes. Consider the case where the fundamental wing bending and torsion modes have equal natural frequencies. When these two modes are excited by wing-root rotation they will respond with equal phase relative to the input. In other words, the wing bending (translation) and wing torsion (rotation) will be in phase. Now suppose that the bending natural frequency is twice the torsional natural frequency. If the wing is excited by an input at the torsional natural frequency then the torsional response will lag the input by 90 deg. However, the excitation frequency is half of the bending natural frequency so that the bending response is nearly in phase with the excitation. The net result is that the torsional (rotation) will lag the bending (translation) by 90 deg. Therefore, by simply tuning the natural frequencies of the wing we are able to achieve a range of lag and lead phase between translation and rotation.

This concept of tailoring the wing articulation through passive, elastic design does not take into account the effects of aerodynamic loading on the bending and torsion response. It also assumes that the bending and torsion modes are uncoupled. It is well known in the field of aeroelasticity that when the aerodynamic loading is significant (relative to the structural stiffness) significant modal coupling will result.[7,8] The impact of these aeroelastic coupling effects on flapping wing performance is one of the primary targets of this investigation. The test results described in the next section demonstrate the impact of these effects.

III. Results

A. Experimental Objectives and Procedures

The primary objective of these experiments was to quantify the importance of various, fundamental nondimensional parameters on the production of thrust by flapping wings. The parameters under investigation included Reynolds number Re, wing frequency ratio γ, and excitation frequency ratio η. One of the objectives was to develop a desirable bending/torsion motion in the wing with a single degree of freedom excitation. This was motivated by the findings of Dickenson et al.,[6] who established that the relative phase between wing translation and rotation is crucial to the optimal development of thrust. However, typical biological aviators have several degrees of freedom with which they articulate their wings. Our goal was to achieve a similar wing articulation with only a single excitation degree of freedom. This was to be accomplished by tuning the wing bending and torsion natural frequencies of the wing as described previously.

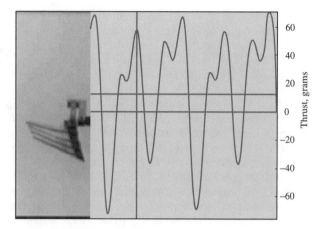

Fig. 4   Synchronized video and data for the $\gamma = 1$ wing.

B.   Experimental Results

As outlined previously, several wings were tested with wing frequency ratios ranging from 0.5 to 2.5. Results for two of these wings will be presented in detail. The first is the wing with equal bending and torsional natural frequencies of 20 Hz, which results in a wing frequency ratio of $\gamma = 1.0$. A snapshot of the synchronized video is shown in Fig. 4 for an excitation frequency ratio of 0.8 and a 20-deg amplitude. The average thrust in this case was 12 g. The oscillating trace shows the instantaneous thrust, the vertical line indicates the time of the frame, and the horizontal line indicates the average thrust produced. Note that the wing is moving down and has just passed the point of producing maximum thrust. The wing articulates such that the bending (or translation) motion is in phase with the torsional (or rotational) motion.

One interesting aspect of the instantaneous thrust data shown in Fig. 4 is the small "dip" in the thrust as the wing approaches the point of maximum thrust. This was at first attributed to nonlinear bending dynamics in the wing. However, recent results from CFD modeling of two-dimensional rigid flapping wings have shown a similar effect, which has been attributed to vortex separation from the wing. Future synchronization of flow visualization with thrust data will seek to verify this effect. A second aspect of the thrust data to note is how the thrust data, which occurs at twice the flapping frequency, exhibits an asymmetry in the negative thrust portion of the data. This is due to a slight asymmetry in the wing construction. The carbon-fiber stiffening ribs were glued to only one side of the wing. Therefore, the wing has a small stiffness asymmetry that results in different behavior on the upstrokes and downstrokes.

Thrust and power data are summarized in Fig. 5 for the $\gamma = 1.0$ wing. This figure shows the average thrust generation and required power over a range of frequencies. Note that the thrust peaks at a value of 12 g and at an excitation frequency that is approximately 80% of the bending and torsion natural frequencies. As expected, the power required to sustain the flapping motion increases linearly with frequency.

The second wing for which detailed data are provided had a bending natural frequency of 40 Hz and a torsional natural frequency of 22 Hz, resulting in a wing

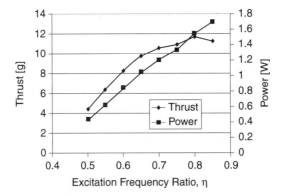

Fig. 5   Thrust and power data for the γ = 1 wing.

frequency ratio of $\gamma = 1.818$. A frame of the synchronized video is shown in Fig. 6 for 20-deg excitation amplitude and an excitation frequency ratio of 0.65. Again, the plot shows the instantaneous thrust, the average thrust, and the time of the wing video frame relative to the thrust data. In this case the wing articulates such that the torsional motion lags the bending motion by 90 deg. This is the type of wing articulation employed by *drosophylia* as noted by Dickenson et al.[6] The fact that the wing is being excited at torsional resonance and well below the bending resonance is apparent in Fig. 6, as indicated by the large twisting deformation of the wing and small spanwise deformation.

The lift and power data as a function of excitation frequency ratio are shown in Fig. 7 for the $\gamma = 1.818$ wing. As in the previous case the maximum thrust is generated somewhat below the torsional natural frequency. Of particular interest is the fact that the maximum thrust generated by this wing is less than that generated by the wing with a wing frequency ratio of $\gamma = 1$. This was unexpected since this is the type of articulation exhibited by *drosophilae*.

This trend held over a range of wing frequency ratios as summarized in Fig. 8, which shows the thrust generated by several wings possessing different wing

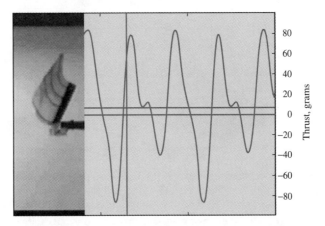

Fig. 6   Synchronized video and data for the γ = 1.818 wing.

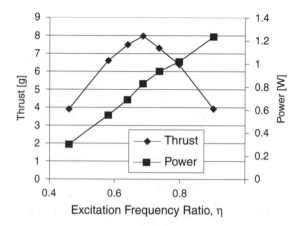

Fig. 7   Thrust and power data for the $\gamma = 1.818$ wing.

frequency ratios. Overall, the wing that produced the maximum thrust was the one with a wing frequency ratio of 1. Two interesting phenomena are illustrated in this plot. One is that the maximum thrust is generated by the $\gamma = 1$ wing for all excitation frequencies. The second is that the maximum thrust is generated at an excitation frequency that is below the torsional resonant frequency.

One plausible explanation for why the *drosophylia* type wing articulation does not generate the maximum thrust is demonstrated in Fig. 9. This figure shows the thrust to power ratio for both the $\gamma = 1$ and $\gamma = 2$ wings. Note that although the $\gamma = 1$ wing produces the most thrust, the $\gamma = 2$ wing (which is similar in

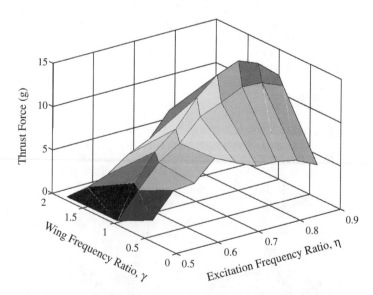

Fig. 8   Thrust force as a function of wing frequency ratio γ and excitation frequency ratio η.

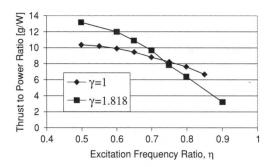

Fig. 9 Thrust to power ratio.

articulation to that of *drosophylia*) exhibits the largest thrust to power ratio. If one considers the thrust to power ratio as a measure of efficiency, it appears that nature has chosen optimum efficiency over optimum performance.

The final result addresses the fact that the maximum thrust generated by all wings occurs at a frequency somewhat below the torsional natural frequency. It was expected that the most thrust would be generated at resonance since this would result in the maximum wing displacement. However, as shown in Fig. 10, the frequency of maximum thrust decreases with increasing flapping amplitude. In fact, the maximum thrust generation is produced at the frequency that results in the maximum torsional displacement. However, as the flapping amplitude increases the aerodynamic loading results in an effective added mass to the wing, which lowers the resonant frequency.

IV. Conclusions

The development of aeroelastic wings for flapping flight has been discussed. A rig for testing the thrust production of flapping wings has been developed. Initial results indicate that the phase between bending motion and torsional motion is critical to the production of thrust. It was noted that a wing with bending and torsional motion in phase creates the largest thrust whereas a wing with the torsional motion lagging the bending motion by 90 deg results in the best efficiency.

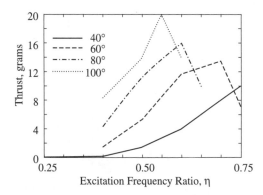

Fig. 10 Effect of flapping amplitude on generated thrust.

Acknowledgments

The authors would like to acknowledge the support of the Defense Advanced Research Projects Agency, and the NASA Langley Research Centers Dynamics and Controls Branch. The authors would also like to acknowledge the contribution of Paul Pao, NASA Langley Research Center, for his contribution toward the explanation of thrust data anomalies.

References

[1]Brodsky, A. K., *The Evolution of Insect Flight*, Oxford Univ. Press, New York, 1994.

[2]Nyborg, U. M., *Vertebrate Flight: Mechanics, Physiology, Morphology, Ecology and Evolution*, Springer-Verlag, New York, 1990.

[3]Ellington, C. P., "The Aerodynamics of Hovering Flight," *Philosophical Transactions of the Royal Society of London*, Vol. 305, No. 1122, 1984, pp. 1–181.

[4]Wooton, R. J., "The Mechanical Design of Insect Wings," *Scientific American*, Vol. 263, No. 5, Nov. 1990, pp. 114–120.

[5]Ennos, A. R., "The Importance of Torsion in the Design of Insect Wings," *Journal of Experimental Biology*, Vol. 140, 1988, pp. 137–160.

[6]Dickenson, M. H., Lehmann, F., and Sane, S. P., "Wing Rotation and the Aerodynamic Basis of Insect Flight," *Science*, Vol. 284, June 1999, pp. 1954–1960.

[7]Dowell, E. H. (ed.), *A Modern Course in Aeroelasticity*, 3rd ed., Kluwer Academic, Dordrecht, The Netherlands, 1994.

[8]Bisplinghoff, R. L., Ashley, H., and Halfman, R. L., *Aeroelasticity*, Addison-Wesley, Cambridge, MA, 1955.

[9]Lan, C. E., "The Unsteady Quasi-Vortex-Lattice Method with Applications to Animal Propulsion," *Journal of Fluid Mechanics*, Vol. 93, 1979, pp. 747–765.

[10]Vest, M. S., and Katz, J., "Unsteady Aerodynamic Model of Flapping Wings," *AIAA Journal*, Vol. 34, No. 7, 1996, pp. 1435–1440.

[11]Liu, H., Ellington, C. P., and Kawachi, K., "A Computational Fluid Dynamic Study of Hawkmoth Hovering," *Journal of Experimental Biology*, Vol. 201, No. 4, 1998, pp. 461–477.

[12]DeLaurier, J. D., "An Aerodynamic Model for Flapping Wing Flight," *Aeronautical Journal*, Vol. 97, 1993, pp. 125–130.

Chapter 22

Shape Memory Alloy Actuators as Locomotor Muscles

Othon K. Rediniotis[*] and Dimitris C. Lagoudas[‡]
Texas A&M University, College Station, Texas

Nomenclature

A = austenite
A_F = austenite finish temperature
A_S = austenite start temperature
A_{0F} = austenite finish temperature at zero stress
A_{0S} = austenite start temperature at zero stress
a_m = gain to determine how fast $\hat{T}$ approaches T_r
b_m = gain to determine how fast $\hat{T}$ approaches T_r
e_H = error
H = hysteresis model
H_{max}^T = maximum transformation strain
H^{-1} = hysteresis model inverse
i = current signal
i_{max} = predefined safety limit on current
M = martensite
M_F = martensite finish temperature
M_S = martensite start temperature
M_{0F} = martensite finish temperature at zero stress
M_{0S} = martensite start temperature at zero stress
$r(t)$ = reference displacement signal
T_a = annealing temperature
T_r = reference temperature
$\hat{T}$ = estimated temperature
u = required heating
y = actual displacement
$\hat{y}$ = estimate of the displacement output

---

[*]Associate Professor, Aerospace Engineering Department.
[‡]Ford Professor, Aerospace Engineering Department.

α = physical parameter based on heat convection from the SMA wire
β = physical parameter based on resistance of the SMA wire
$\varepsilon_{\max}^T$ = maximum transformation strain
$\varepsilon_{\min}^T$ = minimum transformation strain
θ_{ij} = hysteresis model parameters
λ^0 = mean percentage of transformation strain
λ^a = amplitude of transformation strain percentage

I. Introduction

B IRD flight has inspired and guided the design and development of aircraft since its inception. However, it is striking how primitive these man-made machines are compared to their natural counterparts in terms of efficiency, agility, coordination, autonomy, and adaptability. In birds and insects, nature has evolved into a great synergy between muscles (actuators), wings (structure), and brain (control) that has not been approached in the most advanced aircraft and rotorcraft. In nature, every living organism that can generate lift does so through the flapping of wings. A significant advantage of flapping wing propulsion is that lift can be generated with little or no forward velocity and with small wing size.

In the area of underwater vehicle design, the development of small highly maneu-verable vehicles is also presently of interest. Their design is based on the undulatory body motion, swimming techniques, and anatomic structure of fish, primarily the highly controllable fins and the large aspect ratio lunate tail.[1] These pursuits in air and underwater vehicles triggered the emergence of the science of biomimetics, which is the study of natural systems in order to improve the design and function-ality of synthetic systems (data available online at http://www.graylab.ac.uk/cgi-bin/omd? biomimetics [cited 1998]). The tailoring and implementation of the accumulated knowledge into biomimetic vehicles is obviously a task of multidis-ciplinary nature.

In biomimetics, the use of conventional systems of gears and servomotors to provide the actuation power leaves little interior room in the device for control systems and payload (Barrett, D., "Design of the MIT Robo Tuna," MIT RoboTuna Home Page, http://web.mit.edu/towtank/www/tuna/brad/design.html [cited Sept. 1999]; Kumph, J. M., "The Robot Pike Project," MIT Robot Pike Home Page, http://web.mit.edu/towtank/www/pike/stryrp.html [cited Jan. 2000]). Especially in micro air vehicles, power density of the actuation system is a crucial design parameter. In actuator technology, active or "smart" materials have opened new horizons in terms of actuation simplicity, compactness, and miniaturization po-tential. With the current advances in shape memory alloy (SMA) actuation tech-nology, SMAs are probably the most suited actuators to a range of biomimetic applications.[1]

As an illustration of the benefits of SMA actuation over other active materi-als such as piezoceramics and magnetostrictives, let us consider two biomimetic examples: a flapping wing micro air vehicle (FWMAV), with a wingspan of 6 to 12 in. with large wingtip excursions, and a fishlike autonomous underwater vehicle (AUV) on the order of 1-m long.

For the FWMAV, during flapping, actuators with large actuation stroke/strain are preferable. Also, the small available onboard space dictates the use of high energy-density actuation systems. Piezoceramic and magnetostrictive materials have relatively low energy densities and high actuation frequencies, which are not required in this application. SMAs have energy densities much higher than piezoceramic materials. Furthermore, piezoceramic and magnetostrictive materials are only capable of producing small strains and displacements (small fractions of 1%) compared to those attained by SMA materials (as high as 8% for one-way trained and 4% for two-way trained SMAs).[2] At first glance, the frequencies (on the order of 10 Hz) may seem too high for SMA actuation. As we will show later, this is a misconception. We will demonstrate that SMA actuation frequencies even higher than 10 Hz are possible and accurately controllable. Our previous work also indicates that thin SMA layers (~6 μm thick) under partial transformation are capable of delivering frequencies of about 30 Hz at peak stress of 145 MPa.[3] Energy considerations for such an application are covered in a later section.

Consider a second example, a fishlike AUV, on the order of 1-m long. The vehicle is to be propelled by a caudal-fin equivalent system, with the ratio (caudal-fin trailing-edge excursion)/(vehicle length) maintained equal to an average value observed in the propulsion of its aquatic counterparts[4,5] (around 0.22). For the above data and for underwater vehicle speeds between 3 and 7 knots (1.5 to 3.5 m/s), caudal-fin oscillation frequencies will range from 1.5 to 5 Hz. The frequency bandwidth, combined with double-amplitude trailing-edge excursions on the order of 0.25 m, makes SMA-based actuation ideal for the particular application. A six-segment biomimetic hydrofoil[1] has been developed and is currently under testing and modifications to evaluate the use of SMAs to power such an AUV. Results are discussed in a later section.

The small sizes and high force densities of SMAs (compared to conventional actuators) result in availability of much larger percentage of vehicle internal volume. This greater internal volume allows for a larger payload for comparably sized vehicles. Additional SMA advantages include:

a) *Simplicity of actuation mechanism.* SMAs can be all-electric devices and can be used as direct-drive linear actuators requiring little or no additional gear reduction or motion amplification hardware. These merits permit the realization of small or even miniature actuation systems to overcome space limitations.

b) *Silent actuation.* Since no acoustic signature is associated with such a propulsion system, acoustic detectability will be reduced.

c) *Low driving voltages, when powered electrically.* Nitinol (NiTi) SMAs can be actuated with very low voltages (10 to 20 V), thus requiring very simple power supply hardware. In contrast, piezoceramic materials typically require voltages on the order of 100 V for their actuation. However, for SMAs large currents may be required for resistive heating, depending on SMA actuator dimensions, leading to heavy power supply hardware.

d) *SMAs are heat engines.* This means that they do not necessarily need to be powered electrically. The electrical power required to actuate SMAs with loads, frequencies, and fatigue life practical for dynamic applications demand the use of relatively heavy power supplies/batteries, which in turn results in high system weights and poor energy densities. But as heat engines, they can utilize onboard parasitic heat, if available, to produce useful mechanical work. If parasitic heat is not available, they can readily use the chemical energy of fuels, which have outstanding energy densities (order of 13,000 W · h/kg or 46,800 kJ/kg) compared to compact batteries (125–300 W · h/kg or 450–1080 kJ/kg). As an example, schematics with heating and cooling cycles for such a configuration are shown in Fig. 13 and presented in a later section.

SMA actuation is not free of drawbacks, however. Some of them, such as actuation loss and fatigue over repetitive cyclic loading, are in the frontier of SMA research. Many researchers have investigated the effect of both mechanical and thermal cyclic loading on the thermomechanical response of NiTi.[6–11] However, these tests were primarily concerned with the development and stability of two-way strain, the evolution of plastic strain, or hardening in the stress–strain response.

Extension of the cyclic loading path until specimen failure occurs has been studied by a limited number of researchers, with the primary emphasis on mechanical transformation fatigue (i.e., a stress-induced phase transformation). Melton and Mercier[12] gave the earliest detailed report on fatigue properties for NiTi specimens. Constant stress amplitude fatigue results (S–N curves) showed that fatigue limits reached 10^7 cycles for stress levels that did not induce a phase transformation. The fatigue life for stress amplitudes that induced phase transformations (i.e., mechanical transformation fatigue) was reduced to approximately 10^3 to 10^4 cycles. Constant strain amplitude fatigue results that included a phase transformation showed that low cycle fatigue obeyed the Coffin–Manson law relating plastic strain and cycles to failure.[13]

Miyazaki,[14] Miyazaki et al.,[15] and Tobushi et al.[16,17] have performed some of the most recent work on stress-induced transformation fatigue. Miyazaki performed an isothermal cyclic tensile loading at constant stress amplitude, similar to that of Melton and Mercier,[12] for different test temperatures. Owing to the lower percentage of martensitic transformation associated with elevated test temperatures, a higher fatigue life was identified, approaching 10^6 cycles. Study of the thermal transformation fatigue of NiTi SMA was performed by McNichols and Brooks.[18] Results from this study showed a thermal transformation fatigue life between 10^4 and 10^5 cycles for constant strain amplitude levels between 4.4 and 8.3%. It appears that the thermally-induced phase transformation fatigue of SMAs has not been studied systematically and the effect of the magnitude of transformation strain on the fatigue life of SMAs needs to be explicitly examined.

The remaining sections of this chapter start with an overview of SMA actuators, followed by sections on thermomechanical transformation fatigue, adaptive control, and energy considerations for SMA actuators. In the end, actual implementation of SMA actuators for an AUV are briefly reviewed and are followed by conclusions and acknowledgments.

II. Brief Overview of SMA Actuators

Our research has focused on NiTi, which is a nickel-titanium based SMA alloy. In fact, Nitinol was the first shape memory alloy to be discovered. The name comes from the two element symbols, Ni and Ti, and the abbreviation of the lab where the alloy was discovered, the Naval Ordinance Laboratories (NOL).[19] Shape memory properties of Nitinol become evident once a specimen has been properly annealed.

The shape memory effect comes from the differences in the crystalline structure between the martensitic and austenitic states. Figure 1 shows the difference between the two crystal structures. Properties of the SMA vary depending on the amounts of martensite and austenite present. The percentages of each state depend on the stress and temperature of the specimen ("An Introduction to Shape Memory Alloys," SMART Home Page, http://smart.tamu.edu/pageIntro.htm

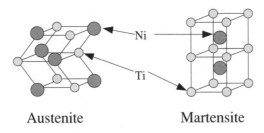

Austenite   Martensite

Fig. 1   Martensite and austenite crystal.

[cited Nov. 2000]). Figure 2 shows a qualitative SMA stress–temperature phase diagram. To the left of martensite finish M_F, only martensite is present. To the right of austenite finish A_F, only austenite is present. In between M_F and A_F, there is a combination of martensite and austenite. When the wire is deformed in the martensitic state and the temperature is raised above austenite start A_S, the wire will begin to return to its original shape and will be at the original shape when the temperature is above A_F. This process is called one-way training.[20] If this process is repeated (that is, if a specimen is deformed in martensite, heated to austenite, and cooled to martensite), the wire will develop what is called two-way training.[20] It will expand by the trained amount when the temperature drops below M_F without a load on the wire.

A qualitative schematic of the SMA phase transformation diagram is shown in Fig. 3 to clarify the actuator functions. The phase change in NiTi is accompanied by a significant exchange of heat with the surroundings. The time rate of the transformation is controlled solely by the time rate of the heat transfer. Typically, SMA heating, which triggers the martensite-to-austenite (M-to-A) phase change, is achieved electrically (by resistive heating). The heating rate, and thus the transformation rate, can be controlled by regulating the applied voltage and thus the electrical current. To trigger the reverse transformation (A-to-M), heat removal from the SMA is necessary. Conventional heat-removal mechanisms are inherently slower than electrical heat addition and usually rely on forced convection. Therefore, in a complete heating–cooling cycle (transformation/actuation, M-to-A and then A-to-M), the cooling process usually limits the speed of the cycle (i.e., the maximum attainable frequency response). An effort has been made to address this problem in a later section.

Another important consideration is that unless SMA actuators are properly employed, plastic strains may develop and the required force or displacement may

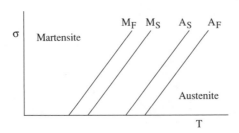

Fig. 2   Typical stress–temperature phase diagram for NiTi SMA.

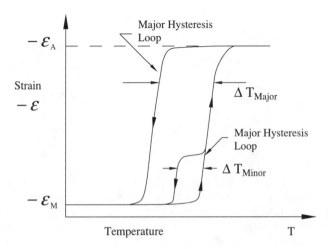

Fig. 3   Strain–temperature SMA transformation diagram.

not be attained on actuation. This variation in thermomechanical response requires further understanding of its influence for successful implementation of SMA actuators. To develop a better understanding of the thermomechanical response a discussion on the thermomechanical fatigue of the NiTi SMA actuator is provided in the following section.

III.   Thermomechanical Transformation Fatigue of SMA Actuators

In many applications of SMA actuators, the shape memory effect (SME) is used many times by heating and cooling and/or loading and unloading. A test frame and experimental setup developed in earlier work[21] was used to determine the thermal transformation fatigue life of SMA wires. The loading paths shown in Fig. 4 represent the various ways in which SMAs were loaded. Loading paths 1 and 3 do not contain a phase transformation of any kind. Therefore, in terms of

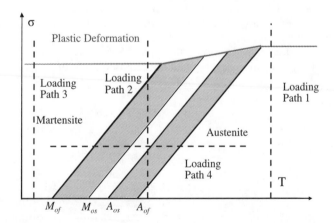

Fig. 4   Representation of fatigue loading paths in stress vs temperature space.

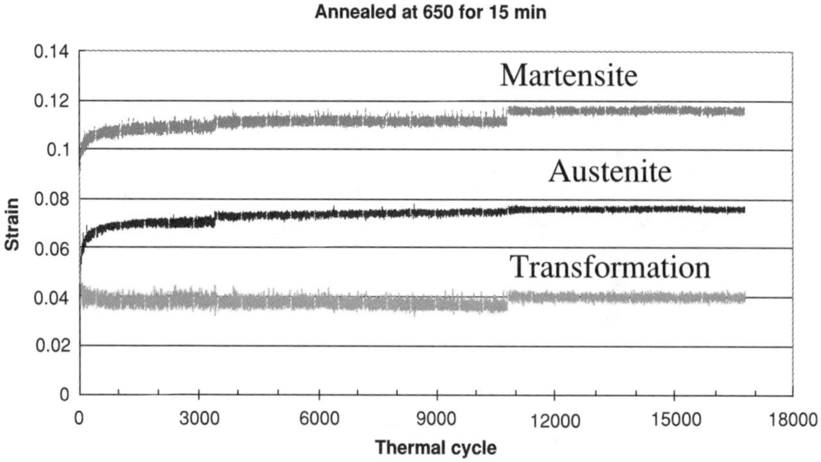

Annealed at 650 for 15 min

Fig. 5   A typical curve of thermomechanical fatigue of NiTiCu annealed at 650°C for 15 min.

transformation fatigue characterization, it is assumed that these tests would result in a response typical of mechanical fatigue seen in common engineering materials. Loading path 2 shows a stress-induced martensitic phase transformation and represents the mechanical transformation fatigue loading path studied by previous researchers.[12,14] Loading path 4 shows a thermally induced phase transformation and best describes the conditions seen by SMA actuators.

Two important experimental findings have been revealed that are related to the fatigue response of SMA wire actuators. The first finding relates the transformation fatigue life of NiTiCu to the heat treatment temperature, which was studied by Miller and Lagoudas[22] and Lagoudas et al.[23] It is found that M_S and A_S temperatures increase slightly as annealing temperature (T_a) increases when the wire is annealed below 400°C, whereas a drastic increase is found in the interval of 450–550°C. Figure 5 shows typical curves of different strains developed in the cycling process. A common feature in all samples is the large plastic strain developed during the first 300–500 cycles, which saturated to a nearly constant value during higher cycles.

The second experimental finding is the effect of stress level on the fatigue life of SMA wires. Cyclic thermal transformations were performed on NiTiCu wires that had a nominal diameter of 0.6 mm and an initial cold work of 30%. Prior to testing, the wires were heat-treated at 600°C for 30 min and at 550°C for 15 min. Results for thermal transformation fatigue tests under a complete phase transformation and varying levels of constant applied stress are shown in Fig. 6. The results show that, for higher stress levels, fewer cycles were needed to cause the specimen to fail and subsequently, as the stress level decreases, the number of cycles to failure increases substantially.

Besides these fully transformed (major loop) fatigue tests, partially transformed (minor loop) fatigue tests were performed by Lagoudas et al.[23] Partial transformation cycles were created by an incomplete martensitic transformation. The percentage of the transformation was controlled by choosing different levels of

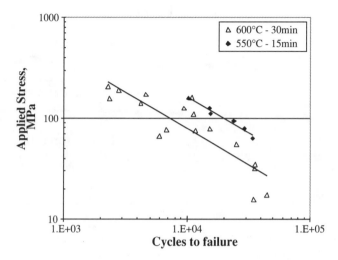

Fig. 6   Applied stress vs thermal cycles to failure for NiTiCu SMA for a complete transformation.

the mean percentage of transformation strain λ^0 and the amplitude of transformation strain percentage λ^a (shown in Fig. 7), which were percentages of the complete transformation. Tests were performed for partial thermal transformation fatigue under a constant applied stress of 78.8 MPa. Prior to testing, the wires were annealed at 600°C for 30 min. A summary of the results is shown in Fig. 8. Error estimates for strain values for both Figs. 5 and 8 are on the order of ±0.075%.

$$\lambda^a = \frac{\varepsilon^T_{max} - \varepsilon^T_{min}}{2H^T_{max}} = \frac{\lambda^{max} - \lambda^{min}}{2} \qquad \lambda^0 = \frac{\varepsilon^T_{max} + \varepsilon^T_{min}}{2H^T_{max}} = \frac{\lambda^{max} + \lambda^{min}}{2}$$

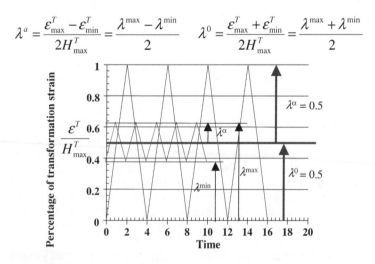

Fig. 7   Schematic of partial transformation loops showing physical meaning of the mean percentage of transformation strain λ^0 and the amplitude of transformation strain percentage λ^a.

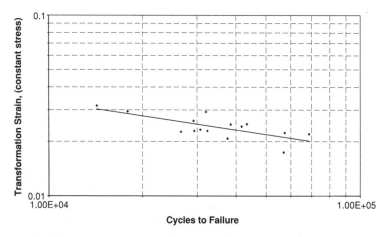

Fig. 8   The effect of transformation strain on the thermal transformation fatigue life for minor hysteresis loops.

Results so far indicate that, for partial thermally induced phase transformation, the fatigue life can be drastically increased with respect to full transformation cycles. Hence the idea of using multiple actuators with partial transformations may provide better actuation response. The following section discusses the developed control scheme for SMA actuators along with a method of achieving higher cooling rates.

IV.   Adaptive Control of SMA Actuator Wires

Exposure of the SMA actuators to the ambient medium has adverse effects on the SMA control. In warm environments, the SMA cooling rate may not be sufficient, and in cold environments the energy expended to heat up the SMAs to austenite may be too large, depending on actuator applications. To achieve SMA actuation frequencies high enough to generate propulsive and maneuvering forces for FWMAVs and AUVs, forced convection cooling of the SMAs must be used. It was therefore decided to thermally insulate the SMA actuators from the ambient environment and embed them inside closed-system channels containing a temperature-controlled cooling medium.

In situations where electrical heating for M-to-A transformation and forced cooling for A-to-M transformation is utilized for actuation, one difficulty imposed by forced convection is that temperature measurements of the SMA cannot be achieved via thermocouples. A thermocouple is usually attached to the SMA wire/strip with a small "ball" of thermal paste, which is thermally, but not electrically, conductive. However, the strong heat transfer caused by active convection of the cooling medium means that the thermal paste dissipates the heat from the wire to the ambient medium much faster than in a pure conduction environment. This causes the resulting temperature measurement to not be an exact representation of the SMA temperature.

We have pursued actuation control schemes that rely only on actuator displacement feedback and do not require temperature feedback. The control approach in this work implements an adaptive hysteresis model for feedback compensation for control of SMA wire actuators.[24,25] In feedback compensation, the tracking error was not directly used in the control law. Rather, it was used to update (identify on-line) the hysteresis model (H) to account for discrepancies between the model and the actual input–output relationship. The inverse maps a desired reference trajectory into a temperature signal. When temperature can be measured, an adaptive thermal model then commands a current to control the temperature of the SMA to track the desired temperature. For cases where temperature measurements are unavailable, integration of a simplified thermal model (based only on rough estimates of the thermal parameters) was used in place of actual temperature measurements.

The control algorithm used for a single-wire test setup can best be described by referring to the schematic in Fig. 9. The reference displacement signal $r(t)$ was mapped into a reference temperature T_r via the hysteresis model inverse H^{-1}. Here u is the heating required to force $\hat{T}$ to track T_r. The current signal i is the input to the SMA wire. The current was also limited by strict bounds on u given by $u \in [0, i_{\max}{}^2]$, where $i_{\max}$ was a predefined safety limit on the current. The lower limit on u essentially meant that the fastest cooling that could be achieved was when the current to the SMA wire was turned off. Gains that determined how fast $\hat{T}$ would approach $\hat{T}_r$ were a_m and b_m. Physical parameters based on the convection and the resistance of the wire were α and β respectively.

To adaptively update the hysteresis model to represent the y-vs-$\hat{T}$ relationship, the estimate $\hat{y}$ of the displacement output was generated from $H(\hat{T})$. The estimation error was calculated using the error between $\hat{y}$ and the actual displacement y. The hysteresis model parameters θ_{ij} were updated using the gradient adaptive law from Section 3.1.2 of Ref. 24, in turn updating the inverse model.

In this control method, the thermal parameters remain constant. Although it may be possible to adaptively update the thermal parameters to represent the heating and convection more accurately for a time-varying cooling environment, for the results presented, the burden of accounting for these variations lies solely with the adaptive hysteresis model. Reasonable estimates of the thermal parameters served to ensure that $\hat{T}$ remained in the defined input region of the hysteresis model.[24]

Figure 10 is a schematic of the experimental setup[1] used to test the SMA control theory. The C shape of the test stand provided support for the wire while allowing

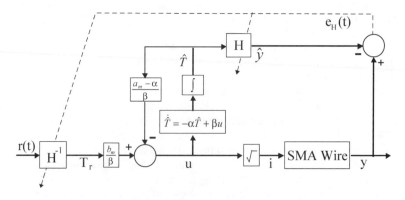

Fig. 9   SMA control scheme.

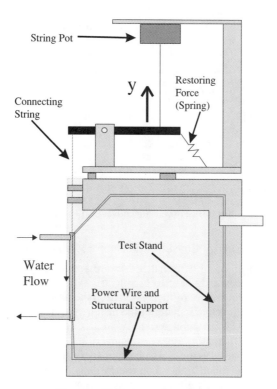

Fig. 10   SMA control test stand.

easy access to the wire. The wire was embedded in a cooling channel. The SMA wire was connected to power leads by brass connectors, one on each end. The free end (top) of the wire was attached to the lever arm by a Kevlar® string. A spring attached to the opposite end of the lever arm both held the SMA wire in tension and provided a restoring force for the SMA. The spring was used instead of a constant weight (mass) as inertial forces created by a dead mass at higher frequencies caused the SMA wire to bounce, resulting in compressive forces on the SMA wire. Displacements were then measured using a Celesco string-potentiometer. The resolution of the string-potentiometer, in conjunction with a 16-bit data acquisition board (National Instruments AT-MIO-16XE-50), was 0.0008 in. The total worst-case uncertainty was 0.011 in. in absolute position. The average wire used in the test frame was 12-in. long, so at 2% strain, with a lever motion amplification of 4 to 1, this resulted in a displacement uncertainty of 1.15%.

On application of our adaptive control scheme to water-cooled SMA actuators for this experimental setup, good tracking for actuation frequencies as high as 20 Hz was achieved. Figure 11 presents the control results for the case of using cool air as the convection medium. As shown, the tracking was very good, but because of the limited convection rates of air, only low actuation frequencies could be achieved. Figure 12 presents the tracking control results for the case of using water as the convection medium. Tracking was very good, and actuation frequencies as high as 20 Hz were achieved.

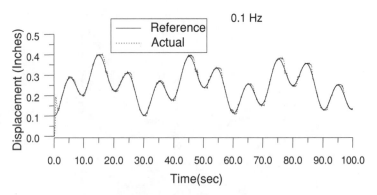

Fig. 11   Forced-air convection cooling.

V.   Energy Considerations for SMA Actuators

It should be kept in mind here that SMAs are actuated by heat, which means that one does not necessarily have to use electricity to actuate them. One would rather use the chemical energy of fuels. We are in the process of developing small combustors along with active transfer of the heat to and from the SMA. Figure 13 shows a compact SMA actuator system, which is comprised of minipumps, a combustor, the SMA actuator, a heat exchanger, and fuel as the energy source. Propane or gasoline is used as the fuel since their energy densities are on the order of 50,000 kJ/kg, which is much higher than energy densities of batteries or fuel cells. Ethylene glycol is considered as the heating fluid because of its high boiling point. It is heated in a combustor, which acts like a high-efficiency boiler. The hot ethylene glycol heats the SMA actuator convectively above austenite finish temperature, causing strain recovery and hence actuation. Ethylene glycol is circulated in the heating circuit through a pump. Another pump is used to circulate the coolant in the cooling circuit for cooling the SMA down to the martensite finish temperature. A small battery operates these two pumps. The coolant is also circulated through a heat exchanger to dissipate heat removed from the SMA actuator.

Let us consider the following example: A typical, generic fuel has an energy density (enthalpy) of 13,000 W · h/kg (48,800 kJ/kg). The full transformation latent heat per unit volume for a typical SMA is 0.10 J/mm^3. For a flapping wing MAV with a wingspan on the order of 0.3 m, approximately 300 mm (length) of NiTi wire with a diameter of 0.6 mm (0.023 in.) (keeping in mind that a NiTi wire of that diameter can pull, when actuated, as much as 70 N or 15.7 lbf) is considered. For an average flapping frequency of 20 Hz, this amount of wire (a total of 85 mm^3) would need a total of 170 J/s. Assuming full transformation per cycle and a combustion efficiency of approximately 85% (typical of high-efficiency boilers) 1 g of fuel will be sufficient to maintain the vehicle in cruise for about 4 min. The same calculations for a wingspan of 0.15 m (6 in.) yields a flying time of approximately 8 min/g of fuel. Note that this analysis is based on full transformations. Partial transformation with multiple thin SMA actuators would give higher frequency response[3] and less fuel consumption per SMA layer, yielding greater flying time for a given wingspan.

To get an idea about work done per unit muscle mass (J/kg), we compare typical work done by an SMA muscle to that done by a typical dragonfly and a

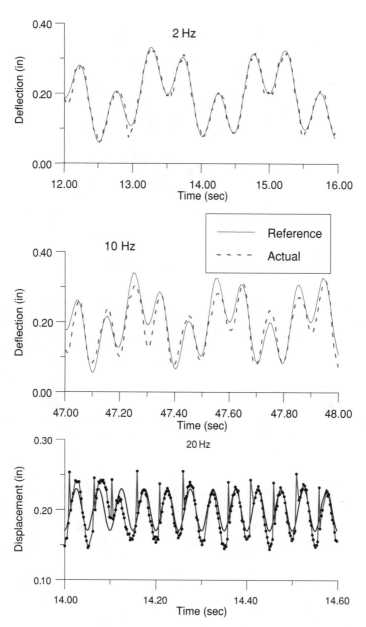

Fig. 12   SMA actuator control using an adaptive hysteresis model with active cooling.

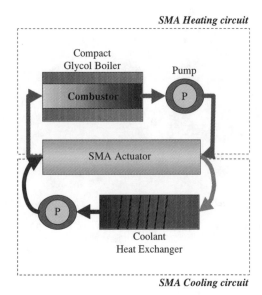

Fig. 13   Compact actuator system.

hummingbird wing muscle as shown in Table 1. The reason for choosing humming-birds and dragonflies is in accordance with the general understanding that devel-opment of an FWMAV based on biomimetics requires study of flight mechanisms and characteristics as found in hummingbirds, dragonflies, and other flapping wing insects. Based on conservative estimates, the work done by a typical SMA actuator per unit mass surpasses the estimates for hummingbirds[27] and dragonflies.[26] Of course, this fact alone does not justify usage of SMA for FWMAVs. However, this combined with recent developments—such as smaller system weight/size, high actuation frequencies, high energy densities, active heating/cooling, and relatively high fatigue life—provides enough incentive for SMA actuators to be considered as an alternative.

VI.   SMA Actuators as Locomotor Muscles for a Biomimetic Hydrofoil

The profile shape of the six-segment biomimetic hydrofoil[1] was based on a NACA 0009 airfoil section with a 30-in. (76.2-cm) chord and minor modifications in the aft portion of the profile. There are three main components to the biomimetic

Table 1   Work done per muscle mass

Muscle type	Work done per muscles mass (J/kg)
Dragonfly (*S. sanguineum*)[26]	4
Hummingbird[27]	2–7.41
SMA	1000

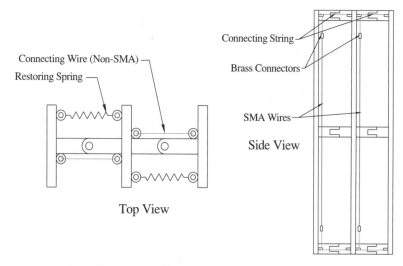

Fig. 14 Schematic of SMA muscles on biomimetic hydrofoil.

hydrofoil: a) the aluminum backbone and rib structure, b) the SMA muscles, and c) the scales or skin.

The ribs are seven 0.25-in.-thick (0.64-cm-thick) aluminum plates. The seven ribs are connected with six sets of hinge pairs (vertebrae). The vertebrae are machined aluminum, with bearings at the pivot points to ensure smooth motion between the ribs. Mounted on each vertebra of each pair is a position encoder. The encoder gives the relative angle between adjacent ribs.

The SMA muscle has two parts that pull in opposing directions. The opposing force is provided by two springs. The SMA wires run parallel to the segments, as shown in Fig. 14. This orientation allows the wires to be longer, and with a pulley system a larger angular displacement of the vertebrae is achieved.

The skin is the interface between the vehicle and the ambient fluid and allows the hydrofoil to efficiently transfer energy to the fluid, thus generating thrust. The skin is set up as six overlapping scales on each side, with nose and tail sections added at the leading and the trailing edge respectively. Figure 15 shows the fully assembled hydrofoil.

Fig. 15 Assembled biomimetic hydrofoil.

Fig. 16 Photograph of a tail position during an actuation cycle in the water tunnel.

In-air actuation tests were performed to track various types of reference signals. Water tunnel actuation was performed to track several types of reference signals and to measure the thrust generated by the resulting hydrofoil shapes. Sinusoidal deflection patterns were also tested for thrust-producing capabilities. Figure 16 shows a picture from one such actuation cycle.

VII. Conclusions

Controllable actuation of SMAs up to 20 Hz has been demonstrated. Increasing actuation bandwidth would require using an active temperature-controlled cooling medium. We are in the process of implementing SMA actuators that utilize partial transformation and active cooling and are capable of actuation bandwidths of 50 to 100 Hz.

It has been shown that fatigue life of SMAs depends upon dislocation density, applied load, and the extent of phase transformation. Fatigue life increases under reduced load and partial phase transformation. Areas such as low energy efficiency and relatively high current requirements for SMA actuators, if electric SMA heating is utilized, need to be addressed.

Small fuel combustors coupled with active transfer of heat to and from the SMA are being researched as high energy-density sources for SMAs. An effort has been made to provide a general proof of concept of using SMAs as locomotor muscles for FWMAVs and AUVs. Actual implementation of SMAs as locomotor muscles for a biomimetic hydrofoil has been discussed and acts as a stepping stone toward further development and understanding of SMA-actuated FWMAVs and AUVs.

Acknowledgments

The authors wish to acknowledge the financial support for the research on the Biomimetic Hydrofoil by the Office of Naval Research, Contract No. N00014-97-1-0840 and Aeroprobe Corporation, Project No. 96-386. The sponsorship by the Texas Higher Education Coordination Board, TD&T Grant No. 000512-0278-1999, is also gratefully acknowledged. The authors would also like to acknowledge the help of Mughees Khan in the preparation of the manuscript.

References

[1] Rediniotis, O. K., Lagoudas, D. C., and Wilson, L. N., "Development of a Shape Memory Alloy Actuated Biomimetic Hydrofoil," AIAA Paper 2000-521, Jan. 2000.

[2] Tobushi, H., Ohashi, Y., Saida, H., Hori, T., and Shirai, S., *JSME International Journal*, Vol. I, 1992, p. 84.

[3]Lagoudas, D. C., and Bhattacharyya, A., "Modeling of Thin Layer Extensional Thermo-electric SMA Actuators," *International Journal of Solids and Structures*, Vol. 35, No. 3–4, 1998, pp. 331–362.

[4]Kudva, J., Jardine, P., Martin, C., and Appa, K., "Overview of the ARPA/WL 'Smart Structures and Materials Development—Smart Wing' Contract," *Proceedings of the 3rd SPIE Symposium on Smart Structures and Materials: Industrial and Commercial Applications of Smart Structures Technologies*, The International Society for Optical Engineering, Vol. 2721, 1996, pp. 10–16.

[5]Gilbert, W. W., "Mission Adaptive Wing System for Tactical Aircraft," *Journal of Aircraft*, Vol. 18, No. 7, 1981, pp. 597–602.

[6]Hebda, D., and White, S. R., "Effect of Training Conditions and Extended Thermal Cycling on Nitinol Two-Way Shape Memory Behavior," *Smart Materials and Structures*, Vol. 4, 1995, pp. 298–304.

[7]Lim, T. J., and McDowell, D. L., "Degradation of an NiTi Alloy During Cyclic Loading," *Proceedings of the 1994 North American Conference on Smart Structures and Materials*, The International Society for Optical Engineering, Orlando, FL, 1994, pp. 153–165.

[8]Miyazaki, S., Imai, T., Igo, Y., and Otsuka, K., "Effect of Cyclic Deformation on the Pseudoelasticity Characteristics of Ti-Ni Alloys," *Metallurgical Transactions A*, Vol. 17, 1986, pp. 115–120.

[9]Tanaka, K., Nishimura, F., Hayashi, T., Tobushi, H., and Lexcellent, C., "Phenomenological Analysis on Subloops and Cyclic Behavior in Shape-Memory Alloys Under Mechanical and or Thermal Loads," *Mechanics of Materials*, Vol. 19, 1995, pp. 281–292.

[10]Lagoudas, D. C., and Bo, Z., "Thermomechanical Modeling of Poly-crystalline SMAs Under Cyclic Loading, Part II: Material Characterization and Experimental Results for a Stable Transformation Cycle," *International Journal of Engineering Science*, Vol. 37, 1999, pp. 1141–1173.

[11]McCormick, P. G., and Liu, Y., "Thermodynamic Analysis of the Martensitic Transformation in NiTi. 1. Effect of Heat-Treatment on Transformation Behavior," *Acta Metallurgica el Materialia*, Vol. 42, No. 7, 1994, pp. 2401–2406.

[12]Melton, K. N., and Mercier, O., "Fatigue of NiTi Thermoelastic Martensites," *Acta Metallurgica*, Vol. 27, 1979, pp. 137–144.

[13]Suresh, S., *Fatigue of Metals*, Cambridge Univ. Press, Cambridge, U.K., 1991.

[14]Miyazaki, S., "Thermal and Stress Cycling Effects and Fatigue Properties of Ni-Ti Alloys," *Engineering Aspects of Shape Memory Alloys*, edited by T. W. Duerig, K. N. Melton, D. Stockel, and C. M. Wayman, Butterworth-Heinemann, London, 1990, p. 394.

[15]Miyazaki, S., Mizukoshi, K., Ueki, T., Sakuma, T., and Liu, Y., "Fatigue Life of Ti-50 at. % Ni and Ti-40Ni-10Cu (at. %) Shape Memory Alloy Wires," *Materials Science and Engineering A*, Vol. 273–275, 1999, pp. 658–663.

[16]Tobushi, H., Hachisuka, T., Yamada, S., and Lin, P.-H., "Rotating-Bending Fatigue of a TiNi Shape-Memory Alloy Wire," *Mechanics of Materials*, Vol. 26, 1997, pp. 35–42.

[17]Tobushi, H., Hachisuka, T., Hashimoto, T., and Yamada, S., "Cyclic Deformation and Fatigue of a TiNi Shape Memory Alloy Wire Subjected to Rotating Bending," *Journal of Engineering Materials and Technology*, Vol. 120, 1998, pp. 64–70.

[18]McNichols, J. L., and Brooks, P. C., "NiTi Fatigue Behavior," *Journal of Applied Physics*, Vol. 52, 1981, pp. 7442–7444.

[19]Buehler, W. J., and Wiley, R. C., "Nickel-Based Alloys," U.S. Patent 3,174,851, 1965.

[20]Wayman, C. M., *Phase Transformations, Nondiffusive*, Elsevier, New York, 1983, pp. 1031–1075.

[21]Lagoudas, D. C., and Miller, D. A., "Experiments of Thermomechanical Fatigue of SMAs," *Proceedings of the 1999 Conference on Smart Structures and Materials*, The International Society for Optical Engineering, 1999, pp. 275–282.

[22]Miller, D. A., and Lagoudas, D. C., "Influence of Cold Work and Heat on the Shape Memory Effect and Plastic Strain Development of NiTi Wire Actuators," *Journal of Materials Science and Engineering A*, Vol. 308, 2000, pp. 161–175.

[23]Lagoudas, D. C., Miller, D. A., Rong, L., and Li, C., "Thermomechanical Transformation Fatigue of SMA Actuators," *Proceedings of the 2000 Conference on Smart Structures and Materials*, The International Society for Optical Engineering, Vol. 3992, 2000, pp. 420–429.

[24]Webb, G., Wilson, L., Rediniotis, O., and Lagoudas, D., "Adaptive Control for Shape Memory Alloys Wires in Underwater Applications," *AIAA Journal*, Vol. 37, No. 11, 1999.

[25]Webb, G. V., "Adaptive Identification and Compensation for a Class of Hysteresis Operators," Ph.D. Dissertation, Dept. of Aerospace Engineering, Texas A&M Univ., College Station, TX, 1998.

[26]Wakeling, J. M., and Ellington, C. P., "Dragon Fly Flight: III. Lift and Power Requirements," *Journal of Experimental Biology*, Vol. 200, 1997, pp. 583–600.

[27]Chai, P., "Hummingbird Hovering Energetics during Moult of Primary Flight Feathers," *Journal of Experimental Biology*, Vol. 200, 1997, pp. 1527–1536.

Part III.   Micro Air Vehicle Applications

Chapter 23

Mesoscale Flight and Miniature Rotorcraft Development

Ilan Kroo[*] and Peter Kunz[†]

Stanford University, Stanford, California

Nomenclature

C_L	=	lift coefficient, lift/$(\frac{1}{2}\rho V^2 S)$
$C_{l\max}$	=	maximum section lift coefficient
L/D	=	lift-to-drag ratio
l/d	=	section lift-to-drag ratio
M	=	rotor figure of merit
P	=	power required to maintain flight
P'	=	ratio of power required for hover to power required for forward flight
S	=	wing area
T	=	thrust
V	=	forward speed
V_h	=	average induced velocity in hover
W	=	weight
η_p	=	propeller efficiency
ρ	=	mass density of atmosphere

I. Introduction

THE rapid development of microelectronics and microelectromechanical systems (MEMS) has made it possible to incorporate a variety of computing, communications, and sensing functions on air vehicles with masses of less than 100 g. Micro air vehicles (MAVs) with 15-cm spans are now flying with real-time video, GPS, and sophisticated autopilot functions. With the idea that progress in these areas will continue, we have looked at what might be possible for future MAVs, focusing on airframe technologies that would enable flight at even smaller scales.

---

[*]Professor, Department of Aeronautics and Astronautics. Fellow AIAA.

[†]Doctoral Student, Department of Aeronautics and Astronautics. Member AIAA.

Such vehicles would have many unique capabilities, including the ability to fly indoors or in swarms to provide sensor information over a wide area at a specific time. The very low mass of these devices might make them attractive for planetary exploration, of Mars or Titan for example, because of the high cost of transporting objects into space.[1] Although subgram imaging systems are not available, miniature aerial robots might be used in the near term for simple atmospheric sensing tasks.

Our investigation focuses on mesoscale systems—devices larger than microscopic, yet significantly smaller than conventional air vehicles. Such rotorcraft, termed mesicopters, are centimeter-scale devices with masses of 3 to 15 g, powered by DC motors (Fig. 1). To achieve this goal, the program has begun with the development of somewhat larger vehicles with masses up to 60 g. This provides near-term payload capabilities while reducing the cost of the development program.

A. Hovering vs Forward Flight

That fixed wing aircraft dominate aeronautical development may be attributed to two factors. First, aircraft have generally been employed for transportation of people or goods. The goal is not just to remain airborne, but to get from point A to point B. Not all applications require the high speed capability of aircraft, but this has been an obvious feature to exploit. Although surveillance, communications,

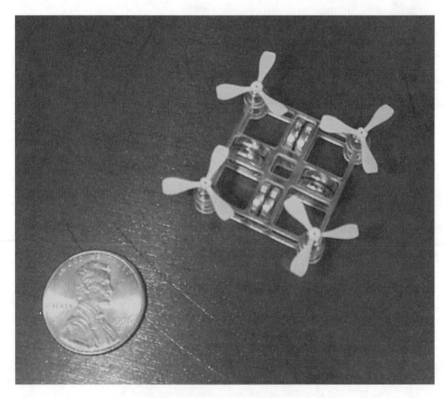

Fig. 1   The mesicopter: a mesoscale flying device.

and imaging (information-related) applications have existed for some time, only recently has this started to become possible with very low mass systems. Second, and most importantly, hovering requires substantially more power than does forward flight—at least for conventionally sized systems. Even if the rotor and wing dimensions are similar, the aircraft in forward flight requires a thrust of $W/(L/D)$ to maintain level flight, whereas the rotor requires a thrust equal to the vehicle weight. However, the required power for a fixed wing airplane increases with speed when the thrust is given. So if the wing L/D is small and the speed that would be required to sustain flight is large, the discrepancy between the required power for forward flight and hover is less significant. In some cases, especially at very low Reynolds numbers, the two flight modes require very similar power inputs. This can be seen more quantitatively as follows:

The power required by a fixed wing aircraft to maintain level flight is

$$P = TV/\eta_p = [W/(L/D)](2W/S\rho C_L)^{\frac{1}{2}}/\eta_p$$

where T is the thrust, V is the forward speed, η_p is the propeller efficiency, W is the weight, S is the wing area, and C_L is the wing lift coefficient. The power required for hover is

$$P = TV_h/M = W(W/2S\rho)^{\frac{1}{2}}/M$$

where V_h is the induced velocity in hover and M is the rotor figure of merit. If we compare the power required for hover with that required for level flight of a propeller-powered fixed wing aircraft, we get

$$P' = (P/W)_{\text{hover}}/(P/W)_{\text{fixed}} = \left[L/D\,\eta_p\,(W/2S\rho)^{\frac{1}{2}}\right]/\left[M(2W/S\rho C_L)^{\frac{1}{2}}\right]$$

Now if, just for comparison purposes, we set the disk loading equal to the wing loading and operate the vehicles at the same density, we get

$$P' = L/D(CL/4)^{\frac{1}{2}}\,\eta_p/M$$

So the difference between rotorcraft and fixed wing aircraft is diminished as the L/D and C_L of the fixed wing vehicle become small, as in the case of low Reynolds number flight. If we consider a 15-cm span MAV with a L/D of 5 at a C_L of 0.2 (and assume, arguably, that the rotor figure of merit and propeller efficiency are similar), a hovering vehicle would require only 12% more power than the fixed wing device. For larger aircraft this is not at all the case. For a high-altitude, long-endurance UAV with $L/D = 35$ and $C_L = 1.0$ the power ratio is 17.5. Of course, the fixed wing MAV flies forward at rather high speeds, whereas the rotorcraft hovers. This may be an advantage or disadvantage, depending on the intended mission, but the point is that at these scales the often-assumed efficiency disadvantage of rotorcraft is not apparent. As the scale is further reduced, and the L/D and optimal C_L of the fixed wing airplane are further reduced, the comparison is even more favorable. Furthermore, the rotor weight for a given disk area may be significantly lower than that of a similarly sized wing together with a propeller and tail surfaces.

Rotorcraft may also be desirable for certain missions because of their compact form factor and ability to maintain their position in hover. In many imaging applications, the conventional aircraft's minimum speed limitations are problematic. Current designs for a Mars aircraft indicate that to avoid excessive vehicle dimensions, flight speeds of Mach 0.5 to 0.6 are required, limiting low-altitude, high-resolution imaging options. Finally, with a rotorcraft design of this size we can provide sufficient control for a four-rotor vehicle using motor speed control, avoiding problems with control surface aerodynamics and actuation that plague small aircraft of conventional design.

B. Rotating vs Flapping

It has been noted that the only truly successful examples of flight at such small scales are insects, which achieve flight using wing flapping, rather than rotary motion. Indeed Ellington,[2] Dickinson et al.,[3] and others have argued that unsteady aerodynamic effects may be significant features of insect flight, increasing the achievable maximum lift at very low Reynolds numbers. A number of successful powered ornithopters have been developed and, despite the fact that these have generally achieved poor efficiencies, there is little question that the approach can be used. It is most attractive at small scales where inertial loads do not dominate and where unsteady phenomena may be helpful. Still, the complexity of the flow-field, the required wing motion, and the mechanism itself lead one first to ask not whether a flapping device can be used, but whether it must be used. Just as automobiles depart from the paradigm suggested by walking animals, we have started by considering the simple steady motion of a rotor, eliminating mechanical complexity and simplifying stability and control issues. Rotary motion does not preclude the possibility of exploiting unsteady effects and we are currently considering the potential for increased maximum lift with rotor speed modulation.

C. Fundamental Scaling Issues

The obvious success of small flying animals, in fact the absence of very large flying animals, suggests that flight at very small scales may be more easily accomplished than flight at larger scales. Some of the relevant scaling laws (e.g., increased strength and stiffness with smaller size) have a beneficial effect on the design of small flight vehicles, whereas others (e.g., aerodynamics) make the design of small-scale aircraft more difficult. Tennekes[4] shows that the wing loading (W/S) of flying devices increases with size. Although this result is likely related to various versions of a square–cubed law, it is surely more complicated than the simple constant density scaling suggested in this reference. One part of the explanation for this trend in nature is that smaller, lighter creatures can more feasibly contend with relatively larger wings because of improved structural properties (strength and stiffness/weight) at small scales. This is also a feature that can be exploited with a small rotorcraft. Rotor weight and aeroelasticity are not problems for tiny rotors with a disk loading of less than 25 N/m^2 (0.5 lb/ft^2), whereas large helicopters have disk loadings of 500 N/m^2 (10 lb/ft^2) or more and would become unmanageable with rotors 4–5 times as large. The lower attainable disk loading is one favorable attribute of small rotorcraft since the *required* power-to-weight ratio scales as the square root of the disk loading.

The *available* power-to-weight ratio depends on the motor and power source. Although scaling of devices such as gas turbines may be quite different from permanent magnet motors, out initial devices consist of electrically driven rotors with energy stored in batteries. Available small DC motors show relatively little effect

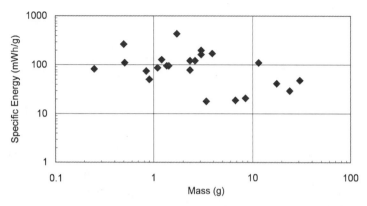

Fig. 2   Battery performance. Some of the batteries here have power output limitations that make them unsuitable. Plot includes several chemistries.

of size on their power-to-weight ratio, although the efficiencies of the very small motors are usually smaller than larger versions. The specific energy of batteries depends more on chemistry than on size, although very small batteries are often dominated by case weight. A study of commercial DC motors and batteries suggested that scaling effects on available power-to-weight ratio were not significant over the 4–5 orders of magnitude considered (see Figs. 2 and 3). If it were not for Reynolds number effects, a miniature electric rotorcraft would have some fundamental advantages over larger devices since the required power-to-weight ratio is reduced by low disk loading, whereas the available power-to-weight is little changed with size.

The reduction of Reynolds number with vehicle size poses one of the greatest challenges for mesoscale flight. Whereas section lift-to-drag ratios of 200 and above are common for large airfoils, the increased skin friction at small scales leads to values in the range of 5 to 15 for Reynolds numbers in the thousands (Fig. 4).

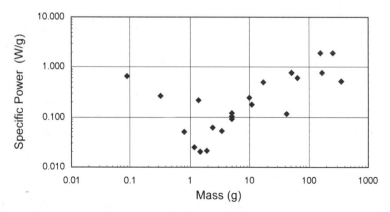

Fig. 3   Specific power of several commercially available dc motors. Efficiency varies and motors include brushed, coreless, and brushless configurations.

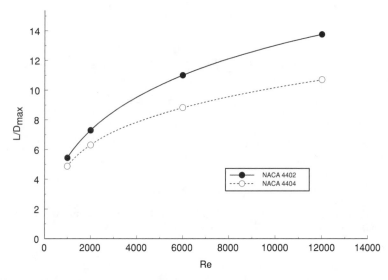

Fig. 4   Reduction in section L/D with Reynolds number. Results are from INS2D with fully laminar flow. Note importance of section thickness.

II.   Approach

The development of the mesoscale rotorcraft is ongoing and has involved challenges in many disciplines. Some of the primary issues are summarized below.

Insect-Scale Aerodynamics. The Reynolds number of the mesicopter rotors lies in the range of 1000 to 6000 where aerodynamics are dominated by viscous considerations and few design tools are available. This is one of the areas in which scaling laws are unfavorable, with lower lift-to-drag ratios and limited rotor lift capabilities. Some of the aerodynamic features are poorly understood in this size regime and means by which improved performance may be realized have been little explored. Because the flow is viscous, some of the simpler tools used for propeller and rotor design are not applicable and basic design rules (e.g., nearly constant inflow) are not appropriate.

Three-dimensional Micromanufacturing. To achieve high lift-to-drag ratios smooth rotors with three-dimensional surfaces at microscale dimensions must be built. Traditional microfabrication techniques can generate features at and below the desired size scales. Yet the need to produce smooth three-dimensional surface features requires rethinking processing steps commonly used for the building of IC and MEMS structures. Traditional three-dimensional machining methods are not normally employed for the fabrication of parts and devices as thin as 50 μm, yet their resolution of a few micrometers makes them attractive candidates for shaping surfaces within this size regime.

Integration of Power and Control Systems. Although many types of batteries with high specific energy are becoming available, identifying very small batteries suitable for the mesicopter, with good specific energy and high current rates, is not easy. The control of these small devices is also a problem. Because of their size, stability time constants are very short and the mass budget for motor/flight control sensors and processing is limited.

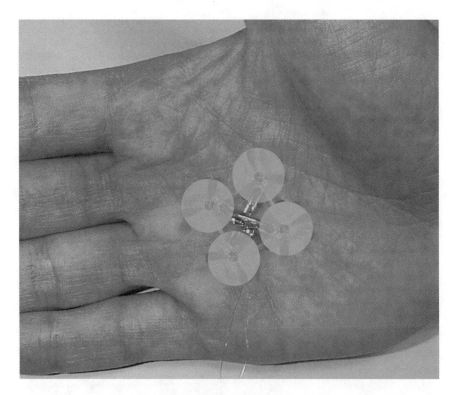

Fig. 5 Initial prototype.

The basic approach has been to develop scalable design and fabrication methods and to start with devices that are larger than the eventual goal. The (super)scale model prototypes are sufficiently large that commercial motors, batteries, and electronics can be employed. The first such prototype is shown in Fig. 5 with a maximum takeoff weight of about 3 g. This device was used to gather data on aerodynamic performance and required an external power supply since the planned Li-ion batteries were not yet available. A second prototype with a maximum weight of 10–15 g is currently being tested and can utilize existing batteries. Finally, a stability and control test bed with even larger dimensions and with a mass of 60 g is also under development. As these systems are refined, the scale will be reduced to explore the limits of this technology.

A. Aerodynamic Design

The operating regime of the mesoscale helicopter poses difficulties for aerodynamic analysis and design. Current sizing and motor parameters result in a rotor tip Reynolds number of approximately 5000. Little experimental or computation work has been published on aerodynamic lifting surfaces operating at such low Reynolds numbers[5] and it is unclear to what extent classical airfoil and rotor analysis and design methods are applicable in this flow regime. The highly viscous nature of the flowfield, large increases in the boundary layer thickness, and the potential for large regions of separated flow create the potential for large discrepancies in performance from what might be expected based on experience at higher Reynolds numbers. The present approach involves a simplified three-dimensional rotor analysis and optimization code, coupled with more complete two-dimensional rotor section analysis. Results from the viscous section analyses are combined with the three-dimensional design code using regression-based models of the two-dimensional results. Although this approach is similar to that used for larger-scale rotorcraft design, the successful implementation of the approach was not straightforward and some surprising results were obtained.

1. Two-Dimensional Analysis and Design

It has been suggested that at such low Reynolds numbers airfoil geometry is not important and that most any inclined plate is as good as an airfoil section. Our initial looks at section performance using the Navier–Stokes solver INS2D[6] indicated that this was not correct. In addition to important effects of thickness (as shown in Fig. 4), details of the section edge shape are significant as is the camber distribution. Figure 6 shows a sample result from the CFD computation, illustrating the very thick "boundary layer" and in Fig. 7 the effect of Reynolds number and thickness form on section drag polars. The large influence of viscosity on the section pressure distribution changes some of the usual approaches to airfoil design. The influence

Fig. 6   Contours of constant total pressure from CFD analysis illustrate thick boundary layer flow at these conditions ($Re = 5000$, $\alpha = 8$ deg).

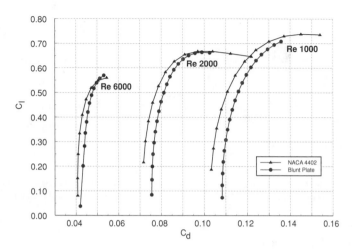

Fig. 7   NACA and 2% cambered plate polars from CFD analysis. Results are from INS2D with fully laminar flow.

of Reynolds number and airfoil geometry on drag is illustrated in Fig. 7. These considerations are described further in a companion paper[7] on airfoil design for the mesicopter, which includes the numerical optimization of section camber that leads to the geometry shown in Fig. 8.

2.   Three-Dimensional Design

The three-dimensional rotor design was based on classical blade element methods with inflow computed using momentum and vortex theory.[8] The analysis incorporates viscous effects in several ways, including an estimate of the swirl introduced by blade profile drag. This simple method was reasonably successful in predicting the rotor performance in hover, but an improved analysis using three-dimensional Navier–Stokes modeling is currently underway.

Because the blade lift-to-drag ratio is low and it and $C_{l\text{max}}$ depend strongly on Reynolds number, some of the simpler approaches to design (e.g., minimum induced loss concepts) lead to less than optimal solutions. Nonlinear optimization was therefore employed to determine the blade chord and twist distribution along with rotor diameter and rpm. Models of the section drag polars were constructed from the Navier–Stokes computations, and motor performance models, based on tests of the brushless DC motors, were incorporated directly into the optimization. Optimization results for the larger device show that the rotor is more strongly constrained by maximum solidity. The second prototype requires approximately four times the lift on each rotor and is constrained to 2.2-cm rotor diameter if ungeared commercial DC motors are used. This leads to the geometry shown on the right of Fig. 9.

Fig. 8   Airfoil optimized for maximum L/D at $Re = 6000$.

Fig. 9   Optimized blade geometries differ greatly owing to different motor characteristics. (Left) 1.5-cm rotors. (Right) 2.2-cm rotors designed for four times the thrust.

B.   Rotor Fabrication

One of the more challenging aspects of creating an efficient mesicopter is the fabrication of the rotor. The optimally designed blades are very thin three-dimensional
structures, with minimum strength and stiffness properties for operation and handling. For the 1.5-cm rotors significant aerodynamic performance penalties were
predicted for thicknesses in excess of 50 μm. Three material categories were
considered—polymers, metals, and ceramics—and a variety of manufacturing
processes for these material categories were explored. The process selected and
implemented by Stanford's Rapid Prototyping Laboratory is known as Shape
Deposition Manufacturing (SDM), a sequence of additive and subtractive processing steps for the fabrication of complex three-dimensional parts. Mold SDM
is a variation of this process for the creation of complex shaped fugitive wax molds.
A spectrum of castable polymer and ceramic materials have been used to make
parts from these molds.[9]

The sequence of manufacturing steps is illustrated in Fig. 10 and involves the
following:

1) *CAD modeling based on the design parameters.* Chord length, twist angle,
and cross-section shapes are given at several stations along the radius. As the result
of manufacturing and strength considerations, the parts close to the center hub are
modified to avoid weak connections and stress concentrations.

2) *CNC code generation.* After the model is created, CNC machining code is
generated using a commercial CAD/CAM package.

3) *Substrate preparation.* Support material is machined to obtain the geometry
of the bottom surface of the rotor by three-axis CNC mill (Step 1).

4) *Polymer casting.* Part material (i.e., polymer) is cast to fill cavity (Step 2).

5) *Surface flattening.* Excess polymer on top of the wax surface is removed (Step 3).

Fig. 10   Steps in rotor fabrication.

6) *Material shaping to net shape.* CNC machines geometry of top rotor surface (Step 4).

7) *Substrate removal.* If the rotors cannot be pulled out of the substrate directly, wax is melted at 150°C; remaining traces can be removed with BioAct (Step 5).

Rotor testing revealed that the actual thrust produced was less than that predicted by the aerodynamic analysis. Although the aerodynamic approximations made in the interest of reasonable computation times might account for this, it is also possible that the as-built parts did not conform to the intended design. To verify this, detailed studies of the rotor shape were conducted using scanning electron microscopy and laser validation of the actual rotor. An example image of the section shape at 75% of the rotor radius is shown in Fig. 11. On the right side of the figure, optical sensing shows significant discrepancies in the as-built incidence. This was corrected by using a wax substrate with lower melting temperatures for subsequent rotors.

In fact, the rotor section shapes did not well approximate the initially designed sections as the desired thicknesses dropped considerably below the minimum 50 μm that could be machined using this process. Subsequent computational fluid dynamics (CFD) analysis showed that although maintaining small maximum thickness was important to good aerodynamic performance, sections with more uniform thickness distributions were acceptable.

C. Power Systems

Initial prototypes use commercially available brushless DC motors (made by RMB in Switzerland). These motors achieve very high efficiencies (60–67%) for their small size (mass as low as 325 mg). Of course brushless motors require motor control electronics and to achieve the rated power and efficiency, rather sophisticated closed-loop controllers are required. The motor manufacturer sells a closed-loop controller, but this weighs hundreds of grams. For this project, the control electronics have been replicated using small components with a total weight of much less than 1 g. A more difficult problem is associated with the voltage requirements for the motors and controllers. Because an input of 4–9 V is required, a rather large number of cells is necessary to drive the motors, using NiCd or AgO$_2$ chemistries.

Fig. 11 **(Left) SEM image of section shape at 0.75R. Chord is approximately 3 mm. (Right) Laser scanning measurements showing error in as-built geometry.**

Lithium batteries are a natural choice, but small, high-current lithiums are not available in the sizes required. Stanford and SRI researchers have explored new lithium polymer technologies that will eventually provide an ideal power source for these devices, but this system is still evolving and is not currently available for the mesicopter prototypes. A more convenient approach involves the use of fewer cells (perhaps as few as one) and a voltage multiplier to achieve the required voltage levels for the available motors. This is our current approach for the smaller devices and electronics development is proceeding in parallel with the system testing.

D. Control

The basic concept of the four-rotor design is that vehicle control can be achieved using the motor controllers described in the previous section. This is convenient as it requires no additional electronics and avoids problems associated with additional actuators. By varying the torque applied to the four motors one can achieve roll, pitch and yaw control, and overall thrust. This strategy for control is not feasible for large rotorcraft, but because of their small size, the mesicopter rotor inertia is very low and the control bandwidth is high. Such configurations have been flown on a larger scale and several successful hobby models achieve stable controllable flight with some additional damping provided by gyro-based feedback.

To determine if the configuration could be made passively stable or to design gains that would be required, a linear model of the rotor aerodynamics was developed and combined with a nonlinear simulation of the vehicle dynamics. This analysis suggested that the vehicle was unstable but could be stabilized with a moderate amount of rate feedback from a MEMS gyro. Subsequent studies showed that by carefully positioning the center of gravity and canting the thrustline inward, natural stability might be achieved. Figure 12 shows the results of this analysis in the form of a root-locus diagram. The figure indicates that certain combinations of vertical center of gravity (c.g.) position and rotor cant angle produce well-damped

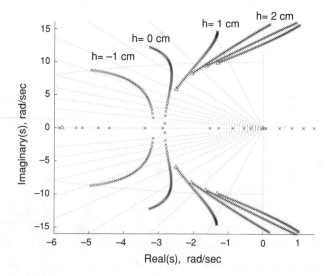

Fig. 12   Roots of linearized mesicopter characteristic equations of motion vs vertical c.g. position and rotor cant angle.

designs. Whether this is sufficient for commanded control of the vehicles is being addressed with the development of a stability and control test bed. This 60-g rotorcraft flies using commercial lithium-ion batteries and is commanded using a conventional pulse-width modulated signal from a radio-controlled transmitter.

E. Sensors

Work on performance and stability has remained the focus of this research to date. One of the next areas for study includes possible sensors for improved flight control. Although rate gyros may be used to provide stability augmentation, very small scale magnetometers and air data systems have been developed for the Defense Advanced Research Projects Agency's (DARPA) MAV program and may be integrated into these devices at some point. Even extremely small scale GPS is a possibility. New concepts for centimeter-level position sensing using carrier-phase differential GPS with a flight system weight of order 1 g are currently being considered. Less exotic sensors, including imaging systems, can more easily be incorporated on the stability and control test bed and this is planned in the next few months.

III. Testing

The testing program to date has included motor, battery, and controller characterization, rotor testing to determine thrust and torque, and complete four-rotor constrained vehicle tests. Figure 13 shows initial single rotor and four-rotor tests on a pivoted arm that constrained the motion to a single degree of freedom before stability and control issues were addressed. This approach has been superceded by more accurate force and moment tests on a test stand constructed for this purpose.

Results of these tests suggested that the aerodynamic design approach was appropriate at this scale, with maximum thrusts of about 80% of the predicted values, despite departures from the assumed section shape. Rotors for the larger prototype have also been fabricated and tested, but they show substantially less lift than predicted. This has been attributed to differences between the designed and as-built geometries, although more significant three-dimensional viscous effects (e.g., viscous-induced swirl) than have been modeled remains a possibility under current study.

Initial tests of a Miniature rotorcraft designed to fly in the Martian atmosphere have recently been completed in collaboration with researchers at the Jet Propulsion Laboratory (JPL). The basic configuration is similar to the other mesicopter

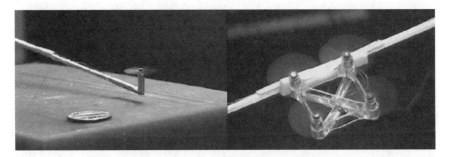

Fig. 13   Initial single rotor and four-rotor lift tests. Rotor diameter: 25 mm.

Fig. 14   Rotor and experimental test setup at JPL for miniature Mars rotorcraft

designs, but owing to the very low atmospheric density and the desire to carry payloads of at least 10 g, the required rotor size was 10 times as large. In addition, the Reynolds numbers are even smaller than for the 1.5-cm-diam mesicopter, ranging from 1000 to 2000. A very lightweight carbon blade was fabricated and tested in a Mars atmosphere simulation chamber at JPL. A view of the test from the observation window and an image of the rotor are shown in Fig. 14. The rotor did produce lift despite the near vacuum conditions, but like the previous high solidity design, the lift produced was somewhat less than predicted based on the simple three-dimensional model. Additional design studies and rotor analyses are planned for this application and related missions such as described in Ref. 1.

IV.   Conclusions

A set of analysis, design, and fabrication methods has been applied to investigate the feasibility of very small rotorcraft. Studies have included a range of vehicle sizes and suggest that mesicopters as small as 1.5 cm are possible, although devices that can carry 10 g of payload may be more easily realized and are of greater current interest. These devices may be used in the near future to carry very simple sensors and may, in the more distant future, be controlled in groups that can provide unique information-gathering capabilities.

Continuing work in three-dimensional, low Reynolds number aerodynamics will be pursued in parallel with a focused effort on stability, control, and communication. Free flights of our prototypes are imminent. These prototypes will provide an excellent test bed for work on distributed control concepts, aerodynamics, and miniature systems development.

Acknowledgments

This work was supported by the NASA Institute for Advanced Concepts, established by the Universities Space Research Association, and funded by NASA. This work is a partnership between the Aeronautics and Astronautics Department and the Mechanical Engineering Department at Stanford. Professor Fritz Prinz leads the fabrication aspects of the project and directs the Rapid Prototyping Lab where the mesicopters are built. Important aspects of this work were accomplished by doctoral students in Aeronautics and Astronautics and in Mechanical Engineering, and researchers at NASA Ames, Langley, and JPL have provided much appreciated technical assistance.

References

[1] Young, L., Chen, R., Aiken, E., and Briggs, G., "Design Opportunities and Challenges in the Development of Vertical Lift Planetary Aerial Vehicles," *Proceedings of the American Helicopter Society International Vertical Lift Aircraft Design Specialists' Meeting*, San Francisco, Jan. 2000.

[2] Ellington, C. P., "The Aerodynamics of Hovering Insect Flight," *Philosophical Transactions of the Royal Society of London Series B*, Vol. 305, 1984, p. 145.

[3] Dickinson, M. H., Lehmann, F. O., and Sane, S. P., "Wing Rotation and the Aerodynamic Basis of Insect Flight," *Science*, Vol. 284, No. 5422, June 1999, pp. 1954–1960.

[4] Tennekes, H., *The Simple Science of Flight*, MIT Press, Cambridge, MA, 1997.

[5] Miki, N., and Shimoyama, I., "Analysis of the Flight Performance of Small Magnetic Rotating Wings for Use in Microrobots," *Proceedings of the 1998 IEEE International Conference on Robotics and Automation, Leuven, Belgium*, IEEE, May 1998, pp. 3065–3070.

[6] Rogers, S. E., and Kwak, D., "An Upwind Differencing Scheme for the Time Accurate Incompressible Navier–Stokes Equations," AIAA Paper 88-2583, June 1988.

[7] Kunz, P., and Kroo, I., "Analysis and Design of Airfoils for Use at Ultra-Low Reynolds Numbers," *Conference on Fixed, Flapping, and Rotary Wing Vehicles at Very Low Reynolds Numbers*, edited by T. J. Mueller, Univ. of Notre Dame, Notre Dame, IN, June 2000.

[8] Johnson, W., *Helicopter Theory*, Princeton Univ. Press, Princeton, NJ, 1980.

[9] Cooper, A. G., Kang, S., Kietzman, J. W., Prinz, F. B., Lombardi J. L., and Weiss, L., "Automated Fabrication of Complex Molded Parts Using Mold SDM," *Proceedings of the Solid Freeform Fabrication Symposium*, Univ. of Texas, Austin, TX, Aug. 1998.

Chapter 24

Development of the Black Widow
Micro Air Vehicle

Joel M. Grasmeyer* and Matthew T. Keennon[†]

AeroVironment, Inc., Simi Valley, California

I. Introduction

T HE first feasibility study for micro air vehicles (MAVs) was performed by the RAND Corporation in 1993.[1] The authors indicated that the development of insect-size flying and crawling systems could help give the United States a significant military advantage in the coming years. During the following two years, a more detailed study was performed at Lincoln Laboratory.[2] This study resulted in a Defense Advanced Research Projects Agency (DARPA) workshop on MAVs in 1995. In the fall of 1996, DARPA funded further MAV studies under the Small Business Innovation Research (SBIR) program. AeroVironment performed a Phase I study, which concluded that a 6-in. MAV was feasible. In the spring of 1998, AeroVironment was awarded a Phase II SBIR contract, which resulted in the current Black Widow MAV configuration.

Several universities have also been involved in MAV research. Competitions have been held since 1997 at the University of Florida and Arizona State University (http://www.aero.ufl.edu/~issmo/mav/mav.htm; http://www.eas.asu.edu/~uav/main.html cited March 2001). The goals of the competitions have been to observe a target located 600 m from the launch site and to keep a 2-oz payload aloft for at least 2 min.

II. Early Prototypes

In the early stages of the Black Widow MAV program, several prototypes were built to explore the 6-in. aircraft design space, which was largely unknown at the time. About 20 balsa wood gliders with different wing configurations were built, and glide tests were performed to determine the lift-to-drag ratios. These tests showed that the disk configuration had some promise, and so a powered version was built next. The powered 6-in. disk performed a 9-s flight in the spring of 1996.

---

*Aeromechanical Engineer. Member AIAA.
[†]Program Manager.

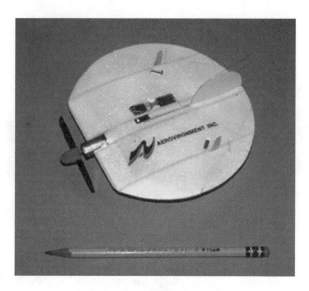

Fig. 1   Early MAV prototype.

The endurance was gradually increased using the disk configuration, culminating in a 16-min flight using lithium batteries in November 1997. This MAV weighed 40 g, was manually controlled by elevons, and did not carry a payload (Fig. 1).

III.   Multidisciplinary Design Optimization

The early prototypes demonstrated that a 6-in. aircraft was feasible. However, the MAV still required a video camera and an advanced control system that would allow operation by an unskilled operator. Since the prototype MAVs were clearly not capable of handling the extra weight and power of these additional systems, a more rigorous design approach was required to continue evolving the system toward maturity.

For this reason, a multidisciplinary design optimization (MDO) methodology was developed in the summer of 1998 to maximize the performance of the MAVs. The goal of the MDO methodology was to create a simulated environment in which the optimum MAV configuration could evolve. The simulated environment consists of physics-based models of the key aspects of the MAV design space, as shown in Table 1. The design variables used for the first-generation MAV optimization study are shown in Table 2. The wingspan was fixed at 6 in. and the wing root chord was fixed at 5.4 in. to accommodate the motor and propeller within the 6-in.-cube constraint. The objective function was to maximize the endurance of the MAV. Because there was no up-front mission requirement, the MAV was optimized around a single cruise design point. The only constraint was that thrust must be greater than drag, and this was implemented with a penalty function. In the MAV MDO code, the optimization is performed by a genetic algorithm.

The vehicle aerodynamics model was validated by performing wind tunnel tests on a variety of wing configurations. This allowed validation of the induced drag and friction drag parts of the code. The protuberance and interference drag components were then added individually.

Table 1   Subsystem model descriptions

Subsystem model	Description
Vehicle aerodynamics	Lifting line theory for induced drag, Blasius skin friction formulas for friction drag, Hoerner equations for interference drag, and equivalent parasite areas for protuberance drag
Propeller aerodynamics	Minimum induced loss methodology[3]
Motor performance	Analytic motor model with coefficients adjusted to match experimental data for each motor
Battery performance	Curve fits of battery endurance vs power draw based on experimental data
Weight buildup	Mass budget of all fixed components, and simple weight equations for variable masses

The propeller aerodynamics model was validated by testing four different propellers in the wind tunnel over a range of velocities, air speeds, and power levels. Both direct-drive and geared props were tested.

The motor model was validated by using the Solver in Microsoft® Excel to adjust the coefficients in the analytic motor model equations to minimize the error between the model predictions and experimental performance measurements over a range of shaft loads and power levels. Most of the motor models show less than 5% error from the experimental data.

The battery model is simply a curve fit through experimental discharge data at different power loads, and so no validation was necessary.

Most of the early MAV prototype vehicles were flown with direct-drive propellers. A few geared propeller configurations were built, but they had marginal performance. However, it was felt that the geared prop concept still had enough potential to merit further study for the first generation of the Black Widow. Therefore two optimum configurations—a direct-drive prop and a geared prop—were created as candidates for the final configuration.

Even though the entire vehicle was optimized for each of the drivetrain types, the tip chords of both configurations were roughly the same, and the loiter velocities

Table 2   MAV design variables

Design variable	Range/options
Battery type	2
Motor type	9
Gearbox type	4
Motor power draw	1–5 W
Propeller diameter	2–4 in.
Wingtip chord	0–6 in.
Loiter velocity	20–40 mph

Table 3 Optimum wing shape parameters (common
to both direct drive and geared propeller designs)

Wingspan	6.0 in.
Wing centerline chord	5.4 in.
Chord at spanwise breakpoint	5.4 in.
Wingtip chord (in)	3.9 in.
Spanwise position of breakpoint	1.5 in.
Wing thickness/chord ratio	8.4%

were roughly the same. Additionally, a series of wind tunnel tests showed that the configuration planform was not a major design consideration. Most of the configurations that were tested had similar performance characteristics. The optimizer also selected the same battery and the same motor for both direct-drive and geared configurations. Therefore these parameters were all frozen to the same values, to allow a normalized comparison between the direct-drive and geared propulsion systems. Table 3 shows the wing shape parameters, and Table 4 shows a comparison of the direct-drive and geared propulsion systems.

The optimization code predicts that the endurance of the geared prop configuration is 30.2 min, whereas the direct-drive prop configuration has an endurance of 33.4 min. The direct-drive prop configuration achieves about 10% greater endurance than the geared prop configuration. This is mainly due to the efficiency loss in the gearbox, the added weight of the gearbox, and the larger and heavier propeller.

To increase confidence in this prediction, both propulsion systems were tested in the AeroVironment wind tunnel over a range of operating conditions. Figure 2 shows the propeller efficiency, motor/gearbox efficiency, and the combined motor/gearbox/propeller efficiency vs thrust at a freestream velocity of 25 mph for the geared propeller configuration. The vehicle drag at 25 mph is 9.9 g for the geared prop configuration. Therefore 9.9 g of thrust is required for level flight. A total

Table 4 Propulsion system parameters for direct drive
and geared propeller configurations

	Geared prop (4:1)	Direct drive prop
Required thrust	9.9 g	9.4 g
Propeller diameter	3.81 in.	2.67 in.
Propeller rpm	5365	22,400
Propeller efficiency	80%	68%
Gearbox efficiency	81%	N/A
Motor rpm	21,460	22,400
Motor efficiency	62%	63%
Total propulsion efficiency	40%	43%
Power draw from batteries	4.65 W	4.35 W
Battery endurance	30.2 min	33.4 min

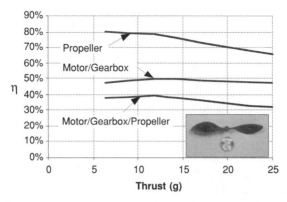

Fig. 2   Propeller, motor/gearbox, and combined efficiencies vs thrust at 25 mph for geared propeller.

propulsion system efficiency of 40% was achieved for this level of thrust. Figure 2 shows the experimental data, which agrees well with the predicted performance shown in Table 4.

Figure 3 shows the propeller, motor, and combined propeller/motor efficiencies vs thrust at 25 mph for the direct-drive prop. Because the direct-drive prop configuration is slightly lighter than the geared prop configuration, it has less induced drag, and the required thrust is 9.4 g. In the direct-drive case, the total propulsion system efficiency is 43%. The propeller wind tunnel tests have increased our confidence in the validity of the optimization code, and they have shown that the direct-drive propeller configuration outperforms the geared prop configuration. Therefore we chose to use a direct-drive prop for the first-generation Black Widow configuration. Table 5 presents a performance summary for the first-generation Black Widow configuration. Figure 4 shows the mass breakdown.

After arriving at the optimum configuration, a sensitivity analysis was performed. This analysis showed that an additional 1 g of drag would decrease the endurance by 3 min, and an additional 1 g of mass would decrease the endurance by 30 s.

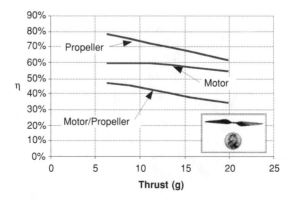

Fig. 3   Prop, motor, and combined efficiencies vs thrust at 25 mph for direct-drive propeller.

**Table 5 Performance summary for the
first-generation Black Widow MAV**

Total mass	56.5 g
Loiter drag	9.4 g
Lift/drag ratio	6.0
Loiter velocity	25 mph
Loiter lift coefficient	0.42
Loiter throttle setting	70%
Endurance	33.4 min

The first-generation MAV configuration performed a 22-min. flight with a black and white video camera on 3 March 1999. This vehicle weighed 56 g and had a cruise speed of 25 mph. The next step was to add color video and increase the endurance to our goal of 30 min. In the summer of 1999, we performed another design iteration and further refined the Black Widow design. The final vehicle is shown in Fig. 5. The vehicle is controlled by a rudder on the central fin and a small elevator in the middle of the trailing edge. The pitot-static tube can be seen extending forward from the right wingtip.

IV. Energy Storage

In the beginning of the MAV program, we evaluated a wide range of power sources, including internal combustion engines, fuel cells, microturbines, and solar power, but the best source of energy among currently available technologies turned out to be modern lithium batteries. Fossil fuels have a much higher energy density than batteries, but the currently available small internal combustion engines are extremely inefficient, difficult to throttle, and generally quite unreliable. Small fuel cell technology looks promising, but practical working systems are not yet available. Microturbines also look promising, but their technology may take even longer to mature. Solar cells cannot supply enough energy to sustain level flight, but they could be used to recharge the batteries while the MAV is parked somewhere.

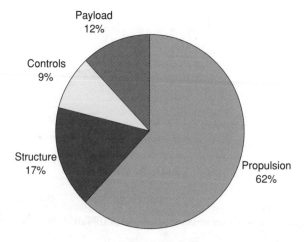

Fig. 4   Mass breakdown for the first-generation Black Widow configuration.

Fig. 5   Final Black Widow MAV configuration.

Batteries and electric motors are extremely reliable, inexpensive, and quiet. However, there are tradeoffs among different battery chemistries. Nickel-cadmium (NiCd) and nickel-metal-hydride (NiMH) batteries have very high power densities but very low energy densities. Lithium (Li) batteries are generally designed to have high energy densities, but they have relatively low power densities. Also, NiCd and NiMH batteries are rechargeable, whereas most lithium batteries are not. We chose to use NiCd and NiMH batteries for flight testing and lithium batteries for demonstration flights.

As with any aircraft, the energy source is a primary design driver. Therefore it is critical to have quantitative data for many different batteries in order to select the best battery for the vehicle. During the MAV program, we characterized a variety of small batteries using discharge tests over a range of temperatures. We reduced these data to a series of curve fits, and we used the curve fits in the MAV optimization code.

V.   Motors

Throughout the MAV program, we tested and evaluated several electric motor candidates. We built a dynamometer specifically for small motors, and we tested each motor over a wide range of operating conditions. The motor test data were used to create an analytic math model of each motor. These motor models were then integrated into the MDO code.

The dynamometer tests showed that efficiencies as high as 70% can be achieved with small motors (less than 10 g mass). The trends are that larger motors have higher efficiency to power ratios, and higher voltage motors have higher efficiency to power ratios. Unfortunately, the motor voltage is limited by the battery supply voltage unless a power converter is used.

VI. Micropropeller Design

The early MAV prototypes used plastic propellers developed for small model airplanes. Some of these propellers were modified by cutting and sanding commercially available props. Because the propeller performance is critical to the success of the MAV, we developed a propeller design methodology that allows us to significantly increase the efficiency of small propellers.

The nominal mission profile for the Black Widow is to climb to about 200 ft above ground level and cruise around at the optimum loiter velocity gathering video data. Therefore, at least 90% of the flight occurs at a single flight condition. This greatly simplifies the propeller optimization, since the off-design conditions do not strongly affect the overall performance. It also allowed us to use the minimum induced loss propeller design methodology to optimize the twist and chord distribution for the loiter flight condition.[3] The prop diameter was optimized by the genetic algorithm along with the other vehicle design variables. The optimum twist distribution is determined by the minimum induced loss propeller design methodology.[3]

The propeller shown in Fig. 6 was optimized for a 4:1 gearbox and a 7-g DC motor. The pitch at the 75% radius station is 6.04 in., and the propeller diameter is 3.81 in.

A three-dimensional model of the propeller geometry was created using the SolidWorks® solid modeling software. Stereolithography models of the upper and lower mold halves were then created from the virtual solid model. Figures 7 and 8 show the prop mold geometry. The propeller was fabricated from unidirectional and woven carbon-fiber composites.

To validate the propeller design code, a series of tests were performed in the AeroVironment wind tunnel. The wind tunnel is an open-circuit, suction design with a test section that is 20-in. wide, 20-in. high, and 40-in. long. The tunnel is capable of producing velocities between 5 and 80 mph in the test section. The torque and thrust were measured using a balance that was constructed using three load cells from commercially available lab scales. The load cells have a 0.1-g accuracy, and they are insensitive to offset loads and moments. This was validated during calibration tests with the completed balance.

Fig. 6   Micropropeller designed for 4:1 gearbox.

Fig. 7 Upper half of propeller mold.

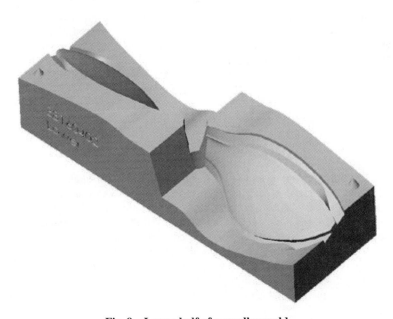

Fig. 8 Lower half of propeller mold.

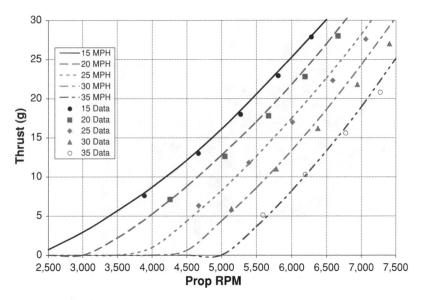

Fig. 9   Thrust vs rpm and freestream velocity for 3.81-in.-diam propeller.

Figure 9 shows the thrust vs rpm and freestream velocity for the propeller. The plot shows excellent agreement between the experimental data and the code predictions. The propeller was designed to produce 10 g of thrust at 25 mph and 5250 rpm.

Figure 10 shows the propeller efficiency vs rpm and velocity for the propeller. The best measured efficiency was 83%, whereas the code predicted a peak efficiency of 82%. The propeller actually operates at 78% efficiency, even though the peak efficiency at 25 mph is 81%. Because the motor efficiency is higher at higher speeds, a slight sacrifice in propeller efficiency increased the efficiency of the total propulsion system. Also note that the peak efficiency increases with increasing freestream velocity owing to higher blade Reynolds numbers.

Figure 11 shows the efficiency vs advance ratio, and Fig. 12 shows the thrust and power coefficients vs advance ratio.

VII.   Airframe Structural Design

The structural design of an MAV presents several unique challenges. Because of the square–cube law, the inertial loads induced on an MAV during accelerations and decelerations (such as takeoff and landing) are quite small relative to those of larger aircraft. During a typical landing, the MAV flies into the ground at a shallow angle and survives a few bounces with no damage. In fact, the worst case design loads for many parts of the structure are the handling loads imposed by people.

The first-generation Black Widow design used a solid foam wing structure with some internal reinforcements in high-stress areas. The wing structure consists of the basic wing, the internal rigid structure, and the vertical fin assembly, as shown in Fig. 13.

The wings are fabricated from expanded polystyrene (EPS) foam. The foam has many desirable qualities, including ease of shaping, light weight, strength, and ease of bonding.

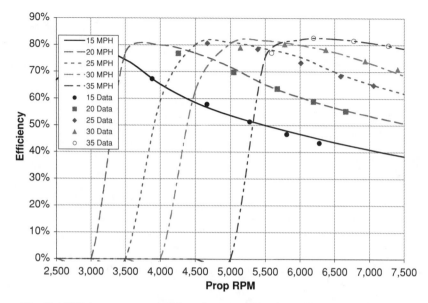

Fig. 10   Efficiency vs rpm and freestream velocity for 3.81-in.-diam propeller.

The main pieces in Fig. 13 are shaped by cutting using a hot-wire tool. The ends of the wire are moved by computer-controlled stepper motors, such that precision cuts can be made from CAD drawings.

A cavity is cut from the leading edge of the center wing section. This is the area where the internal rigid structure is embedded. The internal rigid structure is designed to hold the most massive parts of the MAV (batteries and motor) together and tie into the high load points of the MAV, such as the launch lug. The rigid structure is mainly fabricated from fiberglass sheet.

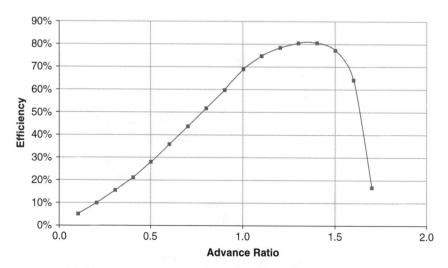

Fig. 11   Efficiency vs advance ratio for 3.81-in.-diam propeller.

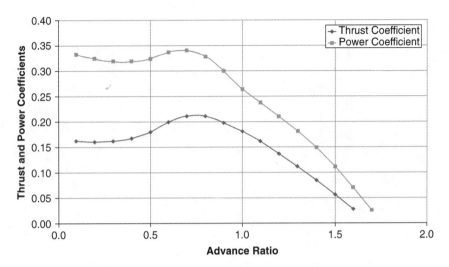

Fig. 12   Thrust and power coefficients vs advance ratio for 3.81-in.-diam propeller.

The vertical fin assembly is hand fabricated from balsa wood. The rudder is also made of balsa wood and is hinged with Kevlar® cloth.

VIII.   Avionics

One of the objectives for the Black Widow MAV is to achieve autonomous flight so that the vehicle can be easily operated by an unskilled operator. The first step toward autonomous flight is to sense the state of the vehicle and pass the data to the flight computer. For this reason, the Black Widow has a two-axis magnetometer to sense compass heading, a pitot-static tube connected to an absolute pressure sensor to sense altitude, and a differential pressure sensor to sense dynamic pressure. The vehicle also uses a piezoelectric gyro to sense the turn rate.

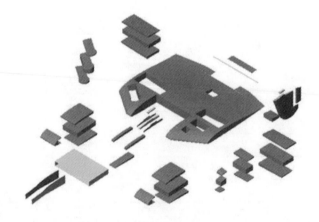

Fig. 13   Solid foam wing structure.

The MAV must also receive commands from the ground station and translate the commands into control surface movements and throttle changes. This requires a command uplink receiver, a flight computer, and control actuators. The uplink receiver has a mass of 2 g and is about the size of two postage stamps. It operates at 433 MHz. The aircraft uses two microprocessors to perform onboard computations. The rudder and elevator control surfaces are moved with custom-developed 0.5-g actuators.

IX.   Video Camera Payload

The Black Widow MAV was developed as a platform to deliver live color video images in real time to an observer on the ground. The video payload evolved from a current off-the-shelf (COTS) video transmitter and a modified COTS black and white complementary metal-oxide semiconductor (CMOS) camera into a custom video transmitter and a custom color CMOS camera, as shown in Fig. 14.

There are a wide variety of micro video cameras on the market today. The challenge for an MAV is to find a good balance of high image quality, low weight, low power, and small size. We found a good compromise with the CMOS video cameras. Table 6 shows the specifications for the CMOS cameras used on the Black Widow.

To get the video from the onboard camera to the ground, we used a radio frequency (RF) transmitter operating at 2.4 GHz. The transmitter takes the analog video stream as an input, modulates it using frequency modulation (FM), and outputs it as an RF signal. Because commercial video receivers and antennas are readily available, 2.4 GHz was used. The first-generation COTS video transmitter had moderate performance because of low power conversion efficiency in the RF amplifier section. For the final MAV system, an improved transmitter with higher output power, smaller size, and lighter weight was developed. Table 7 shows the video downlink transmitter specifications.

Fig. 14   Black and white camera, COTS transmitter (top); custom color camera, custom transmitter (bottom).

Table 6   CMOS video camera specifications

	Black & white	Color
Mass (g)	2.2	1.7
Power (mW)	50	150
Resolution (pixels)	320×240	510×488

X.   Stability and Control

The small size of an MAV creates several unique stability and control challenges. Because the mass moment of inertia scales as the fifth power of the characteristic dimension, small vehicles tend to have high natural frequencies of rotational oscillation. Obtaining a stable video image requires an actively stabilized camera mount or an actively stabilized aircraft. Therefore high oscillation frequencies require a control system with fast processors and fast actuators to stabilize the camera or the entire MAV. Since wing loading decreases with decreasing size, small air vehicles are quite susceptible to gusts. Even small birds (with highly evolved active control systems) have trouble maintaining steady flight in extremely turbulent conditions.

The main stability augmentation system used on the MAV is a yaw damper. Many of the early prototype MAVs showed a 3-Hz Dutch roll oscillation. The addition of more vertical tails and the yaw damper significantly increased the damping ratio. The MAV has three autopilot modes: dynamic pressure hold, altitude hold, and heading hold. More autopilot modes may be added in the future as the system becomes more advanced and as new sensors, such as a GPS receiver, are added. We also developed a data logging system that can sample 16 channels of data at 20 Hz for 4 min. This was used to evaluate and refine the control system dynamics.

XI.   Performance

On 10 August 2000 the Black Widow MAV performed a flight that most likely established several world records for the MAV class of aircraft. Table 8 summarizes the performance on this flight. The pilot flew about 90% of the flight "heads-down," which means he was looking only at the video image and downlinked sensor data from the MAV.

Because the Black Widow uses an electric propulsion system, it is extremely difficult to observe in the air. It cannot be heard above ambient noise at 100 ft, and

Table 7   Video downlink transmitter specifications

	First generation	Final
Mass (g)	3.3	1.4
Power input (mW)	550	550
Power output (mW)	50	100
Frequency (GHz)	2.4	2.4

Table 8 Performance summary for the Black Widow flight on 10 August 2000

Endurance	30 min
Maximum communications range	1.8 km
Maximum altitude	769 ft
Mass	80 g

unless you are specifically looking for a 6-in.-square black dot in the sky directly overhead, you cannot see it. It looks more like a bird than an airplane. In fact, we have seen sparrows and seagulls flocking around the MAV several times.

XII. Ground Control Unit

In addition to the MAV itself, a fully functional MAV system requires a user-friendly, rugged, and compact ground control unit (GCU). The Black Widow GCU evolved through three stages to reach its final form. The first-generation GCU was a collection of off-the-shelf equipment that was quite bulky and had to be assembled at the field. The second-generation GCU was a 15-lb briefcase that contained the MAV, a pneumatic launcher, a removable pilot's control unit with a 4-in. LCD display for the downlinked video, and an automatic tracking antenna. The final GCU (Fig. 15) is built around a plastic case, which is extremely rugged, compact, and waterproof. The MAV is stored in a separate cassette box, which also serves as the launcher. To fly an MAV, the user simply connects the GCU to the launch cassette with a cord, aims the box at the sky, and presses the launch button on the pilot's controller.

XIII. Conclusions

The Black Widow MAV program has been quite successful in proving that a 6-in. aircraft is not only feasible but can perform useful missions that were previously deemed impossible. Additionally, the micro air vehicle concept has opened the doors to many new avenues of research in the fields of aerodynamics, propulsion, stability and control, multidisciplinary design optimization, microelectronics, and artificial intelligence. Some of the specific conclusions resulting from the Black Widow development are:

1) A direct-drive propulsion system appears to be more efficient than a geared propulsion system at the MAV scale.

2) Propeller efficiencies of 80% or greater are possible at the MAV scale.

3) Motor efficiencies of 70% or greater are possible at the MAV scale.

4) An electric propulsion system appears to be the simplest, cheapest, stealthiest, and most reliable option with today's technology.

5) The multidisciplinary design optimization methodology was extremely helpful for maximizing the efficiency of the entire MAV propulsion system.

6) The total propulsion system efficiency is the key parameter in maximizing the endurance of an MAV.

Fig. 15   Black Widow ground control unit and cassette launcher shown deployed (above) and stowed (below).

7) Because the structural weight of an MAV does not vary significantly with configuration variations, the structures subsystem is weakly coupled to the other aircraft subsystems at the MAV scale.

8) It is possible to build a basic avionics suite with data logging capability at the MAV scale.

9) A color video camera with downlink transmitter can be built with a mass of about 3 g using current technology.

10) A 6-in. span, electric MAV of 80-g mass is capable of downlinking live color video from a range of 1.8 km, with an endurance of 30 min.

Acknowledgments

This work was supported by the DARPA Tactical Technology Office, under Contract DAAH01-98-C-R084. The authors would like to express thanks to the rest of the Micro Air Vehicle team at AeroVironment for their tremendous contributions to this project. The team includes Alex Andriukov, Ken Carbine, Craig Foxgord, Dave Ganzer, Mark Levoe, Les Littlefield, Scott Newbern, Stuart Sechrist, and Carrie Sundra.

References

[1]Hundley, R. O., and Gritton, E. C., "Future Technology-Driven Revolutions in Military Operations," RAND Corporation, Document No. DB-110-ARPA, 1994.

[2]Davis, W. R., "Micro UAV," presentation to 23rd AUVSI Symposium, 15–19 July 1996.

[3]Adkins, C. N., and Liebeck, R. H., "Design of Optimum Propellers," AIAA Paper 83-0190, Jan. 1983.

Chapter 25

Computation of Aerodynamic Characteristics of a Micro Air Vehicle

Ravi Ramamurti[*] and William Sandberg[†]

Naval Research Laboratory, Washington, DC

Nomenclature

C_D = coefficient of drag $(F_D/q_\infty L_{\text{ref}}^2)$
C_L = coefficient of lift $(F_L/q_\infty L_{\text{ref}}^2)$
C_M = coefficient of moment $(M/q_\infty L_{\text{ref}}^3)$
C_T = coefficient of thrust
C_Q = coefficient of torque
c = chord length
D = propeller diameter
F = force
J = advance ratio (V/nD)
L = length
M = momentum
n = angular velocity
p = pressure
q_∞ = freestream dynamic pressure $[(1/2)\,\rho_\infty V_\infty^2]$
Re = chord Reynolds number
u = velocity in x direction
v = velocity in y direction
V_∞ = freestream velocity
$\boldsymbol{v}$ = flow velocity vector
$\boldsymbol{w}$ = mesh velocity vector
w = velocity in z direction

Greek Symbols
α = pitch angle of attack
β = yaw angle of attack

---

[*]Aerospace Engineer, Laboratory for Computational Physics and Fluid Dynamics. Senior Member AIAA.
[†]Deputy Director, Laboratory for Computational Physics and Fluid Dynamics.

η = propeller efficiency, $\frac{1}{2\pi}\frac{C_T}{C_Q}J$

θ = elevon angle

ρ = density

σ = stress tensor

ϕ = cant angle of the control surface

Subscripts

a = refers to advection

D = refers to drag

L = refers to lift

ref   = reference

x = refers to x axis

z = refers to z axis

∞ = freestream value

I. Introduction

UNMANNED aerial vehicles (UAVs) play an important role in reconnaissance and war fighting.[1-5] Recently, micro air vehicles (MAVs)[6] have been of significant interest to the war fighting as well as the intelligence collection communities. MAVs have the potential for providing a new technological capability that can be used not only by the military but also in commercial applications including law enforcement, environmental hazard detection and assessments, and inspection of the interior of large buildings. There is more to designing micro air vehicles than just scaling down the dimensions of UAVs. The aerodynamics of the MAVs in the low Reynolds number regime differs significantly from the aerodynamics of mini vehicles, such as the UAVs. The objective of the computations discussed in this chapter is to investigate the inviscid and low-*Re* aerodynamics of novel MAV designs.

In this study, a finite element based incompressible flow solver is employed. The simple elements enable the flow solver to be as fast as possible, reducing the overhead in building element matrices, residual vectors, etc. The governing equations are written in arbitrary Lagrangian–Eulerian (ALE) form, which enables simulation of flow with moving bodies. For viscous flow cases, the mesh requirement for resolving the boundary layer is met by employing arbitrary semistructured grids close to wetted surfaces and wakes. An actuator disk model is employed for modeling the propeller flow. Turbulent flow simulations are performed using the Baldwin–Lomax turbulence model. The details of the flow solver, the rigid body motion, and adaptive remeshing are given by Ramamurti et al.[7] and are summarized next.

II. The Incompressible Flow Solver

The governing equations employed are the incompressible Navier–Stokes equations in ALE formulation, which are written as

$$\frac{\partial \mathbf{v}}{\partial t} + \mathbf{v}_a \cdot \nabla \mathbf{v} + \nabla p = \nabla \cdot \sigma \tag{1}$$

$$\nabla \cdot \mathbf{v} = 0 \tag{2}$$

where p denotes the pressure, $\mathbf{v}_a = \mathbf{v} - \mathbf{w}$ is the advective velocity vector (flow velocity v minus mesh velocity w), and both the pressure p and the stress ten-sor σ have been normalized by the (constant) density ρ and are discretized in time using an implicit time-stepping procedure. It is important for the flow solver to be able to capture the unsteadiness of a flowfield, if such exists. The present flow solver is time-accurate, allowing local time stepping as an option. The result-ing expressions are subsequently discretized in space using a Galerkin procedure with linear tetrahedral elements. To be as fast as possible, the overhead in building element matrices, residual vectors, etc. should be kept to a minimum. This require-ment is met by employing simple, low-order elements that have all the variables (u, v, w, and p) at the same node location. The resulting matrix systems are solved iteratively using a preconditioned conjugate gradient algorithm (PCG). The pre-conditioning is achieved through linelets as described by Martin and Löhner.[8] The flow solver has been successfully evaluated for both two-dimensional and three-dimensional, laminar and turbulent flow problems by Ramamurti and Löhner,[9] and Ramamurti et al.[10]

III.   Description of the Micro Air Vehicle Model

The design of the basic MAV model examined in this study was developed at the Naval Research Laboratory (NRL) and is called the Micro Tactical Expendable (MITE). The wingspan of the vehicle is 6 in. with a span to wing chord aspect ratio of 1.25. The MITE model is driven by a single engine with two counterrotating propellers. The main airfoil is a NACA 0006 section and the end plates are NACA 0015 sections. The end plates are blended into a body of revolution that serves as the engine propeller nacelles. The control surface for the configuration is a wedge airfoil that combines the functions of a rudder and an aileron. A schematic of the

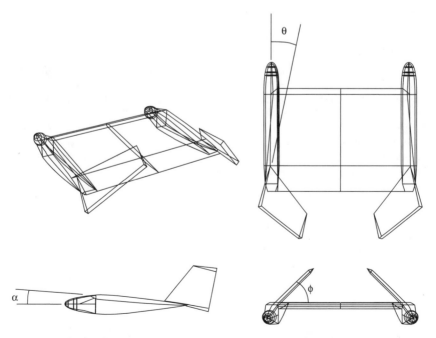

Fig. 1   Schematic of MITE showing the control surface angles.

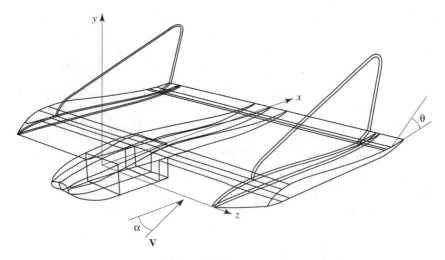

Fig. 2   Schematic of the MITE2 configuration.

configuration[11] is shown in Fig. 1. Another MAV configuration that was tested is called the MITE2. This is a larger model with a 14.5-in. wingspan and a chord of 10 in. This configuration was successfully flown at NRL. This configuration (Fig. 2) differs from the MITE in that it has two tail fins acting as roll stabilizers and elevons, which combine the functions of the elevators and ailerons, as the primary control surface, and is shown in Fig. 2.

IV.   Discussion of Results

A.   MITE

Three-dimensional flow past the MITE was computed at various angles of attack α, ranging from 0 to 14.5 deg, of the main foil. These results were obtained using inviscid calculations on a grid consisting of approximately 160,000 points and 860,000 tetrahedra. Computations were performed over half the vehicle assuming symmetry about then $z = 0$ plane. The length scale was nondimensionalized with the chord length and the inflow velocity with the freestream velocity. The aerodynamic characteristics of the vehicle were also evaluated for several cant angles ϕ (see Fig. 1) and are shown in Fig. 3.

The lift, drag, and moment coefficients are obtained by nondimensionalizing with the dynamic head and the area and are given by

$$C_L = \frac{F_L}{\left(\frac{1}{2}\rho_\infty V_\infty^2 L_{ref}^2\right)}, \qquad C_D = \frac{F_D}{\left(\frac{1}{2}\rho_\infty V_\infty^2 L_{ref}^2\right)} \qquad (3a,b)$$

and

$$C_{M_x} = \frac{M_x}{\left(\frac{1}{2}\rho_\infty V_\infty^2 L_{ref}^3\right)} \qquad (3c)$$

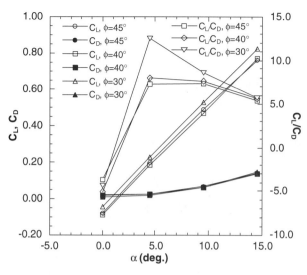

Fig. 3   Aerodynamic characteristics of MITE.

Here, F_L and F_D are the lift and drag forces, M_x is the moment about x axis, V_∞ is the inflow velocity, and L_{ref} is the reference length. The lift force C_L increases linearly with α, as expected. The calculations show that the lift-to-drag ratio (C_L/C_D) attains a maximum around $\alpha = 4.5$ deg, for ϕ values ranging from 30 to 45 deg. The results show that C_L/C_D increases as ϕ is decreased from 45 to 30 deg. This is due to the increase in lift at $\phi = 30$ deg. The drag is almost the same for all the cant angles tested. The surface pressure distribution on the configuration and on the symmetry plane, for $\phi = 45$ deg and $\alpha = 0$ deg is shown in Fig. 4. It

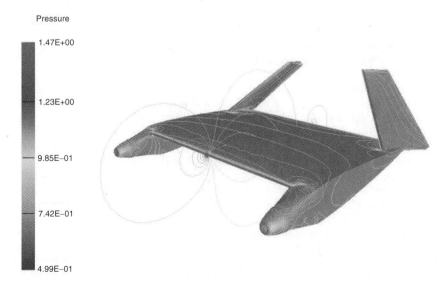

Fig. 4   Surface pressure distribution on MITE at $\phi = 45$ deg and $\alpha = 0$ deg.

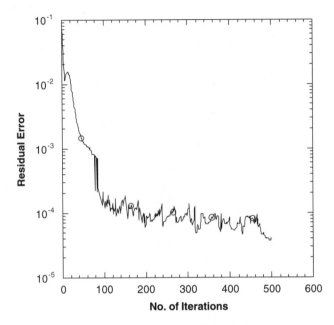

Fig. 5   Convergence history for inviscid flow.

can be seen that the maximum pressure is at the stagnation regions at the nose of the end plate and at the leading edge of the main airfoil, and the minimum value occurs at the junction of the main airfoil and the control surface. The convergence history for the case of $\alpha = 0$ deg is shown in Fig. 5. The grid employed for this case consisted of 157,000 points and 856,000 tetrahedral elements. It can be seen that the residual in pressure dropped by more than 3 orders of magnitude in 500 iterations.

To validate the computational results, a grid refinement study was performed. The resolution of the mesh on the surface and in the vicinity of the model is doubled. The resulting mesh consisted of 290,000 points and 1.6×10^6 tetrahedral elements. Figure 6 shows the variation of the computed lift and drag for various angles of attack using the coarse and the fine grids. It can be seen that a grid-independent solution has been achieved and that the grid consisting of 157,000 points is sufficient for the computations.

Laminar and turbulent flow were computed past this configuration for Reynolds number based on the chord of the main airfoil. The Baldwin–Lomax model[12] was employed to model the eddy viscosity. The contribution of the pressure and the viscous forces to the lift and drag are shown in Table 1. The C_L/C_D for the laminar case, at $Re = 50,000$, is reduced from the inviscid result of 7.3 to 1.46, primarily because of the viscous drag. As Re is increased further, the C_L/C_D ratio increases to 3.06, owing to the increase in lift and a slight drop in the drag force. For the turbulent case, the entire model was assumed to be in a fully turbulent flow. The drag attributed to the pressure and the viscous components increases by a modest amount, compared to the laminar case. The lift force is also slightly reduced, resulting in a decrease of C_L/C_D, to a value of 1.89. The convergence history for

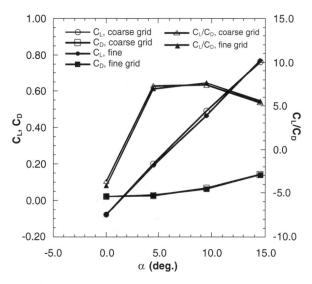

Fig. 6 Effect of grid refinement.

the laminar case is shown in Fig. 7. Again, the pressure converges by more than 3 orders of magnitude in 700 iterations.

B. Effect of Fuselage

The initial MITE configuration shown in Fig. 1 was modified to include a fuse-lage at the center with the entire fuselage located below the leading edge of the main airfoil and the control surface cant angle at $\phi = 45$ deg. The grid that was employed for this configuration consisted of approximately 187,000 points and 1.03×10^6 tetrahedral elements. The third configuration that was evaluated was the MITE configuration with the fuselage smoothly joined on the upper surface of the main airfoil. A grid with 211,000 points and 1.16×10^6 tetrahedral elements was employed for this case. The effect of the fuselage on the C_L/C_D characteristic of the vehicle was also studied at selected angles of attack and the results are plotted in Fig. 8. The addition of the fuselage lowers the lift and the C_L/C_D ratio at $\alpha = 4.5$ deg. The maximum C_L/C_D ratio occurs at an angle of attack $\alpha = 9.5$ deg.

Table 1 Variation of lift and drag for the MITE

Re	C_D(pr.)	C_D(visc.)	Total C_D	C_L(pr.)	C(visc.)	Total C_L	C_L/C_D
Euler	2.72×10^{-2}	0.00	2.72×10^{-2}	1.99×10^{-1}	0.00	1.99×10^{-1}	7.31
50,000 (laminar)	3.70×10^{-2}	5.78×10^{-2}	9.48×10^{-2}	1.41×10^{-1}	$-1.78e\text{-}03$	1.43×10^{-1}	1.46
100,000 (laminar)	3.19×10^{-2}	2.08×10^{-2}	5.27×10^{-2}	1.62×10^{-1}	$-4.89e\text{-}03$	1.61×10^{-2}	3.06
100,000 (turbulent)	3.66×10^{-2}	3.30×10^{-2}	6.95×10^{-2}	1.32×10^{-1}	$-1.06e\text{-}03$	1.31×10^{-1}	1.89

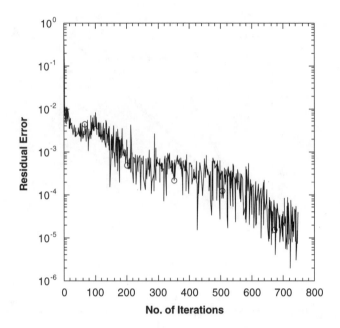

Fig. 7 Convergence history for laminar flow.

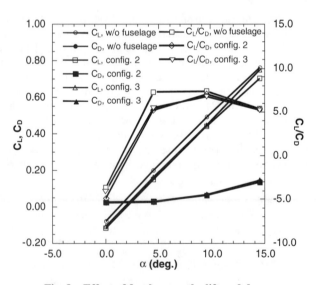

Fig. 8 Effect of fuselage on the lift and drag.

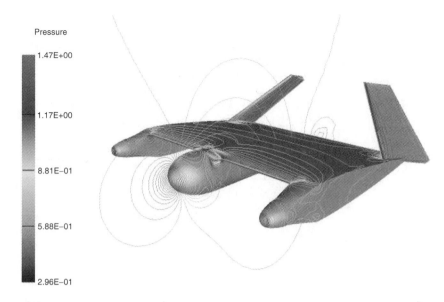

Pressure

1.47E+00

1.17E+00

8.81E−01

5.88E−01

2.96E−01

Fig. 9 Surface pressure distribution over the MITE configuration with fuselage; ϕ = 45 deg and α = 0 deg.

The drags for all three configurations considered are nearly the same. The surface pressure distribution for configuration 3 is shown in Fig. 9.

C. Effect of Propeller

To study the effect the propeller has on the flow, we could employ the adaptive remeshing flow solver to simulate the flow past the moving blades and the interaction of this flow with the vehicle. This computation would be long because of the unsteady nature of the flow, and the remeshing around moving bodies makes it computationally expensive. Since the objective of this study was to evaluate several configurations at various flow angles and control surface orientations, a simpler model to mimic this flow, the actuator disk model, was employed.

The region where the propeller is situated, shown in Fig. 10, is marked. In this region, source terms are added as body forces in the momentum equation, Eq. (1). This model has been successfully employed by Oh et al.[13] for modeling the flow in a ducted propulsor. The source strength for the propeller region is determined from the thrust per unit volume. First, a constant body-force term in the x direction is added. The strength of the source was computed to compensate for the drag force. The lift force increased from 0.038 to 0.039, but the drag force also increases from a value of 0.007 to 0.010. Hence, the source strength was increased by an order of magnitude. For this case, although the lift increased to 0.045, the drag also increased to 0.015, reducing the C_L/C_D to approximately 3.06.

D. MITE2 Configuration

The elevon deflection angle θ of the MITE2 configuration, shown in Fig. 2, was varied in the range from −15 to +15 deg; positive angles are for the elevon deflected

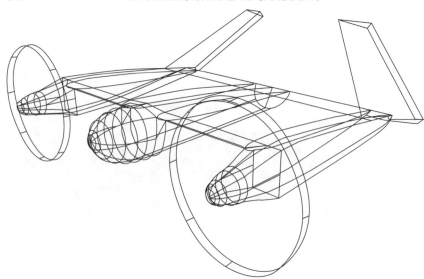

Fig. 10   MITE configuration with the propeller.

in the counterclockwise direction from the original position. The variation of the coefficient of lift with change in angle of attack for various elevon angles is shown in Fig. 11. The lift and drag coefficients are computed using Eq. (3). The L_{ref} is taken to be 10 in. for this configuration. It is clear from Fig. 11a that C_L varies linearly with α for all the elevon angles tested, and $dC_L/d\alpha$ is approximately 3.43/radian. For the MITE configuration, with L_{ref} being unity, at an angle of attack $\alpha = 14.5$ deg and $\phi = 30$ deg, the lift force is approximately 0.21, resulting in a C_L of 0.41. For the MITE2 configuration, the C_L exceeds this value for all elevon angles $\theta \geq 5$ deg. The cases where the two elevons of the vehicle are deflected asymmetrically are denoted by two values for the angle θ. In Fig. 11a, the variation of lift for a differential deflection of 5 deg falls between the curve corresponding to the case where both the elevons are deflected by 0 and 5 deg. The variation of the lift coefficient with angle of attack when the elevons are deflected asymmetrically is shown in Fig. 11b. For all the deflections studied, the $dC_L/d\alpha$ is almost constant, except for the case when one of the elevons is deflected by +15 deg. In this case, the pressure on the lower side of the elevon is reduced, resulting in a loss of lift at high angles of attack.

The variation of C_L with respect to elevon angle is shown in Fig. 12. It is clear that the variation is almost linear with $dC_L/d\theta$ of approximately 1.36/radian. The variation of the drag coefficient is shown in Fig. 13. For all the cases where the elevons are deflected by a positive angle, the drag coefficient is a minimum at $\alpha = 5$ deg and increases with further increase in α. For the negative deflection cases the minima occurs at 0 deg and increases with increase in α. The drag coefficient for the case of the split elevon falls between the values corresponding to $\theta = 0$ deg and 5 deg. The variation of the ratio C_L/C_D is shown in Fig. 14. At $\alpha = 0$ deg, C_L/C_D is negative except for $\theta = -15$ deg. For all the cases with negative elevon deflection, C_L/C_D achieves a maximum at $\alpha = 5$ deg; for positive deflection angles, C_L/C_D drops when α is increased to 5 deg and recovers with further increase in α. It is interesting to note that at $\alpha = 15$ deg, all the cases come together to a value of around 5.0, perhaps because of the dominance of the wing–fuselage flow at high

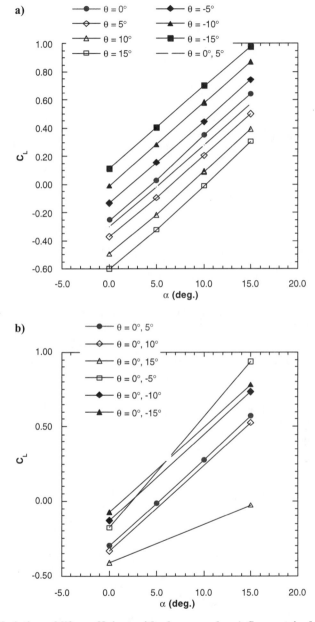

Fig. 11   Variation of lift coefficient with elevon angle: a) Symmetric deflection of elevons, b) Asymmetric deflections of elevons.

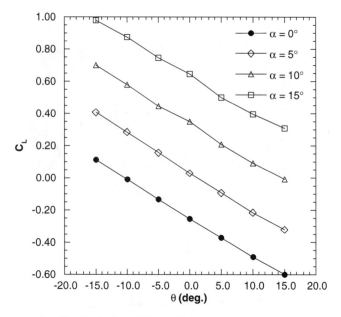

Fig. 12   Variation of lift coefficient with elevon angle.

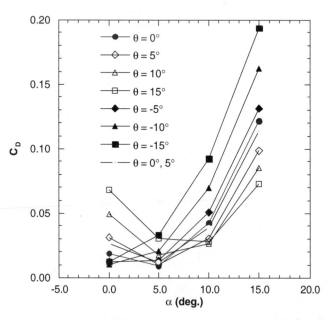

Fig. 13   Variation of drag coefficient with angle of attack for the MITE2 configuration.

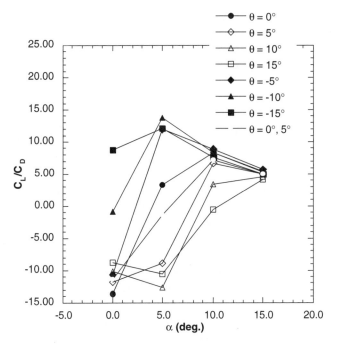

Fig. 14   Variation of C_L/C_D with angle of attack.

angles of attack. The variation of the roll moment is shown in Fig. 15. It can be seen that for both angles of attack tested, the roll moment varies linearly with elevon deflection angle. The value of $dC_{M_x}/d\theta$ is approximately 0.16/radian. The variation of the pitch moment about the center of gravity of the vehicle is shown in Fig. 16. From Fig. 16a, it can be seen that the pitch moment varies linearly with θ, and the slope is approximately 0.91/radian. The variation with the angle of attack is shown in Fig. 16b. As α is increased, C_{M_z} increases for all elevon angles.

E.   Effect of Propeller

As in the case of the MITE, the region where the propeller is situated is first marked. In this region source terms are added as body forces in the momentum equations. The strengths of the source terms were obtained from the power characteristics of the propeller. The propeller for this case was 7 in. in diameter, and at 3000 rpm the shaft power consumed was approximately 4 W. Assuming a nominal efficiency $\eta = 0.65$, the thrust generated for a forward flight velocity of 24 ft/s is approximately 0.08 lb. The propeller efficiency is given by

$$\eta = \frac{1}{2\pi} \frac{C_T}{C_Q} J$$

$$J = \frac{V}{nD}$$

(4a, b)

were J is the advance ratio, n is the angular velocity of the propeller (in revolutions per second), D is the diameter of the propeller, and C_T and C_Q are the thrust and

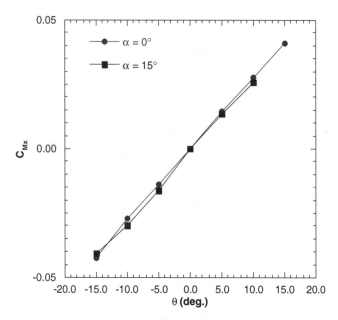

Fig. 15   Variation of roll moment with elevon angle.

the torque coefficients, respectively. From Eq. (4) the ratio C_T/C_Q can be found to be approximately 5.0. Knowing the thrust required and incoming velocity, the momentum source needed in the axial direction can be computed. The momentum source in the tangential direction is then computed using the C_T/C_Q ratio. Three cases were run for the MITE2 configuration at angles of attack of 0 and 15 deg. In the first two cases, the source was distributed uniformly in the propeller region. In the first case, the tangential source was in the direction of the wingtip vortex, and in the second case the swirl opposed the wingtip vortex. In the third case, the source was distributed in the radial direction. A blade element analysis was performed to obtain this distribution. Figures 17a and 17b show the effect of the propeller on the coefficients of lift and pitch moment, respectively. It is clear that the slope of these curves are increased marginally.

F.   Trajectory Simulation

The derivatives obtained from these computations were also used as input to a six-degree-of-freedom trajectory simulation. The characteristics of the vehicle were supplied together with simple control laws for maintaining vehicle altitude and direction. An initial trajectory simulation showed that when the vehicle was subjected to a 90 deg turn, it underwent a roll instability. It was subsequently found that the yaw–roll coupling was incorrect. To get a better estimate of this coupling, computations were performed for this vehicle at a yaw angle of attack $\beta = 5$ deg, and with $\alpha = 8$ and 15 deg. The effect of yaw on the transverse force coefficient C_N and the roll moment C_{M_x} are shown in Fig. 18. With these additional characteristics for the vehicle, the simulated trajectory showed that the vehicle was able to execute a 90-deg horizontal plane turn with no roll instability.

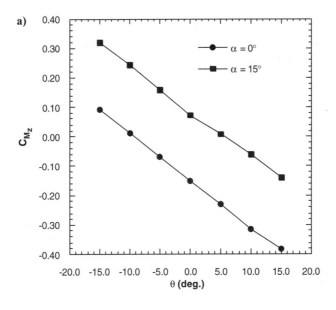

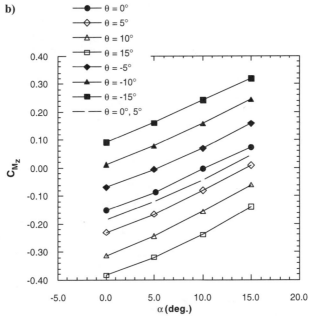

Fig. 16   Variation of coefficient of pitch moment: a) Effect of elevon angle, b) Effect of angle of attack.

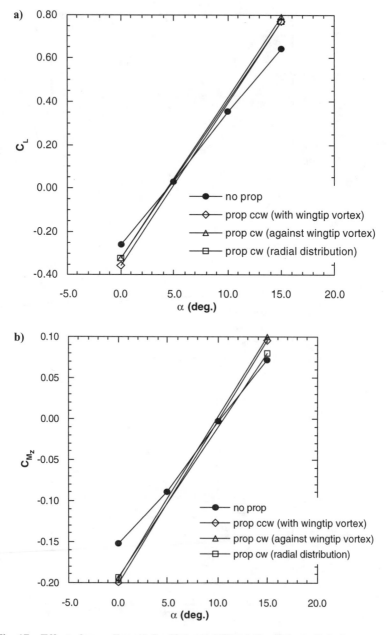

Fig. 17   Effect of propeller: a) Coefficient of lift, b) Coefficient of pitch moment.

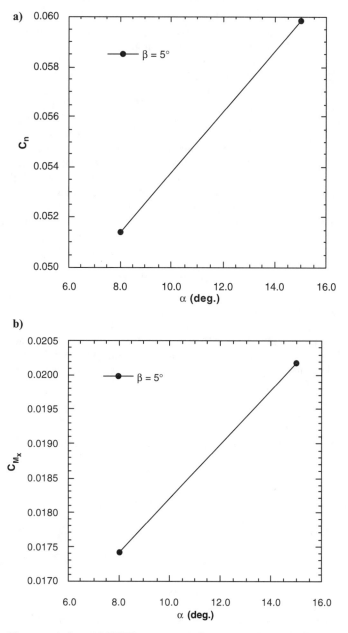

Fig. 18   Characteristics of MITE2 at yaw: a) Transverse force coefficient, b) Roll moment coefficient.

V. Summary and Conclusions

A finite element flow solver based on unstructured grids has been employed to study the aerodynamic characteristics of micro air vehicles. Several configurations developed at NRL, the MITE, with and without the fuselage, and the MITE2, were evaluated. Both laminar and turbulent flows past these configurations were computed for selected cases. For the MITE configuration without the fuselage, the L/D ratio attains a maximum value of 12.5 for a control surface cant angle $\phi = 30$ deg and at an angle of attack $\alpha = 4.5$ deg. The addition of fuselage lowers the L/D ratio at $\alpha = 4.5$ deg. An actuator disk model was employed to simulate the effect of the propellers. For the MITE2 configuration, several computations were performed in which the angle of attack and the elevon angle were varied. From these results, the characteristic derivatives were computed. These derivatives were supplied as an input for a trajectory simulation. For miniature air vehicles, the conventional propeller may not be an efficient mechanism for propulsion. For these vehicles, flapping foil propulsion is an attractive alternative. The lift and thrust enhancement in this mode of propulsion was studied in two dimensions by Ramamurti et al.[14] and is being extended to three dimensions.

Acknowledgments

This work was supported by the Office of Naval Research through the Tactical Electronic Warfare Division of the Naval Research Laboratory. The authors would like to thank Professor Rainald Löhner of George Mason University and Mr. John Gardner of LCP&FD; NRL for their helpful discussions and assistance; and Mr. Kevin Ailinger for his support throughout the course of this work. This work was supported in part by a grant of HPC time from the DoD HPC Centers, ARL MSRC SGI-O2K and NRL SGI-O2K.

References

[1]Cook, N., "USA's Revolutionary Plan for Air Vehicles Unveiled," *Jane's Defense Weekly*, March 1997, p. 5.

[2]Evers, S., "'Culture of Innovation' Return to UAV Projects," *Jane's Defense Weekly*, March 1997, p. 14.

[3]Hewish, M., "Building a Bird's-Eye View of the Battlefield," *Jane's International Defense Review*, Feb. 1997, pp. 55–61.

[4]Canan, J. W., "Seeing More, and Risking Less, with UAVs," *Aerospace America*, Oct. 1999, pp. 26–31.

[5]Wilson, R., "Mini Technologies for Major Impact," *Aerospace America*, May 1998, pp. 36–42.

[6]Hewish, M., "Rucksack Reece Takes Wing," *Jane's International Defense Review*, Vol. 30, No. 2, Feb. 1997, p. 63.

[7]Ramamurti, R., Sandberg, W. C., and Löhner, R., "Simulation of a Torpedo Launch Using a 3-D Incompressible Finite Element Flow Solver and Adaptive Remeshing," AIAA Paper 95-0086, Jan. 1995.

[8]Martin, D., and Löhner, R., "An Implicit Linelet-Based Solver for Incompressible Flows," AIAA Paper 92-0668, 1992.

[9]Ramamurti, R., and Löhner, R., "Evaluation of an Incompressible Flow Solver Based on Simple Elements," *Advances in Finite Element Analysis in Fluid Dynamics*, FED Vol. 137, edited by M. N. Dhaubhadel et al., ASME, New York, 1992, pp. 33–42.

[10]Ramamurti, R., Löhner, R., and Sandberg, W. C., "Evaluation of Scalable 3-D Incompressible Finite Element Solver," AIAA Paper 94-0756, 1994.

[11]Hewish, M., "A Bird in Hand, Miniature and Micro Air Vehicles Challenge Conventional Thinking," *Jane's International Defense Review*, Vol. 32, No. 11, Nov. 1999, pp. 22–28.

[12]Baldwin, B. S., and Lomax, H., "Thin Layer Approximation and Algebraic Model for Separated Turbulent Flows," AIAA Paper 78-257, 1978.

[13]Oh, C. K., Löhner, R., Ramamurti, R., and Sandberg, W. C., "An Actuator Disc Simulation of Body-Ducted Propulsor Flow," APS Fluid Meeting, New Orleans, LA, 21–23 Nov. 1999.

[14]Ramamurti, R., Sandberg, W. C., and Löhner, R., "Simulation of Flow about Flapping Foils Using a Finite Element Incompressible Flow Solver," AIAA Paper 99-0652, 1999; *AIAA Journal*, Vol. 39, No. 2, 2001, pp. 253–260.

Chapter 26

Optic Flow Sensors for MAV Navigation

Geoffrey L. Barrows,* Craig Neely, and Kurt T. Miller
Naval Research Laboratory, Washington, DC

I. Introduction

T HERE are several major efforts to build and fly so-called micro air vehicles
(MAV), which are loosely defined as aircraft with 15-cm or smaller wingspans.
A desired, but currently unobtained, capability is for such an MAV to be able to
fly toward its destination autonomously. This includes "small-scale navigation,"
which refers to collision avoidance, altitude control, landing, and other similar
tasks. Furthermore, it is desirable for all processing and control, including any collision avoidance, to be performed onboard. Because of the MAV's small size, any
onboard collision avoidance system needs to weigh no more than a few grams. Early
successes in the field of optic flow chips[1,2] led the authors to consider optic flow
sensors as a means of providing basic small-scale navigation capability to MAVs.

II. Optic Flow

A. Optic Flow for Navigation

The term "optic flow"[3] refers to the speed at which texture moves in an image
focal plane as a result of relative motion between the observer and objects in the
environment. Optic flow is typically formulated as a vector field over an image,
in which the vectors define the velocity at which the texture is moving in the
image plane. Figure 1 depicts how a micro air vehicle might experience optic flow
while in flight and how the optic flow could be used for small-scale navigation and
collision avoidance. Shown are two optic flowfields as seen by two sensors: one
aimed downward and one aimed forward. In this example, we consider an MAV
flying in a straight line over ground.

The optic flow in the downward direction moves backward. This is simply
because the ground appears to be moving backward when the MAV is flying
forward. The optic flow will be slightly faster for objects whose top surfaces are
above the ground and hence closer to the MAV. The magnitude of the optic flow will

*Currently a founder of Centeye, Inc.

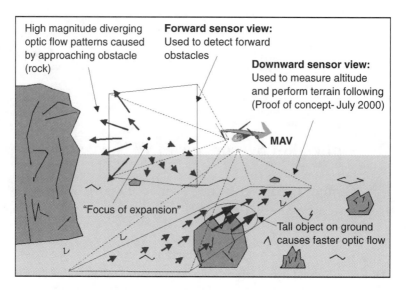

Fig. 1   Optic flow as seen by an MAV flying above ground.

be inversely proportional to the distance between the MAV and such top surfaces. Thus a tree or a rock whose height is just below the flight altitude of the MAV will have a corresponding large optic flow vector.

The optic flow in the forward direction is more complex and depends on the presence of any obstacles in the front. The focus of expansion (FOE) is the point from which the optic flow vectors radiate. The FOE indicates the MAV's direction of heading. If the MAV is approaching an obstacle, then the forward optic flowfield will have a large divergence or expansion quality, and the FOE will be inside the obstacle in the visual field. If the MAV has a flight path that will cause it to fly near an obstacle but without collision, the associated optic flowfield will have a large expansion on the side of the obstacle. In this case the FOE will not be located inside the obstacle. If there are no obstacles in the flight path then there will be only a small divergence in the optic flow, resulting from the ground and the horizon. Figure 1 depicts the scenario in which the MAV will pass a large rock on the left. Thus the optic flow on the right side has a large divergence quality, but the focus of expansion is not within this region of high divergence. The optic flow on the upper left portion of this visual field is zero, since this region of the visual field includes the sky and other distant objects.

Consider another example, not shown, in which the MAV is flying down a tunnel and has two additional sensors looking in the left and right directions. The MAV will be able to tell where it is in relation to the two walls by measuring the ratio of the optic flow on both sides. If the two measured optic flows are equal, then the MAV is in the center of the tunnel. This is independent of the tunnel's actual width. In fact, it is believed that honey bees use such a rule to fly down the center of a tunnel or hallway:[4] When one of the walls is shifted forward or backward at a constant rate, the honey bee shifts over appropriately so that the two optic flows appear equal.

The above discussion gives a greatly simplified and intuitive understanding of how optic flow can be used for basic navigation and collision avoidance tasks.

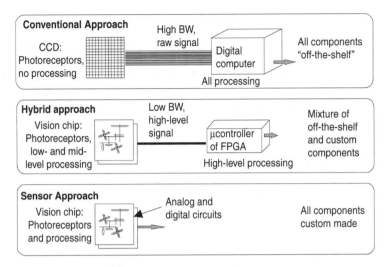

Fig. 2   Different approaches for implementing machine vision.

Examples of how more complex optic flow patterns can be used to analyze increasingly complex scenarios can be found in Aloimones's book[5] and elsewhere in the literature.

B.   Optic Flow Sensors

Next we describe the "sensor" approach of measuring optic flow. Figure 2 depicts three abstracted hardware approaches, ranging from conventional machine vision to a fully integrated sensor. First consider the "conventional machine vision" approach, which is loosely defined as approaches in which a charge-coupled device (CCD) or similar imager digitizes the visual field according to some regular space and time pattern, and a digital computer processes this raw information. The output of such an imager is essentially a video signal. The measurement of optic flow using this approach is known to be a computationally intensive problem. Qualitatively, this is because such video signals tend to have a large bandwidth. Furthermore, the type of processing that computes optic flow searches for features that extend in both space and time. Algorithms that compute optic flow thus search for such features that span the three-dimensional video signal and consequently require many CPU instructions. Arguably this computational complexity problem is due to some basic aspects of conventional machine vision: The image is sampled by a CCD or other imager according to some arbitrary space–time grid, which generates a large amount of raw data. This raw data form is appropriate for collecting imagery for, say, image understanding or image interpretation applications, but is not of the form that is convenient for computing optic flow. Thus a large amount of processing needs to be performed. However, the conventional machine vision approach has the advantage that all the parts are commercial off-the-shelf (COTS) and that the digital computer is reprogrammable, thus allowing easy modifications to the system. This "conventional approach" represents almost all machine vision systems.

In contrast is the "sensor" approach for measuring optic flow, in which the entire machine vision system is placed on a single integrated circuit or multichip module (MCM). First, the sensor only computes the part of the optic flowfield that is necessary for the given task. This reduces the required number of computations. Second, the sensor includes analog and digital circuitry, as well as specially shaped photoreceptors, that "naturally" perform many of the computations in a compact implementation. This allows for the implementation of faster and much more compact hardware to compute optic flow. There are many examples of integrated visual sensors (also called "vision chips"), including those that measure optic flow, that were developed with this philosophy. The interested reader is referred to a review paper[1] and a book.[2] Included in these documents are several examples of single-chip optic flow sensors. The strengths of these sensor chips are that they are compact and that they consume very little power (some on the order of microwatts). The weakness of many of these sensors is that they have limited performance in that they often need artificially high contrast textures (such as vertical black and white bars) to function. Furthermore, they are all hard-wired. To implement a new function, a new chip must be designed and fabricated. This process requires a significant amount of time and money. This contrasts with conventional machine vision systems, where a new sensor can be implemented by simply reprogramming the computer. Obviously, this approach for implementing machine vision systems is in the minority, accounting for much less than 1% of all machine vision efforts.

A compromise is the "hybrid sensor" approach, which is essentially a "middle of the road" approach between the sensor approach and the conventional approach. The goal is to retain some of the compact size, low-power consumption, and high-speed characteristics of analog circuitry while maintaining some of the re-programmability of digital circuits. In this approach, the front-end processing is performed in the analog domain using specially shaped photoreceptors and ana-log circuits. The hard-wired "sensor head" produces a low-bandwidth but high information content signal, which can then be processed by a simple digital com-puter such as a microcontroller or even a field-programmable gate array (FPGA). Essentially the sensor head performs many of the brute force computations up front. The use of a microcontroller allows the entire sensor to be reprogrammed or reconfigured to some extent without requiring a large digital computer. This hybrid approach for building machine vision systems results in a somewhat larger sensor than a single-chip sensor and consumes more power. However, the flexibi-lity afforded by the microcontroller facilitates optimization and therefore increases the chance that the sensor will solve a practical problem.

It should be noted that pursuing the hybrid approach does not commit one to abandon the development of a single-chip sensor. In fact, the hybrid sensor can be considered a stepping stone: Once the digital algorithm has been worked out, standard electronic design CAD tools can be used to design digital circuitry that performs the same function, and then this circuitry can be placed on the same chip as the analog circuitry.

III. Description of the Optic Flow Sensor

It should be noted that one of the challenges associated with optic flow sensors is to be able to correctly measure motion when the visual environment is "realistic." This means that the sensor must be able to function in environments that occur in the real world, indoors or outdoors, rather than in the laboratory and exposed to artificial textures such as black and white bars or similar patterns. This chapter describes an optic flow sensor, developed by the author, that satisfies these conditions and is thus practical in the above sense. First we describe the competitive feature tracker (CFT) algorithm, a new algorithm for measuring optic flow that exhibits robust performance with real-world textures.[6,7] Then a fusion algorithm is described that merges the outputs of multiple CFT sensors to provide more robust measurements.[8] Recently these sensors have been integrated with gliders and aircraft, in a control loop. This chapter gives an overview of the above work. For further details, the reader is referred to Refs. 6–11.

A. Generalized Model for Implementing Optic Flow Algorithms

Figure 3 depicts the general model of optic flow measuring architectures that have guided the research described here. This model was inspired form the basic known architecture of animal visual systems. A lens focuses an image of the environment onto a focal plane chip, which contains photoreceptor circuits and other circuits necessary to compute optic flow. Low-level feature detectors respond to different spatial and temporal entities in the visual field, such as edges, spots, and corners. The elementary motion detector (EMD) is the most basic structure or entity that senses visual motion, though its output may not be in a form easily used. Fusion circuitry fuses information from the EMDs to reduce errors, increase robustness, and produce a meaningful representation of the optic flow for specific applications. For implementing hybrid sensors, one engineering decision that must be made is placement of the boundary in this figure that separates processing on the vision chip sensor head from processing in the microcontroller.

B. The CFT Algorithm

Here we briefly describe the CFT EMD algorithm as introduced in a previous paper.[6] For a detailed description of a single-chip implementation of CFT, refer to an earlier paper.[10] Figure 4 depicts the basic CFT EMD architecture. Figure 5 depicts the edge detection kernel implemented by a differential amplifier and two photoreceptors. Figure 6 shows sample traces of feature signals and feature location signals. Many of these EMD structures would be implemented, scattered across the visual field, in a full sensor. The discussion here covers how optic flow is measured in one tiny part of the visual field.

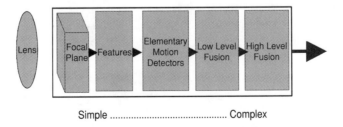

Fig. 3   Overall sensor architecture.

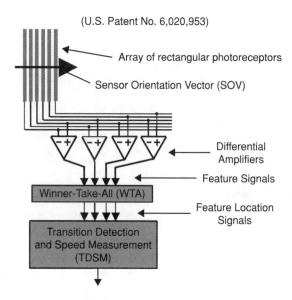

(U.S. Patent No. 6,020,953)

Array of rectangular photoreceptors

Sensor Orientation Vector (SOV)

Differential Amplifiers

Feature Signals

Winner-Take-All (WTA)

Feature Location Signals

Transition Detection and Speed Measurement (TDSM)

Fig. 4   The competitive feature tracker elementary motion detector.

Functionally there are four sections, as shown in Fig. 4: photoreceptors, feature detectors (shown here as differential amplifiers), a winner-take-all (WTA), and a transition detection and speed measurement (TDSM) section. A section of the focal plane is sampled with an array of elongated rectangular photoreceptors laid out so that the array is positioned along the sensor orientation vector (SOV). The photoreceptor rectangles are arranged so that their long axes are perpendicular to the SOV. This geometry filters out visual information perpendicular to the SOV direction while retaining information in the parallel direction. One effect of these rectangular photoreceptors is that the optic flow measurement produced by the EMD is

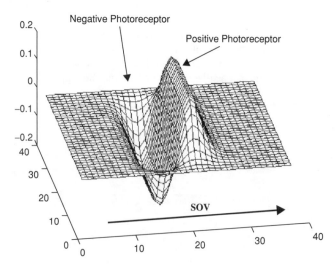

Fig. 5   Edge detection kernel.

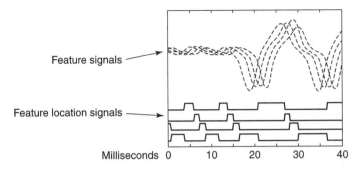

Fig. 6   Sample signal traces.

actually a measurement of the projection of the two-dimensional optic flow vector onto the SOV vector. In other words, if the optic flow is parallel to the SOV, then the magnitude (and sign) of the measured optic flow corresponds to the actual optic flow measurement. If the optic flow is perpendicular to the SOV, then the measured optic flow is zero. If the optic flow direction is, say, 45 deg off the SOV, then the magnitude of the measured optic flow is that of the actual optic flow divided by $\sqrt{2}$. A proof of this statement is contained in the first author's dissertation.[7]

The outputs from the photoreceptors are sent to an array of four feature detectors that output four analog feature signals. A feature detector circuit attains its highest output value when the feature to which it is "tuned" appears on its input photoreceptors. For example, suppose the feature detectors are differential amplifiers. Then their effective response function is the edge detection kernel, shown in Fig. 5. A feature signal will have a high value when a vertical edge is located between the input photoreceptors with the brighter side of the edge illuminating the positively connected photoreceptor.

A wide variety of other linear feature detectors can be implemented, as shown in Fig. 7. An array of weights called the "configuration vector" can be used to

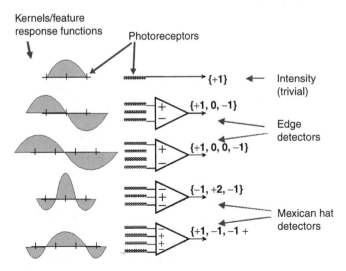

Fig. 7   Different configurations of feature detectors.

describe how the photoreceptor signals are linearly combined. For example, the configuration vector $\{-1, 0, 1\}$ describes the edge detector used in Fig. 4. This edge detector has a "response function" shown in Fig. 5. Another edge detector is described by the $\{-1, 0, 0, 1\}$ configuration. Another type of feature detector is a second derivative or "linear Mexican hat" detector implemented by the $\{-1, 2, -1\}$ and $\{-1, 1, 1, -1\}$ configurations. Such a feature detector is similar to a one-dimensional difference-of-Gaussian (DOG) filter.

The four analog feature signals are then sent to WTA. The WTA has four analog inputs and four digital outputs. The WTA determines which input has the highest value and sets the corresponding output to a digital high (or "1") and all the other outputs to a low (or "0"). The location of the high value indicates where on the photoreceptor array the image is most like the feature defined by the configuration vector. The WTA outputs are thus called feature location signals. As an edge moves across the photoreceptors shown in Fig. 4, the high value will move sequentially across the WTA outputs. This is easily visualized with the aid of the signal traces in Fig. 6. Shown are four feature signals and their corresponding feature location signals when the photoreceptors are exposed to a moving black and white bar pattern.

The TDSM circuit converts the movement of the high WTA output into a velocity measurement. Essentially this circuit interprets the $1{\to}2{\to}3{\to}4$ motions of the high WTA output as visual motion. Whenever the high feature location signal moves in a manner that indicates visual motion, the EMD generates a measurement of the optic flow. The direction of the visual motion is determined by the direction of travel of the high feature location signal. Likewise the speed is obtained from the lag time from one feature location to the next. The actual optic flow (in degrees or radians per second) can be determined from the physical geometry of the photoreceptor array and the sensor optics. More details on how this speed measurement is performed can be found in Refs. 6 and 7.

It should be noted that this basic EMD architecture can be modified in different ways. First, more than four feature detectors can be used. The number of feature detectors (and thus the number of inputs and outputs of the WTA) is called the "width" of the EMD. Second, it is possible to construct two-dimensional versions of this EMD.[7]

C. Fusion

Above it was described how the basic CFT EMD can be reconfigured to track different features by changing the configuration vector of the feature detectors. In practice it has been observed that a wide variety of feature detector functions can be used to implement reliable EMDs. Each of these EMDs will occasionally make erroneous measurements, owing to spatial aliasing effects caused by the photoreceptors. However, it has been observed that EMDs using different configurations tend to make these errors at different times. This is not surprising given that different feature detectors effectively transform the focal plane image in different ways. This suggests that if several different EMDs share the same photoreceptor array, then their measurements could be fused somehow to produce a more reliable answer.

Several different methods of fusing the EMD outputs could be postulated. For example, a median or a mean could be taken over all measurements taken within a time window. Such a method is plausible for implementation on a digital computer but is not amenable to implementation in mixed-signal very large scale integration (VLSI) circuitry. Another method utilizing majority-vote principles is to quantize the range of possible speed measurements into velocity ranges, and then keep track of how many measurements fall within each velocity range over a time period. The velocity range that has received the most stimulation provides the sensor output. Alternately, a leaky integrator can be used to record the activity of each velocity range. This latter method is the method favored by us. More details are located in Refs. 7 and 8.

D. Implementation of a First-Generation Hybrid Sensor

Two generations of the sensor architecture described earlier were fabricated and tested. Both generations used the "hybrid" approach. Figure 8 shows one of the first-generation sensors. A custom analog integrated circuit sensor head with two photoreceptor arrays and four configurations of feature detectors was fabricated in a 1.2-μm N-well process via the MOSIS service.[12] The sensor head was approximately 2-mm square. A small 3-mm focal length glass lens was mounted above the chip surface to project an image of the environment onto the chip. A PIC microcontroller was used to implement the winner-take-all and TDSM sections of the CFT algorithm with software. Details on the implementation can be found in the first author's dissertation.[7] These particular microcontrollers digitized the feature signals at approximately a 1-kHz rate using an on-chip 8-bit A/D converter. The fusion algorithm was also implemented in software on the microcontroller. A total of eight velocity ranges were implemented, four in the forward direction and four in the backward direction. Two first-generation sensors were prototyped: The first sensor used a PIC 16c715 microcontroller and only one of the photoreceptor arrays. This sensor measured optic flow in one direction only. The second sensor used a PIC 16c76 microcontroller and both photoreceptor arrays to measure optic flow in two directions. A mirror was used to split the visual field of the sensor head so that optic flow in two opposite directions could be measured.

A more powerful processor than the 16c715 or the 16c76 can certainly be used for this task. However, this particular microcontroller is compact and light.

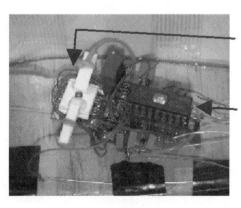

• **Vision chip front-end:**
Photoreceptors and feature detectors

• **PIC 16C715 microcontroller:**
(2kB PROM, 128B RAM, 1MIPS) Remainder of CFT algorithm and glider control

Fig. 8   First-generation hybrid sensor.

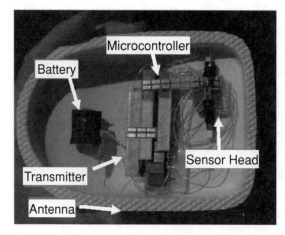

Fig. 9   Second-generation hybrid sensor.

Using this microcontroller, the custom analog chip, and a few discrete components, it was possible to fabricate a full compact optic flow sensing device weighting approximately 25 g. With such a low mass, it was possible to experiment with integrating the sensor into a glider for use in a control loop.

E.   Implementation of a Second-Generation Hybrid Sensor

A second-generation hybrid sensor was fabricated and tested (Fig. 9). In this sensor, the sensor head vision chip had two photoreceptor arrays and four configurations of feature detectors for each array. Also on-chip was the WTA circuit. This particular sensor had a width of six—there were six, rather than four, feature signals for each configuration, and the WTA circuit had six inputs and six outputs. A single WTA circuit whose inputs were multiplexed between the feature detector signals was used. A PIC 16c76 microcontroller was again used for back-end processing. This sensor was an improvement over the first generation because the sensor head outputted digital signals (WTA outputs) rather than analog signals. Thus the PIC did not have to perform analog to digital conversion, which sped up the entire sensor. Also, the higher width allowed the definition of more complex transition rules to improve the accuracy and reliability of the sensor.[11]

IV.   Use of Optic Flow for Navigation

Provided that one has linear optic flow sensors that are capable of measuring optic flow robustly with real-world textures, such sensors should be useful for a variety of MAV navigation and control tasks. Many of these control functions can be described as simple rules or heuristics that directly (or reflexively) map an optic flow feature to a behavior or an interpretation of the current state and environment.

By mounting several of these sensors aimed in different directions, it is possible to measure the roll, pitch, and yaw rates of an aircraft. Such information would provide stabilization information to the aircraft in much the same manner as a gyroscope. In the case of a fixed wing aircraft, both yaw and pitch rates can be estimated with as little as one sensor aimed directly forward, with photoreceptor arrays oriented both vertically and horizontally. Assuming the aircraft is flying forward and not undergoing sideways motion (due to, say, air currents coming in from the side), optic flow in the forward direction can be attributed directly to pitch or yaw motion. Additional sensors aimed in the backward direction will increase the robustness of the measurement and also handle the scenario in which the aircraft is undergoing sideways motion. The use of additional sensors mounted backward is thus also appropriate for rotary-wing aircraft. The weakness of this approach is that the presence of other moving objects in the environment could "corrupt" the optic flowfield, causing erroneous measurements. The best sensor configuration for measuring roll, pitch, and yaw motion would have optic flow measured in all directions, so that global optic flow features can be measured. Such a configuration is, of course, what is implemented in insect vision, as insect eyes tend to have a panoramic view of the world.

The measurement of altitude could be performed with as little as one sensor aimed downward, as depicted in Fig. 1. Assuming the aircraft is traveling in a straight line, the optic flow moving backward in the downward direction is a function of both the altitude and ground speed of the aircraft. Hence if one of these variables is known, and the optic flow is measured, then the other variable can be determined. If the aircraft is undergoing pitch motion as well then this rotational component would need to be removed from the downward optic flow measurement to obtain optic flow resulting from translation only. An improved version of altitude control could be obtained by looking in a variety of directions downward. The shape of the terrain could be interpreted, allowing for an improved control of altitude as well as terrain following.

Flanking a wall is a similar problem to that of maintaining an altitude. From the perspective of optic flow sensing, the only difference is that the sensor is aimed sideways instead of downward. The most significant difference that would be encountered in practice is that the control laws for guiding the aircraft would be different. Similar to flanking a wall is flying down the center of a tunnel or hallway. This can be performed with two sensors, one aimed left and one aimed right. A flight path down the middle would be obtained by simply selecting the path that equalizes the optic flow on both sides. This particular strategy has been observed in experiments with honey bees.[4]

A more difficult set of tasks would be the detection of imminent collisions with obstacles. Collisions can be detected by looking for a strong divergence of the optic flowfield that results when the colliding object appears to grow rapidly in size. A simple control rule would be to then steer away from regions in which the divergence is the greatest. An approximate measurement of divergence can be performed using several linear optic flow sensors arranged in a cross pattern, as shown in Fig. 10, using theory developed by Ancona and Poggio.[13] Experiments in using such optic flow divergence to perform obstacle avoidance in aircraft are the subject of current research efforts.

V. Initial In-Flight Experiments

Several sets of experiments were performed with the above-described sensors to validate the concept that such sensors can provide small-scale navigation capability to aircraft.

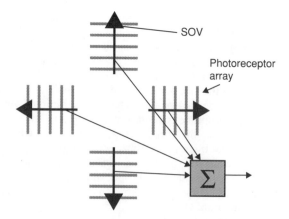

Fig. 10 Measuring optic flow divergence with four optic flow sensors.

A. Experiments with First-Generation Sensor on a Glider

The first-generation sensors were mounted onto small balsa wood and Styrofoam gliders (Fig. 11) to experiment with the use of the sensor in flight in a control loop. The goal was to use the optic flow sensor in several simple navigation scenarios. These experiments were intended to serve as an initial proof-of-principle of the use of optic flow as a means of guiding an MAV. The sensor was mounted on the

Fig. 11 Glider with first-generation sensor.

glider's fuselage and pointed in the appropriate direction. Additional code was added to the microcontroller that caused the microcontroller to provide power to a flap-type control surface that could serve as a rudder or an elevator.

The first experiment explored the use of downward-aimed sensors for altitude measurement. The sensor utilizing one photoreceptor array to measure optic flow in one direction was used for this experiment (e.g., the first of the first-generation sensors). The control surface was mounted horizontally on the tail as an elevator. The control surface was programmed to turn upward whenever the measured optic flow reached a threshold velocity and to provide neutral control otherwise. Thus when the glider was released it should first glide gently toward the ground. The measured optic flow should increase as the glider approached the ground. Then at an altitude that corresponds to the sensor's threshold optic flow value, the control surface should turn upward and cause the glider to flare up into a stall. The glider should "porpoise" up and down one or more times until it runs out of momentum and lands. These behaviors were indeed observed whenever the glider was released.

In the second experiment both the sensor and the control surface were rotated 90 deg. The purpose was to demonstrate the use of optic flow to steer the glider away from a wall. The glider was tossed toward the wall at a shallow angle. At first, the control surface provided neutral control and did not modify the glider's path, allowing the glider to approach the wall. Just before the glider crashed into the wall, the sensor detected an optic flow rate that exceeded the predetermined threshold. The sensor modified the control surface appropriately and caused the glider to turn away from the wall. Repetition of this experiment gave the same result. It is acknowledged by the author that such simple use of a rudder is not the most efficient method of causing yaw motion; however, the effect was still strong enough for the purpose of demonstration.

The goal of the third experiment was to demonstrate the use of optic flow to avoid both walls. For this experiment the second first-generation sensor (e.g., the one looking in two directions) was used. The microcontroller was programmed so that when the optic flow on one side was substantially higher than that on the other side, the glider would steer in the opposite direction. When the measured optic flow on the two sides was balanced, the control surface provided neutral control. Using this simple setup, it was possible to demonstrate a limited form of wall avoidance: When thrown toward one wall the glider generally steered away into the opposite direction. As the glider approached the other wall, the sensor generally detected it and initiated a second turn. However, in this particular setup, the second wall was usually not detected soon enough to be avoided. This is a result of the combined dynamics of the glider and the sensor: There is lag in both the sensor and in the glider. Thus the rudder would turn on closer to the wall than it would turn off when heading away. Therefore the glider would leave the wall at a higher angle of incidence than when traveling toward the wall. So after a few wall avoidances, the glider travels at too steep an angle toward the wall for the glider to be able to steer away. This effect could be reduced by modifying the control rule to cause the glider to correct its path after avoiding one wall. Likewise, one could put memory into the sensor to cause it to estimate its angle of incidence, and modify the control surface appropriately. (It is known in control theory that to control x, one needs to have information about both x and dx/dt.) However, this route was not explored in detail for these experiments.

B. Experiments with the Second-Generation Sensor on an Aircraft

The second-generation sensor was used to explore altitude measurement and control on a small (1-m wingspan) RC (remote control) type aircraft in two experiments. The sensor was mounted underneath the wing of the aircraft, next to the fuselage, and aimed toward the ground. In the first experiment, the aircraft was flown entirely by a pilot, and the optic flow measurement was obtained by radio telemetry. The test environment was a grassy field with trees and a parking lot area to the side. An inverse relationship between the flight altitude and the measured optic flow was observed. Crossing between the grassy field area and the parking lot area did not alter the obtained optic flow measurement. In addition, when the aircraft was flown over the trees, the optic flow increased abruptly. These behaviors are as expected.

In the second experiment, the sensor was integrated in a control loop to demonstrate altitude control. Figure 12 contains an image of the sensor mounted on the aircraft. The integration was performed as follows: The sensor had a separate and small control surface for controlling aircraft pitch. This way the human pilot had overriding control of the aircraft for takeoff, landing, and general safety. The sensor controlled this surface via a standard pulse-width-modulated (PWM) servo connected directly to the sensor. A simple, but predictable, control rule was implemented: If the measured optic flow is faster than a set threshold, then pitch up. Otherwise, pitch down.

The flight experiments were performed as follows: The aircraft was launched and the human pilot guided the aircraft to a desired altitude at one end of a grassy field. The pilot then trimmed the hand-held RC controller to provide neutral control that would cause the aircraft to fly forward in a straight line when the controls were released. Once the trimming was performed, the aircraft was again guided to one end of the field. The controls were released, and it was observed whether the sensor was able to keep the aircraft stable at a set altitude. When the other end of the field was reached, the pilot regained control of the aircraft and turned it around for another pass.

When the pilot released the controls, the sensor indeed gave pitch feedback to the aircraft. However (as easily seen in Fig. 12), the sensor's control surface was off

Fig. 12   RC type aircraft with second-generation sensor.

Fig. 13   Four images, taken approximately 1 s apart, showing altitude control being performed with pitch control by second-generation sensor.

center. Thus the aircraft would roll whenever the sensor's control surface moved. It was found that the human pilot had to provide roll correction to keep the aircraft flying. (This roll correction did not noticeably affect the aircraft pitch, and thus it was the sensor alone that controlled the pitch for this purpose.) When this corrective feedback, and only this feedback, was provided by the human, the sensor was able to control the aircraft's flight altitude. An altitude of approximately 5 m was obtained, with a flight speed of approximately 20 m/s. Figure 13 shows four frames, about a second apart, of one flight pass. Of course, with the zero-order control rule provided, the aircraft did not stay at a constant altitude but varied up and down around that altitude as expected. Over a flight time of about 15 min, the human pilot "rescued" the aircraft three times, when the aircraft dropped too close to the ground. This experiment proved that aircraft altitude can be controlled by optic flow.

VI.   Next-Generation Sensors

Described earlier were several prototype hybrid optic flow sensors that were integrated to provide in-flight control to two types of aircraft. The sensors were very limited in that they measured optic flow in just one or two locations of

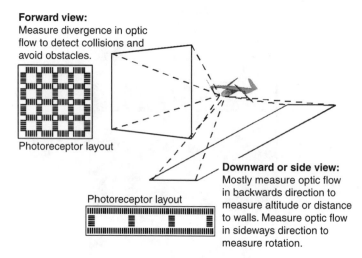

Forward view:
Measure divergence in optic flow to detect collisions and avoid obstacles.

Photoreceptor layout

Photoreceptor layout

Downward or side view:
Mostly measure optic flow in backwards direction to measure altitude or distance to walls. Measure optic flow in sideways direction to measure rotation.

Fig. 14   Possible photoreceptor layouts defining downward- and forward-looking sensors.

the visual field and their mass was on the order of 25 g or larger. To be truly practical, the coverage of the visual field must be expanded, the sensor weight must be decreased to several grams, and more advanced control rules must be used. This discussion will focus on the first two requirements. The size of the sensor can easily be reduced using standard electronic construction practices. As visible in Fig. 8, a dual in-line packaged (DIP) version of the microcontroller was used, and larger through-the-hole discrete components were used for interfacing the sensor head with the microcontroller. By using a more compact board layout and using the most compact surface mount components, the mass can be reduced to the required weight. The mass can actually be reduced further by moving beyond a hybrid sensor to a true single-chip sensor. All the necessary electronics to do so have been demonstrated.[7] Thus it would be an exercise in engineering to create the chip layout.

Figure 14 depicts how the visual field can be sampled with greater coverage, using the same sensor views as depicted in Fig. 1. Shown for the two directions are photoreceptor array layouts to be implemented on-chip. Not shown are the layouts for feature detectors and the remainder of the algorithm, which will take up comparatively smaller amount of on-chip "real estate" and whose layout is less critical.

In the downward direction, the main interest is in measuring the backward-moving optic flow from the ground. Thus the majority of the photoreceptors should be oriented with their long axis in the left–right direction to sense this motion. That way the outputs of these photoreceptors will be less sensitive to roll in the MAV. A smaller number of photoreceptors should be implemented in the orthogonal direction to measure the roll and/or any horizontal drift in the MAV. Such a sensor layout will be able to measure the altitude in a number of directions ranging from underneath the aircraft to the forward direction, thus providing a robust estimation of aircraft altitude and the terrain curvature.

In the forward direction, the main interest is in measuring divergence in the optic flow field for detecting collisions. Thus the photoreceptor layouts should be equally distributed between both horizontal and vertical directions. This will allow the two-dimensional optic flow and hence the divergence to be computed.

It should be noted that the above sensor heads were fabricated in an integrated circuit process that represents early 1990s technology: A complete integrated sensor[10] with one photoreceptor array and three EMDs was fabricated on a chip in a 1.2-μm process. This particular design took up a 1.1×1.1 mm area in a layout that was not optimized for size. Thus if one extrapolates to 0.15 μm processes that are available at the time of this chapter's writing, it should be possible to squeeze a 30 by 30 array of such sensors onto a single 5-mm square chip.

VII. Conclusion

Research is being performed to develop compact optic flow sensors for use in a variety of small-scale navigation tasks for MAVs. By using a combination of the CFT EMD architecture and the fusion algorithm, optic flow sensors have advanced to the point of being practical in real-world textures. A proof of principle has been obtained in a set of simple experiments in which a glider avoided collisions with a wall or a floor by using optic flow and where an aircraft had its altitude controlled using optic flow.

The above proof of principle is clearly not sophisticated enough for practical applications. However, it does demonstrate rudimentary small-scale navigation with a very simple sensor and a "zero-order" control rule. As implied above and as discussed in the first author's dissertation,[7] all the building blocks are available for the fabrication of next-generation sensors that will measure optic flow as shown in Fig. 1. At this point the fabrication of such sensors is an exercise in engineering. This is the subject of current efforts by the authors.

Acknowledgments

The help of Don Srull of Quest, Inc. is acknowledged in assembling the gliders. The help of Kevin Ailinger, Rudolph Montoleon, Eric Petticord, and Rick Foch of NRL Code 5710 in providing advice is greatly acknowledged. Steve Tayman of NRL Code 5710 was the human pilot for the demonstration involving the second generation sensor.

This work was funded by the Office of Naval Research.

References

[1] Sarpeshkar, R., Kramer, J., and Koch, C., "Analog VLSI Motion Detection: From Fundamental Limits to System Applications," *Proceedings of the IEEE, Special Issue on Parallel Architectures for Image Processing.* Vol. 84. No. 7, IEEE, Piscataway, NJ, 1996.

[2] Moini, A., *Vision Chips*, Kluwer Academic, Boston, 2000.

[3] Gibson, J. J., *The Ecological Approach to Visual Perception.* Houghton Mifflin, Boston, 1950.

[4] Srinivasan, M. V., Zhang, S. W., Lehrer, M., and Collett, T. S., "Honeybee Navigation En-Route to the Goal: Visual Flight Control and Odometry," *Journal of Experimental Biology, Special Issue on Navigation*, Vol. 199. No. 1, 1996, pp. 237–244.

[5] Aloimones, Y. (ed.), *Visual Navigation, From Biological Systems to Unmanned Ground Vehicles*, Lawrence Erlbaum, Mahwah, NJ, 1997.

[6]Barrows, G., "Feature Tracking Linear Optic Flow Sensor for 2-D Optic Flow Measurement," *1998 International Conference on Control, Automation, Robotics, and Vision (ICARCV'98)*, Nanyang Technological University, Singapore, 1998, pp. 732–736.

[7]Barrows, G. L., "Mixed-Mode VLSI Optic Flow Sensors for Micro Air Vehicles," Ph.D. Dissertation, Department of Electrical Engineering, Univ. Maryland, College Park, MD, Dec. 1999.

[8]Barrows, G. L., Miller, K. T., and Krantz, B., "Fusing Neuromorphic Motion Detector Outputs for Robust Optic Flow Measurement," *1999 International Joint Conference on Neural Networks (IJCNN'99)*, Washington, DC, IEEE, Piscataway, NJ, 1999.

[9]Barrows, G. L., "Mixed-Mode VLSI Optic Flow Sensors for In-Flight Control of a Micro Air Vehicle," *1999 Robotics and Applications Conference (RA'99)*, Santa Barbara, CA, IASTED, 1999, pp. 306–311.

[10]Miller, K., and Barrows, G., "Feature Tracking Linear Optic Flow Sensor Chip," *1999 IEEE International Symposium on Circuits and Systems (ISCAS'99)*, Vol. V, Orlando, FL, IEEE, Piscataway, NJ, May 30–June 2, 1999, pp. 116–119.

[11]Barrows, G., and Neely, C., "Mixed-Mode VLSI Optic Flow Sensors for In-Flight Control of a Micro Air Vehicle," *Critical Technologies for the Future of Computing*, SPIE Vol. 4109, Society of Photo-Optical Instrumentation Engineers, July 2000.

[12]MOSIS: The Metal Oxide Semiconductor Integration Service, at the Information Sciences Institute, 4676 Admiralty Way, 7th Floor, Marina Del Rey, CA 90292-6695.

[13]Ancona, N., and Poggio, T., "A Real-Time Miniaturized Optical Sensor for Motion Estimation and Time-to-crash Detection," SPIE Vol. 2950, Society of Photo-Optical Instrumentation Engineers, 1996, pp. 75–85.

PROGRESS IN ASTRONAUTICS AND AERONAUTICS
SERIES VOLUMES

*1. Solid Propellant
Rocket Research (1960)
Martin Summerfield
Princeton University

*2. Liquid Rockets and
Propellants (1960)
Loren E. Bollinger
Ohio State University
Martin Goldsmith
The Rand Corp.
Alexis W. Lemmon Jr.
Battelle Memorial Institute

*3. Energy Conversion
for Space Power (1961)
Nathan W. Snyder
*Institute for Defense
Analyses*

*4. Space Power Systems
(1961)
Nathan W. Snyder
*Institute for Defense
Analyses*

*5. Electrostatic
Propulsion (1961)
David B. Langmuir
*Space Technology
Laboratories, Inc.*
Ernst Stuhlinger
*NASA George C. Marshall
Space Flight Center*
J. M. Sellen Jr.
*Space Technology
Laboratories, Inc.*

*6. Detonation and
Two-Phase Flow (1962)
S. S. Penner
*California Institute of
Technology*
F. A. Williams
Harvard University

*7. Hypersonic Flow
Research (1962)
Frederick R. Riddell
AVCO Corp.

*8. Guidance and Control
(1962)
Robert E. Roberson
Consultant
James S. Farrior
*Lockheed Missiles and
Space Co.*

*9. Electric Propulsion
Development (1963)
Ernst Stuhlinger
*NASA George C. Marshall
Space Flight Center*

*10. Technology of Lunar
Exploration (1963)
Clifford I. Cumming
Harold R. Lawrence
Jet Propulsion Laboratory

*11. Power Systems for
Space Flight (1963)
Morris A. Zipkin
Russell N. Edwards
General Electric Co.

*12. Ionization in High-
Temperature Gases (1963)
Kurt E. Shuler, Editor
*National Bureau of
Standards*
John B. Fenn,
Associate Editor
Princeton University

*13. Guidance and
Control–II (1964)
Robert C. Langford
General Precision Inc.
Charles J. Mundo
Institute of Naval Studies

*14. Celestial Mechanics
and Astrodynamics (1964)
Victor G. Szebehely
Yale University Observatory

*15. Heterogeneous
Combustion (1964)
Hans G. Wolfhard
*Institute for Defense
Analyses*
Irvin Glassman
Princeton University
Leon Green Jr.
*Air Force Systems
Command*

*16. Space Power Systems
Engineering (1966)
George C. Szego
*Institute for Defense
Analyses*
J. Edward Taylor
TRW Inc.

*17. Methods in
Astrodynamics and
Celestial Mechanics
(1966)
Raynor L. Duncombe
U.S. Naval Observatory
Victor G. Szebehely
*Yale University
Observatory*

*18. Thermophysics and
Temperature Control of
Spacecraft and Entry
Vehicles (1966)
Gerhard B. Heller
*NASA George C. Marshall
Space Flight Center*

---

*Out of print.

---

*Out of print.

*53. Experimental
Diagnostics in Gas Phase
Combustion Systems
(1977)
Ben T. Zinn, Editor
*Georgia Institute of
Technology*
Craig T. Bowman,
Associate Editor
Stanford University
Daniel L. Hartley,
Associate Editor
Sandia Laboratories
Edward W. Price,
Associate Editor
*Georgia Institute of
Technology*
James G. Skifstad,
Associate Editor
Purdue University
ISBN 0-915928-18-3

*54. Satellite
Communication: Future
Systems (1977)
David Jarett
TRW Inc.
ISBN 0-915928-18-3

*55. Satellite
Communications:
Advanced Technologies
(1977)
David Jarett
TRW Inc.
ISBN 0-915928-19-1

*56. Thermophysics of
Spacecraft and Outer
Planet Entry Probes
(1977)
Allie M. Smith
ARO Inc.
ISBN 0-915928-20-5

*57. Space-Based
Manufacturing from
Nonterrestrial Materials
(1977)
Gerald K. O'Neill, Editor
Brian O'Leary,
Assistant Editor
Princeton University
ISBN 0-915928-21-3

*58. Turbulent
Combustion (1978)
Lawrence A. Kennedy
*State University of
New York at Buffalo*
ISBN 0-915928-22-1

*59. Aerodynamic
Heating and Thermal
Protection Systems (1978)
Leroy S. Fletcher
University of Virginia
ISBN 0-915928-23-X

*60. Heat Transfer and
Thermal Control Systems
(1978)
Leroy S. Fletcher
University of Virginia
ISBN 0-915928-24-8

*61. Radiation Energy
Conversion in Space
(1978)
Kenneth W. Billman
*NASA Ames Research
Center*
ISBN 0-915928-26-4

*62. Alternative
Hydrocarbon Fuels:
Combustion and Chemical
Kinetics (1978)
Craig T. Bowman
Stanford University
Jorgen Birkeland
Department of Energy
ISBN 0-915928-25-6

*63. Experimental
Diagnostics in Combustion
of Solids (1978)
Thomas L. Boggs
Naval Weapons Center
Ben T. Zinn
*Georgia Institute of
Technology*
ISBN 0-915928-28-0

*64. Outer Planet Entry
Heating and Thermal
Protection (1979)
Raymond Viskanta
Purdue University
ISBN 0-915928-29-9

*65. Thermophysics and
Thermal Control (1979)
Raymond Viskanta
Purdue University
ISBN 0-915928-30-2

*66. Interior Ballistics of
Guns (1979)
Herman Krier
*University of Illinois at
Urbana–Champaign*
Martin Summerfield
New York University
ISBN 0-915928-32-9

*67. Remote Sensing of
Earth from Space: Role of
"Smart Sensors" (1979)
Roger A. Breckenridge
*NASA Langley Research
Center*
ISBN 0-915928-33-7

*68. Injection and Mixing
in Turbulent Flow (1980)
Joseph A. Schetz
*Virginia Polytechnic
Institute and State
University*
ISBN 0-915928-35-3

*69. Entry Heating and
Thermal Protection (1980)
Walter B. Olstad
NASA Headquarters
ISBN 0-915928-38-8

*70. Heat Transfer,
Thermal Control, and
Heat Pipes (1980)
Walter B. Olstad
NASA Headquarters
ISBN 0-915928-39-6

*71. Space Systems and
Their Interactions with
Earth's Space
Environment (1980)
Henry B. Garrett
Charles P. Pike
Hanscom Air Force Base
ISBN 0-915928-41-8

---

*Out of print.

---

*Out of print.

---

*Out of print.

148. Metallurgical Technologies, Energy Conversion, and Magneto-hydrodynamic Flows (1993)
Herman Branover
Yeshajahu Unger
Ben-Gurion University of the Negev
ISBN 1-56347-019-5

149. Advances in Turbulence Studies (1993)
Herman Branover
Yeshajahu Unger
Ben-Gurion University of the Negev
ISBN 1-56347-018-7

150. Structural Optimization: Status and Promise (1993)
Manohar P. Kamat
Georgia Institute of Technology
ISBN 1-56347-056-X

151. Dynamics of Gaseous Combustion (1993)
A. L. Kuhl
Lawrence Livermore National Laboratory
J.-C. Leyer
Universite de Poitiers
A. A. Borisov
USSR Academy of Sciences
W. A. Sirignano
University of California
ISBN 1-56347-060-8

152. Dynamics of Heterogeneous Gaseous Combustion and Reacting Systems (1993)
A. L. Kuhl
Lawrence Livermore National Laboratory
J.-C. Leyer
Universite de Poitiers
A. A. Borisov
USSR Academy of Sciences
W. A. Sirignano
University of California
ISBN 1-56347-058-6

153. Dynamic Aspects of Detonations (1993)
A. L. Kuhl
Lawrence Livermore National Laboratory
J.-C. Leyer
Universite de Poitiers
A. A. Borisov
USSR Academy of Sciences
W. A. Sirignano
University of California
ISBN 1-56347-057-8

154. Dynamic Aspects of Explosion Phenomena (1993)
A. L. Kuhl
Lawrence Livermore National Laboratory
J.-C. Leyer
Universite de Poitiers
A. A. Borisov
USSR Academy of Sciences
W.A. Sirignano
University of California
ISBN 1-56347-059-4

155. Tactical Missile Warheads (1993)
Joseph Carleone
Aerojet General Corporation
ISBN 1-56347-067-5

156. Toward a Science of Command, Control, and Communications (1993)
Carl R. Jones
Naval Postgraduate School
ISBN 1-56347-068-3

*****157. Tactical and Strategic Missile Guidance Second Edition (1994)**
Paul Zarchan
Charles Stark Draper Laboratory, Inc.
ISBN 1-56347-077-2

158. Rarefied Gas Dynamics: Experimental Techniques and Physical Systems (1994)
Bernie D. Shizgal
University of British Columbia
David P. Weaver
Phillips Laboratory
ISBN 1-56347-079-9

*****159. Rarefied Gas Dynamics: Theory and Simulations (1994)**
Bernie D. Shizgal
University of British Columbia
David P. Weaver
Phillips Laboratory
ISBN 1-56347-080-2

160. Rarefied Gas Dynamics: Space Sciences and Engineering (1994)
Bernie D. Shizgal
University of British Columbia
David P. Weaver
Phillips Laboratory
ISBN 1-56347-081-0

161. Teleoperation and Robotics in Space (1994)
Steven B. Skaar
University of Notre Dame
Carl F. Ruoff
Jet Propulsion Laboratory, California Institute of Technology
ISBN 1-56347-095-0

162. Progress in Turbulence Research (1994)
Herman Branover
Yeshajahu Unger
Ben-Gurion University of the Negev
ISBN 1-56347-099-3

180. Advances in Missile Guidance Theory (1998)
Joseph Z. Ben-Asher
Isaac Yaesh
Israel Military Industries— Advanced Systems Division
ISBN 1-56347-275-9

181. Satellite Thermal Control for Systems Engineers (1998)
Robert D. Karam
ISBN 1-56347-276-7

182. Progress in Fluid Flow Research: Turbulence and Applied MHD (1998)
Yeshajahu Unger
Herman Branover
Ben-Gurion University of the Negev
ISBN 1-56347-284-8

183. Aviation Weather Surveillance Systems (1999)
Pravas R. Mahapatra
Indian Institute of Science
ISBN 1-56347-340-2

184. Flight Control Systems (2000)
Rodger W. Pratt
Loughborough University
ISBN 1-56347-404-2

185. Solid Propellant Chemistry, Combustion, and Motor Interior Ballistics (2000)
Vigor Yang
Pennsylvania State University
Thomas B. Brill
University of Delaware
Wu-Zhen Ren
China Ordnance Society
ISBN 1-56347-442-5

186. Approximate Methods for Weapons Aerodynamics (2000)
Frank G. Moore
ISBN 1-56347-399-2

187. Micropropulsion for Small Spacecraft (2000)
Michael M. Micci
Pennsylvania State University
Andrew D. Ketsdever
Air Force Research Laboratory, Edwards Air Force Base
ISBN 1-56347-448-4

188. Structures Technology for Future Aerospace Systems (2000)
Ahmed K. Noor
NASA Langley Research Center
ISBN 1-56347-384-4

189. Scramjet Propulsion (2000)
E. T. Curran
Department of the Air Force
S. N. B. Murthy
Purdue University
ISBN 1-56347-322-4

190. Fundamentals of Kalman Filtering: A Practical Approach (2000)
Paul Zarchan
Howard Musoff
Charles Stark Draper Laboratory, Inc.
ISBN 1-56347-455-7

191. Gossamer Spacecraft: Membrane and Inflatable Structures Technology for Space Applications (2001)
Christopher H. M. Jenkins
South Dakota School of Mines
ISBN 1-56347-403-4

192. Theater Ballistic Missile Defense (2001)
Ben-Zion Naveh
Azriel Lorber
WALES Ltd.
ISBN 1-56347-385-2

193. Air Transportation Systems Engineering (2001)
George L. Donohue
George Mason University
Andres G. Zellweger
Embry-Riddle Aeronautical University
ISBN 1-56347-474-3

194. Physics of Direct Hit and Near Miss Warhead Technology (2001)
Richard M. Lloyd
Raytheon Electronics Systems
ISBN 1-56347-473-5

195. Fixed and Flapping Wing Aerodynamics for Micro Air Vehicle Applications (2001)
Thomas J. Mueller
University of Notre Dame
ISBN 1-56347-517-0